Food Structures, Digestion and Health

Food Structures, Digestion and Health

Edited by

Mike Boland, Matt Golding and Harjinder Singh

Riddet Institute, Massey University, Palmerston North, New Zealand

AMSTERDAM • BOSTON • HEIDELBERG • LONDON
NEW YORK • OXFORD • PARIS • SAN DIEGO
SAN FRANCISCO • SINGAPORE • SYDNEY • TOKYO

Academic Press is an Imprint of Elsevier

ELSEVIER

Academic Press is an imprint of Elsevier
32 Jamestown Road, London NW1 7BY, UK
225 Wyman Street, Waltham, MA 02451, USA
525 B Street, Suite 1800, San Diego, CA 92101-4495, USA

Notice
No responsibility is assumed by the publisher for any injury and/or damage to persons or property as a matter of products liability, negligence or otherwise, or from any use or operation of any methods, products, instructions or ideas contained in the material herein. Because of rapid advances in the medical sciences, in particular, independent verification of diagnoses and drug dosages should be made

British Library Cataloguing-in-Publication Data
A catalogue record for this book is available from the British Library

Library of Congress Cataloging-in-Publication Data
A catalog record for this book is available from the Library of Congress

ISBN: 978-0-12-404610-8

For information on all Academic Press publications
visit our website at www.store.elsevier.com

Typeset by TNQ Books and Journals
www.tnq.co.in

Printed and bound by CPI Group (UK) Ltd, Croydon, CR0 4YY

14 15 16 17 18   10 9 8 7 6 5 4 3 2 1

Working together
to grow libraries in
developing countries

www.elsevier.com • www.bookaid.org

Contents

SECTION 4 FOOD DEVELOPMENTS TO MEET THE MODERN CHALLENGES OF HUMAN HEALTH

CHAPTER 13 Applying Structuring Approaches for Satiety: Challenges Faced, Lessons Learned

CHAPTER 14 Technological Means to Modulate Food Digestion and Physiological Response

All figures that have been provided in color are available online at
www.booksite.elsevier.com/9780124046108

List of Contributors

Timothy R. Angeli
Auckland Bioengineering Institute, The University of Auckland, Auckland, New Zealand

M.S. Anokhina
N.M. Emanuel Institute of Biochemical Physics of Russian Academy of Sciences, Moscow, Russian Federation

A.S. Antipova
N.M. Emanuel Institute of Biochemical Physics of Russian Academy of Sciences, Moscow, Russian Federation

Thierry Astruc
INRA Clermont-Ferrand Theix, "Quality of Animal Products" Research Unit, Saint Genès Champanelle, France

S. Bassett
Food Nutrition & Health Team, Food & Bio-based Products Group, AgResearch Grasslands, Palmerston North, New Zealand

L.E. Belyakova
N.M. Emanuel Institute of Biochemical Physics of Russian Academy of Sciences, Moscow, Russian Federation

Thilo Berg
Riddet Institute, Massey University, Palmerston North, New Zealand

Mike J. Boland
Riddet Institute, Massey University, Palmerston North, New Zealand

Leo K. Cheng
Auckland Bioengineering Institute, The University of Auckland, Auckland, New Zealand

Milena Corredig
Department of Food Science, University of Guelph, Guelph, Ontario, Canada

Douglas G. Dalgleish
Department of Food Science, University of Guelph, Guelph, Ontario, Canada

Eric Dickinson
School of Food Science and Nutrition, University of Leeds, Leeds, UK

L. Donato-Capel
Food Science and Technology Department, Nestec Ltd, Nestlé Research Center, Lausanne, Switzerland

Peng Du
Auckland Bioengineering Institute, The University of Auckland, Auckland, New Zealand

Jolon M. Dyer
Food & Bio-Based Products, AgResearch Lincoln Research Centre, Christchurch, New Zealand, Biomolecular Interaction Centre, University of Canterbury, Christchurch, New Zealand, Wine, Food & Molecular

Biosciences, Lincoln University, Canterbury, New Zealand, Riddet Institute, based at Massey University, Palmerston North, New Zealand

A. Erkner
Nutrition and Health Department, Nestec Ltd, Nestlé Research Center, Lausanne, Switzerland

M.J. Ferrua
Riddet Institute, Massey University, Palmerston North, New Zealand

Sophie Gallier
Riddet Institute, Massey University, Palmerston North, New Zealand

Danone Nutricia Research, Uppsalalaan, Utrecht, The Netherlands

C.L. Garcia-Rodenas
Nutrition and Health Department, Nestec Ltd, Nestlé Research Center, Lausanne, Switzerland

Manohar Garg
School of Biomedical Sciences & Pharmacy, University of Newcastle, Callaghan, NSW, Australia; Riddet Institute, Massey University, Palmerston North, New Zealand

Matt Golding
Institute of Food, Nutrition & Human Health and Riddet Institute, Massey University, Palmerston North, New Zealand

N.V. Grigorovich
N.M. Emanuel Institute of Biochemical Physics of Russian Academy of Sciences, Moscow, Russian Federation

Anita Grosvenor
Food & Bio-Based Products, AgResearch Lincoln Research Centre, Christchurch, New Zealand

Anilda Guri
Department of Food Science, University of Guelph, Ontario, Canada
Canadian Research Institute for Food Safety, University of Guelph, Ontario, Canada

Allan Hardacre
Institute of Food Nutrition and Human Health, Massey University, Palmerston North, New Zealand

E. Hughes
Food Science and Technology Department, Nestec Ltd, Nestlé Research Center, Lausanne, Switzerland

Rafael Jiménez-Flores
Dairy Products Technology Center, California Polytechnic State University, San Luis Obispo, CA, USA

E. Kolodziejczyk
Food Science and Technology Department, Nestec Ltd, Nestlé Research Center, Lausanne, Switzerland

Andrea Laubscher
Dairy Products Technology Center, California Polytechnic State University, San Luis Obispo, CA, USA

U. Lehmann
Food Science and Technology Department, Nestec Ltd, Nestlé Research Center, Lausanne, Switzerland

W.C. McNabb
Riddet Institute, Massey University, Palmerston North, New Zealand, Gravida, National Centre for Growth and Development, The University of Auckland, Auckland, New Zealand, AgResearch Grasslands, Palmerston North, New Zealand

David J. Mela
Unilever R & D Vlaardingen, AC Vlaardingen, The Netherlands

D.V. Moiseenko
N.M. Emanuel Institute of Biochemical Physics of Russian Academy of Sciences, Moscow, Russian Federation

Paul J. Moughan
Riddet Institute, Massey University, Palmerston North, New Zealand

Niranchan Paskaranandavadivel
Auckland Bioengineering Institute, The University of Auckland, Auckland, New Zealand

Melinda Phang
University of Newcastle, Nutraceuticals Research Group, Newcastle, NSW, Australia

Yu.N. Polikarpov
N.M. Emanuel Institute of Biochemical Physics of Russian Academy of Sciences, Moscow, Russian Federation

E. Pouteau
Nutrition and Health Department, Nestec Ltd, Nestlé Research Center, Lausanne, Switzerland

N.C. Roy
Food Nutrition & Health Team, Food & Bio-based Products Group, AgResearch Grasslands, Palmerston North, New Zealand, Riddet Institute, Massey University, Palmerston North, New Zealand, Gravida, National Centre for Growth and Development, The University of Auckland, Auckland, New Zealand

L. Sagalowicz
Food Science and Technology Department, Nutrition and Health Department, Nestec Ltd, Nestlé Research Center, Lausanne, Switzerland

M.G. Semenova
N.M. Emanuel Institute of Biochemical Physics of Russian Academy of Sciences, Moscow, Russian Federation

P.R. Shorten
Bioinformatics, Maths & Stats Team, Knowledge & Analytics Group, AgResearch, Ruakura Research Centre, Hamilton, New Zealand, Riddet Institute, Massey University, Palmerston North, New Zealand, Gravida, National Centre for Growth and Development, The University of Auckland, Auckland, New Zealand

Harjinder Singh
Riddet Institute, Massey University, Palmerston North, New Zealand

Jaspreet Singh
Riddet Institute, Massey University, Palmerston North, New Zealand

R. Paul Singh
Riddet Institute, Massey University, Palmerston North, New Zealand, Department of Biological and
Agricultural Engineering, University of California, Davis, CA, USA

S. Srichuwong
Food Science and Technology Department, Nestec Ltd, Nestlé Research Center, Lausanne, Switzerland

C. Thum
Food Nutrition & Health Team, Food & Bio-based Products Group, AgResearch Grasslands, Palmerston
North, New Zealand, Riddet Institute, Massey University, Palmerston North, New Zealand

E.N. Tsapkina
N.M. Emanuel Institute of Biochemical Physics of Russian Academy of Sciences, Moscow,
Russian Federation

A.S. Van Wey
Bioinformatics, Maths & Stats Team, Knowledge & Analytics Group, AgResearch, Ruakura Research
Centre, Hamilton, New Zealand, Riddet Institute, Massey University, Palmerston North, New Zealand

T.J. Wooster
Food Science and Technology Department, Nestec Ltd, Nestlé Research Center, Lausanne, Switzerland

Z. Xue
Department of Biological and Agricultural Engineering, University of California, Davis, CA, USA

W. Young
Food Nutrition & Health Team, Food & Bio-based Products Group, AgResearch Grasslands, Palmerston
North, New Zealand

Preface

Over the past two decades or so, the emphasis on nutrition has moved beyond simple assessment of the amounts of nutrients in a diet, to take into account the way in which those nutrients are delivered. This involves a consideration of both the rates at which nutrients are taken up by the body (a consideration in nutrition akin to what pharmacokinetics is to drug delivery) and the sites in the gastrointestinal tract where the nutrients are released and are taken up by the body. The rate of release of glucose from carbohydrates and uptake by the body was one of the earliest aspects of this, manifest in the glycemic index, a measure that is particularly important in the management of diabetes, but also an important consideration in the development of foods for weight management. Indeed, the terms used throughout this volume, such as glycemic index, resistant starch, and satiety, are now entering the public stream of consciousness, and consumers are increasingly aware of the nutritional value of the foods they eat. The role of food structure in modifying digestion and release of nutrients and bioactives builds on understandings of food structure derived from recent developments in material science and nanotechnology. Natural foods contain important structural components at the molecular, nano-, micro-, and meso-structural scale. Processing usually modifies and often destroys these structures and thus modifies the digestion profile of nutrients. In today's food processing industry, it is increasingly important to be able to manufacture foods that release nutrients in ways that mimic natural foods, thus providing a more natural flow of nutrients following consumption. Food structure is also important for the delivery of bioactives: some bioactives are acid labile, for example, and need to be protected from stomach acids and released in the neutral pH of the small intestine, and appropriate structures can achieve this.

The Riddet Institute, a Centre of Research Excellence in New Zealand, was set up to lead research into the relationship between food structure and health. In 2012, it hosted the inaugural conference on food structures, digestion, and health, to bring together experts from around the world to discuss and present on this important topic. A second such conference is in preparation at the time of writing. The present volume has evolved from a selected range of those conference presentations.

In this volume, we have covered a broad range of approaches in different disciplines to understand the interactions between food structures and digestion and health, with offerings from those involved at the forefront of research in their particular areas. The book is structured around four sections:

1. Understanding food structures in natural and processed foods and their behavior during physiological processing
2. Impact of food structures and matrices on nutrient uptake and bioavailability
3. Modeling the gastrointestinal tract
4. Food developments to meet the modern challenges of human health

In producing the volume, we have aimed for simplicity and clarity of language, so that the work of an expert in one particular area is accessible to readers from all areas. It is particularly pleasing to have a wide range of authors, not only across disciplines and across different food types, but also a spectrum from basic university-based research through to applied work by multinational food companies.

In preparing this volume, we would like to thank all of the authors and all others who have helped in this project. Particular thanks must go to Ansley Te Hiwi for secretarial support.

Understanding Food Structures in Natural and Processed Foods and their Behavior During Physiological Processing

Understanding Food Structures: The Colloid Science Approach

Eric Dickinson

School of Food Science and Nutrition, University of Leeds, Leeds, UK

CONTENTS

INTRODUCTION

As diet-related health problems continue to increase globally, there is recognition within the research community of the need for more detailed knowledge of the behavior of foods as they are processed within the human digestive system. Individual foods differ considerably in their nutrient composition and also in terms of the matrix materials within which the nutrients are embedded. During eating, the breakdown property of the food matrix is a major controlling factor for the perception of texture and flavor in the mouth. After swallowing, the processing of the disrupted food matrix in the gastrointestinal tract influences the perception of

Food Structures, Digestion and Health. http://dx.doi.org/10.1016/B978-0-12-404610-8.00001-3

postprandial satiety and bioavailability of nutrients. It seems reasonable to assert that, in order for food technologists to continue to be able to develop nutritious foods from healthier combinations of ingredients, there is an underlying requirement to understand more fully the changing structural behavior of foods during eating and digestion.

The challenges posed by the complex dietary health issues are made more extreme by the potentially conflicting demands of consumers that food should be simultaneously tasty, wholesome, healthy, and cheap. According to the food industry, it is generally necessary for processed foods to contain high levels of fat, salt, and sugar in order to meet existing consumer expectations with respect to flavor and texture. Nevertheless, well-founded concern over the adverse health implications of the overconsumption of certain types of lipids has led the industry to develop alternative "low-fat" and "reduced fat" food products. In addition, the identification and widespread public recognition of the health-promoting properties of certain bioactive compounds has generated commercial opportunities for marketing high-value specialist products containing encapsulated bioactives (nutraceuticals). On the downside, however, many of the notionally healthier products containing less fat (or salt or sugar) are often perceived by consumers as being of inferior organoleptic quality. This is because the methods used to modify food composition have effects on other essential food characteristics such as taste, appearance, and texture (Velikov and Pelan, 2008). Furthermore, many of the specialist products containing added health-beneficial nutraceuticals may be regarded as expensive "niche" products of significant benefit only to a small fraction of consumers with specific recognized medical conditions. Overshadowing these commercial trends is one further problem: the available evidence suggests that a large proportion of the consumers in Western societies are not easily persuaded to compromise their eating pleasure, or to increase their grocery shopping expenditure, simply for the sake of some promised long-term health benefits. Hence, the successes of governments and industry in modifying eating habits for the sake of improving long-term well-being remain disappointingly limited.

Against this challenging background, the food technologist aims to develop cheap healthier alternatives to existing processed foods without diminishing the consumer's organoleptic experience. Understanding how this can be done requires detailed insight into the relationship between the composition and processing of the food and its multifaceted properties—nutritional, sensory, and physicochemical. These days it is an implicit belief of most food researchers that one important piece of information, the *food structure*, is a prerequisite to determining how the ingredient composition and processing conditions are mechanistically related to the product properties. There was perhaps once a time when the subject of food structure was solely the specialist domain of the food microscopist, but that time has long since gone. Structural information is now an essential requirement of all those concerned with the control of food ingredient functionality during food manufacture, storage, and digestion.

So what is meant by "food structure"? The answer depends to some extent on the perspective of the observer—as physicist, chemist, biologist, or engineer. The answer is also influenced by the type of food under consideration. Take the category of fruits, plants, and nuts, for example. These are commonly eaten as whole foods in their nearly natural state. Therefore, it is the biological perspective that would seem to be paramount. Natural materials can be regarded as hierarchical fibrous composites composed of a relatively small number of basic components. The spatial organization of the structural units (cells, fibers, membranes, etc.) has its origin in the biological origin and function of the material, and hence the perceived food texture may be systematically interpreted in terms of the structure and properties of the hierarchical fibrous composites (Vincent, 2008).

The structural complexity of much of the food consumed by humans in the modern world is far removed from the fibrous composite character of the living plant or animal materials. Natural structuring agents like cell wall materials are rarely used in their unrefined state (Foster, 2011). Typically, the food is prepared in the kitchen or factory from a recipe involving a multicomponent mixture of separate ingredients, each of which has been subjected to many different stages of mechanical, biochemical, and thermal processing (Aguilera and Lillford, 2008). Under such circumstances, the conventional biological perspective is not an adequate one for describing or understanding the structure. In the first place, this is because most of the natural biological structure has been substantially modified or destroyed during the process of extracting the individual ingredients. But a second, and even more important, reason for dismissing the biological perspective is that these individual ingredients are subsequently reassembled into a complex structure that is completely different from any encountered in the living world. The main challenge in defining and understanding this complex structure is to identify what are the key structural elements that determine its associated textural and sensory properties.

Structure formation within a manufactured food product is commonly approached from an engineering or technological perspective. The traditional discipline of food technology has been elegantly defined as "a controlled attempt to preserve, transform, create or destroy a structure that has been imparted by nature or processing" (Aguilera and Stanley, 1999; Aguilera, 2005). Our objective here, however, is to move beyond mere technological know-how to a state of understanding that would allow systematic control, prediction, and design of food material properties. For a product manufactured from ingredients of known composition, the specification of the relationship between the processing conditions and the material properties requires an analysis of the food structure from the perspective of physical chemistry (Walstra, 2003) or, equivalently, chemical physics (Belton, 2007). This type of analysis is not easily realized, however, because foods are non-equilibrium structures. And these structures change continuously with time and with the external environment during processing, storage, and cooking—and, most importantly, during eating and digestion. Experimental investigation of the structure of a manufactured food material typically reveals the presence of many different coexisting phases organized on a wide range of spatial length scales from molecular to microscopic. When observed from the physicochemical perspective, such multiphase food systems are conveniently described as "food colloids" (Dickinson and Stainsby, 1982; Dickinson, 1992).

The broad multidisciplinary scope of food structure investigations is illustrated by the set of bibliometric data plotted in Figure 1.1. These data were derived from a *Web of Science* search of the subjects of peer-reviewed research papers published since January 2000. The search data indicate that the authors of these papers have separately chosen to identify the topic of "food structure" to a roughly equal extent with the disciplines of biology, engineering, and chemistry. The association with physics is less well developed, but there is recent evidence that this trend is changing (Mezzenga, 2007; Ubbink, 2012). The data in Figure 1.1 also confirm the expected association of food structure with texture, taste, and rheology. More specifically, the search data reveal that author descriptions of food structure are commonly expressed in physicochemical language using the readily recognizable words of colloid science, i.e., gel, particle, emulsion, interface, and foam. We therefore infer that there exists a substantial body of recent food structure research that recognizes and promotes the colloid science approach.

According to Ubbink (2012) the term "structured foods" is really a pleonastic concept since every food is necessarily structured on a continuum of length scales from molecular to macroscopic. Nevertheless, the term is meaningful because of the physicochemical conceptual perspective that lies behind it. That is, the investigation of food structure goes beyond the mere specification of the geometrical organization of the structural elements in the food material. It extends to a consideration of the nature of the *interactions* between

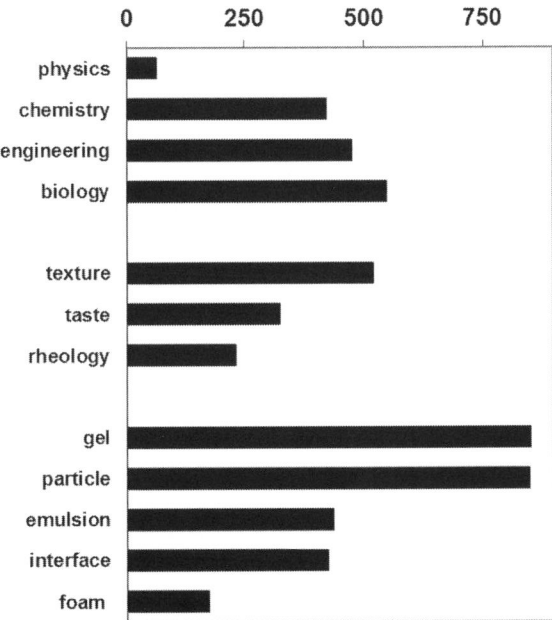

■ **FIGURE 1.1** Multidisciplinary character and technical language of peer-reviewed journal papers on food structure published during the period from January 2000 to September 2012. The data refer to the numbers of hits resulting from online searches in *ISI Web of Science* of the term "food structure" combined separately with each of the other indicated terms.

those elements. In the same way that structural elements may range in size from molecular to macroscopic, so the interactions between those elements may operate over different length scales. Inferences about interactions between structural elements emerge from the physical and mathematical models of food colloidal systems. The essential character of any model may be statistical, thermodynamic, kinetic, or phenomenological—or some combination of all these. The aim of such models is to provide insight into aspects of the relationship between the structure of food systems and their rheological and stability properties.

Without doubt, a major factor influencing progress in the physicochemical investigation of food structure is the availability of advanced experimental instrumentation (Dickinson, 1995a; Aguilera and Stanley, 1999; McClements, 2007). Modern laboratory tools have allowed food researchers to access previously inaccessible information, thereby exposing these complex multicomponent food systems to a more comprehensive and wide-ranging level of investigation. The routine use of various types of commercial microscopes, sensitive rheometers, and reliable particle-sizing equipment has led to major advances in the quantitative characterization of model food systems. In addition, specialized techniques are used to provide more precise information over specific length scales. These techniques include scattering methods such as X-ray diffraction, neutron scattering/reflection, and diffusing wave spectroscopy; microscopy techniques such as atomic force microscopy, confocal laser scanning microscopy, Brewster angle microscopy, and cryoscopic scanning electron microscopy; and numerous other analytical techniques such as mass spectrometry, nuclear magnetic resonance, ultrasonic spectroscopy, and differential scanning calorimetry. In addition, this progress in measurement science has been complemented by developments in image analysis for quantifying microstructural information (Aguilera and Germain, 2007; Pugnaloni *et al.*, 2005) and by advances in the methods and applications of numerical simulations and modeling (Ettelaie, 2003; Euston *et al.*, 2007; van der Sman, 2012).

This chapter outlines recent advances in the structuring of food systems from the perspective of the colloid scientist. While the main emphasis is on protein-based emulsion systems, it is believed that the underlying conceptual approach has direct relevance to a broad range of food systems. However, before addressing the essential principles and applications of the colloid science approach, let us pause briefly to reflect on the historical perspective and the contemporary linguistic landscape.

ON COLLOID TERMINOLOGY IN THE AGE OF "NANO"

As in many other areas of human activity, the development of science is influenced by fashion. This is most clearly reflected in the language (jargon) that practitioners employ to map out the local intellectual territory. So, if I say "colloidal," you may say "nano," and they may say "mesoscopic." Presumably, as specialists, *we* know what *we* mean by these terms. But is everyone really saying the same thing? Or are there subtle differences of meaning involved here?

The online version of the *Oxford English Dictionary* defines a colloid as a "homogeneous non-crystalline substance of large molecules or ultramicro scopic particles of one substance dispersed through a second substance." In addition, it is asserted that "colloids include gels, sols and emulsions" and that "(colloidal) particles do not settle, and cannot be separated out by ordinary filtering or centrifuging like those in a suspension." The dictionary definition is therefore based firmly on experimental observation. The colloid appears homogeneous and non-crystalline to the naked eye. But its inherent heterogeneity is revealed when viewed under a powerful light microscope (ultramicroscope). History tells us that this term "colloid" was used by the early physical chemists to categorize a whole basket of messy systems, usually of biological origin, whose characteristics could not be explained in terms of the then known world of small molecules and elementary states of matter. This attitude prevailed for a long time, as memorably expressed in the dramatic statement of Hedges (1931): "the word 'colloidal' conjures up visions of things indefinite in shape, indefinite in chemical composition and physical properties, fickle in chemical deportment, things infilterable and generally unmanageable."

There is nothing explicit on colloidal dimensions in the *Oxford English Dictionary* definition. But it is asserted that colloidal particles do not settle under gravity and are not readily separable by filtration or centrifugation. Hence, it is a short step to apply current knowledge of the statistical character of Brownian motion and the hydrodynamics of fluid flow to infer that there is an effective upper limit to the colloidal length scale. This analysis leads to a maximum colloidal particle size of around 1 μm (i.e., ≈ 1000 nm), which, by chance, corresponds very roughly to the resolution of the standard optical microscope. This upper limit for the colloidal length scale is rather approximate because the experimentally observed criteria are themselves necessarily subjective in character. Furthermore, the theory of particle sedimentation tells us that the settling rate is dependent on other properties such as the overall particle concentration, the relative densities of the phases, and the viscosity of the dispersion medium, all of which can (and often do) change substantially from one system to another. Despite this variability, one essential point remains: the system's colloidal credentials are established not through measuring its chemical composition, but as a consequence of observing its characteristic experimental behavior.

The prefix "nano" denotes the factor 10^{-9}. Hence, a nanoscale system is one with a length scale of 10^{-9} meters (1 nm). This is the size of one large molecule or a group of small molecules. It therefore follows that the study of nanoscale systems, namely "nanoscience," involves investigating the chemistry and physics of materials from the perspective of structures containing individual molecules (or their assemblies) as the essential primary building blocks. By its very nature, nanoscience research involves the development of new nanoscale materials whose safety within the human biological environment is necessarily uncertain (Magnuson *et al.*, 2011). Hence, the food industry has to be properly cautious about using novel nanoscale materials in its products, and perhaps also even more cautious concerning the risks of possible misconceptions by consumers regarding the dangers of any such use.

Despite the reluctance of the food industry's public face to embrace "nano" terminology, the words "nanoscale" and "nanoscience" have become increasingly familiar to readers of the physical science literature. Moreover, this jargon has been extended into other kinds of nanospeak. New words have been constructed by prefixing "nano" to the established terms of colloid science: "nanoparticle," "nanodroplet," "nanocapsule," "nanogel," "nanoemulsion," etc. (Possibly the culmination of this trend is the apparently tautological nanocolloid.) And while the value of the upper size limit of the nanoscale remains somewhat ill-defined, there is a growing convention that it should be set at around 100 nm. This allows the word "nanoparticle" to be used to distinguish a small colloidal particle (diameter <0.1 μm) from a larger microparticle (≈ 1 μm). That having been said, confusing statements do still persist in the literature concerning nanoparticles (and other nanoscale objects) with apparent dimensions of several hundred nanometers (or more).

Another term, "mesoscopic," is also applied to materials of length scale intermediate between molecular (atomic) and macroscopic. This word comes from a branch of condensed matter physics called mesoscopic physics which deals with the fundamental properties of nanotechnological devices relevant to the microelectronics industry. A normal macroscopic object can be well described in terms of the average properties of the material from which it is made. But a mesoscopic object is so small that the fluctuations around the average bulk material properties are very important. The consequence is that mesoscopic behavior is governed not by the familiar laws of classical (Newtonian) mechanics, but rather by the laws of quantum mechanics. Due to the close overlap of length scales, the methods and terminology of mesoscopic physics would appear to be applicable to colloid science or nanoscale systems. But in practice, because the electronic properties of materials are not really significant for food scientists, there is little overlap between the fields. This contrasts sharply with the area of soft matter physics, which is properly considered to offer a relevant conceptual framework for describing food structure, even though the formal definition of "soft matter" includes no explicit concept of length scale (de Kruif, 2012).

In the broader philosophical context, there is an underlying perspective to nanoscience terminology and language extending beyond the simple length-scale specification or the vagaries of scientific fashion. This perspective is based on the capability to fabricate structures using a precise knowledge of the physics and chemistry underlying the organization of the individual building blocks. This concept of structure formation is known as the "bottom-up" approach (Semenova and Dickinson, 2010). That is, the application of the nanoscale perspective involves building the characteristics of a complex system, as manifest on the microscale through to the macroscale, by means of the control of structure and behavior on the nanoscale, i.e., at the molecular level. With such a philosophical perspective, the term "nanoscience" suggests a more comprehensive and ambitious scientific vision than that implied by traditional "colloid science," whose structure-generating methods are mainly of the so-called "top-down" variety (e.g., particle size reduction by application of brute force). Therefore, while their operational length scales overlap very considerably, the approaches of colloid science and nanoscience do remain conceptually distinct. Traditional food colloid science is grounded in the experimental investigation of observable behavior, whereas the study of food nanoscience implies the design and control of supramolecular assemblies (Leser *et al.*, 2003).

Although essentially benign in its influence on scientific thinking, some of the new "nano" nomenclature does have potentially confusing consequences. The term "nanoemulsion" is rather noteworthy in this context, especially when used by those who appear unaware of the already well-established meaning of the "microemulsion." As carefully explained by McClements (2012a), a nanoemulsion is simply a conventional emulsion with droplets of nanoscale size (up to $\approx 0.1\ \mu m$). That is, it is a thermodynamically unstable dispersion of one liquid in another; the morphology is either oil-in-water (O/W) or water-in-oil (W/O). Therefore, an O/W nanoemulsion is roughly equivalent to what is known in the field of emulsion polymerization as a "miniemulsion," although by convention the latter has a larger upper size limit of $\approx 0.5\ \mu m$ (Landfester *et al.*, 1999).

In contrast to the intense mechanical agitation required to form a nanoemulsion (or miniemulsion), the microemulsion is a thermodynamically stable colloidal system formed spontaneously by mixing oil and water in the presence of a suitable surfactant and cosolvent (Garti and Aserin, 2012). It is a type of "association colloid"; it may be oil-continuous, water-continuous or bicontinuous; and it consists of entities called self-assembled structures (micelles, bilayers, etc.). In fact, the oil (or water) droplets in an O/W (or W/O) microemulsion can be considered to possess the structural character of swollen surfactant micelles (or reverse micelles). Most importantly, microemulsion droplets are typically just a few nanometers in size, i.e., considerably smaller than the average droplet size of a nanoemulsion. Perversely, then, "nano" is bigger than "micro" in the world of emulsion science!

ESSENTIAL PRINCIPLES OF STRUCTURE FORMATION AND STABILIZATION

The two main classes of structural entities found in food colloids are particles and polymers (Dickinson, 1992). These entities exist in a wide range of shapes and sizes, as illustrated schematically in Figure 1.2. Particles may be

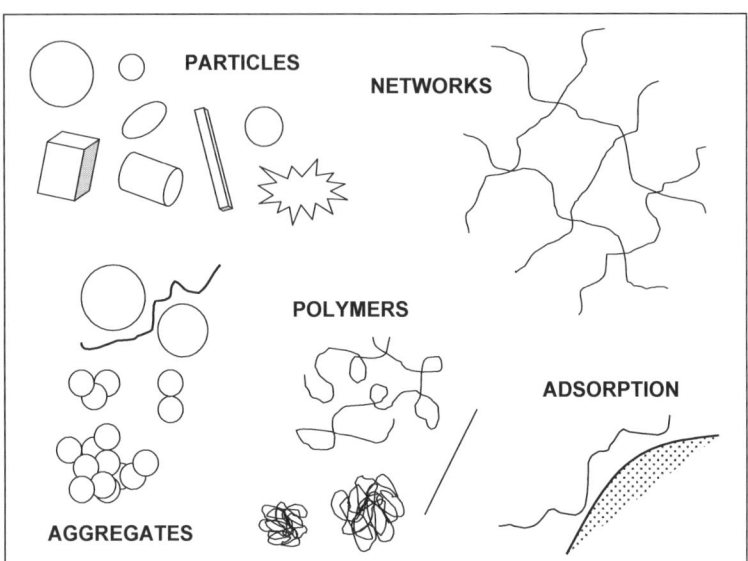

■ **FIGURE 1.2** Schematic representation of the primary structural units of food colloids. Particles tend to exhibit aggregation. Polymers tend to form networks and stick to surfaces. *After Dickinson (1992) with some modifications.*

exactly spherical, like isolated liquid droplets and gas bubbles, or more irregular, like fat crystals, starch granules, and protein aggregates. Polymers may exist as long extended chains (food polysaccharides) or compact organized structures (food proteins). Both particles and polymers may be aggregated to some extent, and these states of aggregation may extend to macroscopic dimensions (gel-like networks). Some of the polymers tend to stick to the surfaces of solid particles, droplets, and bubbles; and some form bridges between these surfaces. Some particles are themselves composed of polymers, and some may be trapped within polymer networks.

Food polysaccharides and food proteins have contrasting functional properties (Dickinson, 2003). Polysaccharides are stiff polydisperse polymers of high molecular weight and predominantly hydrophilic character. Under the technical description of "hydrocolloids," they are routinely used for texture control in food colloids as thickening agents (xanthan gum, guar gum, and carboxymethylcellulose) or gelling agents (alginate, pectin, carrageenan). Gelation of a solution/dispersion of polysaccharide may be induced in various alternative ways, e.g., by heating, cooling, or addition of salts. Beyond their thickening/gelling functionality, some polysaccharides such as modified starch, maltodextrin, and chitosan are also employed in various types of encapsulation technology. Furthermore, in addition to their widespread use in food processing as texture modifying agents, hydrocolloid thickeners have several applications related to digestion and health. The rheological properties of hydrocolloids are exploited in the formulation of liquid foods for patients with swallowing difficulties (Funami, 2011) and in the development of foods with high satiating capacity (Fiszman and Varela, 2013). One specific kind of application involves the controlled gelation of a hydrocolloid under acidic conditions in the stomach, which slows down the process of gastric emptying, leading to an increased feeling of fullness, and hence a potential health benefit in terms of appetite control (Lundin *et al.*, 2008; Ström *et al.*, 2010). In addition, non-starch polysaccharides have an important structural role as dietary fiber, i.e., as food polymers that are resistant to enzymatic breakdown in the mouth and small intestine, but undergo slow fermentation in the colon (Ouwehand *et al.*, 2009).

Proteins exhibit a wide diversity of functional properties in food colloids as a consequence of their complex reactivity and amphiphilic characteristics (Foegeding and Davis, 2011). Animal-derived structural proteins such as casein, whey protein, gelatin, and egg proteins, as well as some plant proteins, are used for food colloid stabilization and texture control. Not only do proteins have the capability to adsorb strongly at oil—water and air—water interfaces and to function as effective stabilizers of emulsions and foams, they also have a strong tendency towards self-assembly, aggregation, and gelation, especially following heating or pH change. The wide diversity of aggregated protein structures that can form is exemplified by the case of the milk protein β-lactoglobulin (Nicolai *et al.*, 2011). Dense protein microspheres may be used as structure-forming particles within protein gels as an alternative to oil droplets in protein-filled emulsion gels (Saglam, 2012). In general, the exploitation of protein aggregation and gelation occurs within bulk aqueous phases of food systems. Nevertheless, it has been demonstrated recently (Iqbal *et al.*, 2013) that controlled aggregation of protein microspheres is useful also for structuring of lipid phases.

In combination with their ligand-binding properties, the self-assembly behavior of proteins provides a powerful method of generating nanoparticle structural units and nanoscale delivery vehicles that are capable of encapsulating bioactive food ingredients (Livney, 2010; Semenova and Dickinson, 2010; Matalanis *et al.*, 2011). Various kinds of protein-based nanoscale structuring are possible:

- micellar protein assembly
- enzymatically cross-linked protein nanogel particles
- protein-based nanotubes/nanospheres/nanofibers/nanocapsules
- complexes of proteins with amphiphilic compounds
- protein—polysaccharide complexes/conjugates/coacervates

- core—shell protein nanoparticles
- protein-stabilized lipid nanoparticles

By exploiting the self-assembly behavior of a major food protein such as casein, nanoscience-based opportunities are emerging for the fabrication of new functional structures with potential applications for nutraceutical encapsulation, e.g., casein nanocapsules (Semo *et al.*, 2007) or hollow casein nanospheres (Liu *et al.*, 2010). In relation to protein gelation and interfacial stabilization, one type of nanostructuring system that has been generating considerable interest recently is the class of long insoluble fibrils arising from the aggregation of a globular protein like β-lactoglobulin into highly ordered amyloid-type assemblies (Adamcik and Mezzenga, 2012; Kroes-Nijboer *et al.*, 2012).

The formation of a protein—polysaccharide complex implies the presence of an attractive nanoscale force between the two kinds of food biopolymers (Turgeon *et al.*, 2007). The interacting macromolecules may be dissolved in the aqueous phase, or they may reside at the surface of a colloidal particle such as a casein micelle (Corredig *et al.*, 2011). The character of the attractive protein—polysaccharide interaction may be strong and long-lasting, or weak and reversible. The presence of a covalent bond between two biopolymers represents an extreme kind of specific interaction, one that is strong and permanent. Non-specific attractive protein—polysaccharide interactions arise from the combination of many different kinds of individual chemical interactions (ionic, hydrogen bonding, hydrophobic, etc.) averaged in time and space over the pair of macromolecules. Depending on the solution conditions, the contribution of the electrostatic interactions may be predominant (Dickinson, 2008a, b). The presence of strong electrostatic interactions between oppositely charged biopolymers (e.g., gum arabic + gelatin) produces complex coacervates with the capability of stabilizing thin encapsulation shells around dispersed oil droplets. Acidic solution conditions of low ionic strength enhance the strength of attractive interactions between positively charged proteins and negatively charged polysaccharides; but weaker, more reversible complexes are formed around neutral pH. Hence the adjustment of acidity (or ionic strength) may cause protein—polysaccharide interactions to be substantially modified, even changing over from net attractive to net repulsive (or vice versa). The main significance of protein—polysaccharide complexation in food colloid systems is that, while most polysaccharides are not themselves surface active, when present in mixed biopolymer complexes they may exhibit a strong tendency to stick to oil—water or air—water interfaces (Dickinson, 2003, 2009a).

The long-standing success of the colloid science approach to food structuring has been most clearly demonstrated in the field of dairy science and technology (Mulder and Walstra, 1974; Walstra *et al.*, 2006). Using a combination of traditional and industrial processing methods, a single natural food ingredient, liquid milk, can be converted into a diverse collection of derived food products possessing a wide range of textures and multiphase structures—gels, emulsions, foams, plastic solids, and powders. The starting point in each of these cases of colloidal processing is a set of just three building blocks—fat globules, casein micelles, and whey proteins. As illustrated in Figure 1.3, the key stage in the transformation of the building blocks into the final structures is the triggering of some kind of "active state" by means of heating, acidification, enzyme activity, or mechanical action (Aguilera and Stanley, 1999; Aguilera, 2006). The textural character of each dairy product is determined by the proportion and distribution of the different phases present (liquid, solid, gas) and the colloidal nature of the stabilizing entities. In particular, desirable solid-like textural characteristics emerge from the formation of aggregated networks of structured colloidal particles: butter has a fat crystal network, whipped cream has a fat globule

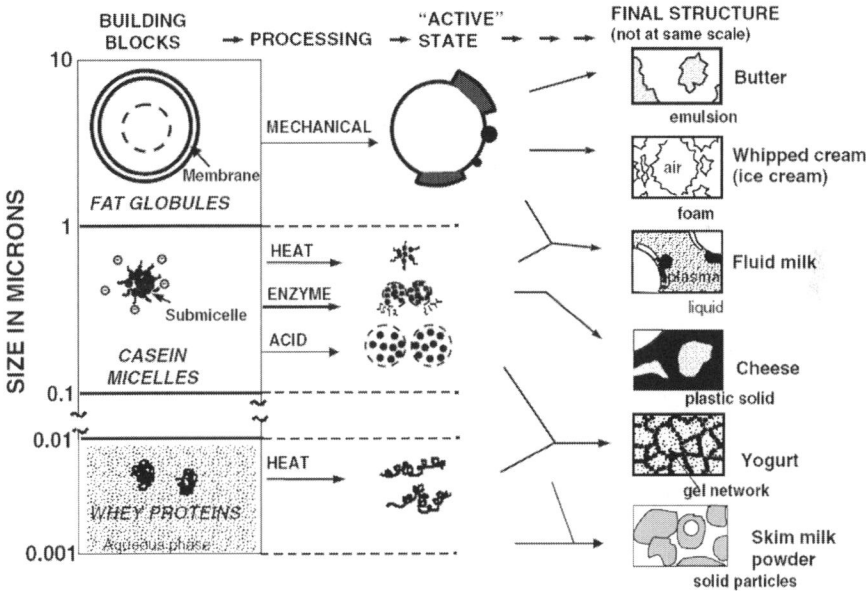

■ **FIGURE 1.3** Schematic representation of the fabrication of dairy products from the colloidal and molecular components of raw milk. Each processing stage transforms the building blocks into an "active state" during the structural transformation to the final product. *Diagram reproduced from Aguilera (2006) with permission.*

network, and cheese and yogurt have networks of partly destabilized casein micelles (Dickinson, 1988, 1992).

Emulsion droplets are extremely important structural entities involved in the fabrication of food products such as ice-cream, mayonnaise, and fatty spreads. As well as being an essential unit operation in product manufacture, emulsification is a key primary step in the encapsulation of hydrophobic nutraceuticals via the commonly used industrial technique of spray drying. The basic principles and practice of emulsification are now fairly well established (McClements, 2005). Conventional technology involves the "top-down" approach: intense mechanical forces are rapidly applied to an oil + water suspension with the aim of disrupting the large liquid drops into dispersed droplets of micrometer dimensions or smaller. A range of laboratory-scale emulsification devices are available based on ultrasonic disruption, high-speed mixing, jet homogenization, or high-pressure micro-fluidization. Large-scale equipment used in the food industry is based on the stirred tank, the colloid mill, or, for making the smallest droplets, the high-pressure valve homogenizer. More recently, the development of advanced emulsification methods has eliminated the need to apply an intense indiscriminate flow field by generating individual droplets as liquids pass through membrane pores or micro-channels (Boom, 2008). These low-intensity methods have potential advantages for the implementation of smart encapsulation technologies and for novel emulsion design involving shear-sensitive structural components. However, the current use of these low energy methods is still mainly restricted to laboratory studies.

Emulsifiers and stabilizers are the essential functional components of food emulsions (Dickinson and Stainsby, 1982). The emulsifier (emulsifying agent) is a surface-active ingredient that adsorbs at the oil—water interface during emulsification and protects the newly formed droplets against immediate recoalescence. The stabilizer provides long-term protection against the combined instability phenomena of flocculation, coalescence, and creaming (sedimentation). Stabilizers that adsorb at the droplet surface achieve their effectiveness through the colloid stability mechanisms of electrostatic and steric stabilization (Dickinson, 1992). Protein ingredients derived from milk or eggs can act as both emulsifiers and stabilizers in many food product formulations. The surface structures of the resulting protein stabilizing layers are sensitive to changes in thermodynamic variables such pH and temperature, which occur during emulsion processing and also during food digestion (Maldonado-Valderrama et al., 2009). The functional properties of proteins in food products are commonly further influenced by molecular interactions with small-molecule emulsifiers (Nylander et al., 2008). The emulsion stabilizing action of hydrocolloids such as carboxymethycellose or xanthan is mainly attributed to the thickening or gelling behavior of the biopolymer in the aqueous continuous phase (Dickinson, 2003, 2004).

Structural and rheological properties of food dispersions (emulsions) are ultimately determined by the nature of the interactions between the constituent colloidal particles (emulsion droplets). For surfaces covered with adsorbed biopolymers, the total free energy of interaction $W_{tot}(r)$ as a function of the separation distance r for a pair of particles (droplets) is composed of a sum of (at least) four separate contributions:

$$W_{tot}(r) = W_{disp}(r) + W_{el}(r) + W_{dep}(r) + W_{steric}(r) \qquad (1.1)$$

The individual free energy terms in Eq. 1.1 are defined as follows (Semenova and Dickinson, 2010):

- W_{disp} is the attractive van der Waals potential arising from the ubiquitous London dispersion forces acting between the fluctuating dipoles of all the polarizable molecules within the interacting particles and their adsorbed biopolymer layers.
- W_{el} is the electrostatic repulsive potential arising from overlap of electrical double layers around the charged particles.
- W_{dep} is the attractive depletion potential induced by the presence of non-adsorbed species (polymers, micelles, nanoparticles, etc.) in the vicinity of the particle surfaces.
- W_{steric} is the steric repulsive potential arising from the entropic interaction between overlapping biopolymer adsorbed layers.

The presence of a low concentration of non-adsorbing polysaccharide in the aqueous phase of a protein-stabilized emulsion may lead to reversible colloidal structuring due to depletion flocculation. This is illustrated in Figure 1.4 for the case a caseinate-stabilized emulsion with added xanthan gum (Moschakis *et al.*, 2005). When the polysaccharide is absent, or present at very low concentration (0.01 wt%), the emulsion appears stable and homogeneous, apart from the presence of some large droplets that cream to the top of the sample (see Figure 1.4A). On increasing the polysaccharide concentration to 0.05 wt%, the system exhibits microscale phase separation into discrete oil-rich and hydrocolloid-rich regions, the latter appearing as dark blobs against the lighter background, as shown in Figure 1.4B. The origin of this microscale phase separation is depletion flocculation of the emulsion droplets due to the presence of non-adsorbed polymer. The morphology of the phase-separated emulsion system is determined by a mechanical balance of the thermodynamic forces (interfacial tension) and the viscoelastic forces associated with the flocculated droplet network. This is an example of a general structure-forming phenomenon in soft matter

(A)   **(B)**

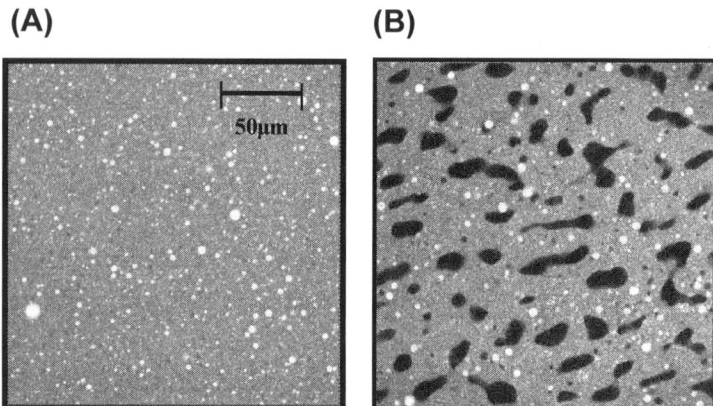

■ **FIGURE 1.4** Influence of non-adsorbing polysaccharide on the structure of a protein-stabilized emulsion (30 vol% oil, 1.4 wt% sodium caseinate, pH = 7) as observed by confocal microscopy: (A) stable emulsion (0.01 wt% xanthan); (B) microscopic phase separation and depletion flocculation (0.05 wt% xanthan). Large oil droplets appear as white particles; phase-separated regions depleted of emulsion droplets appear as dark blobs. *Reproduced from Moschakis et al. (2005) with permission.*

systems known as "viscoelastic phase separation" (Tanaka, 2012). Depletion flocculation of a protein-stabilized emulsion, and any associated viscoelastic phase separation, may also be induced by various kinds of non-adsorbing nanoparticles and micellar species (Dickinson, 2010a).

When the polysaccharide adsorbs at the droplet surface, the colloidal (in) stability behavior is rather different. Starting from a stable emulsion (Figure 1.5A), the addition of a small amount of polymer leads to bridging flocculation (Figure 1.5B). With further polymer addition, the extent of bridging flocculation increases steadily, reaching a maximum at an added polysaccharide concentration corresponding to about half-coverage of the available droplet surface area. When enough adsorbing polymer is present to saturate the whole of the available interface, the system re-establishes colloidal stability due to steric/electrostatic stabilization by the new outer polysaccharide layer (Figure 1.5C). At even higher added polymer concentrations, the dispersed droplets are immobilized in an entangled network of polysaccharide gel (Figure 1.5D) (Dickinson, 2009b). The same kind of emulsion stabilization mechanism, involving a combination of particle adsorption and entanglements, has also been observed with dispersed linear particles such as cellulosic nanorods (Kalashnikova *et al.*, 2013).

Making use of the basic principles of food colloid science, there has been good progress during the last few years in the application of structural design principles to the fabrication of emulsions with novel functional

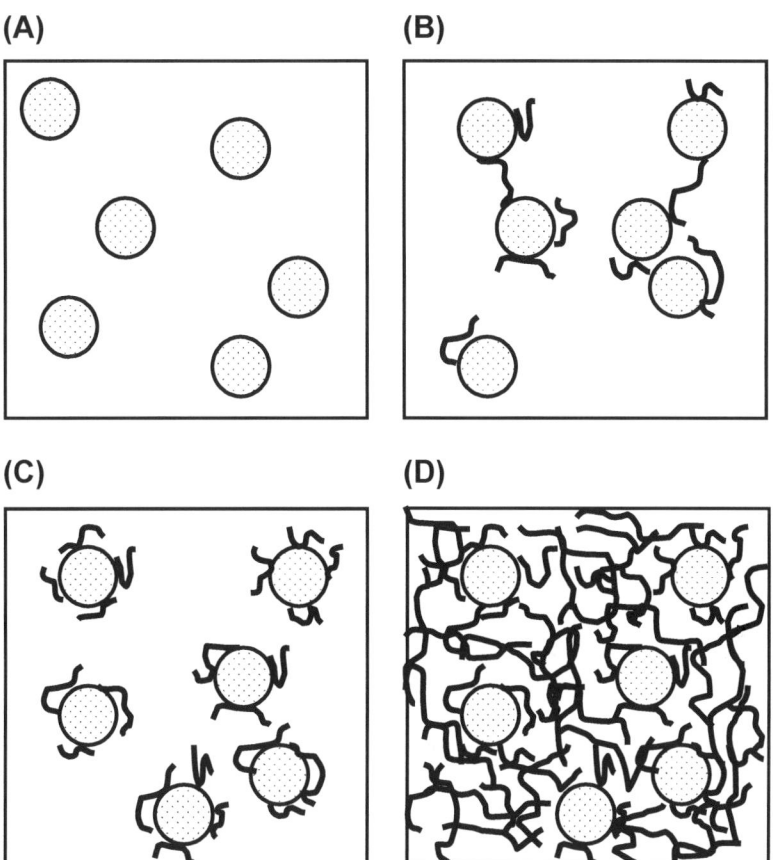

■ **FIGURE 1.5** Schematic representations of the structural states of protein-stabilized emulsions as a consequence of polysaccharide adsorption at the surface of the droplets: (A) original stable emulsion; (B) bridging flocculation at low polymer concentration; (C) steric stabilization at polymer concentration corresponding to saturated coverage; (D) stabilization by droplet entrapment in polymer network at high concentration.

properties. This activity has been driven by an increased interest from the food industry in the use of emulsions as delivery systems in foods (Appelqvist *et al.*, 2007; Velikov, 2012; Bouquerand *et al.*, 2012). A common objective of ongoing emulsion research is the systematic control of biopolymer interactions with the objective of stabilizing well-defined nanoscale structures around and between the dispersed droplets. In a recent overview of developments in this field (McClements, 2012b), it was explained that three kinds of generic approaches are commonly adopted: layering, embedding, and clustering.

The *layering* approach involves tailoring emulsion properties by building laminated coatings around the droplets. For food applications, these coatings typically take the form of adsorbed layers of proteins and polysaccharides produced by the sequential or simultaneous deposition of oppositely charged biopolymers at the emulsion droplet surface. Interpreted more broadly, this layering approach also embraces the concept of fabricating laminated coatings with inorganic particles, biopolymer fibers, or even (nano)emulsion droplets. Once initially formed, the multilayered structures may be further stabilized and strengthened by physical, chemical, or enzymatic cross-linking (McClements, 2010).

In the *embedding* approach, the individual emulsion droplets are trapped within another phase composed of a different material. The outer phase region may be of macroscopic dimensions, as in the case of an emulsion-filled biopolymer gel (Dickinson, 2012b). Or it may take the form of a larger dispersed entity: a double emulsion droplet, a spray-dried powder particle, or a filled hydrogel microsphere. This embedding approach allows the functional performance of the overall system to be controlled by the manipulation of the composition and properties of the matrix surrounding the emulsion droplets (Augustin and Hemar, 2009).

The *clustering* approach involves control of functional properties by modification of the aggregation state of the droplets. For a food protein-stabilized emulsion system, aggregation may be induced in various ways—by addition of polysaccharide, by enzyme action, or by the adjustment of temperature, pH, or ionic strength. Droplet clustering tends to increase the viscoelastic character of the emulsion system, with substantial consequences for colloidal stability (McClements, 2005), for in-mouth texture perception (Sarkar and Singh, 2012), and for lipid digestibility in the gastrointestinal tract (Singh *et al.*, 2009; Golding *et al.*, 2011).

SOME SPECIFIC TYPES OF FOOD EMULSION STRUCTURING

Multilayer Emulsions

Emulsions prepared with small-molecule emulsifiers or food proteins may exhibit loss of stability during long-term storage or when subjected to environmental stresses such as heating, freezing, drying, pH change, salt addition, or mechanical disturbance. A powerful strategy to reduce or eliminate these instability issues is to prepare an emulsion having two or more surface layers, with the sensitive inner layer of emulsifier or protein covered by one or more outer protective layers composed of a sterically stabilizing biopolymer. To be effective, the adjacent layers must be held together by strong molecular interactions. As well as enhancing physical stability, this multilayer technology may provide improved protection of O/W emulsions against oxidation of emulsified lipids and degradation of labile lipophilic compounds. In addition, there is the potential to control the rate of delivery of encapsulated lipophilic compounds by manipulating the permeability of multilayer coatings, and to design emulsion-based release triggers in response to the various changes which occur under gastrointestinal conditions (McClements, 2010).

The essence of this multilayering approach is illustrated schematically in Figure 1.6 for the case of an emulsion system with up to three layers. The primary O/W emulsion is prepared by homogenizing oil and water in the presence of an emulsifier such as an anionic surfactant or protein. Biopolymer A, of opposite net charge to the primary emulsifier, is subsequently adsorbed at the surface of the oil droplets to produce a secondary emulsion with a bilayer structure. Another biopolymer B, of opposite net charge to biopolymer A, is then deposited to produce a tertiary emulsion with a triple layer structure. Prior to introducing each biopolymer ingredient, the experimental protocol would normally comprise, first, a washing step (centrifugation or filtration) to remove any excess unadsorbed emulsifier/biopolymer, and, second, an agitation step (intense stirring or sonication) to disrupt unwanted aggregates formed by bridging flocculation (Guzey and McClements, 2006). Assuming that the primary droplets carry a net negative charge at neutral pH, the secondary emulsion could be prepared

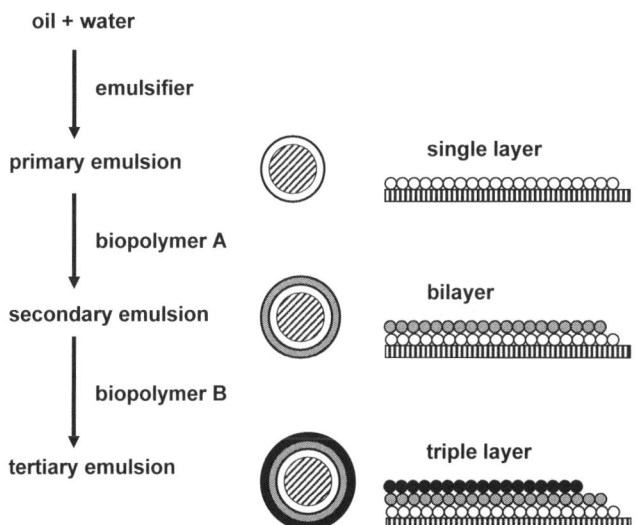

■ **FIGURE 1.6** Conceptual procedure for the fabrication of the nanoscale structure of multilayer O/W emulsions by sequential adsorption of layers of biopolymers onto oil droplet surfaces. The emulsifier used to make the primary emulsion may be a small-molecule surfactant or a biopolymer.

with either a cationic polysaccharide (e.g., chitosan) or a cationic protein (e.g., lactoferrin); and then the tertiary emulsion could be prepared using an anionic polysaccharide (e.g., pectin, carrageenan) or an anionic protein (e.g., casein, β-lactoglobulin) (McClements, 2012b). The performance of electrostatically deposited multilayers may be further enhanced by covalent cross-linking of the biopolymer(s). In such a way, the freeze–thaw stability of a bilayer emulsion of fish gelatin + sugar beet pectin has been found to be substantially improved by cross-linking the outer pectin layer via its ferulic acid groups with the enzyme laccase (Zeeb et al., 2011).

The nanoscale structure of a mixed biopolymer layer at an oil–water (or air–water) interface will depend on many factors, including the molecular conformation of the biopolymers, the aqueous solution conditions, and the method of preparation. For the case of sequential adsorption, the uniformity of the multilayer architecture which appears in Figure 1.6 is too idealized. In reality, there is likely to be a degree of interpenetration of adjacent biopolymer layers as well as heterogeneity in composition parallel to the interface (Dickinson, 1995b, 2011a). Instead of forming the bilayer emulsion by the sequential adsorption of protein and polysaccharide, an alternative method is simply to use the protein–polysaccharide complex as a mixed emulsifying ingredient in a single stage of homogenization. For the case of emulsions based on sodium caseinate + dextran sulfate, it has been shown (Jourdain et al., 2008) that, as well as being more convenient, this one-step preparation method leads to better emulsion stability than the two-step bilayer approach. Together with separate results from studies involving the sequential and simultaneous adsorption of β-lactoglobulin + pectin (Ganzevles et al., 2007), it may be inferred that the more condensed surface structure resulting from mixed biopolymer complex adsorption produces distinctly different colloid stabilizing behavior from that obtained with the equivalent bilayer interface. On the other hand, for extremely long adsorption times, thermodynamic considerations suggest that the nanoscale structures of the two kinds of interfaces should converge towards a single quasi-equilibrium state (Dickinson, 2011a).

Complexation of proteins with polysaccharides in mixed biopolymer layers may affect the accessibility of adsorbed proteins to enzymatic hydrolysis, with obvious implications for in vivo protein digestion kinetics. The essential principle underlying this phenomenon has been demonstrated experimentally (Jourdain, 2008) for the case of caseinate-based emulsions in the presence of dextran sulfate or ι-carrageenan, where the O/W emulsion was prepared by either sequential or simultaneous adsorption of the protein and polysaccharide at the oil–water interface. The data presented in Figure 1.7 indicate that the extent of proteolysis of adsorbed β-casein by trypsin, as measured

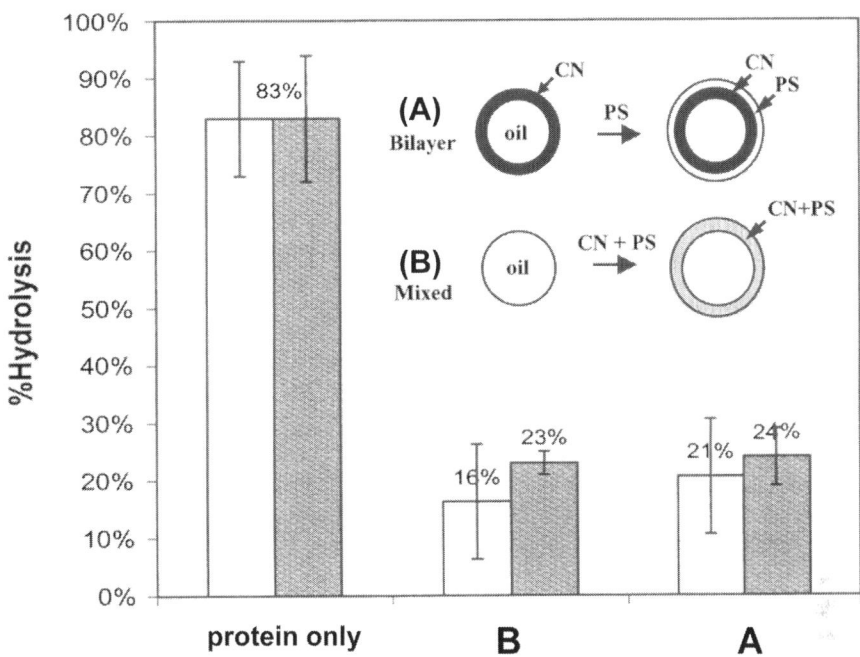

■ **FIGURE 1.7** Effect of interfacial protein–polysaccharide complexation on the extent of hydrolysis of β-casein by trypsin in O/W emulsions (20 vol% oil, 0.5 wt% sodium caseinate, pH = 6) containing either 1 wt% dextran sulfate (filled columns) or 1 wt% ι-carrageenan (white columns): A, bilayer emulsion system (sequential adsorption of CN and PS); B, mixed emulsion system (simultaneous adsorption of CN and PS). Proteolysis was allowed to take place for a period of 60 minutes at 25°C at an enzyme/substrate ratio of 1:2000. *Data taken from Jourdain (2008).*

under controlled laboratory conditions of pH and enzyme/substrate ratio, becomes considerably reduced (by a factor of around 4) on replacing the reference protein-stabilized emulsion by one stabilized by protein (CN) + anionic polysaccharide (PS). The reduced extent of hydrolysis could be attributed to the steric effect of the complexed polysaccharide, which inhibits the close approach of the enzyme to the casein's hydrolysis sites, as well as to indirect effects involving enzyme–polysaccharide electrostatic interactions (Jourdain, 2008).

Pickering Emulsions

Food emulsions commonly contain various kinds of dispersed particles. These may accumulate at the surfaces of liquid droplets and hence contribute to long-term emulsion stability behavior (Dickinson, 2006). In cases where the nanoparticles or microparticles are the primary stabilizing entities, the system is known as a Pickering emulsion. The conventional explanation for Pickering stabilization is that the adsorbed particles form a densely packed layer at the oil–water interface which prevents the close approach of neighboring droplets (Binks and Horozov, 2006). In this way the steric barrier of layered particles protects the emulsion against droplet flocculation and coalescence. The emulsion system may be of the O/W or W/O type depending on whether the Pickering particles are preferentially dispersible in water (i.e., particle surface is predominantly hydrophilic) or in oil (predominantly hydrophobic). Traditional foods exhibiting this Pickering stabilization mechanism include homogenized milk (an O/W emulsion stabilized by casein micelles) and fatty spreads (W/O emulsions stabilized by fat crystals).

The location of a single spherical particle with respect to an oil–water interface is determined by the balance of free energies at the particle–water, particle–oil, and oil–water surfaces, as shown in Figure 1.8A. The resolution of the three forces (per unit length) parallel to the solid surface at the junction of the three phases is expressed by the classical Young equation:

$$\cos \theta = \left(\gamma_{po} - \gamma_{pw}\right)/\gamma_{ow} \tag{1.2}$$

where θ is the contact angle, and γ_{po}, γ_{pw}, and γ_{ow} are the tensions at the particle–oil, particle–water, and oil–water interfaces, respectively. The case illustrated in Figure 1.8A corresponds to a particle that is preferentially wetted by water ($\theta < 90°$), which is a situation favoring the stabilization of an oil droplet in water (Figure 1.8B). Once attached to the oil–water interface with a finite contact angle, a particle of radius r can be regarded as being permanently adsorbed, since the free energy of spontaneous desorption, ΔG_d, is extremely high in relation to the thermal energy (kT):

$$\Delta G_d = \pi r^2 \gamma_{ow}(1 - \cos \theta)^2. \tag{1.3}$$

With typical values taken for γ_{ow} and θ, the predicted desorption energy for a spherical microparticle ($r = 1 \, \mu m$) is enormously large, i.e., $\approx 10^6 \, kT$. Even for the case of a nanoparticle (say $r \approx 5$ nm), assuming θ is not near $0°$ or $180°$, the state of adsorption can be regarded as essentially irreversible ($\Delta G_d >> 10 \, kT$) (Binks and Horozov, 2006).

Pickering stabilization conventionally requires the presence of a complete monolayer of closely packed particles at the oil–water interface, as illustrated in Figure 1.8B. In practice, however, emulsion stabilization by dispersed particles may also be achieved under conditions of reduced or low particle surface coverage if the particles are extensively aggregated (see Figure 1.8C). Under these aggregated conditions there are interconnected steric barriers formed around and among the droplets, and the viscoelastic character of the interconnected particle network protects the

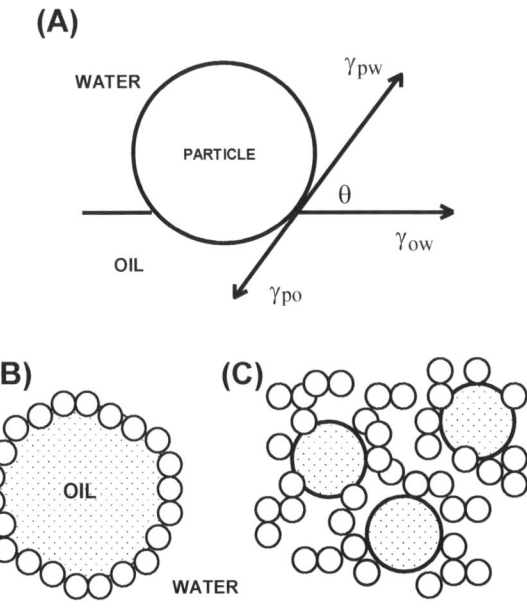

■ **FIGURE 1.8** Schematic representations of emulsion stabilization by spherical particles. (A) Location of predominantly hydrophilic particle at the oil—water interface, where θ is the contact angle and γ_{po}, γ_{pw}, and γ_{ow} are the tensions at the particle—oil, particle—water and oil—water boundaries, respectively. (B) Close-packed monolayer of equal-sized particles around an oil droplet in a Pickering O/W emulsion. (C) Aggregated particles in the aqueous continuous phase conferring emulsion stability through immobilization of oil droplets in a gel-like particle network.

O/W emulsion system, not only against oil droplet coalescence, but also against droplet creaming and macroscopic phase separation. Furthermore, in the case of aerated systems, the adsorption of particles and aggregates at the air—water interface provides excellent long-term protection against gas bubble shrinkage and foam coarsening (Du *et al.*, 2003; Murray, 2007).

Reports from various investigators have demonstrated that a wide range of microparticles and nanoparticles of biological origin can act as Pickering stabilizers (Dickinson, 2010b, 2012a). Food-grade microparticles effective as emulsion stabilizers include hydrophobically modified starch granules and cellulose particles/fibers. Microparticle stabilizers can be prepared from native starch granules by chemical modification with octenyl succinic anhydride (OSA) or by heat treatment (Yusoff and Murray, 2011; Rayner *et al.*, 2012). Nevertheless, these microparticles are relatively inefficient as emulsifying agents owing to their large particle size and inability to lower the oil—water interfacial tension, which means that the resulting Pickering emulsions consist of rather coarse droplets ($\approx 10\ \mu m$ diameter or larger).

In order to make Pickering food emulsions with colloid-sized droplets (i.e., ≈ 1 µm or less), investigators have turned to nanoscale particles of biological origin. The recent literature includes numerous reports of Pickering stabilization by various types of food-grade particulate materials, including systems based on flavonoids (tiliroside, rutin), proteins (zein), and polysaccharides (cellulose, chitin, starch). The successful incorporation of any of these nanoparticle-based stabilizing agents into a real food product is, however, unlikely to be straightforward. This is because it is inevitable that the nanoparticle aggregation behavior and its contact angle at the oil–water interface will be influenced by the presence of other food ingredients, especially surface-active lipids and soluble biopolymers (Dickinson, 2012a, 2013). For instance, it has been recently shown (Zhu *et al.*, 2013) that the presence of just a trace quantity of free fatty acids in a commercial triglyceride oil greatly modifies the effective contact angle of calcium carbonate nanoparticles at the triglyceride oil–water interface, sufficient even to cause emulsion phase inversion (O/W $\rightarrow$ W/O).

Double Emulsions

A double emulsion is an emulsion within an emulsion. There are two main types: water-in-oil-in-water (W/O/W) systems with water droplets inside oil droplets, and oil-in-water-in-oil (O/W/O) systems with oil droplets inside water droplets. Therefore, the double emulsion has three distinct bulk phases and two oil–water interfaces. In order to generate the two separate oil–water interfaces of the most common type of double emulsion, the W/O/W system, a two-stage emulsification procedure is typically employed involving two different emulsifying/stabilizing ingredients. A lipophilic emulsifier is used for preparing the inner droplets (the primary W/O emulsion) and a hydrophilic emulsifier for the outer droplets (the secondary O/W emulsion).

Double emulsions of the W/O/W type are recognized as attractive functional systems for incorporation into food products (Garti and Benichou, 2004; Muschiolik, 2007). There are two main areas of application. First, there is the capability to encapsulate a sensitive ingredient such as a hydrophilic nutrient or flavor compound within the inner aqueous phase, and then to release it at a controlled rate in response to some external trigger during eating/digestion. For instance, a protein-stabilized double emulsion formulation with encapsulation and release of anthocyanins under gastrointestinal conditions has been reported (Frank *et al.*, 2012). The second area of potential application lies in reduced-fat product development where a conventional O/W emulsion is replaced by an equivalent amount of W/O/W emulsion of lower absolute oil content but similarly perceived fat content (Norton and Norton, 2010). Both these areas of application have important commercial implications, although these have yet to be realized in practice. This is because, in attempting to move away from the synthetic surfactants and polymers that are routinely found in cosmetic and pharmaceutical formulations, it has proved to be a considerable challenge for food researchers to produce double emulsions of long shelf-life and good consumer acceptability using food-grade emulsifiers and food biopolymer stabilizers (Dickinson, 2011b).

The conventional method of making a double emulsion employs two separate emulsification steps involving high-intensity mixing or valve homogenization. Such vigorous processing generally leads to a heterogeneous state of ill-controlled structuring with a high level of polydispersity for both inner and outer droplets. However, recent advances in membrane-based methods, with their milder shear flow conditions, are now facilitating the production of double emulsions with outer droplets that are more uniform in size (Pawlik and Norton, 2012), albeit with a rather larger average diameter (tens of micrometers). One particularly elegant way to make monodisperse double emulsions based on large inner and outer droplets is by means of microfluidics (Utada *et al.*, 2005; Nisisako, 2008). Figure 1.9A shows a schematic diagram of a microfluidic device that has been used to prepare highly monodisperse double emulsions in a single step. The coaxial geometry of the device induces hydrodynamic focusing of the inner fluid flow inside a middle fluid, which is pumped down the injection tube in the opposite direction to the outer fluid. This microfluidic technology allows precise control of the number and size of inner droplets within each outer droplet (Utada *et al.*, 2005). With further refinements in the design of this kind of device, there is the possibility to produce a more complex multiple emulsion, e.g., a triple emulsion or "emulsion of double emulsion" (W/O/W/O or O/W/O/W). Figure 1.9B shows some examples of individual monodisperse droplets of double and triple emulsions produced with this coaxial microfluidic technology (Chu *et al.*, 2007).

Compared with equivalent O/W emulsions, the formulation of W/O/W emulsions is a much greater challenge in terms of the maintenance of long-term stability (Dickinson and McClements, 1995). Depending on the precise composition and morphology, the double emulsion may exhibit the following destabilization processes:

- coalescence of outer oil droplets, without any change in inner droplets;
- coalescence of inner water droplets, without any change in outer droplet interface;
- coalescence of inner droplets with outer droplet interface, leading to transfer of internal aqueous phase to external aqueous phase; and

(A)

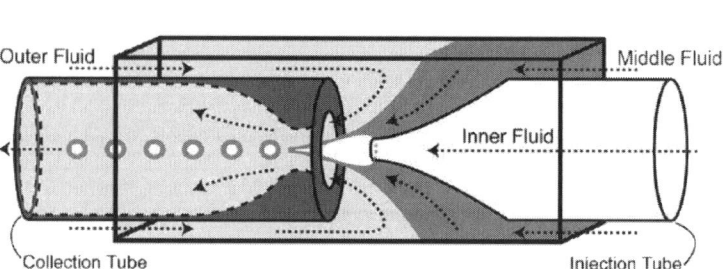

(B)

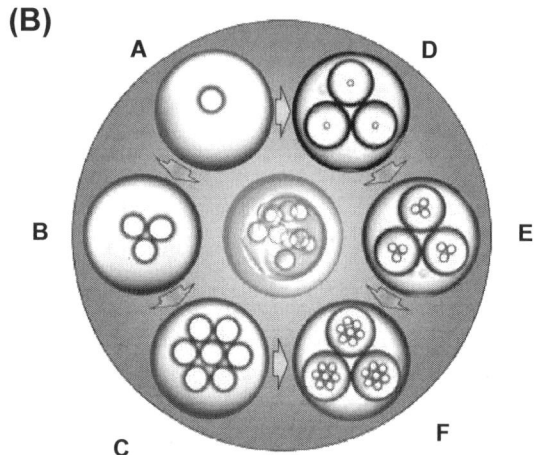

■ **FIGURE 1.9** Microfluidic fabrication of monodisperse multiple emulsions using a coaxial microcapillary device (Utada *et al.*, 2005). (A) Schematic representation of geometrical arrangement of cylindrical glass tubes nested within a square glass tube. The innermost fluid is pumped through a tapered capillary tube (hole diameter 10—50 μm). The middle fluid is pumped through the outer injection tube. The outermost fluid is pumped through the injection tube from the opposite direction. (B) Images of individual multiple droplets: A—C, double emulsion droplets containing 1, 3, or 7 inner droplets; D—F, triple emulsion droplets, each containing 3 intermediate-sized droplets, within which are 1, 3, or 7 inner droplets. *Reproduced from Chu* et al. *(2007) with permission.*

■ shrinkage or swelling of inner water droplets, arising from diffusive exchange of material between internal aqueous phase and external aqueous phase as a result of mass transport through the intervening liquid film.

The middle oil phase of a W/O/W emulsion droplet behaves rather like a miniature liquid "membrane" separating the two aqueous phases. Depending on the osmotic pressure difference, there is a thermodynamic driving force for water to diffuse through the oily membrane from the inner aqueous phase to the outer one (or vice versa). In principle, this instability can be controlled by using an oil phase of low water solubility and by carefully balancing the osmotic pressure difference with simple solutes (sugars) and electrolytes (Mezzenga, 2007). In practice, however, the osmotic balance is difficult to achieve in the context of a manufactured food product.

Another complication with conventional double emulsion formulation is that, in order to ensure that droplets of the primary W/O emulsion are sufficiently fine and stable, it is generally necessary to incorporate a rather high concentration of lipophilic emulsifier (e.g., polyglycerol polyricinoleate) (Dickinson and McClements, 1995). This requirement has commercial disadvantages in terms of adverse taste perception and difficulties of compliance with food ingredient regulations. There are also negative consequences for the double emulsion stability because the excess lipophilic emulsifier, in the form of aggregates and reverse micelles, tends to enhance the solubilization and transport of hydrophilic compounds through the oily membrane. In addition, there is an inevitable migration of some lipophilic emulsifier to the outer oil–water interface, which has the potential to undermine the stabilizing ability of the hydrophilic secondary emulsifier. One way to circumvent all these problems is to replace the molecular emulsifying/stabilizing ingredients by dispersed particles, i.e., to formulate a food-grade Pickering-in-Pickering emulsion. In principle, such a particle-stabilized double emulsion may be prepared with the primary W/O emulsion stabilized by fat crystals (monoglyceride/triglyceride) and the secondary emulsion with silica particles or modified starch particles (Garrec et al., 2012). An appealing feature of this particle-stabilized formulation is the temperature-responsive capability for triggered release of encapsulated hydrophilic compounds (salts, vitamins, etc.) after in-mouth melting of the fat crystals stabilizing the inner oil–water interface.

Emulsion Gels

Many complex food systems exhibit a combination of emulsion-like and gel-like attributes. The expression "emulsion gel" is used to describe an emulsion possessing a network structure and solid-like mechanical properties (Dickinson, 2012b). This type of colloidal state may be generated, for example, by heating a liquid-like whey protein-stabilized emulsion or acidifying a casein-stabilized emulsion. The structuring of the O/W emulsion gel may be characterized by (1) a network of flocculated oil droplets or (2) a network of cross-linked biopolymer molecules in the aqueous phase. These two situations are examples, respectively, of the clustering and embedding approaches of food structure design. An emulsion gel whose textural properties are determined by the interactions between the aggregated droplets is

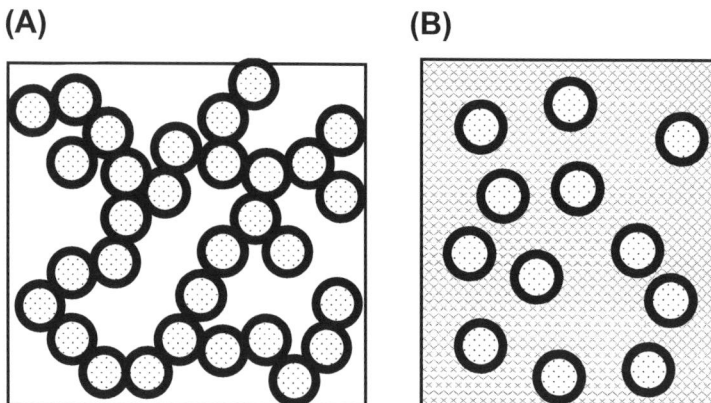

■ **FIGURE 1.10** Schematic presentations of two types of idealized O/W emulsion gel structure: (A) network of flocculated oil droplets; (B) dispersed oil droplets in biopolymer network.

illustrated schematically in Figure 1.10A. A droplet-filled gel with textural properties dominated by the biopolymer network of the gel matrix is displayed in Figure 1.10B. In practice, of course, the structuring of a real food emulsion gel will typically combine features from both these idealized representations.

Emulsion gels can be produced with a broad range of rheological properties and mouthfeel characteristics. Hence, they are convenient model systems for assessing the relationship between the structural characteristics of food colloids and sensory properties such as stickiness and creaminess (Sala *et al.*, 2008). The mechanical properties of emulsion gels are determined by various intrinsic factors such as the volume fraction of droplets and the nature of the filler—matrix interaction. The process of incorporating protein-stabilized emulsion droplets into a heat-set globular protein gel network leads to an enhancement in gel rigidity due to matrix reinforcement by the protein-coated droplets functioning as active filler particles (Chen and Dickinson, 1998). In contrast, when emulsion droplets are stabilized by small-molecule surfactants (e.g., Tween 20), they behave as inactive filler particles which lower gel rigidity (Dickinson and Hong, 1995). The matrix of the model emulsion gel illustrated schematically in Figure 1.10B represents a network of denatured protein molecules or cross-linked polysaccharide chains. Such a network has some of the features of an ideal rubber-like material, insofar as the elasticity is dominated by entropic contributions from flexible polymer chains distributed among the dispersed droplets. For an emulsion gel of moderately high volume fraction, the nature of the droplet—droplet interactions is inevitably more important in determining the rheological behavior; consequently, the aggregated particle model (Figure 1.10A) is a more suitable structural description.

Computer simulation is a powerful tool for exploring the relationship between the nature of the colloidal interactions and the properties of emulsion gels represented as aggregated particle networks. A pragmatic justification for the simulation approach is its success in reproducing the characteristic fractal-like structural features of real particle gel systems (Bijsterbosch *et al.*, 1995). Two alternative models have been used to study particle gels using Brownian dynamics simulation: a bonding model, where particles are joined together by flexible permanent cross-links, and a non-bonding model, in which pairs of particles are linked by reversible attractive colloidal forces (Dickinson, 2000). The shear modulus G of the aggregated particle network is determined by the geometrical structure of the network and the average number of stress-bearing connections (bonds) (van Vliet, 1999):

$$G = nC\left(d^2 F/dx^2\right). \qquad (1.4)$$

In Eq. 1.4, n is the number of bonds (per unit area) carrying the applied stress, C is a parameter related to the network structure, and dF is the change in Gibbs free energy due to a small length change dx between cross-links. For an emulsion gel modeled as an ensemble of flocculated droplets, the effective value of C depends on the manner in which the flocs are linked together. In addition to providing information on the small deformation elastic properties, computer simulations can give insight into the effect of colloidal interactions and microstructure on large deformation rheology and fracture properties of emulsion gels modeled as particle networks (Whittle and Dickinson, 1998; Rzepiela *et al.*, 2004).

A convenient method of encapsulating lipophilic compounds is the entrapment of emulsion oil droplets within the matrix of hydrogel microspheres. Several kinds of experimental methods are available to fabricate these filled hydrogel microspheres, including thermodynamic approaches based on coacervation or incompatibility of biopolymers, and engineering approaches based on injection, spray drying, emulsion templating, or microfluids (Augustin and Hemar, 2009; Matalanis *et al.*, 2011). Using the injection method, for example, emulsion droplets can be mixed into a solution of a hydrocolloid (alginate) and then injected into a solution containing calcium ions to induce rapid gelation. The release of functional components from food after eating may be achieved by controlling the shrinkage/swelling of hydrogel microparticles or by triggering their degradation/dissolution as a consequence of changing pH, temperature, or enzymatic conditions (Wang *et al.*, 2012). For instance, the digestion of protein-stabilized lipid droplets under simulated intestinal conditions has been shown to be appreciably delayed by encapsulating the droplets in hydrogel beads made from complex coacervates of chitosan + calcium alginate (Li and McClements, 2011). Other potential applications of microgel particles are as Pickering-type emulsion stabilizers (Brugger *et al.*, 2008; Destribats *et al.*, 2011), as delivery vehicles for protection of encapsulated lipophilic antioxidants against chemical degradation (McClements, 2012b), and as swollen dispersed particles for viscosity enhancement in low-fat products (Garrec *et al.*, 2012).

Aerated Emulsions

Air is inexpensive and non-fattening. And aerated foods have a smooth and appealing texture for consumers. Consequently, there is strong incentive for the food industry to make effective use of air bubbles and air cells as structural units in the fabrication of low-calorie low-fat products. Small bubbles are especially attractive as structural units because of their known contribution to creaminess perception (Kilcast and Clegg, 2002). The main technical problem is, of course, one of colloidal stability.

Aerated liquid dispersions and foams are more difficult to prepare and to keep stable than other kinds of colloidal structures. Compared with dispersions of fine emulsion droplets, the larger air–water interfacial area of aerated systems increases the probability of film rupture and bubble–bubble coalescence. Furthermore, the appreciable solubility of air in water facilitates steady gas diffusion between bubbles of different sizes (disproportionation) and slow loss of total gas-phase volume (Dickinson, 1992). In addition, the strong buoyancy forces acting on individual bubbles in liquid media cause rapid gravity-induced creaming and phase separation. In principle, this creaming may be inhibited by providing the system with a yield stress sufficient to overcome the buoyancy forces. For instance, in a model aerated dispersion of sugar particles of high solid content (≈ 40 wt%), broadly equivalent to the composition of a fondant confectionery product, it has been shown that phase separation due to gas bubble creaming may be inhibited for a period of several weeks (Lau and Dickinson, 2007). In industrial practice, the conventional wisdom is that, in order to confer long shelf-life on a bubble-filled food product, it is necessary to convert the transient liquid foam into a solid foam. This may be achieved by means of biopolymer gelation (in mousses), thermal treatment (in bakery products), or freezing (in ice-cream).

A saturated monolayer of food protein at the air–water interface generally provides rather effective protection against quiescent bubble coalescence, but not against bubble shrinkage due to gas diffusion (Dickinson *et al.*,

2002). The reason for this is that, while the adsorbed protein layer behaves elastically on short timescales ($<<1$ s), it exhibits viscous flow on long time-scales ($>>1$ s). Hence, the mechanical strength of the protein monolayer is insufficient to resist the slow relentless change in bubble surface area that accompanies the process of disproportionation. For air-filled emulsions, it has been reported (Tchuenbou-Magaia *et al.*, 2011) that one way to over-come the problem of the collapsing protein molecular layer is to surround gas bubbles by nanoscale protein capsules (>10 nm) made from a cysteine-rich protein (bovine serum albumin or egg albumen) subjected to cross-linking by heat treatment. These protein nanocapsules are sufficiently thick and mechanically strong to resist bubble growth and shrinkage. An alternative way of protecting protein-stabilized gas bubbles against disproportionation involves using a unique protein called hydrophobin, which has the special ability to self-assemble into thin rigid layers at air–water interfaces, thereby allowing the generation of bubbles and foams that can remain stable for as long as several months (Cox *et al.*, 2009).

The rheology and perceived texture of an aerated emulsion are determined by how the gas bubbles interact with the emulsion droplets. Whipped dairy cream is the most familiar example of an edible soft solid that is synergis-tically stabilized by a structured combination of bubbles and droplets. Following a few minutes of brisk whipping, a bowl of liquid dairy cream (fat content $\approx 30-35$ wt%) is readily transformed into a smooth aerated colloid having a significant yield stress. The solid-like structure of whipped cream is held together by the continuous network of partially coalesced fat globules generated by the shearing forces during whipping (Anderson and Brooker, 1988). For the requisite degree of orthokinetic clumping to take place, the fat phase has to be semi-crystalline. Maintaining the temperature in the range $5-10°C$ during whipping delivers the optimum solid fat fraction of fat globules in the whipped cream.

While there are certain health-related reasons for aspiring to incorporate a liquid vegetable oil instead of dairy fat into whipped food emulsion prod-ucts, this objective requires the implementation of an alternative kind of colloidal structuring to replace fat globule clumping. In particular, when a protein-stabilized O/W emulsion contains completely liquid oil droplets, the aerated emulsion structure may be stabilized by trapping the air bubbles within a flocculated droplet network. The most convenient way to induce droplet flocculation in a casein-based emulsion is by acidification (Dickinson, 2010a). This approach has been exploited in the stabilization of an imitation whipped cream made by the aeration of an O/W emulsion containing a liquid vegetable oil as the dispersed phase (Allen *et al.*, 2006). The network structure of the resulting acid-induced emulsion gel was generated by whipping a sodium caseinate-stabilized emulsion at the same time as the pH was steadily lowered with glucono-δ-lactone. Due to the presence of the protein-mediated droplet—droplet interactions, this aerated emulsion system exhibits rheological behavior that is distinctly different from that of conventional whipped cream (Allen *et al.*, 2008). Additionally, the texture and long-term stability of the system may be enhanced by incorporating a hydrocolloid stabilizer (pectin).

RELATIONSHIP OF STRUCTURE TO SENSORY PERCEPTION

Once food has entered the mouth, its colloidal structure becomes disrupted by chewing and mastication. The constituent particles and aggregates are reduced in size and lubricated with saliva. Flavor is perceived through the body's detection of small molecules released during structural break-down (Taylor, 2009). As a consequence of hydrodynamic stresses within the oral cavity and interactions with mucus-coated surfaces, a food bolus is generated having structural characteristics essentially different from the original food. According to the (now) classical diagram of Hutchings and Lillford (1988), which is reproduced in Figure 1.11, the food bolus becomes

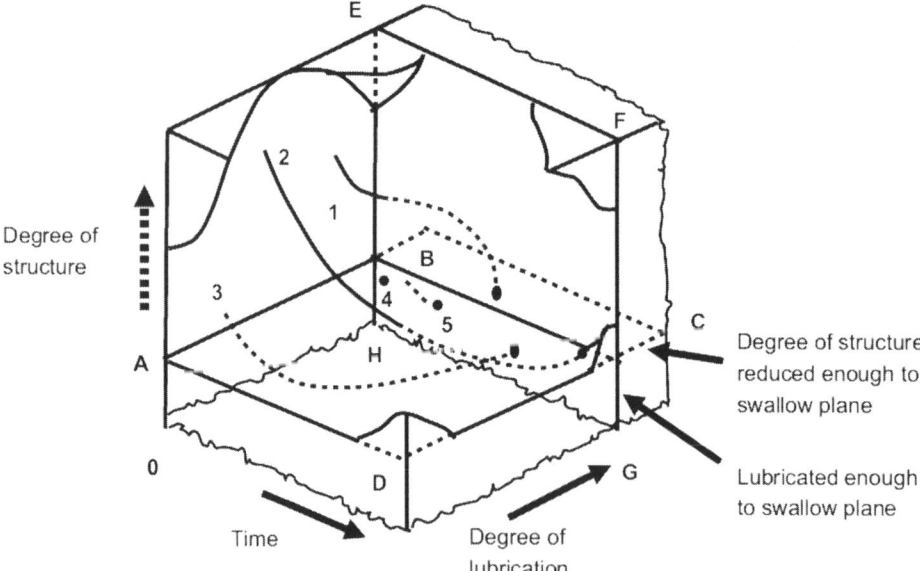

■ **FIGURE 1.11** Breakdown pathway of food in the mouth according to the dynamic model of Hutchings and Lillford (1988). The "degree of structure" and "degree of lubrication" are plotted against time for five representative food types: (1) tender juicy steak, (2) tough dry meat, (3) dry sponge cake, (4) oyster, and (5) liquids. The condition for swallowing of the bolus is defined by planes ABCD (structure) and EFGH (lubrication). *Diagram reproduced from Chen (2009) with permission.*

acceptable for swallowing once the "degree of structure" is sufficiently reduced and an appropriate "degree of lubrication" is reached. The swallowing time therefore depends sensitively on the composition and structure of the food: soft colloidal systems such as emulsions and gels exhibit relatively short swallowing times compared with those of highly structured foods like raw vegetables or tough dry meat.

The perceived texture of a colloidal system during oral processing is determined by its evolving rheological and tribological behavior (Chen, 2009; Foegeding *et al.*, 2011; Foster *et al.*, 2011; Chen and Stokes, 2012). Physiological signals generated during mastication regulate oral muscle activity and jaw movements, as well as providing the sensory input that informs the cognitive view of the texture of the food being eaten. Analyzing the processes taking place in the mouth over colloidal length scales is therefore essential for grasping the complex interplay between sensory perception, oral physiology, and food properties (van Vliet *et al.*, 2009). In relation to understanding the physicochemical basis of texture perception, it seems convenient to distinguish between liquid, solid, and semi-solid (or soft-solid) foods. Flow behavior and colloidal stability are the predominant factors determining the textural properties of liquid-like emulsions, whereas solid food texture perception is more obviously related to the stiffness and fracture behavior of the material.

Complex structural changes are exhibited by emulsion systems during oral processing. Systematic studies of model systems have led to the discovery of several distinct mechanistic processes (van Aken *et al.*, 2007, 2009; Sarkar and Singh, 2012):

- depletion flocculation of oil droplets by unadsorbed mucin;
- bridging flocculation due to associative interactions with mucin;
- release of oil droplets from biopolymer hydrogel particles;
- shear-induced coalescence and oil spreading at air−water interfaces and oral mucosa surfaces;
- breakdown of emulsions made with starch-based emulsifiers/stabilizers as a consequence of α-amylase activity; and
- phase inversion of W/O emulsions stabilized by fat crystals.

The structural changes associated with each of these mechanisms may affect oral sensory perception in various ways—including flavor release (Arancibia *et al.*, 2011). Reversible depletion flocculation leads to an enhancement in perceived thickness and creaminess, whereas bridging flocculation is associated with an increased perception of dryness, roughness, and astringency. Enhancement of fatty and creamy sensations is associated with the release of trapped oil droplets as a result of the fracturing and melting of emulsion gels or hydrogel microspheres. In-mouth coalescence of oil droplets leads to enhanced perception of fatty mouthfeel. This latter effect is due to the release of lipophilic aroma compounds from large coalesced droplets, as well as to the spreading of liquid fat on the oral mucosa, which reduces the perceived friction between tongue and palate. In fatty spread products, the phase inversion of the W/O emulsion during oral processing enhances the flavor perception of hydrophilic compounds (e.g., sodium chloride) previously trapped inside aqueous droplets (Le Révérend *et al.*, 2010).

The sensory attribute of "creaminess" is one of the most important characteristics of semi-solid food systems (van Aken *et al.*, 2009). Consumers invariably consider this attribute to be a multimodal sensation combining both texture ("thick," "smooth") and taste ("fatty," "dairy"). For model protein-stabilized emulsions, a strong correlation exists between perceived creaminess or thickness and the shear viscosity of the emulsion measured under controlled laboratory conditions (Akhtar *et al.*, 2005). Fat content is also a significant factor affecting perceived creaminess of model protein-stabilized emulsions and emulsion-filled gels, with higher scores recorded for creaminess of biopolymer gels containing inactive (unbound) oil droplets and gelatin-based emulsion gels that "melt in the mouth" (Sala *et al.*, 2008). For soft-solid foods like mayonnaise and custard, it has been found (van Vliet *et al.*, 2009) that rheological parameters characterizing structural breakdown in the mouth give the best correlation with scores of creaminess perception. While the word "creaminess" itself implies a strong association with the character of a high-fat dairy cream, it has been demonstrated (Janssen *et al.*, 2007) that bulk rheology and structural breakdown behavior are more significant factors than oil/fat content in determining the perceived creaminess of a non-dairy starch-based custard. This is because the inferred creaminess appears to be inversely related to the extent of shear-induced enzymatic breakdown during oral processing. Nonetheless, it was also established (de Wijk *et al.*, 2006) that the creaminess perceived in sensory tests with this same kind of low-fat starch-based system is further enhanced by the presence of fat which migrates to the surface of the bolus and causes release of fat-soluble flavors.

RELATIONSHIP OF STRUCTURE TO DIGESTION AND HEALTH

The past decade or so has seen significant growth in multidisciplinary research directed towards understanding the physical processes that occur during the passage of food through the various stages of the human gastrointestinal tract. This trend is illustrated by the bibliometric data in Figure 1.12 showing annual numbers of peer-reviewed publications and

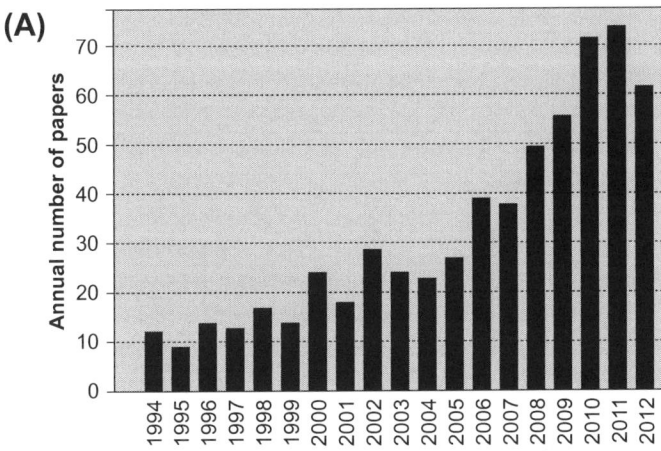

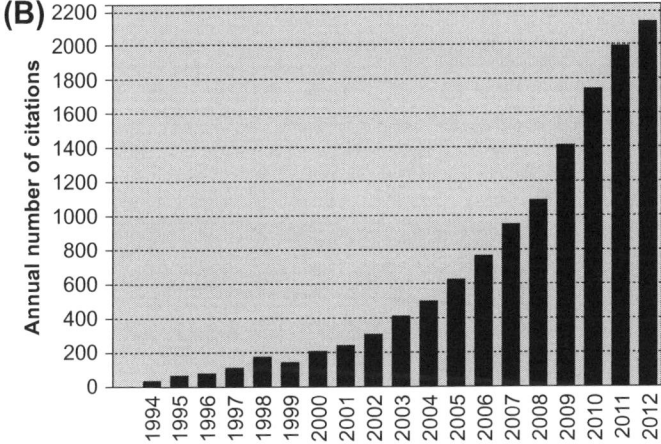

■ **FIGURE 1.12** Published research activity on food structure in relation to digestion based on a search of the database of *ISI Web of Science* over the past 20 years using the combined search topics of "food structure" and "digestion": (A) number of papers published each year, and (B) number of citations received by these papers each year.

citations for the combined topics of "food structure" and "digestion" over the period 1993–2012. We can observe a substantial increase in the annual number of published articles over this period, and an enormous growth rate in annual citations. According to the authors of a large proportion of these publications, the major long-term objective of their research is to address issues of diet-related ill-health. Therefore, it is no surprise to find that plots similar to those shown in Figure 1.12 (but not reproduced here) can also be generated from combining the search topics of "food structure" and "health."

There now exists a considerable body of knowledge concerning the basic types of physiological and biochemical processes taking place during the digestion and absorption of lipids from food (Lairon, 2009; Golding and Wooster, 2010; Lentle and Janssen, 2011). The bulk fats present in the diet, together with the large fat globules from partially destabilized emulsion-based foods, reside initially as a phase-separated cream layer at the top of the stomach (see Figure 1.13). The regular muscle contractions in the stomach and the intense shear forces in the antrum/pylorus region lead to (re)emulsification of this fat in the presence of bile salts and phospholipids. Enzymatic breakdown takes place at the surface of emulsion droplets, and so the total area of the oil–water interface is an important physicochemical parameter affecting the rate of digestion. Partial lipid digestion by gastric lipase occurs in the stomach under acidic conditions, but most of the digestion of the dietary lipids occurs in the small intestine through the action of pancreatic lipase (and colipase).

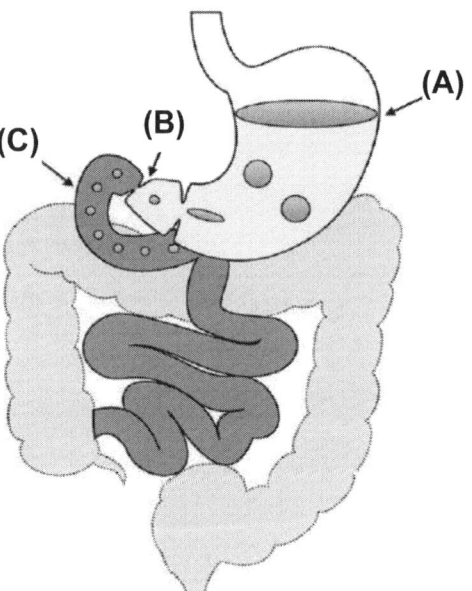

■ **FIGURE 1.13** Specific locations of fat digestion within the human gastrointestinal tract. (A) Emulsion destabilization under acidic gastric conditions produces a structured cream layer in the stomach. (B) The rate of enzyme-induced degradation is enhanced by mixing vortices in the stomach due to muscle contractions, and by the high shearing forces in the antrum/pylorus which cause (re)emulsification of large fat globules. (C) The small intestine is the major site of fat biochemical digestion involving pancreatic lipase, colipase, bile salts, and calcium ions. *Reproduced from Golding and Wooster (2010) with permission.*

While the absorption of lipids from food is undoubtedly a complex process mechanistically, some distinct structure-related factors have been identified as having a substantial influence on lipid digestibility and bioavailability (McClements, 2008; Singh *et al.*, 2009; van Aken, 2010; Wilde and Chu, 2011; Lentle and Janssen, 2011):

- ease of dissolution and breakdown of food matrices within the mouth, stomach, and small intestine;
- lipid composition and fatty acid chain-length distribution;
- initial emulsion droplet size and subsequent changes within the gastrointestinal tract;
- composition and structure of the adsorbed layer at the oil–water interface; and
- lipid solubilization within mixed micelles of bile salts and phospholipids.

Opportunities clearly exist for the exploitation of these structural factors in the design and fabrication of food colloid systems. For instance, matrix materials composed of dietary fiber (non-starch polysaccharides) are expected to remain largely intact until reaching the large intestine, whereas those based on starch or protein would be readily digested by oral or gastric enzymes (amylases and proteases). The rate of proteolysis during digestion is known to be affected by the state of protein aggregation and by the structure and composition of the adsorbed protein layer (Mackie and Macierzanka, 2010; Malaki Nik *et al.*, 2011; Singh and Sarkar, 2011). In a study of caseinate-based O/W emulsions, it was demonstrated (Macierzanka *et al.*, 2012) that cross-linking the interfacial protein with the enzyme transglutaminase prevents emulsion destabilization and retards proteolysis under *in vitro* gastric conditions. In addition to the aggregating effect of acidification, the gastric environment influences emulsion flocculation behavior via the elevated ionic strength and the interactions with mucin (Sarkar *et al.*, 2010). Furthermore, significant differences in emulsion stability under acidic conditions caused by changes in interfacial properties of emulsifier layers can be expected to influence the kinetics of lipid absorption, as well as the rate of gastric emptying and perceptions of postprandial satiety (Marciani *et al.*, 2009).

Despite impressive developments in computational fluid dynamics modeling (Ferrua *et al.*, 2011) and *in vivo* measurement techniques like magnetic resonance imaging (Schulze, 2006; Spiller *et al.*, 2009), most research on food colloid digestion relies on *in vitro* laboratory-scale experiments that purport to mimic the essential physicochemical features of the human digestion process. In such a static *in vitro* model, an attempt is made to replicate the biochemical environment within a single region of the gastrointestinal tract

(McClements and Li, 2010). Temperature, pH, and ionic strength are regulated, and an appropriate cocktail of enzymes and biosurfactants is included in order to imitate the typical aqueous solution conditions of the stomach or small intestine. In a further refinement of this static approach, the investigated sample is subjected to some kind of regulated *in vitro* mechanical agitation in an attempt to replicate the hydrodynamic conditions of the physiological environment (Kong and Singh, 2010; Chen *et al.*, 2011). And in order to mimic the temporal evolution of the entire digestion process, an integrated multi-chamber device is seemingly required (Wickham *et al.*, 2009).

A screening tool for studying the effect of food emulsion composition and structure on lipid digestibility in the small intestine is the so-called "pH stat" method. This simple test involves measuring how much alkali is required to neutralize the free fatty acids released from triglycerides following lipase addition under standardized conditions (McClements and Li, 2010). Mechanistic information derived from the pH stat method includes the effects on lipid digestion of a range of physicochemical factors such as fat composition, initial emulsion droplet size, the composition of the emulsifier/stabilizer layer, the aggregation state of the droplets, and the properties of any encapsulation matrix. For example, it has been demonstrated that lipid digestion is substantially delayed by trapping fat droplets within the gel matrix of alginate beads (Li *et al.*, 2011, 2012b) or by designing the fat composition in such a way that droplets undergo partial coalescence under gastric conditions (Golding *et al.*, 2011). Findings from selected *in vitro* experiments have also been validated against *in vivo* studies.

The adaptability and biochemical redundancy of physiological processing allows for efficient lipid digestion from varied diets and food sources. That is, in common with most biological systems, the human digestion system is "antifragile" towards environmental volatility (Taleb, 2012). So it has been found that the presence of a multilayer biopolymer coating on emulsion droplets does not necessarily delay lipid digestion to any appreciable extent (Hu *et al.*, 2011). In part, this is a direct consequence of the efficiency of the natural surfactants (phospholipids and bile salts) in displacing any adsorbed proteins from the emulsion droplet surface, irrespective of the initial degree of structuring of the interfacial layer. Therefore, while different formulations of protein-stabilized O/W emulsions may well exhibit substantial variations in average droplet size, aggregation state, and degree of proteolysis under simulated gastric conditions, it would appear that these differences do not necessarily influence very much the overall lipid digestion rate under gastric conditions (van Aken *et al.*, 2011) or the kinetics of digestion of lipids and proteins under intestinal conditions (Li *et al.*, 2012a; Macierzanka *et al.*,

2012). Looking further ahead towards the building up of a systematic body of knowledge about structure—function relationships based on these laboratory digestion studies, it would be reassuring to see a greater degree of consistency among researchers in the choices of *in vitro* digestion conditions, as well as more attention given to validation studies involving both *in vivo* animal models and human clinical trials (Hur *et al.*, 2011).

In summary, then, despite ongoing rapid advances, there remain large gaps in our understanding of the physicochemical factors controlling food digestion. In terms of its variability in structure and composition, the human diet is obviously a lot more complicated than the model systems that have so far been investigated under idealized laboratory conditions. Consequently, the precise nature of many of the colloidal interactions that occur between food ingredients under physiological conditions remains largely unknown. Against this background, a tremendous opportunity exists for collaborative investigations by scientists of various backgrounds into the fundamental mechanisms involved in these complex biological processes. In this way, the present tenuous links between food structure, digestion, and health will undoubtedly become more strongly established.

REFERENCES

Adamcik, J., Mezzenga, R., 2012. Protein fibrils from a polymer physics perspective. Macromolecules 45, 1137—1150.

Aguilera, J.M., 2005. Why food microstructure? J. Food Eng. 67, 3—11.

Aguilera, J.M., 2006. Food product engineering: building the right structures. J. Sci. Food Agric. 86, 1147—1155.

Aguilera, J.M., Germain, J.C., 2007. Advances in image analysis for the study of food microstructure. In: McClements, D.J. (Ed.), Understanding and Controlling the Microstructure of Complex Foods. Woodhead Publishing, Cambridge, UK, pp. 261—287.

Aguilera, J.M., Lillford, P.J. (Eds.), 2008. Food Materials Science. Springer, New York.

Aguilera, J.M., Stanley, D.W., 1999. Microstructural Principles of Food Processing and Engineering, second ed. Aspen Publishers, Gaithersburg, MD.

Akhtar, M., Stenzel, J., Murray, B.S., Dickinson, E., 2005. Factors affecting the perception of creaminess of oil-in-water emulsions. Food Hydrocoll. 19, 521—526.

Allen, K.E., Dickinson, E., Murray, B.S., 2006. Acidified sodium caseinate emulsion foams containing liquid fat: a comparison with whipped cream. Lebensmittel Wiss. Technol. 39, 225—234.

Allen, K.E., Dickinson, E., Murray, B.S., 2008. Development of a model whipped cream: effects of emulsion droplet liquid/solid character and added hydrocolloid. Food Hydrocoll. 22, 690—699.

Anderson, M., Brooker, B.E., 1988. Dairy foams. In: Dickinson, E., Stainsby, G. (Eds.), Advances in Food Emulsions and Foams. Elsevier Applied Science, London, pp. 221—255.

Appelqvist, I.A.M., Golding, M., Vreeker, R., Zuidam, N.J., 2007. Emulsions as delivery systems in foods. In: Lakkis, J.M. (Ed.), Encapsulation and Controlled Release Technologies in Food Systems. Blackwell, Oxford, pp. 41−82.

Arancibia, C., Jublot, L., Costell, E., Bayarri, S., 2011. Flavour release and sensory characteristics of o/w emulsions: influence of composition, microstructure and rheological behaviour. Food Res. Int. 44, 1632−1641.

Augustin, M.A., Hemar, Y., 2009. Nano- and micro-structured assemblies for encapsulation of food ingredients. Chem. Soc. Rev. 38, 902−912.

Belton, P. (Ed.), 2007. The Chemical Physics of Food. Blackwell, Oxford.

Bijsterbosch, B.H., Bos, M.T.A., Dickinson, E., van Opheusden, J.H.J., Walstra, P., 1995. Brownian dynamics simulation of particle gel formation: from argon to yoghurt. Faraday Discuss. 101, 51−64.

Binks, B.P., Horozov, T.S. (Eds.), 2006. Colloidal Particles at Liquid Interfaces. University Press, Cambridge, UK.

Boom, R.K., 2008. Emulsions: principles and preparation. In: Aguilera, J.M., Lillford, P.J. (Eds.), Food Materials Science. Springer, New York, pp. 305−339.

Bouquerand, P.-E., Dardelle, G., Erni, P., 2012. An industry perspective on the advantages and disadvantages of different flavour delivery systems. In: Garti, N., McClements, D.J. (Eds.), Encapsulation Technologies and Delivery Systems for Food Ingredients and Nutraceuticals. Woodhead Publishing, Cambridge, UK, pp. 211−251.

Brugger, B., Rosen, B.A., Richtering, W., 2008. Microgels as stimuli-responsive stabilizers for emulsions. Langmuir 24, 12202−12208.

Chen, J., 2009. Food oral processing—a review. Food Hydrocoll. 23, 1−25.

Chen, J., Dickinson, E., 1998. Viscoelastic properties of heat-set whey protein emulsion gels. J. Texture Stud. 29, 285−304.

Chen, J., Gaikwad, V., Holmes, M., Murray, B., Povey, M., Wang, Y., Zhang, Y., 2011. Development of a simple model device for *in vitro* gastric digestion investigation. Food Funct. 2, 174−182.

Chen, J., Stokes, J.R., 2012. Rheology and tribology: two distinctive regimes of food texture sensation. Trends Food Sci. Technol. 25, 4−12.

Chu, L.-Y., Utada, A.S., Shah, R.K., Kim, J.-W., Weitz, D.A., 2007. Controllable monodisperse multiple emulsions. Angew. Chem. Int. Ed. 46, 8970−8974.

Corredig, M., Sharafbafi, N., Kristo, E., 2011. Polysaccharide−protein interactions in dairy matrices: control and design of structures. Food Hydrocoll. 25, 1833−1841.

Cox, A.R., Aldred, D.L., Russell, A.B., 2009. Exceptional stability of food foams using class II hydrophobin HFBII. Food Hydrocoll. 23, 366−376.

de Kruif, C.G., 2012. The future of soft matter and food structure. Faraday Discuss. 158, 523−527.

de Wijk, R.A., Terpstra, M.E.J., Janssen, A.M., Prinz, J.F., 2006. Perceived creaminess of semi-solid foods. Trends Food Sci. Technol. 17, 412−422.

Destribats, M., Lapeyre, V., Sellier, E., Leal-Calderon, F., Schmitt, V., Ravaine, V., 2011. Water-in-oil emulsions stabilized by water-dispersible poly(*N*-isopropyl-acrylamide) microgels: understanding anti-Finkle behaviour. Langmuir 27, 14096−14107.

Dickinson, E., 1988. The structure and stability of emulsions. In: Blanshard, J.M.V., Mitchell, J.R. (Eds.), Food Structure—Its Creation and Evaluation. Butterworths, London, pp. 41−57.

Dickinson, E., 1992. An Introduction to Food Colloids. University Press, Oxford.

Dickinson, E. (Ed.), 1995a. New Physicochemical Techniques for the Characterization of Complex Food Systems. Blackie, Glasgow.

Dickinson, E., 1995b. Mixed biopolymers at interfaces. In: Harding, S.E., Hill, S.E., Mitchell, J.R. (Eds.), Biopolymer Mixtures. University Press, Nottingham, pp. 349–372.

Dickinson, E., 2000. Structure and rheology of simulated gels from aggregated colloidal particles. J. Colloid Interface Sci. 225, 2–15.

Dickinson, E., 2003. Hydrocolloids at interfaces and the influence on the properties of dispersed systems. Food Hydrocoll. 17, 25–39.

Dickinson, E., 2004. Effect of hydrocolloids on emulsion stability. In: Williams, P.A., Phillips, G.O. (Eds.), Gums and Stabilisers for the Food Industry, vol. 12. Royal Society of Chemistry, Cambridge, UK, pp. 394–404.

Dickinson, E., 2006. Interfacial particles in food emulsions and foams. In: Binks, B.P., Horozov, T.S. (Eds.), Colloidal Particles at Liquid Interfaces. University Press, Cambridge, UK, pp. 298–327.

Dickinson, E., 2008a. Interfacial structure and stability of food emulsions as affected by protein–polysaccharide interactions. Soft Matter 4, 932–942.

Dickinson, E., 2008b. Emulsification and emulsion stabilization with protein–polysaccharide complexes. In: Williams, P.A., Phillips, G.O. (Eds.), Gums and Stabilisers for the Food Industry, vol. 14. Royal Society of Chemistry, Cambridge, UK, pp. 221–232.

Dickinson, E., 2009a. Hydrocolloids as emulsifiers and emulsion stabilizers. Food Hydrocoll. 21 3, 1473–1482.

Dickinson, E., 2009b. Hydrocolloids and emulsion stability. In: Phillips, G.O., Williams, P.A. (Eds.), Handbook of Hydrocolloids, second ed. Woodhead Publishing, Cambridge, UK, pp. 23–49.

Dickinson, E., 2010a. Flocculation of protein-stabilized oil-in-water emulsions. Colloids Surf. B. 81, 130–140.

Dickinson, E., 2010b. Food emulsions and foams: stabilization by particles. Curr. Opin. Colloid Interface Sci. 15, 40–49.

Dickinson, E., 2011a. Mixed biopolymers at interfaces: competitive adsorption and multilayer structures. Food Hydrocoll. 25, 1966–1983.

Dickinson, E., 2011b. Double emulsions stabilized by food biopolymers. Food Biophys. 6, 1–11.

Dickinson, E., 2012a. Use of nanoparticles and microparticles in the formation and stabilization of food emulsions. Trends Food Sci. Technol. 24, 4–12.

Dickinson, E., 2012b. Emulsion gels: the structuring of soft solids with protein-stabilized oil droplets. Food Hydrocoll. 28, 224–241.

Dickinson, E., 2013. Stabilizing emulsion-based colloidal structures with mixed food ingredients. J. Sci. Food Agric. 93, 710–721.

Dickinson, E., Hong, S.T., 1995. Influence of water-soluble non-ionic emulsifier on the rheology of heat-set protein-stabilized emulsion gels. J. Agric. Food Chem. 43, 2560–2566.

Dickinson, E., McClements, D.J., 1995. Advances in Food Colloids, chap. 9. Blackie, Glasgow.

Dickinson, E., Stainsby, G., 1982. Colloids in Food. Applied Science, London.

Dickinson, E., Ettelaie, R., Murray, B.S., Du, Z., 2002. Kinetics of disproportionation of air bubbles beneath a planar air–water interface stabilized by food proteins. J. Colloid Interface Sci. 252, 202–213.

Du, Z., Bilbao-Montoya, M.P., Binks, B.P., Dickinson, E., Ettelaie, R., Murray, B.S., 2003. Outstanding stability of particle-stabilized bubbles. Langmuir 19, 3106–3108.

Ettelaie, R., 2003. Computer simulation and modelling of food colloids. Curr. Opin. Colloid Interface Sci. 8, 415–421.

Euston, S.R., Costello, G., Naser, M.A., Nicolosai, M.L., 2007. Modelling and computer simulation of food structures. In: McClements, D.J. (Ed.), Understanding and Controlling the Microstructure of Complex Foods. Woodhead Publishing, Cambridge, UK, pp. 334–386.

Ferrua, M.J., Kong, F., Singh, R.P., 2011. Computational modelling of gastric digestion and the role of food material properties. Trends Food Sci. Technol. 22, 480–491.

Fiszman, S., Varela, P., 2013. The role of gums in satiety/satiation. Food Hydrocoll. 32, 147–154.

Foegeding, E.A., Davis, J.P., 2011. Food protein functionality: a comprehensive approach. Food Hydrocoll. 25, 1853–1864.

Foegeding, E.A., Daubert, C.R., Drake, M.A., Essick, G., Trulsson, M., Vinyard, C.J., van de Velde, F., 2011. A comprehensive approach to understanding textural properties of semi- and soft-solid foods. J. Texture Stud. 42, 103–129.

Foster, K.D., Grigor, J.M.V., Cheong, J.N., Yoo, M.J.Y., Bronlund, J.E., Morgenstern, M.P., 2011. The role of oral processing in dynamic sensory perception. J. Food Sci. 76, R49–R61.

Foster, T.J., 2011. Natural structuring with cell wall materials. Food Hydrocoll. 25, 1828–1832.

Frank, K., Walz, E., Gräf, V., Greiner, R., Köhler, K., Schuchmann, H.P., 2012. Stability of anthocyanin-rich W/O/W emulsions designed for intestinal release in gastrointestinal environment. J. Food Sci. 77, N50–N57.

Funami, T., 2011. Next target for food hydrocolloid studies: texture design of foods using hydrocolloid technology. Food Hydrocoll. 25, 1904–1914.

Ganzevles, R.A., van Vliet, T., Cohen Stuart, M.A., de Jongh, H.H.J., 2007. Manipulation of adsorption behaviour at liquid interfaces by changing protein–polysaccharide interactions. In: Dickinson, E., Leser, M.E. (Eds.), Food Colloids: Self-Assembly and Material Science. Royal Society of Chemistry, Cambridge, UK, pp. 195–208.

Garrec, D.A., Frasch-Melnik, S., Henry, J.V.L., Spyropoulos, F., Norton, I.T., 2012. Designing colloidal structures for micro and macro nutrient content and release in foods. Faraday Discuss. 158, 37–49.

Garti, N., Aserin, A., 2012. Micelles and microemulsions as food ingredient and nutraceutical delivery systems. In: Garti, N., McClements, D.J. (Eds.), Encapsulation Technologies and Delivery Systems for Food Ingredients and Nutraceuticals. Woodhead Publishing, Cambridge, UK, pp. 211–251.

Garti, N., Benichou, A., 2004. Double emulsions for food applications. In: Friberg, S.E., Larsson, K., Sjöblom, J. (Eds.), Food Emulsions, fourth ed. Marcel Dekker, New York, pp. 353–412.

Golding, M., Wooster, T.J., 2010. The influence of emulsion structure and stability on lipid digestion. Curr. Opin. Colloid Interface Sci. 15, 90–101.

Golding, M., Wooster, T.J., Day, L., Xu, M., Keogh, J., Clifton, P., 2011. Impact of gastric structuring on the lipolysis of emulsified lipids. Soft Matter 7, 3513–3523.

Guzey, D., McClements, D.J., 2006. Formation, stability and properties of multilayer emulsions for application in the food industry. Adv. Colloid Interface Sci. 128, 227–248.

Hedges, E.S., 1931. Colloids. Edward Arnold, London.

Hu, M., Li, Y., Decker, E.A., Xiao, H., McClements, D.J., 2011. Impact of layer structure on physical stability and lipase digestibility of lipid droplets coated by biopolymer nanolaminated coatings. Food Biophys. 6, 37–48.

Hur, S.J., Lim, B.O., Decker, E.A., McClements, D.J., 2011. *In vitro* human digestion models for food applications. Food Chem. 125, 1–12.

Hutchings, J.B., Lillford, P.J., 1988. The perception of food texture—the philosophy of the breakdown path. J. Texture Stud. 19, 103–115.

Iqbal, S., Hameed, G., Baloch, M.K., McClements, D.J., 2013. Structuring of lipid phases using controlled heteroaggregation of protein microspheres in water-in-oil emulsions. J. Food Eng. 115, 314–321.

Janssen, A.M., Terpstra, M.E.J., de Wijk, R.A., Prinz, J.F., 2007. Relations between rheological properties, saliva-induced structure breakdown and sensory texture attributes of custards. J. Texture Stud. 38, 42–69.

Jourdain, L.S., 2008. Physical and enzymatic stability of emulsions containing complexes of casein and anionic polysaccharide. Ph.D. thesis, University of Leeds.

Jourdain, L., Leser, M.E., Schmitt, C., Michel, M., Dickinson, E., 2008. Stability of emulsions containing sodium caseinate and dextran sulfate: relationship to complexation in solution. Food Hydrocoll. 22, 647–659.

Kalashnikova, I., Bizot, H., Bertoncini, P., Cathala, B., Capron, I., 2013. Cellulosic nanorods of various aspect ratios for oil-in-water Pickering emulsions. Soft Matter 9, 952–959.

Kilcast, D., Clegg, S., 2002. Sensory perception of creaminess and its relationship with food structure. Food Qual. Pref. 13, 609–623.

Kong, F., Singh, R.P., 2010. A human gastric simulator (HGS) to study food digestion in human stomach. J. Food Sci. 75, E627–E635.

Kroes-Nijboer, A., Venema, P., van der Linden, E., 2012. Fibrillar structures in food. Food Funct. 3, 221–227.

Lairon, D., 2009. Digestion and absorption of lipids. In: McClements, D.J., Decker, E.A. (Eds.), Designing Functional Foods. Woodhead Publishing, Cambridge, UK, pp. 68–93.

Landfester, K., Bechthold, N., Tiarks, F., Antonietti, M., 1999. Formulation and stability mechanisms of polymerizable miniemulsions. Macromolecules 32, 5222–5228.

Lau, C.K., Dickinson, E., 2007. Stabilization of aerated sugar particle systems at high sugar particle concentrations. Colloids Surf. A 301, 289–300.

Le Révérend, B.J.D., Norton, I.T., Cox, P.W., Spyropoulos, F., 2010. Colloidal aspects of eating. Curr. Opin. Colloid Interface Sci. 15, 84–89.

Lentle, R.G., Janssen, P.W.M., 2011. The Physical Processes of Digestion. Springer, New York.

Leser, M.E., Michel, M., Watzke, H.J., 2003. "Food goes nano"—new horizons for food structure research. In: Dickinson, E., van Vliet, T. (Eds.), Food Colloids, Biopolymers and Materials. Royal Society of Chemistry, Cambridge, UK, pp. 3–13.

Li, J., Ye, A., Lee, S.J., Singh, H., 2012a. Influence of gastric digestive reaction on subsequent *in vitro* intestinal digestion of sodium caseinate-stabilized emulsions. Food Funct. 3, 320–326.

Li, Y., McClements, D.J., 2011. Controlling lipid digestion by encapsulation of protein-stabilized lipid droplets within alginate—chitosan complex coacervates. Food Hydrocoll. 25, 1025—1033.

Li, Y., Hu, M., Du, Y., Xiao, H., McClements, D.J., 2011. Control of lipase digestibility of emulsified lipids by encapsulation within calcium alginate beads. Food Hydrocoll. 25, 122—130.

Li, Y., Kim, J., Park, Y., McClements, D.J., 2012b. Modulation of lipid digestibility using structured emulsion-based delivery systems: comparison of *in vivo* and *in vitro* measurements. Food Funct. 3, 528—536.

Liu, C., Yao, W., Zhang, L., Qian, H., Wu, W., Jiang, X., 2010. Cell-penetrating hollow spheres based on milk protein. Chem. Commun. 46, 7566—7568.

Livney, Y.D., 2010. Milk proteins as vehicles for bioactives. Curr. Opin. Colloid Interface Sci. 15, 73—83.

Lundin, L., Golding, M., Wooster, T.J., 2008. Understanding food structure and function in developing food for appetite control. Nutr. Diet. 65, S79—S85.

Macierzanka, A., Böttger, F., Rigby, N.M., Lille, M., Poutanen, K., Mills, E.N.C., Mackie, A.R., 2012. Enzymatically structured emulsions in simulated gastrointestinal environment: impact on interfacial proteolysis and diffusion in intestinal mucus. Langmuir 28, 17349—17362.

Mackie, A., Macierzanka, A., 2010. Colloidal aspects of protein digestion. Curr. Opin. Colloid Interface Sci. 15, 102—108.

Magnuson, B.A., Jonaitis, T.S., Card, J.W., 2011. A brief review of the occurrence, use, and safety of food-related nanomaterials. J. Food Sci. 76, R126—R133.

Malaki Nik, A., Wright, A.J., Corredig, M., 2011. Impact of interfacial composition on emulsion digestion and rate of lipid hydrolysis using different *in vitro* digestion models. Colloids Surf. B 83, 321—330.

Maldonado-Valderrama, J., Gunning, A.P., Ridout, M.J., Wilde, P.J., Morris, V.J., 2009. The effect of physiological conditions on the surface structure of proteins: setting the scene for human digestion of emulsions. Eur. Phys. J. E 30, 165—174.

Marciani, L., Faulks, R., Wickham, M.S.J., Bush, D., Pick, B., Wright, J., et al., 2009. Effect of intragastric acid stability of fat emulsions on gastric emptying, plasma lipid profile and postprandial satiety. Brit. J. Nutr. 101, 919—928.

Matalanis, A., Jones, O.G., McClements, D.J., 2011. Structured biopolymer-based delivery systems for encapsulation, protection, and release of lipophilic compounds. Food Hydrocoll. 25, 1865—1880.

McClements, D.J., 2005. Food Emulsions: Principles, Practice and Techniques, second ed. CRC Press, Boca Raton, FL.

McClements, D.J. (Ed.), 2007. Understanding and Controlling the Microstructure of Complex Foods. Woodhead Publishing, Cambridge, UK.

McClements, D.J., 2008. Designing food structure to control stability, digestion, release and absorption of lipophilic components. Food Biophys. 3, 219—228.

McClements, D.J., 2010. Design of nano-laminated coatings to control bioavailability of lipophilic food components. J. Food Sci. 75, R30—R42.

McClements, D.J., 2012a. Nanoemulsions versus microemulsions: terminology, differences, and similarities. Soft Matter 8, 1719—1729.

McClements, D.J., 2012b. Advances in fabrication of emulsions with enhanced functionality using structural design principles. Curr. Opin. Colloid Interface Sci. 17, 235—245.

McClements, D.J., Li, Y., 2010. Review of *in vitro* digestion models for rapid screening of emulsion-based systems. Food Funct. 1, 32–59.

Mezzenga, R., 2007. Equilibrium and non-equilibrium structures in complex food systems. Food Hydrocoll. 21, 674–682.

Moschakis, T., Murray, B.S., Dickinson, E., 2005. Microstructural evolution of viscoelastic emulsions stabilized by sodium caseinate and xanthan gum. J. Colloid Interface Sci. 284, 714–728.

Mulder, H., Walstra, P., 1974. The Milk Fat Globule. Pudoc Publishing, Wageningen.

Murray, B.S., 2007. Stabilization of bubbles and foams. Curr. Opin. Colloid Interface Sci. 12, 232–241.

Muschiolik, G., 2007. Multiple emulsions for food use. Curr. Opin. Colloid Interface Sci. 12, 213–220.

Nicolai, T., Britten, M., Schmitt, C., 2011. β-Lactoglobulin and WPI aggregates: formation, structure and applications. Food Hydrocoll. 25, 1945–1962.

Nisisako, T., 2008. Microstructured devices for preparing controlled multiple emulsions. Chem. Eng. Technol. 31, 1091–1098.

Norton, J.E., Norton, I.T., 2010. Designer colloids—towards healthy everyday foods. Soft Matter 6, 3735–3742.

Nylander, T., Arnebrant, T., Bos, M., Wilde, P., 2008. Protein/emulsifier interactions. In: Hasenhuettl, G.L., Hartel, R.W. (Eds.), Food Emulsifiers and their Applications, second ed. Springer, New York, pp. 89–171.

Ouwehand, A.C., Tiihonen, K., Mäkeläinen, H., Rautonen, N., Hasselwander, O., Sworn, G., 2009. Non-starch polysaccharides in the gastrointestinal tract. In: McClements, D.J., Decker, E.A. (Eds.), Designing Functional Foods. Woodhead Publishing, Cambridge, UK, pp. 126–147.

Pawlik, A.K., Norton, I.T., 2012. Encapsulation stability of duplex emulsions prepared with SPG cross-flow membrane, SPG rotating membrane and rotor–stator technques—a comparison. J. Membr. Sci. 415–416, 459–468.

Pugnaloni, L.A., Matia-Merino, L., Dickinson, E., 2005. Microstructure of acid-induced caseinate gels containing sucrose: quantification from confocal microscopy and image analysis. Colloids Surf. B 42, 211–217.

Rayner, M., Sjöö, M., Timgren, A., Dejmek, P., 2012. Quinoa starch granules as stabilizing particles for production of Pickering emulsions. Faraday Discuss. 158, 139–155.

Rzepiela, A.A., van Opheusden, J.H.J., van Vliet, T., 2004. Large shear deformation of particle gels studied by Brownian dynamics simulations. J. Rheol. 48, 863–880.

Saglam, D., 2012. Design and functionality of dense protein particles. Ph.D. thesis, Wageningen University.

Sala, G., de Wijk, R.A., van de Velde, F., van Aken, G.A., 2008. Matrix properties affect the sensory perception of emulsion-filled gels. Food Hydrocoll. 22, 353–363.

Sarkar, A., Singh, H., 2012. Oral behaviour of food emulsions. In: Chen, J., Engelen, L. (Eds.), Food Oral Processing. Wiley, Oxford, pp. 111–137.

Sarkar, A., Goh, K.K.T., Singh, H., 2010. Properties of oil-in-water emulsions stabilized by β-lactoglobulin in simulated gastric fluid as influenced by ionic strength and presence of mucin. Food Hydrocoll. 24, 534–541.

Schulze, K., 2006. Imaging and modelling of digestion in the stomach and the duodenum. Neurogastroenterol. Motil. 18, 172–183.

Semenova, M., Dickinson, E., 2010. Biopolymers in Food Colloids: Thermodynamics and Molecular Interactions. Brill, Leiden.

Semo, E., Kesselman, E., Danino, D., Livney, Y.D., 2007. Casein micelle as a natural nanocapsular vehicle for neutraceuticals. Food Hydrocoll. 21, 936–942.

Singh, H., Sarkar, A., 2011. Behaviour of protein-stabilized emulsions under various physiological conditions. Adv. Colloid Interface Sci. 165, 47–57.

Singh, H., Ye, A., Horne, D., 2009. Structuring food emulsions in the gastrointestinal tract to modify lipid digestion. Prog. Lipid Res. 48, 92–100.

Spiller, R., Gowland, P., Marciani, L., 2009. Techniques for assessing the functional response to food of the stomach and small and large intestine. In: McClements, D.J., Decker, E.A. (Eds.), Designing Functional Foods. Woodhead Publishing, Cambridge, UK, pp. 362–386.

Ström, A., Boers, H.M., Koppert, R., Melnikov, S.M., Wiseman, S., Peters, H.P.F., 2010. Physicochemical properties of hydrocolloids determine their appetite effects. In: Williams, P.A., Phillips, G.O. (Eds.), Gums and Stabilisers for the Food Industry, vol. 15. Royal Society of Chemistry, Cambridge, UK, pp. 341–355.

Taleb, N.N., 2012. Antifragile. Allen Lane, London.

Tanaka, H., 2012. Viscoelastic phase separation in soft matter and foods. Faraday Discuss. 158, 371–406.

Taylor, A.J., 2009. Measurement and simulation of flavour release from foods. In: McClements, D.J., Decker, E.A. (Eds.), Designing Functional Foods. Woodhead Publishing, Cambridge, UK, pp. 294–313.

Tchuenbou-Magaia, F.L., Al-Rifai, N., Ishak, N.E.M., Norton, I.T., Cox, P.W., 2011. Suspensions of air cells with cysteine-rich protein coats: air-filled emulsions. J. Cell. Plastics 47, 217–232.

Turgeon, S.L., Schmitt, C., Sanchez, C., 2007. Protein–polysaccharide complexes and coacervates. Curr. Opin. Colloid Interface Sci. 12, 166–178.

Ubbink, J., 2012. Soft matter approaches to structured foods: from "cook-and-look" to rational food design? Faraday Discuss. 158, 9–35.

Utada, A.S., Lorenceau, E., Link, D.R., Kaplan, P.D., Stone, H.A., Weitz, D.A., 2005. Monodisperse double emulsions generated from a microcapillary device. Science 308, 537–541.

Velikov, K.P., 2012. Colloidal emulsions and particles as micronutrient and nutraceutical delivery systems. In: Garti, N., McClements, D.J. (Eds.), Encapsulation Technologies and Delivery Systems for Food Ingredients and Nutraceuticals. Woodhead Publishing, Cambridge, UK, pp. 317–391.

van Aken, G.A., 2010. Relating food emulsion structure and composition to the way it is processed in the gastrointestinal tract and physiological responses: what are the opportunities? Food Biophys. 5, 258–283.

van Aken, G.A., Bomhof, E., Zoet, F.D., Verbeek, M., Oosterveld, A., 2011. Differences in *in vitro* gastric behaviour between homogenized milk and emulsions stabilized by Tween 80, whey protein, or whey protein and caseinate. Food Hydrocoll. 25, 781–788.

van Aken, G.A., de Hoog, E.H.A., Vingerhoeds, M.H., 2009. Oral processing and perception of food emulsions: the relevance for fat reduction in food. In: McClements, D.J., Decker, E.A. (Eds.), Designing Functional Foods. Woodhead Publishing, Cambridge, UK, pp. 481–501.

van Aken, G.A., Vingerhoeds, M.H., de Hoog, E.H.A., 2007. Food colloids under oral conditions. Curr. Opin. Colloid Interface Sci. 12, 251–262.

van der Sman, R.G.M., 2012. Soft matter approaches to food structuring. Adv. Colloid Interface Sci. 176–177, 18–30.

van Vliet, T., 1999. Factors determining small-deformation behaviour of gels. In: Dickinson, E., Rodriguez Patino, J.M. (Eds.), Food Emulsions and Foams: Interfaces, Interactions and Stability. Royal Society of Chemistry, Cambridge, UK, pp. 307–317.

van Vliet, T., van Aken, G.A., de Jongh, H.H.J., Hamer, R.J., 2009. Colloidal aspects of texture perception. Adv. Colloid Interface Sci. 150, 27–40.

Velikov, K.P., Pelan, E., 2008. Colloidal delivery systems for micronutrients and neutraceuticals. Soft Matter 4, 1964–1980.

Vincent, J.F.V., 2008. The composite structure of biological tissue used for food. In: Aguilera, J.M., Lillford, P.J. (Eds.), Food Materials Science. Springer, New York, pp. 11–20.

Walstra, P., 2003. Physical Chemistry of Foods. Marcel Dekker, New York.

Walstra, P., Wouters, J.T.M., Geurts, T.J., 2006. Dairy Science and Technology, second ed. CRC Press, Boca Raton, FL.

Wang, Y., Bamdad, F., Song, Y., Chen, L., 2012. Hydrogel particles and other novel protein-based methods for food ingredient and nutraceutical delivery systems. In: Garti, N., McClements, D.J. (Eds.), Encapsulation Technologies and Delivery Systems for Food Ingredients and Nutraceuticals. Woodhead Publishing, Cambridge, UK, pp. 412–450.

Whittle, M., Dickinson, E., 1998. Large deformation rheological behaviour of a model particle gel. J. Chem. Soc. Faraday Trans. 94, 2453–2462.

Wickham, M., Faulks, R., Mills, C., 2009. *In vitro* digestion methods for assessing the effect of food structure on allergen breakdown. Mol. Nutr. Food Res. 53, 952–958.

Wilde, P.J., Chu, B.S., 2011. Interfacial and colloidal aspects of lipid digestion. Adv. Colloid Interface Sci. 165, 14–22.

Yusoff, A., Murray, B.S., 2011. Modified starch granules as particle stabilizers of oil-in-water emulsions. Food Hydrocoll. 25, 42–55.

Zeeb, B., Fischer, L., Weiss, J., 2011. Cross-linking of interfacial layers affects the salt and temperature stability of multilayered emulsions consisting of fish gelatin and sugar beet pectin. J. Agric. Food Chem. 59, 10546–10555.

Zhu, Y., Lu, L.-H., Gao, J., Cui, Z.-G., Binks, B.P., 2013. Effect of trace impurities in triglyceride oils on phase inversion of Pickering emulsions stabilized by $CaCO_3$ nanoparticles. Colloids Surf. A 417, 126–132.

Chapter 2

Processing of Food Structures in the Gastrointestinal Tract and Physiological Responses

Harjinder Singh[1] and Sophie Gallier[1,2]

[1]*Riddet Institute, Massey University, Palmerston North, New Zealand*, [2]*Danone Nutricia Research, Uppsalalaan, Utrecht, The Netherlands*

CONTENTS

INTRODUCTION

With the growing obesity crisis and the rapid increase in the number of people affected by type 2 diabetes, metabolic syndrome, and cardiovascular diseases, food manufacturers are challenged to develop more nutritious and healthier food products while retaining satisfactory sensory and textural characteristics. Foods are made up of several macro-components (water,

Food Structures, Digestion and Health. http://dx.doi.org/10.1016/B978-0-12-404610-8.00002-5

proteins, lipids, carbohydrates) and micro-components (minerals, vitamins, enzymes, nutraceuticals). The assembly and the interactions of these components define the structure of food materials. This structure at the molecular, nano, micro, and macro levels has a large influence on the sensory properties of foods as well as on the bioavailability of nutrients. The structure is provided by nature to meet specific functionality requirements for biological systems. For example, proteins in muscle fibers and polysaccharides in plant cell walls are involved in maintaining the integrity of the systems, whereas starch granules in tubers and oil bodies in seeds act as energy reserve units (Aguilera, 2005). Various kinds of structures, including fibrous structures (e.g., meat), fleshy materials (e.g., fruits and vegetables), encapsulated embryos (e.g., grains and pulses), and complex fluids (e.g., milk), can be found in natural foods (Aguilera, 2005).

Processing of foods is designed to preserve, transform, destroy, and re-create food structures. In fresh foods, such as meat, fish, fruits, and vegetables, the natural structures are largely preserved through minimal processing. In other food materials, such as grains and legumes, the controlled destruction of natural structures results in a variety of ingredients, including flours, starches, oils, and protein isolates (Aguilera, 2005). These ingredients are then transformed into a variety of manufactured/fabricated foods (e.g., baked products, confectionery, sauces, snacks, and desserts) through mixing, shearing, heating, and cooling processes. New structures are created via interactions of macromolecules during the processing of these foods. The design and the creation of structures in fabricated foods have been the subject of great scientific interest in food science and product development during the last decade. The major focus has been on improving the shelf-life, physical and textural properties, and sensory traits of these fabricated foods, which depend on the architecture of the food matrix. Until recently, the role of food structure/matrix as a variable in nutrient delivery has generally not been considered in the conception, design, or consumer appreciation of these products.

Food is consumed to sustain life as it contains key nutrients that are required to maintain human physiological processes. The most important nutrients are those that supply energy, i.e., lipids (9 kcal/g), and carbohydrates and proteins (both 4 kcal/g) (Jacobs and Tapsell, 2007). These are often called macronutrients and are also the main building blocks of the food structure. Micronutrients in food include vitamins and minerals, but there are many other physiologically active substances in foods (such as antioxidants, polyphenols, hormones, enzymes) that are not produced by human biological systems. The micronutrients and other bioactive compounds are often contained within the food structure and must be liberated from the matrix and converted into a form that can be easily absorbed by the human body. The initial processes of chewing (mastication) and mechanical and enzymatic digestion in the gastrointestinal tract break down the food matrix and release nutrients into the small intestine, the major site of nutrient absorption. In recent years, there has been increased interest in the understanding of how food structure and matrix design influence the rates of nutrient digestion and bioavailability. This research is being undertaken with a view to developing novel foods that regulate calorie intake, provide increased satiety responses, provide controlled digestion, and/or deliver bioactive molecules (Singh and Sarkar, 2011). This chapter provides an overview of current research into how proteins, lipids, and carbohydrates in different structural formats respond to gastrointestinal conditions and how this may affect the kinetics of digestion of macronutrients. Possible effects on postprandial metabolic responses are discussed where appropriate. Two case studies, milk and nuts, are presented to illustrate the impact of the food matrix on digestion, absorption, and physiological responses.

THE PROCESSES OF FOOD DIGESTION

The nutrients in food are contained in a variety of food structures and matrices. Digestion is the process of breaking down these structures to allow the release of individual nutrients that can be absorbed through the wall of the gastrointestinal tract.

The human gastrointestinal tract (GIT) or the alimentary canal is essentially an open-ended tube that passes through the body's ventral cavity. It comprises the oral cavity, the esophagus, the stomach, the small intestine, including duodenum, jejunum, and ileum, the large intestine, including ascending, transverse, and descending colon, rectum, and anus (Figure 2.1). The wall of the GIT consists of four layers throughout its length; a mucous layer, the submucosa, muscularis, and serosa, but their structures vary in different regions of the tract. The glands lining the tract supply the water, enzymes, and chemical environment required for digestion and the movement of material through the tract. The motor functions of the GIT are of two types; mixing movements and propelling movements. Mixing movements occur when smooth muscles contract rhythmically in small sections of the tract. Propelling movements include a wavelike motion called peristalsis, which is caused by contraction behind a mass of food as relaxation allows the mass to enter the next segment of the tract. Here, we provide a brief overview of the basic physicochemical and physiological processes that occur during the digestion of food.

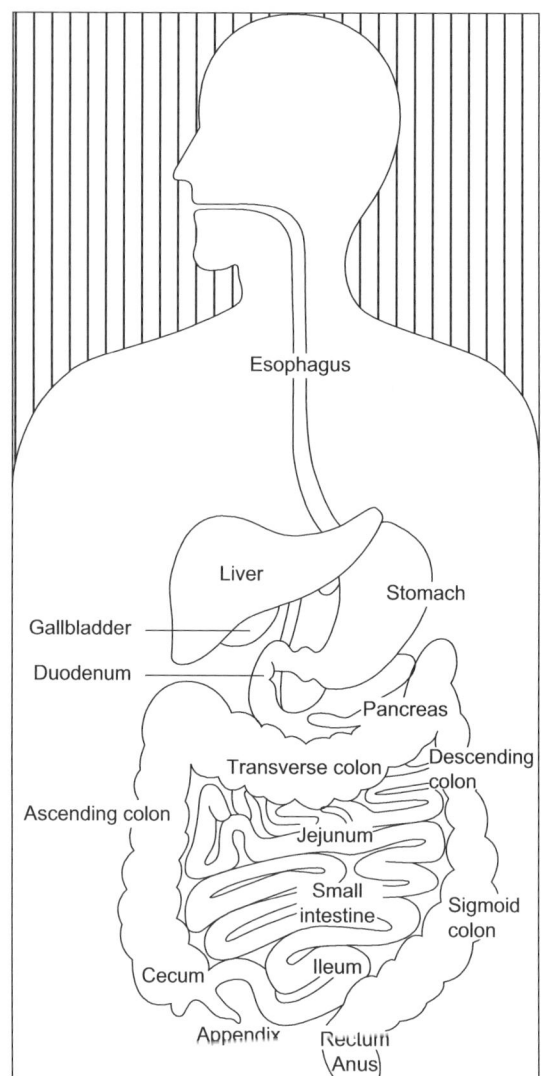

■ **FIGURE 2.1** Human digestive system.

ORAL PROCESSING

The mouth is the first unit of the gastrointestinal tract, through which food enters the digestive system; here, food is manipulated and processed both mechanically and biochemically, to allow its passage through the pharynx and esophagus to the stomach. This first step affects mostly the structure of fresh or processed solid or semi-solid foods. Through the mechanical action of chewing, the food matrix is disintegrated into small pieces/particles; mixing of the food particles with saliva allows the formation of a bolus and cooling or heating of the food to reach body temperature (van Aken, 2010; Turgeon and Rioux, 2011; Benjamin *et al.*, 2012). The residence time of food in the mouth depends on the nature of the food. Solid and semi-solid foods are broken down until they reach a size that is small enough for easy swallowing. Fluids tend to be swallowed rapidly, leading to a brief mixing with saliva. Mastication also allows the release of flavors, which can influence food intake (van Aken, 2010).

Saliva lubricates food to facilitate swallowing. It is a complex biological fluid (pH around 6.8) and contains salivary enzymes (such as α-amylase and carbonic anhydrase) and several other proteins (immunoglobulins, antibacterial proteins, proline-rich proteins, lysozyme, and lactoferrin) (Amado *et al.*, 2005). When a starchy food is consumed, salivary amylase can partially hydrolyze amylose and amylopectin molecules, decreasing the viscosity of these foods. The presence of mucins is important in emulsion-based food products, as mucins are known to bind to oil droplet surfaces, resulting in bridging flocculation of lipid droplets (Sarkar *et al.*, 2009a). When bulk fat/oil is consumed, emulsification of the oil may occur in the mouth, with the adsorption of dietary surfactants or salivary mucins at the oil—water interfaces (Zúñiga and Troncoso, 2012). The presence of a lingual lipase in human saliva is controversial; however, it has been identified in some animals such as rodents (N'Goma *et al.*, 2012).

GASTRIC PROCESSING

The stomach is a J-shaped muscular organ that receives and mixes food with digestive juices, and propels food to the small intestine. The stomach is divided into cardiac, fundic, body, and pyloric regions and a pyloric canal. A pyloric sphincter controls release of food from the stomach into the small intestine. The gastric movements induce a grinding effect between food particles, reducing their size further to less than 1—2 mm, as well as mixing with gastric juices to create a mixture called chyme. The movements of the antrum allow mixing and pumping of the chyme to the small intestine (Thomas, 2006; van Aken, 2010).

In the fasted state, the gastric juice has a basal pH between 1 and 3. After food intake, the pH increases to 5.5—7, decreases to 4—5 after about 60 min, which corresponds to the half stomach emptying time, and then returns to its basal value once food has left the stomach, which takes about 150 min (N'Goma *et al.*, 2012). The acidic environment of the stomach ensures that microbial growth is inhibited but also causes considerable changes to food structures. For example, at low acidic pH, aggregation of proteins may occur and lipid droplets may be destabilized through flocculation, resulting in coalescence (Gallier *et al.*, 2013b).

The gastric juice has an ionic strength of approximately 100 mM and contains pepsin and gastric lipase. Because of the action of pepsin, proteins are partially digested in the stomach, although some proteins are resistant to pepsin action. Physiological surfactants, such as phospholipids from the gastric mucosa, are known to affect the digestion of proteins (Mandalari *et al.*, 2009; Dupont *et al.*, 2010). Mandalari *et al.* (2009), using an *in vitro* model, showed that the presence of physiological phospholipids did not affect the resistance of β-lactoglobulin to pepsinolysis, but protected the protein from subsequent pancreatic proteolysis. The proteolysis of β-lactoglobulin by pepsin is more efficient when it is adsorbed at the surface of oil droplets than when it is in solution (Sarkar *et al.*, 2009b). This has been attributed to a possible change in the conformation of the β-lactoglobulin molecules upon adsorption at the oil−water interface, which exposes the peptic cleavage sites for proteolysis. Interestingly, adsorbed α-lactalbumin in oil-in-water emulsions appears to be more resistant to hydrolysis by pepsin, compared with native α-lactalbumin in solution (Nik *et al.*, 2010). Proteolytic products modulate the secretion of acid and pepsinogen, and stomach emptying.

About 10−30% of lipids are hydrolyzed into free fatty acids and diacylglycerols by the gastric lipase. The gastric emptying rate depends on the viscosity and structure of the food material remaining under the prevailing conditions (Turgeon and Rioux, 2011). In some food systems, phase separation between a water phase and a lipid phase may occur; the former is emptied faster than the latter (Chang *et al.*, 1968).

INTESTINAL PROCESSING

The small intestine is the main site for the digestion and release of nutrients and their conversion into an absorbable form. The secretion of pancreatic juice containing bicarbonate ions induces a drastic change in pH, between 5 and 7, with a mean value at 6.25 (Carriere *et al.*, 2005; N'Goma *et al.*, 2012). The pancreatic secretions contain many enzymes, such as proteases and peptidases (trypsin, chymotrypsin, carboxypeptidases, etc.), lipases, and esterases (pancreatic lipase, cholesterol esterase, phospholipase A_2, etc.) and pancreatic amylases (Singh *et al.*, 2009). Gastric-resistant proteins and peptides are further hydrolyzed by trypsin, chymotrypsin, and other proteases (from the brush border) into amino acids; the latter and small peptides are taken up by the enterocytes of the small intestine (Boirie *et al.*, 1997).

Carbohydrates are digested by amylase and glucosidase into monosaccharides. The digestion of disaccharides and some other oligosaccharides is carried out by the enzymes of the small intestinal brush border. Lipids are further digested by lipases such as pancreatic lipase (forming a complex with colipase), phospholipase A_2, cholesterol esterase, and pancreatic-lipase-related protein. Pancreatic lipolysis results in the release of two free fatty acids and one 2-monoacylglycerol. Biliary secretions contain conjugated bile salts, phospholipids, cholesterol, bilirubin, and electrolytes (Gallier and Singh, 2012b). Bile salts stabilize oil droplets and form micelles, which, in addition to unilamellar vesicles, are the transport vehicles of the lipolytic products to the enterocytes. 2-Monoacylglycerol molecules are easily absorbed by the enterocytes; however, long-chain fatty acids released as free fatty acids can form insoluble soaps with calcium and, to a certain extent, magnesium within the intestine (Knutson *et al.*, 2010). The formation of these soaps reduces the absorption of long-chain fatty acids and calcium and increases their fecal excretion. The peristaltic movements of the small intestine allow the digesta to be mixed with pancreatic and biliary secretions and progress towards the large intestine.

Any undigested material reaches the colon, where it is fermented by the gut microbiota (Wong *et al.*, 2012). Undigested carbohydrates and proteins are fermented, producing short-chain fatty acids (mainly acetate, propionate, and butyrate) that are absorbed in the colon.

FOOD MATRIX, NUTRIENT BIOAVAILABILITY, AND PHYSIOLOGICAL RESPONSES

Bioavailability is generally defined as the amount of an ingested nutrient that is absorbed and available for physiological functions. The bioavailability of a particular nutrient depends on a number of physicochemical processes, including the fraction of nutrient that is released from the food matrix (bioaccessibility), absorption by intestinal cells, and transport to body cells. As discussed above, nutrients are made bioaccessible by the processes of mastication, mixing with acid and complex enzymatic digestion that continues to break down the food matrix. In addition, bioaccessibility is aided by cooking and other food processing methods. This effect is especially prominent in plant foods. For example, micronutrients from fruits and vegetables are better absorbed when the fruits and vegetables are puréed or thermally processed, as the cell walls in the raw fruits and vegetables limit the bioaccessibility of micronutrients. In fruits and vegetables, antioxidants, such as phenolic acids, are easily absorbed when weakly bound to the cell walls (Palafox-Carlos *et al.*, 2011), whereas the absorption of lipid-soluble antioxidants, such as carotenoids, depends on their release from the food matrix during mastication and digestion and their solubilization into bile salt micelles. Because of the lipophilic properties of carotenoids, their absorption is improved in the presence of dietary lipids, similar to phytosterols and lipid-soluble vitamins (A, E, and D). Only 10% of carotenoids are bioavailable in raw carrots whereas 50% are bioavailable from oil and processed products (Parada and Aguilera, 2007). Mastication is a critical step for the release of nutrients from fruits and vegetables, as it allows reduction of the particle size, thus increasing the surface area available for the activity of digestive enzymes. In the small intestine, fibers can form soluble polymer chains, insoluble assemblies, or a swollen, sponge-like network. These forms, such as fruit tissues, can trap antioxidants or affect the mixing process in the upper digestive tract, thus reducing the rate of absorption of antioxidants (Palafox-Carlos *et al.*, 2011).

Macronutrients, i.e., carbohydrates, proteins, and fat, most often occur in foods in combination in a complex whole-food or processed food matrix, and are usually consumed as part of a meal together with other foods. Interactions among all these components form the basis of the food matrix that is seen by the digestive system. The release of macronutrients during digestion is dependent on their native structures and on multiple interactions that occur during food processing operations and in the gastrointestinal tract.

Starch

The principal carbohydrate, i.e., starch (found in plant seeds and tubers), exists in the form of granules, which consist of amylopectin (70–80%) and amylose (20–30%), both of which are glucose polymers. Depending on the botanical origin, physicochemical characteristics, and the type of processing, starch-based carbohydrates are hydrolyzed at different rates and to different extents *in vitro* and *in vivo* (Cummings *et al.*, 1997; Singh *et al*, 2010). Generally, there are three key factors that influence the rate at which starch is digested: the extracellular structures of the plant, the nature of the surface of the granules (i.e., the porosity and the specific interfacial area), and the molecular packing of the amylopectin and amylose molecules (Mishra *et al.*, 2012). Almost all types of native granules contain surface pores, which may allow penetration of water and amylase into the center of the granules (Copeland *et al.*, 2009). Native starches are digested relatively slowly and only to a limited extent when the starch is organized in its native (ungelatinized) state and some native starches, e.g., from potato, are virtually indigestible. Heating or cooking native starches in the presence of excess water causes starch to gelatinize, which dramatically increases its rate of digestion. During gelatinization, the granules swell to many times their original volume, allowing amylose and amylopectin to leach out into the aqueous phase. Amylopectin contributes greatly to the physicochemical and functional characteristics of starch. A higher content of amylose reduces starch hydrolysis and an increased amount of resistant starch, composed of recrystallized amylose, is found in high amylose starch after heating/cooling cycles (Marze, 2013).

In formulated foods, the composition of the food matrix also affects the digestion of starch significantly (Singh *et al.*, 2010). The presence of a protein network surrounding starch granules can affect the gelatinization and subsequent retrogradation of starch, thus slowing down the digestion of starch by α-amylase within the digestive tract (Zúñiga and Troncoso, 2012). Soluble dietary fiber decreases the rate of glucose release in foods through increased viscosity and/or delaying gastric emptying (Jenkins *et al.*, 1978; Holt *et al.*, 1979). Lipids can affect starch hydrolysis by starch–lipid complexation (Marze, 2013). These effects are dependent on the nature of the lipids and starches and the processing (gelatinization or retrogradation) conditions. However, there is a competitive trend between amylose–lipid complexation and amylose retrogradation (Marze, 2013).

The rate of starch hydrolysis during digestion determines the degree to which blood glucose loading exceeds blood glucose clearance. The net increase in blood glucose concentrations consequently determines the intensity of the insulin response that is required to remove the glucose overload and restore normal glucose concentrations. The changes in blood glucose and hormonal responses after the ingestion of carbohydrate foods vary widely, because of the differences in structures and matrices, as discussed above. It is increasingly being recognized that low-glycemic-index (as a measure of the blood-glucose-raising ability of the available carbohydrate in foods) diets improve metabolic variables in diabetes and hyperlipidemia (Jenkins and Jenkins, 1987).

Lipids

More than 95% of lipids are consumed as triacylglycerols, the rest being mainly phospholipids and cholesterol. Triacylglycerols in natural foods (e.g., meat, dairy, nuts) are often a part of structures in which triacylglycerol particles are coated with a solubilizing layer or multilayer of membrane phospholipids and proteins. In processed foods, oils or fats are extracted from animal or plant materials and then incorporated within the food matrix in the form of emulsions (such as yoghurt, cheese, spreads, imitation creams, salad dressings, gravies, sauces, ice-creams, confectionary products, and chocolate). Diacylglycerols, monoacylglycerols, phospholipids, and proteins are commonly used to stabilize these complex emulsion systems. The initial step in the digestion of lipids involves breaking down the surrounding structures and releasing the lipid droplets from the cells, seed bodies, or wherever else they are stored. The physical transformation of food matrices in the gastrointestinal tract allows the release of triacylglycerols in various formats. As triacylglycerols are insoluble under aqueous conditions in the gastrointestinal tract and lipases are water-soluble proteins, lipid digestion takes place at the oil—water interfaces. Therefore, any factor that affects the binding of lipase to the oil—water interface, e.g., the surface area of the interface, the nature of the emulsifier, or the molecular structure of triacylglycerol molecules, would be expected to have an impact on the rate and extent of lipid digestion. Several studies have shown that the ability of lipases to digest emulsified oil droplets is affected by the composition of the interfacial layer and the droplet size of emulsions (see reviews Singh, 2011; Singh and Ye, 2013). Food emulsions formed with different surface-active components, such as small molecule surfactants, phospholipids, and proteins, react differently to lipase action, because of differences in the nature of the adsorbed layer (such as composition, charge, thickness, and rheology). The presence of solid lipids has been shown to slow down the lipid digestion rate (Bonnaire et al., 2008). Fats and oils can be solidified by interesterification (shuffling of the fatty acids within or among the triacylglycerol molecules) or partial hydrogenation (Berry, 2009). Thus, these processes, common in the manufacture of spreads to simulate the texture of butter, influence the digestion rate of lipids. Limited studies on the relationship between structure and lipid digestibility in natural foods, such as milk and nuts, have been carried out. Some of the recent findings in this area are discussed later in this review.

Following the consumption of lipid foods, the concentration of blood-circulating triacylglycerol increases within 1 h and can remain elevated for several hours (Lairon, 2009); this is referred to as postprandial lipemia. Postprandial lipemia is due to an increase in both intestine-derived chylomicrons and liver-derived very-low-density lipoproteins (Cohn et al., 1993) and varies as a function of the quantity and type of lipids ingested and the food matrix. Saturated-fat-induced lipemia is more pronounced and is of longer duration than that induced by polyunsaturated fats (Weintraub et al., 1988). A sharp and prolonged postprandial lipemia is associated with increased risk of cardiovascular heart diseases (Clemente et al., 2003). Short- and medium-chain fatty acids are quickly absorbed by the enterocytes and transported by the portal vein to the liver for β-oxidation, whereas long-chain fatty acids are absorbed by the small intestinal enterocytes and transported through the lymphatic system as chylomicrons (Borgstrom and Patton, 1991). The absorption of long-chain fatty acids in the duodenum and jejunum stimulates the secretion of cholecystokinin, promoting satiety (van Aken, 2010).

Marciani *et al.* (2007, 2009) compared the behaviors of acid-stable and acid-unstable emulsions under gastric conditions in humans and showed that an acid-stable emulsion slowed down the gastric emptying rate and increased satiety, postprandial release of cholecystokinin, and gallbladder contraction in comparison with an acid-unstable emulsion. In addition to the composition and structure of the interface, the size of lipid droplets is an important parameter in the digestion of lipids. Indeed, Armand *et al.* (1999) studied the digestion of two emulsions with different lipid droplet sizes in humans and showed that the emulsion with the smaller lipid droplets was digested to a greater extent, because of the increased surface area, and slowed down the rate of gastric emptying.

Clemente *et al.* (2003) showed that, in type 2 diabetic subjects, ingestion of an isoenergetic diet including milk (liquid), butter (solid), or cheese (semi-solid) had no effect on the duration of postprandial lipemia. However, the triacylglycerol peak was delayed after ingestion of the butter-based diet, probably because of the presence of smaller fat globules in milk and cheese, which were digested at a faster rate than butter fat. The gastric emptying rate was greater with the cheese-based diet than with the milk-based diet. Studies in rats showed that the ingestion of milk fat with skim milk resulted in a faster appearance of plasma triacylglycerol and a sharper triacylglycerol peak than the ingestion of homogenized or non-homogenized cream (Michalski, 2009). Thus, the matrix structure and the oil—water interface have an impact on the physiological response after the ingestion of milk fat. In humans, the daily consumption of butter led to higher fasting total and low-density lipoprotein cholesterol than the daily consumption of cheese (Michalski, 2009).

Vors *et al.* (2013) introduced the concept of "slow fat versus fast fat" in their study on the effect of the emulsification of milk fat, consumed in a mixed meal, on the absorption and postprandial handling of fatty acids in normal-weight and obese subjects. Emulsification enhanced fatty acid absorption and consequently led to a more rapid and sharper postprandial chylomicron peak. Emulsification also increased lipid β-oxidation. Because smaller fat droplets increase the rate of lipolysis (Armand *et al.*, 1999), homogenized milk should result in greater lipolysis than fresh non-homogenized milk; however, the nature of the interface also needs to be considered in the assessment of lipolysis. Indeed, gastric lipolysis of human milk fat globules, covered by the native milk fat globule membrane, is greater than that of infant formula globules with a smaller droplet size and a milk protein interface (Armand *et al.*, 1996).

Berry *et al.* (2008) studied the impact of almond lipid bioaccessibility on the postprandial lipemia of healthy male subjects. Postprandial lipemia was lower after the ingestion of almond nut macroparticles (in a muffin) than after the ingestion of almond oil mixed with defatted almond flour (in a muffin) (Figure 2.2). Indeed, the presence of cell walls, as in almond nut macroparticles, limits the bioaccessibility of trapped almond lipids, as the cell walls are not efficiently degraded in the upper gastrointestinal tract. The cell walls of almonds are rich in non-starch polysaccharides and the coat cell wall is rich in phenolic compounds (Ellis *et al.*, 2004). The mastication of raw almonds releases an estimated 8–12% of the lipids (Berry *et al.*, 2008) and produces particles smaller than 2 mm (Mandalari *et al.*, 2008). Thus, when consuming whole almonds, a large proportion of the lipids are not available for lipases in the first steps of digestion. Ellis *et al.* (2004) studied the influence of cell walls on the bioaccessibility of almond lipids in humans. They

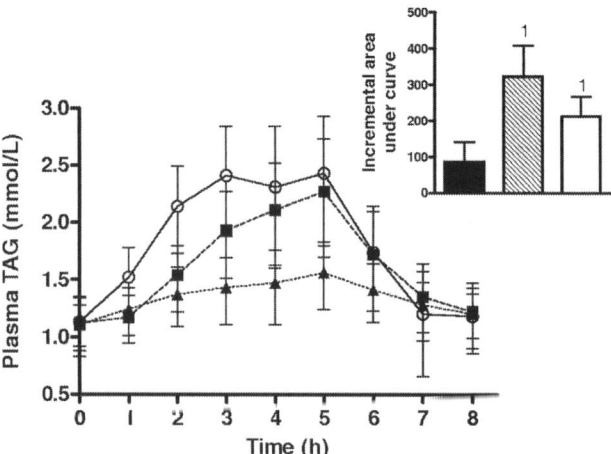

■ **FIGURE 2.2** Geometric mean plasma triacylglycerol (TAG) concentrations and 95% CIs in healthy men ($n = 20$) after test meals containing 50 g of test fat from whole almond seed macroparticles (▲), almond oil and flour (○), or sunflower oil (control, ■). Deviations from fasting values were analyzed by ANOVA, with the three diets and time (0–8 h) as factors: diet effect ($P = 0.002$), time effect ($P < 0.001$), and diet × time interaction ($P < 0.001$). Inset: incremental area under the curve (0–8 h) values are inset and presented as geometric means and 95% CIs ($n = 20$) for whole almond seed macroparticles (■; 86.8; 53.3, 141.1), almond oil and flour (▨; 323.3; 238.2, 438.9), and control oil (□; 212.1; 156.7, 287.1). [1]Significantly different from whole almond seed macroparticles, $P < 0.01$ (Bonferroni multiple comparison test). *Reproduced by permission from The American Journal of Clinical Nutrition (Berry et al., 2008).*

followed the release of lipids during chewing and digestion and collected fecal samples. After mastication, only the surface of the almond particles was broken, leading to minimal release of oil bodies, and underlying cells were intact. Fecal samples showed an abundance of structurally intact cell walls encapsulating undigested lipids (Ellis *et al.*, 2004). These cell walls had escaped from enzymatic activity and microbial fermentation.

Proteins

Dietary proteins are derived from both animal sources (e.g., milk, meat, fish) and plant sources (e.g., cereals, rice, pulses, nuts), in which they are often an integral part of the food matrix. Proteins can be extracted from these sources and used as food ingredients in formulated foods for specific functional properties, such as gelling, foaming, water binding, and emulsification. Food proteins are often denatured and aggregated during processing or cooking operations. Depending on the temperature, pH, ionic strength, and shear, denaturation can lead to the formation of fibrils, micro-particles, spherical micro-gels, and fractal gel aggregates (Kaufmann and Palzer, 2011). Therefore, depending on the source and the processing treatment, proteins may exhibit a wide range of heterogeneous, complex structures in the foods we consume. Such structures may influence the accessibility of protein molecules to the proteolytic enzymes of the gastrointestinal tract. In addition to the matrix effects, it is well known that the native molecular structures of proteins have a profound effect on their susceptibility to proteolysis, as proteins have different tertiary conformations and possess regions with different affinities for hydrophobic and hydrophilic environments. For example, highly disordered proteins, such as caseins, undergo rapid hydrolysis, but proteins with highly folded, compact conformations, such as β-lactoglobulin, show resistance to digestion in the native state. Physical processing, such as heating and emulsification, can change molecular flexibility and accessibility and consequently can markedly enhance digestibility (Zeece *et al.*, 2008; Peram *et al.*, 2013).

The presence of fiber and other polysaccharides affects protein digestion (Turgeon and Rioux, 2011). Polysaccharides may prevent access of digestive proteases to proteins by increasing the viscosity of the gastrointestinal contents or to the cleavage sites on proteins and peptides via electrostatic interactions. Some phytochemicals, such as folate and phenolic compounds, can bind to proteins and thus reduce their bioavailability (Parada and Aguilera, 2007).

Dietary proteins and their digestive products have an effect on physiological functions such as regulation of metabolism and food intake, and on protein synthesis. Proteins are considered to be more satiating than lipids and carbohydrates and induce higher energy expenditure (Kaufmann and Palzer, 2011). It has been recognized that the rate of protein digestion and consequently amino acid absorption can affect postprandial protein synthesis, breakdown, and deposition (Dangin *et al.*, 2003). As discussed above, the rate of protein digestion varies according to the types and structures of ingested food proteins, the presence of other macronutrients and the physicochemical behavior of proteins in the gastrointestinal tract. Some proteins are known to be digested at a slower rate than others in humans. For example, whey proteins remain soluble in the gastric environment whereas caseins form a clot, which delays its gastric emptying and thus probably results in a slower release of amino acids (Mahe *et al.*, 1996). It has been suggested that slowly digested proteins, such as caseins, induce a higher postprandial protein gain than rapidly digested proteins, such as whey proteins, which stimulates protein synthesis but also oxidation (Boirie *et al.*, 1997). Ingestion of micellar casein slows down gastric emptying by forming a clot in the acidic environment in the stomach, whereas ingestion of whey and soy proteins and gluten modulates the release of satiety hormones (Turgeon and Rioux, 2011).

ILEAL BRAKE

When nutrients in their absorbable form reach the small intestine, the intestinal mucosa induces the stimulation of the vagal nerve and the secretion of endocrine hormones, slowing down the stomach emptying rate (van Aken, 2010). The ileal brake mechanism is activated by the presence of unabsorbed nutrients in the ileum, which stimulates the secretion of the satiety hormones, polypeptide tyrosine—tyrosine and glucagon-like peptide-1. Polypeptide tyrosine—tyrosine secretion is higher when unabsorbed proteins rather than unabsorbed carbohydrates or fats are detected and suppresses hunger (van Aken, 2010). The activation of the ileal brake prolongs the gut transit time (see Chapter 13).

CASE STUDIES

BEHAVIOR OF MILK LIPIDS IN THE GASTROINTESTINAL TRACT

The milk lipids are predominantly composed of triacylglycerols that are dispersed in milk in the form of fat globules ranging from 0.1 to 15 μm in diameter (Walstra, 1995). The fat globules are surrounded by the milk fat globule membrane (MFGM), which prevents the milk fat globules from coalescence and protects them against the action of the milk and bacterial lipases (Dewettinck *et al.*, 2008). It has also been speculated that the MFGM has profound effects on the accessibility of the triacylglycerols for lipase-catalyzed digestion (Patton and Keenan, 1975). The MFGM is composed of a trilayer that contains phospholipids, various glycoproteins, enzymes, and cholesterol (Dewettinck *et al.*, 2008). The phospholipids are segregated between liquid-ordered domains (particularly rich in sphingomyelin and cholesterol), analogous lipid rafts, and coexisting liquid-disordered phases (Gallier *et al.*, 2010; Lopez *et al.*, 2010). Milk processing dramatically affects the structure of the milk fat globule and its membrane. Dairy processing operations, such as heat treatment and homogenization, alter the structure and composition of the MFGM. Heat treatment of milk causes the denaturation of the MFGM proteins and association with the whey proteins through sulfydryl—disulfide interactions (Kim and Jimenez-Flores, 1995; Corredig and Dalgleish, 1996; Ye *et al.*, 2004). Homogenization causes disruption of the fat globules into smaller droplets (<2.0 µm). Consequently, skim milk proteins are adsorbed at the oil—water interface as there is insufficient MFGM material to fully cover this additional surface. These alterations in the MFGM may influence the digestion and the bioavailability of the milk fat (Michalski and Januel, 2006).

The digestion of bovine milk and human milk has received considerable research attention over the past 50 years (Kobylka and Carraway, 1973; Berendsen, 1982; Hamosh *et al.*, 1985; Armand, 2008; Gallier *et al.*, 2012). However, relatively few studies have focused on understanding the changes in the structures of fat globules and proteins within the gastrointestinal tract. Berendsen (1982) observed that milk formed a clot in the stomach of suckling rats, with the milk fat globules trapped in a network of casein proteins. Buchheim (1984) observed lamellar structures of monoacylglycerols at the surface of the milk fat globules in the stomach of minipigs fed pasteurized (homogenized and non-homogenized) milk and UHT milk.

Recent studies have focused mainly on the intestinal lipolysis of milk fat globules to understand the role of MFGM structure and composition in the lipolysis of milk fat (Blackberg *et al.*, 1981; Borgstrom and Erlanson-Albertsson, 1982; Berton *et al.*, 2009; Ye *et al.*, 2010). Using an *in vitro* intestinal model, Ye *et al.* (2010) compared the lipid digestion behavior of native fat globules (stabilized by natural MFGM as in raw milk) with that of artificial fat globules (stabilized with milk proteins as in recombined milk) (Figure 2.3). In the absence of bile extract, the rate of lipid digestion was slower in raw milk than in recombined milk, suggesting that the

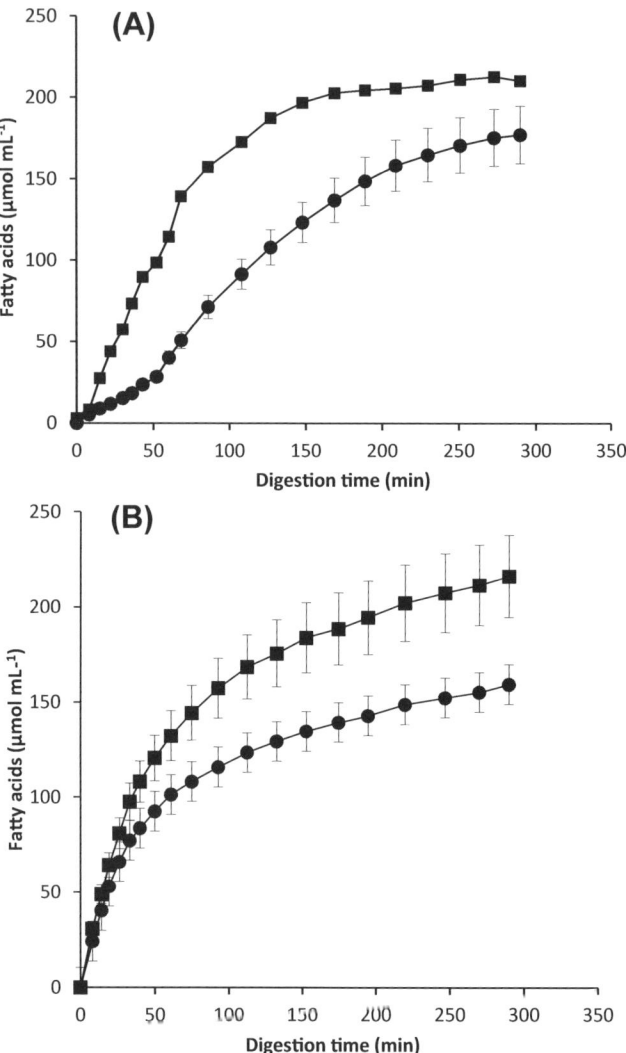

■ **FIGURE 2.3** Level of fatty acids released from fat globules in raw milk (●) and recombined milk ($d_{32} = 2.10$ μm) (■) after digestion with pancreatic lipase in the absence (A) and the presence (B) of bile extract as a function of time. *Adapted from Ye et al. (2010). Reproduced with permission from Elsevier.*

composition of the MFGM influences the rate of lipid hydrolysis in milk. When bile extract was added to the samples, the rate of lipolysis was promoted and the difference between the raw milk samples and the recombined milk samples, in terms of the rate and extent of lipolysis, became less significant. Bile extract probably displaced the materials that inhibit the activity of lipase from the MFGM (Patton *et al.*, 1986). In the later stages, lipid digestion was dependent on the amount of bile extract in the samples, and was less influenced by the surface properties.

In addition to lipolysis by pancreatic enzymes in the small intestine, lipolysis of the milk fat globules by gastric lipase plays an important role in their digestion (Bernback *et al.*, 1989, 1990; Gallier *et al.*, 2013a). The free fatty acids released during gastric lipolysis accumulate at the surface of the fat globules, which may stimulate intestinal lipolysis by facilitating the binding of the pancreatic lipase—colipase complex to the oil—water interface. The MFGM helps the milk fat globules to keep their integrity under the gastric conditions, as some MFGM glycoproteins are resistant to pepsin digestion and phospholipids are not digested by the gastric lipase (Hamosh *et al.*, 1999; Ye *et al.*, 2010).

We have recently studied the structural changes in the MFGM of different milk systems during *in vitro* digestion (Ye *et al.*, 2010, 2011; Gallier *et al.*, 2012). Under *in vitro* gastric conditions, fat globules did not coalesce or aggregate (Figure 2.4) because of their high zeta-potential (Gallier *et al.*, 2012). As phospholipids and some of the highly glycosylated MFGM proteins were not affected by pepsin, the fat globule integrity was maintained to a large extent in the gastric environment (Hamosh *et al.*, 1999). Under *in vitro* intestinal conditions, a lamellar phase of lipolytic products accumulated at the surface of the milk fat globules, until bile salt micelles formed, which solubilized the lipolytic products (Gallier *et al.*, 2012). The intestinal lipolysis did not present a lag phase and proceeded readily, which indicated that bile salts displace easily some of the MFGM components to allow the pancreatic lipase—colipase complex to anchor at the surface of the milk fat globule (Gallier *et al.*, 2012).

In our recent work, we studied the behavior of fat globules and MFGM during *in vivo* gastrointestinal digestion in rats (Gallier *et al.*, 2013a, c). Rats were gavaged with 2 mL of cream (obtained from fresh whole milk) and the gastric contents were collected after 30 min, 2 h, and 3 h. Fat globules appeared to increase in size under gastric conditions (Figure 2.5A). Spherical protrusions rich in lipolytic products were observed at the surface of fat globules in the gastric chyme (Figure 2.5B and C). Under low pH gastric conditions, free fatty acids, and especially long-chain fatty acids, are poorly water soluble and thus accumulate at the oil—water interface (Gallier *et al.*, 2013a). Pafumi *et al.* (2002) associated this interfacial accumulation of free fatty acids with the process of inhibition of gastric lipolysis.

After gavaging rats with 2 mL of cream hourly for 5 h, upper (duodenum and jejunum) and lower (ileum) small intestinal digesta were collected

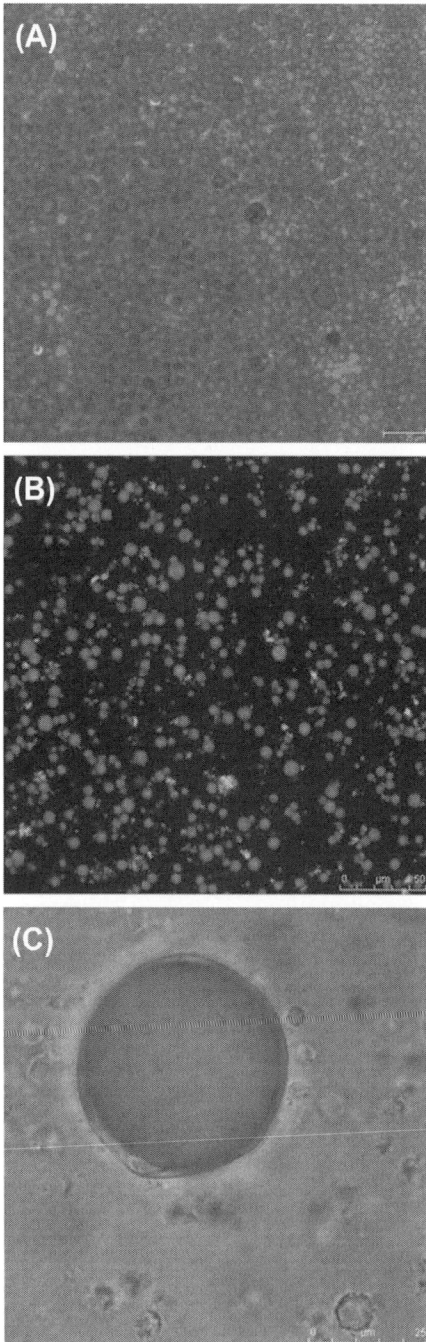

■ **FIGURE 2.4** Confocal laser scanning microscopic images of bovine milk under native conditions (A), after 60 min of gastric digestion (B) and after 60 min of gastric digestion followed by 30 min of intestinal digestion (C). *Adapted from Gallier* et al. *(2012). Reproduced with permission from Elsevier.* [A color version of this figure is available online at www.booksite. elsevier.com/9780124046108].

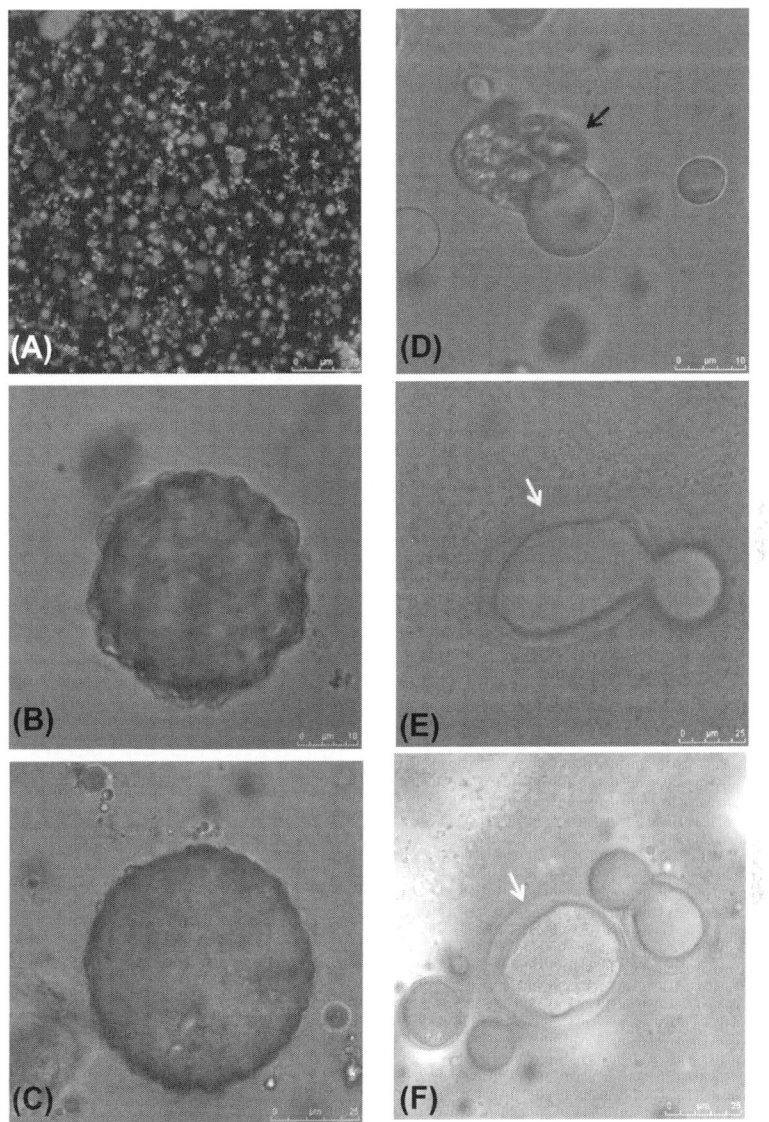

■ **FIGURE 2.5** Differential interference contrast and confocal laser scanning microscopic images of small intestinal digesta of rats gavaged with cream prepared from pasteurized and homogenized bovine milk (A and D), pasteurized bovine milk (B and E), and raw bovine milk (C and F). The white arrows point at liquid—crystalline lamellar phases and the black arrow points at oil coming out of the cracked liquid—crystalline lamellae, which are left behind. Scale bars = 75 μm (A), 10 μm (B and D), and 25 μm (C, E, and F). *Adapted from Gallier et al. (2013c).* [A color version of this figure is available online at www.booksite.elsevier.com/9780124046108].

(Gallier *et al.*, 2013c). During intestinal lipolysis, the fat globules retained the size acquired in the stomach, which was similar to that observed *in vitro* (Gallier *et al.*, 2012). The fat globules were surrounded by a liquid—crystalline lamellar phase (Figure 2.4D—F), formed by the accumulation of lipolytic products and calcium (Patton and Carey, 1979) and possibly bile salts. This lamellar phase was then solubilized as multilamellar vesicles, and, in contact with bile salts, the vesicles formed smaller mixed micelles composed of bile salts, phospholipids, cholesterol, and lipolytic products. The lamellar shell surrounding the fat globules eventually cracked or solubilized, releasing the undigested triacylglycerol core and leaving the empty crystalline shell behind (Figure 2.5D—F).

Crystals of lipolytic products, likely to be made up of long-chain saturated fatty acids and soaps with a melting point higher than the body temperature, were detected within the aqueous phase of the digesta (Figure 2.4E and F). The presence of crystallized fatty acids was also observed by Knutson *et al.* (2010) after ingestion of a yoghurt containing emulsified vegetable oil and, interestingly, this was associated with the activation of the ileal brake mechanism. Long-chain saturated fatty acids can form soaps with calcium within the intestinal lumen, reducing the absorption of both fatty acids and calcium, which are then excreted in stools (Michalski, 2009).

BEHAVIOR OF NUT-DERIVED LIPIDS IN THE GASTROINTESTINAL TRACT

Oil bodies are the storage form of lipids in plants and oilseeds. They are stabilized by a layer of phospholipids and embedded structural proteins, oleosins, and caleosins (Purkrtova *et al.*, 2008). Much research has been carried out on the formation, structure, and purification of oil bodies (Tzen and Huang, 1992; Huang, 1994; Beisson *et al.*, 2001b). There is increasing interest in using oil bodies as natural emulsions and novel carriers for the delivery of lipophilic bioactive molecules (White *et al.*, 2009; Bonnegna *et al.*, 2011). In the last decade, some studies have investigated the digestion of oil bodies (Beisson *et al.*, 2001a; White *et al.*, 2009; Wu *et al.*, 2012). Most *in vivo* studies on nuts have dealt with the physiological responses induced by a diet rich in nuts (Berry *et al.*, 2008; Damasceno *et al.*, 2011) and *in vitro* studies have investigated the digestion of whole nuts and nut oil (Mandalari *et al.*, 2008; Kong and Singh, 2009). Kong and Singh (2009) showed that roasting almonds facilitated gastric digestion and penetration of gastric juices in almond pieces by loosening the structure of the cell walls. A large part of the oil bodies remains trapped within the cell walls after mastication (Ellis *et al.*, 2004), but the bolus contains some oil bodies in suspension in the aqueous phase. In our laboratory, we have been interested in looking at the *in vitro* digestion of nut oil bodies (Gallier and Singh, 2012a; Gallier *et al.*, 2013b), in particular the fate of "free" oil bodies that are released during mastication within the gastrointestinal tract.

Almond and walnut dispersions of oil bodies (Figure 2.6A and B) were obtained by crushing the nuts in water and filtering the solution (Gallier and Singh, 2012a; Gallier *et al.*, 2013b). The dispersions were subjected to a two-step *in vitro* digestion protocol. Almond oil bodies flocculated under gastric conditions (Figure 2.6D), as detected by an increase in the average particle size (Figure 2.7) (Gallier and Singh, 2012a). Some pepsin-resistant peptides remained at the oil–water interface (Figure 2.5D) but presented a low zeta-potential (Figure 2.8), leading to flocculation of the oil bodies. The particle size decreased under intestinal conditions (Figure 2.7) because of lipolysis and micellar formation (Gallier and Singh, 2012a). Pepsin-resistant peptides were digested rapidly by pancreatic proteases (Figure 2.5F) and the zeta-potential of the oil bodies decreased (Figure 2.8) because of the adsorption of bile salts and the accumulation of partially ionized free fatty acids at the oil–water interface.

A short time lag phase in the release of free fatty acids under intestinal conditions was observed, which was considered to be associated with the delay in the absorption of bile salts onto the oil body surface coated by phospholipids and pepsin-resistant peptides (Gallier and Singh, 2012a). Interestingly, this lag phase was not observed when oil bodies were first subjected to gastric lipolysis, producing free long-chain fatty acids accumulating at the oil–water interface and aiding the onset of intestinal lipolysis (Beisson et al., 2001a). A similar lag phase was observed by others during the intestinal lipolysis of sunflower seed oil bodies (White et al., 2009).

Walnut oil bodies coalesced and flocculated under gastric conditions (Figure 2.6C). The average surface-mean diameter of walnut oil bodies was greater than that of almond oil bodies during gastric and intestinal digestion (Figure 2.7). The zeta-potential of the walnut oil bodies under gastric and intestinal conditions (Gallier *et al.*, 2013b) followed the same pattern as that of the almond oil bodies (Figure 2.8). Similar to almond oil bodies (Gallier and Singh, 2012a), pepsin-resistant peptides were also observed at the surface of walnut oil bodies under gastric conditions (Figure 2.6C). Under intestinal conditions and after the start of the lipolysis of walnut oil bodies (Figure 2.6E), Gallier *et al.* (2013b) observed the formation of "spontaneous biological multiple-phase emulsions" (SBMPE).These emulsions were not formed during the *in vitro* digestion of almond oil bodies (Gallier and Singh, 2012a) and bovine milk fat globules (Gallier *et al.*, 2012) using a similar protocol. The

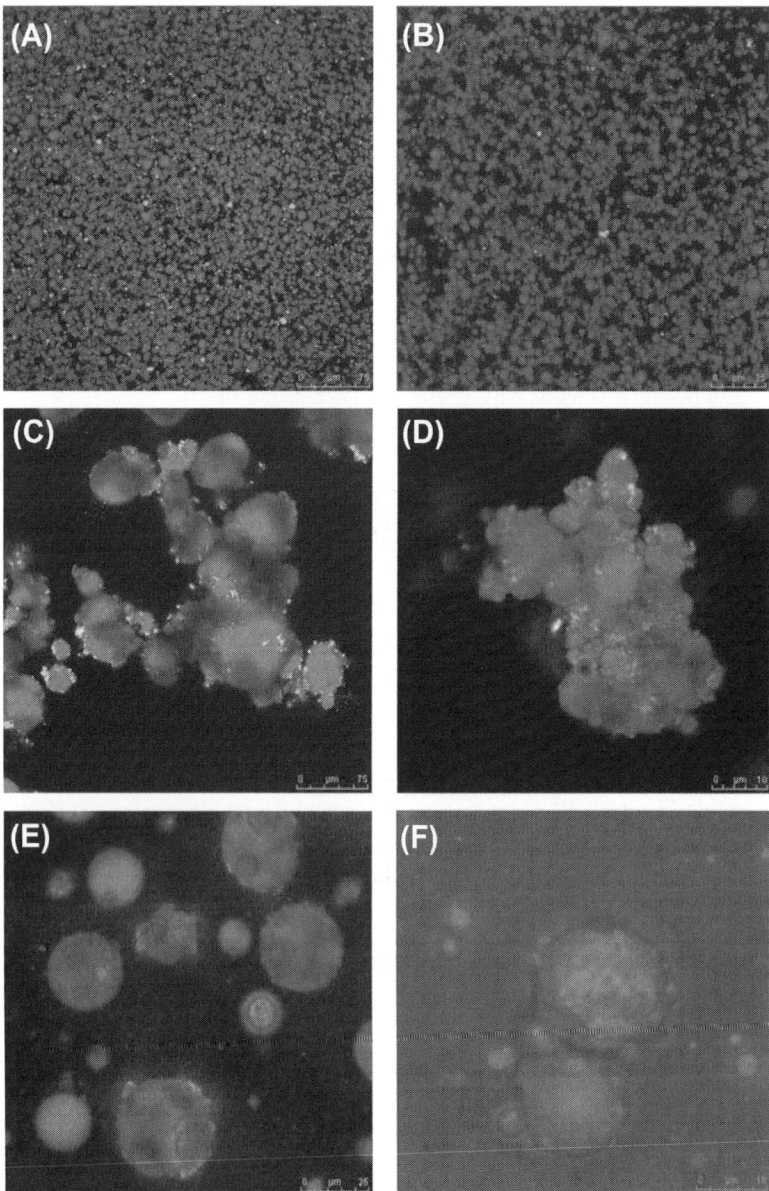

■ **FIGURE 2.6** Confocal laser scanning microscopic images of walnut (A, C, and E) and almond (B, D, and F) oil body dispersions under native conditions (A and B), after 60 min of gastric digestion (C and D) and after 60 min of gastric digestion followed by 60 min of intestinal digestion (E and F). *Adapted by permission from Gallier et al. (2013b) and Gallier and Singh (2012a). Copyright 2013 American Chemical Society and reproduced with permission from the Royal Society of Chemistry, respectively.* [A color version of this figure is available online at www.booksite.elsevier.com/9780124046108].

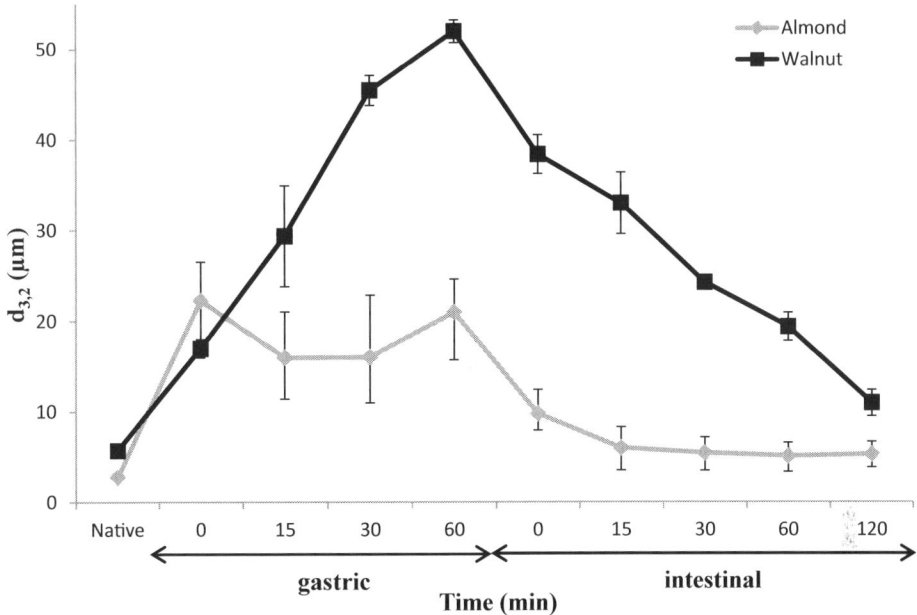

■ **FIGURE 2.7** Surface-mean ($d_{3,2}$) diameters of native and digested walnut oil bodies (black) and native and digested almond oil bodies (gray) during 60 min of gastric digestion followed by 120 min of intestinal digestion. *Adapted by permission from Gallier et al. (2013b) and Gallier and Singh (2012a). Copyright 2013 American Chemical Society and reproduced with permission from the Royal Society of Chemistry, respectively.*

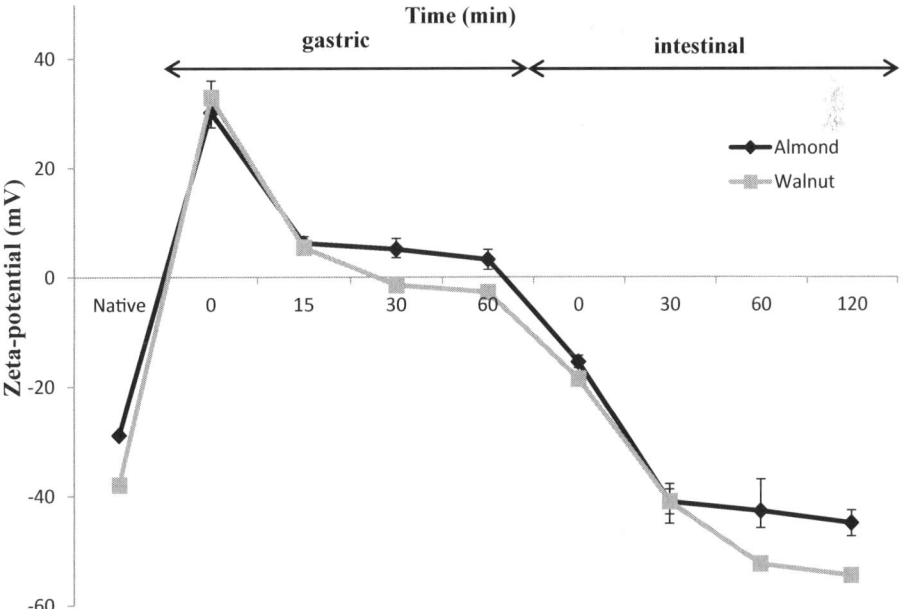

■ **FIGURE 2.8** Zeta-potential of native and digested walnut oil bodies (black) and native and digested almond oil bodies (gray) during 60 min of gastric digestion followed by 120 min of intestinal digestion. *Adapted by permission from Gallier et al. (2013b) and Gallier and Singh (2012a). Copyright 2013 American Chemical Society and reproduced with permission from the Royal Society of Chemistry, respectively.*

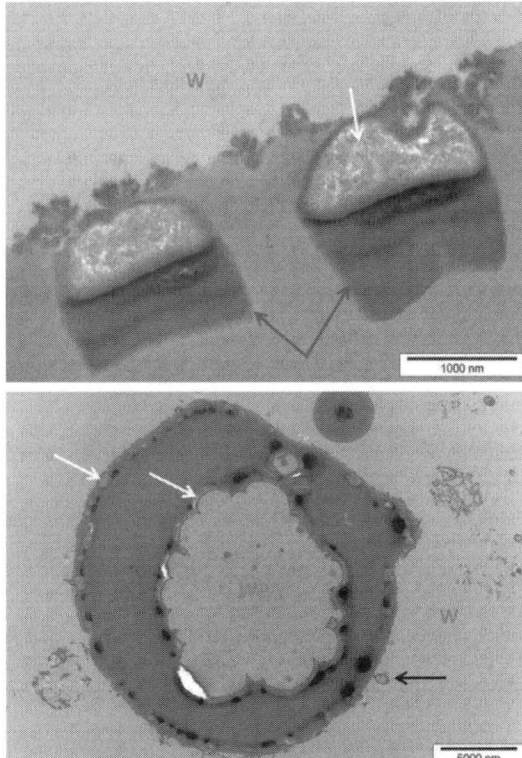

■ **FIGURE 2.9** Transmission electron microscopic images of walnut oil body dispersions after 60 min of gastric digestion followed by 120 min of intestinal digestion. Black arrows point at vesicles and micelles transporting lipolytic products. White arrows point at bile salt crystals. Gray arrows point at lipolytic product crystals. W, water phase; L, oil phase; WP, water—protein-rich phase. *Adapted by permission from Gallier et al. (2013b). Copyright 2013 American Chemical Society.* [A color version of this figure is available online at www.booksite.elsevier.com/9780124046108].

inner droplets seemed to contain proteins or peptide material (Figure 2.6E). Transmission electron microscopy of SBPME (Figure 2.9) revealed the presence of crystals stabilizing the inner and outer droplets. The crystals were likely to be a combination of bile salt crystals, calcium-fatty acid soaps and monoacylglycerol soaps (Gallier *et al.*, 2013b). The fatty acid profile of walnuts, rich in polyunsaturated fatty acids, was probably responsible for the formation of SBPME. The significance of SBPME in the digestion of lipids is unclear and needs to be investigated further.

CONCLUSIONS

The processing of foods within the gastrointestinal tract is extremely complex, involving many different mechanical, physical, and biochemical processes. Our understanding of how various foods with different compositions, physical properties, and structures are affected by these processes remains incomplete. The development of *in vitro* digestion models has generated new knowledge on the changes in food structures within the gastrointestinal tract and has extended our understanding of the role of the food structure/matrix in the release and digestion of nutrients. However, there is a clear need for the standardization of *in vitro* digestion methods, as different systems with a range of biochemical conditions are being used by different research groups. More advanced *in vitro* models that can precisely simulate gastric mobility, gastric emptying, and variations in dilution, pH, and enzyme concentration will allow a clearer picture of the fate of foods within the gastrointestinal tract. Moreover, the *in vitro* models need to be validated in *in vivo* animal/human studies to further understand how different food structures are processed in the human gut. In addition, more sophisticated analytical and microscopy techniques need to be developed to study the nature of food matrices and the location and release of nutrients under physiological conditions.

In the last few years, there has been an increase in research on understanding the link between food structure and physiological outcome in humans. This has led to the recognition that the rate of delivery of a nutrient is as important to human health as the overall food composition, although the fundamental mechanisms involved in these biological processes remain unresolved. To fully exploit the relationship between food structure, digestion, and health, an integrated approach involving scientists from different disciplines (nutrition, biomedical sciences, food science, materials science) will be required. By characterizing the structuring of food within the gastrointestinal tract and the impact of the food matrix and the interactions between food nutrients on their bioavailability, innovative food systems with tailored properties and specific impacts on human health could be created.

REFERENCES

Aguilera, J.M., 2005. Why food microstructure? J. Food Eng. 67, 3—11.

Amado, F.M., Vitorino, R.M., Domingues, P.M., Lobo, M.J.C., Duarte, J.A.R., 2005. Analysis of the human saliva proteome. Expert Rev. Proteom. 2, 521—539.

Armand, M., 2008. Milk fat digestibility. Sciences des Aliments 28, 84—98.

Armand, M., Hamosh, M., Mehta, N.R., Angelus, P.A., Philpott, J.R., Henderson, T.R., et al., 1996. Effect of human milk or formula on gastric function and fat digestion in the premature infant. Pediat. Res. 40, 429—437.

Armand, M., Pasquier, B., Andre, M., Borel, P., Senft, M., Peyrot, J., et al., 1999. Digestion and absorption of 2 fat emulsions with different droplet sizes in the human digestive tract. Am. J. Clin. Nutr. 70, 1096—1106.

Beisson, F., Ferte, N., Bruley, S., Voultoury, R., Verger, R., Arondel, V., 2001a. Oil-bodies as substrates for lipolytic enzymes. Biochim. Biophys. Acta—Mol. Cell Biol. Lipids 1531, 47−58.

Beisson, F., Ferté, N., Voultoury, R., Arondel, V., 2001b. Large scale purification of an almond oleosin using an organic solvent procedure. Plant Physiol. Biochem. 39, 623−630.

Benjamin, O., Silcock, P., Kieser, J.A., Waddell, J.N., Swain, M.V., Everett, D.W., 2012. Development of a model mouth containing an artificial tongue to measure the release of volatile compounds. Innov. Food Sci. Emerg. Technol. 15, 96−103.

Berendsen, P.B., 1982. Ultrastructural studies of milk digestion in the suckling rat. Food Microstr. 1, 83−90.

Bernback, S., Blackberg, L., Hernell, O., 1989. Fatty-acids generated by gastric lipase promote human-milk triacylglycerol digestion by pancreatic colipase-dependent lipase. Biochim. Biophys. Acta 1001, 286−293.

Bernback, S., Blackberg, L., Hernell, O., 1990. The complete digestion of human-milk triacylglycerol *in vitro* requires gastric lipase, pancreatic colipase-dependent lipase, and bile salt stimulated lipase. J. Clin. Invest. 85, 1221−1226.

Berry, S.E.E., 2009. Triacylglycerol structure and interesterification of palmitic and stearic acid-rich fats: an overview and implications for cardiovascular disease. Nutr. Res. Rev. 22, 3−17.

Berry, S.E.E., Tydeman, E.A., Lewis, H.B., Phalora, R., Rosborough, J., Picout, D.R., Ellis, P.R., 2008. Manipulation of lipid bioaccessibility of almond seeds influences postprandial lipemia in healthy human subjects. Am. J. Clin. Nutr. 88, 922−929.

Berton, A., Sebban-Kreuzer, C., Rouvellac, S., Lopez, C., Crenon, I., 2009. Individual and combined action of pancreatic lipase and pancreatic lipase-related proteins 1 and 2 on native versus homogenized milk fat globules. Mol. Nutr. Food Res. 53, 1592−1602.

Blackberg, L., Hernell, O., Olivecrona, T., 1981. Hydrolysis of human-milk fat globules by pancreatic lipase—role of colipase, phospholipase-A2, and bile salts. J. Clin. Invest. 67, 1748−1752.

Boirie, Y., Dangin, M., Gachon, P., Vasson, M.P., Maubois, J.L., Beaufrere, B., 1997. Slow and fast dietary proteins differently modulate postprandial protein accretion. Proc. Natl Acad. Sci. USA 94, 14930−14935.

Bonnaire, L., Sandra, S., Helgason, T., Decker, E.A., Weiss, J., McClements, D.J., 2008. Influence of lipid physical state on the in vitro digestibility of emulsified lipids. J. Agric. Food Chem. 56, 3791−3797.

Bonsegna, S., Bettini, S., Pagano, R., Zacheo, A., Vergaro, V., Giovinazzo, G., et al., 2011. Plant oil bodies: novel carriers to deliver lipophilic molecules. Appl. Biochem. Biotechnol. 163, 792−802.

Borgstrom, B., Erlanson-Albertsson, C., 1982. Hydrolysis of milk fat globules by pancreatic lipase − role of colipase, phospholipase A2, and bile salts. J. Clin. Invest. 70, 30−32.

Borgstrom, B., Patton, J.S., 1991. Luminal events in gastrointestinal lipid digestion. In: Schultz, S.G. (Ed.), Handbook of Physiology, Section 6: The Gastrointestinal System. American Physiological Society, Bethesda, MD, pp. 475−504.

Buchheim, W., 1984. Zum Einfluss unterschiedlicher technologischer Behandlung von Milch auf die Verdauugsvorgange im Magen. IV. Elektronenmikroskopische Charakterisierung des Koagulums und lipolytischer Vorgange im Magen. [Influence of different technological treatments of milk on digestion in the stomach. IV. Electron microscopic characterization of the coagulum and of lipolytic processes in the stomach]. Milchwissenschaft 39, 271−275.

Carriere, F., Grandval, P., Renou, C., Palomba, A., Prieri, F., Giallo, J., et al., 2005. Quantitative study of digestive enzyme secretion and gastrointestinal lipolysis in chronic pancreatitis. Clin. Gastroenterol. Hepatol. 3, 28−38.

Chang, C.A., McKenna, R.D., Beck, I.T., 1968. Gastric emptying rate of the water and fat phases of a mixed test meal in man. Gut 9, 420−424.

Clemente, G., Mancini, M., Nazzaro, F., Lasorella, G., Rivieccio, A., Palumbo, A.M., et al., 2003. Effects of different dairy products on postprandial lipemia. Nutr. Metab. Cardiovasc. Dis. 13, 377−383.

Cohn, J.S., Johnson, E.J., Millar, J.S., Cohn, S.D., Milne, R.W., Marcel, Y.L., et al., 1993. Contribution of apoB-48 and apoB-100 triglyceride-rich lipoproteins (TRL) to post-prandial increases in the plasma concentration of TRL triglycerides and retinyl esters. J. Lipid Res. 34, 2033−2040.

Copeland, L., Blazek, J., Salman, H., Tang, M.C.M., 2009. Form and functionality of starch. Food Hydrocoll. 23, 1527−1534.

Corredig, M., Dalgleish, D.G., 1996. Effect of different heat treatments on the strong binding interactions between whey proteins and milk fat globules in whole milk. J. Dairy Res. 63, 441−449.

Cummings, J.H., Roberfroid, M.B., Members of the Paris Carbohydrate Group, 1997. A new look at dietary carbohydrates: chemistry, physiology and health. Eur. J. Clin. Nutr. 51, 417−423.

Damasceno, N.R.T., Perez-Heras, A., Serra, M., Cofan, M., Sala-Vila, A., Salas-Salvado, J., Ros, E., 2011. Crossover study of diets enriched with virgin olive oil, walnuts or almonds. Effects on lipids and other cardiovascular risk markers. Nutr. Metab. Cardiovasc. Dis. 21, S14−S20.

Dangin, M., Guillet, C., Garcia-Rodenas, C., Gachon, P., Bouteloup-Demange, C., Reiffers-Magnani, K., et al., 2003. The rate of protein digestion affects protein gain differently during aging in humans. J. Physiol.—Lond. 549, 635−644.

Dewettinck, K., Rombaut, R., Thienpont, N., Le, T.T., Messens, K., Van Camp, J., 2008. Nutritional and technological aspects of milk fat globule membrane material. Int. Dairy J. 18, 436−457.

Dupont, D., Mandalari, G., Molle, D., Jardin, J., Leonil, J., Faulks, R.M., et al., 2010. Comparative resistance of food proteins to adult and infant in vitro digestion models. Mol. Nutr. Food Res. 54, 767−780.

Ellis, P.R., Kendall, C.W.C., Ren, Y.L., Parker, C., Pacy, J.F., Waldron, K.W., Jenkins, D.J.A., 2004. Role of cell walls in the bioaccessibility of lipids in almond seeds. Am. J. Clin. Nutr. 80, 604−613.

Gallier, S., Cui, J., Olson, T.D., Rutherfurd, S.M., Ye, A., Moughan, P.J., Singh, H., 2013a. In vivo digestion of bovine milk fat globules: effect of processing and interfacial structural changes. I. Gastric digestion. Food Chem. 141, 3273−3281.

Gallier, S., Gragson, D., Jimenez-Flores, R., Everett, D., 2010. Using confocal laser scanning microscopy to probe the milk fat globule membrane and associated proteins. J. Agric. Food Chem. 58, 4250−4257.

Gallier, S., Singh, H., 2012a. Behavior of almond oil bodies during in vitro gastric and intestinal digestion. Food Funct. 3, 547−555.

Gallier, S., Singh, H., 2012b. The physical and chemical structure of lipids in relation to digestion and absorption. Lipid Technol. 24, 271–273.

Gallier, S., Tate, H., Singh, H., 2013b. In vitro gastric and intestinal digestion of a walnut oil body dispersion. J. Agric. Food Chem. 61, 410–417.

Gallier, S., Ye, A., Singh, H., 2012. Structural changes of bovine milk fat globules during in vitro digestion. J. Dairy Sci. 95, 3579–3592.

Gallier, S., Zhu, X.Q., Rutherfurd, S.M., Ye, A., Moughan, P.J., Singh, H., 2013c. In vivo digestion of bovine milk fat globules: effect of processing and interfacial structural changes. II. Upper digestive tract digestion. Food Chem. 141, 3215–3223.

Hamosh, M., Bitman, J., Wood, D.L., Hamosh, P., Mehta, N.R., 1985. Lipids in milk and the first steps in their digestion. Pediatrics 75, 146–150.

Hamosh, M., Peterson, J.A., Henderson, T.R., Scallan, C.D., Kiwan, R., Ceriani, R.L., et al., 1999. Protective function of human milk: the milk fat globule. Sem. Perinatol. 23, 242–249.

Holt, S., Carter, D.C., Heading, R.C., Prescott, L.F., Tothill, P., 1979. Effect of gel fiber on gastric-emptying and absorption of glucose and paracetamol. Lancet 1, 636–639.

Huang, A.H.C., 1994. Structure of plant seed oil bodies. Curr. Opin.Struct. Biol. 4, 493–498.

Jacobs, D.R., Tapsell, L.C., 2007. Food, not nutrients, is the fundamental unit in nutrition. Nutr. Rev. 65, 439–450.

Jenkins, D.J.A., Wolever, T.M., Leeds, A.R., Gassull, M.A., Haisman, P., Dilawari, J., et al., 1978. Dietary fibres, fibre analogues, and glucose tolerance: importance of viscosity. BMJ 1, 1392–1394.

Jenkins, D.J.A., Jenkins, A.L., 1987. The glycemic index, fiber, and the dietary treatment of hypertriglyceridemia and diabetes. J. Am. Coll. Nutr. 6, 11–17.

Kaufmann, S.F.M., Palzer, S., 2011. Food structure engineering for nutrition, health and wellness, 11th International congress on engineering and food (ICEF11). Procedia Food Science, Athens, Greece, 1479–1486.

Kim, H.H.Y., Jimenez-Flores, R., 1995. Heat-induced interactions between the proteins of milk fat globule membrane and skim milk. J. Dairy Sci. 78, 24–35.

Knutson, L., Koenders, D., Fridblom, H., Viberg, A., Sein, A., Lennernas, H., 2010. Gastrointestinal metabolism of a vegetable-oil emulsion in healthy subjects. Am. J. Clin. Nutr. 92, 515–524.

Kobylka, D., Carraway, K.L., 1973. Proteolytic digestion of proteins of the milk fat globule membrane. Biochim. Biophys. Acta—Biomembranes 307, 133–140.

Kong, F., Singh, R.P., 2009. Digestion of raw and roasted almonds in simulated gastric environment. Food Biophys. 4, 365–377.

Lairon, D., 2009. Digestion and absorption of lipids. In: McClements, D.J., Decker, E.A. (Eds.), Designing Functional Foods: Measuring and Controlling Food Structure Breakdown and Nutrient Absorption. CRC Press, pp. 68–93.

Lopez, C., Madec, M.-N., Jimenez-Flores, R., 2010. Lipid rafts in the bovine milk fat globule membrane revealed by the lateral segregation of phospholipids and heterogeneous distribution of glycoproteins. Food Chem. 120, 22–33.

Mahe, S., Roos, N., Benamouzig, R., Davin, L., Luengo, C., Gagnon, L., et al., 1996. Gastrojejunal kinetics and the digestion of [N-15]beta-lactoglobulin and casein in humans: the influence of the nature and quantity of the protein. Am. J. Clin. Nutr. 63, 546–552.

Mandalari, G., Faulks, R.M., Rich, G.T., Lo Turco, V., Picout, D.R., Lo Curto, R.B., et al., 2008. Release of protein, lipid, and vitamin E from almond seeds during digestion. J. Agric. Food Chem. 56, 3409–3416.

Mandalari, G., Mackie, A.M., Rigby, N.M., Wickham, M.S.J., Mills, E.N.C., 2009. Physiological phosphatidylcholine protects bovine beta-lactoglobulin from simulated gastrointestinal proteolysis. Mol. Nutr. Food Res. 53, S131–S139.

Marciani, L., Faulks, R., Wickham, M.S.J., Bush, D., Pick, B., Wright, J., et al., 2009. Effect of intragastric acid stability of fat emulsions on gastric emptying, plasma lipid profile and postprandial satiety. Br. J. Nutr. 101, 919–928.

Marciani, L., Wickham, M., Singh, G., Bush, D., Pick, B., Cox, E., et al., 2007. Enhancement of intragastric acid stability of a fat emulsion meal delays gastric emptying and increases cholecystokinin release and gallbladder contraction. Am. J. Physiol.—Gastrointest. Liver Physiol. 292, G1607–G1613.

Marze, S., 2013. Bioaccessibility of nutrients and micronutrients from dispersed food systems: impact of the multiscale bulk and interfacial structures. Crit. Rev. Food Sci. Nutr. 53, 76–108.

Michalski, M.C., 2009. Specific molecular and colloidal structures of milk fat affecting lipolysis, absorption and postprandial lipemia. Eur. J. Lipid Sci. Technol. 111, 413–431.

Michalski, M.C., Januel, C., 2006. Does homogenization affect the human health properties of cow's milk? Trends Food Sci. Technol. 17, 423–437.

Mishra, S., Hardacre, A., Monro, J., 2012. Food structure and carbohydrate digestibility, carbohydrates. In: Chang, C.-F. (Ed.), Comprehensive Studies on Glycobiology and Glycotechnology. InTech, Rijeka, Croatia, pp. 289–316.

N'Goma, J.-C.B., Amara, S., Dridi, K., Jannin, V., Carrière, F., 2012. Understanding the lipid-digestion processes in the GI tract before designing lipid-based drug-delivery systems. Therapeut. Deliv. 3, 105–124.

Nik, A.M., Wright, A.J., Corredig, M., 2010. Surface adsorption alters the susceptibility of whey proteins to pepsin-digestion. J. Colloid Interface Sci. 344, 372–381.

Pafumi, Y., Lairon, D., de la Porte, P.L., Juhel, C., Storch, J., Hamosh, M., Armand, M., 2002. Mechanisms of inhibition of triacylglycerol hydrolysis by human gastric lipase. J. Biol. Chem. 277, 28070–28079.

Palafox-Carlos, H., Ayala-Zavala, J.F., Gonzalez-Aguilar, G.A., 2011. The role of dietary fiber in the bioaccessibility and bioavailability of fruit and vegetable antioxidants. J. Food Sci. 76, R6–R15.

Parada, J., Aguilera, J.M., 2007. Food microstructure affects the bioavailability of several nutrients. J. Food Sci. 72, R21–R32.

Patton, J.S., Carey, M.C., 1979. Watching fat digestion. Science 204, 145–148.

Patton, S., Borgstrom, B., Stemberger, B.H., Welsch, U., 1986. Release of membrane from milk-fat globules by conjugated bile-salts. J. Pediatr. Gastroenterol. Nutr. 5, 262–267.

Patton, S., Keenan, T.W., 1975. The milk fat globule membrane. Biochim. Biophys. Acta—Rev. Biomembr. 415, 273–309.

Peram, M.R., Loveday, S.M., Ye, A., Singh, H., 2013. In vitro gastric digestion of heat-induced aggregates of β-lactoglobulin. J. Dairy Sci. 96, 63–74.

Purkrtova, Z., Jolivet, P., Miquel, M., Chardot, T., 2008. Structure and function of seed lipid body-associated proteins. C. R. Biol. 331, 746–754.

Sarkar, A., Goh, K.K.T., Singh, H., 2009a. Colloidal stability and interactions of milk-protein-stabilized emulsions in an artificial saliva. Food Hydrocoll. 23, 1270−1278.

Sarkar, A., Goh, K.K.T., Singh, R.P., Singh, H., 2009b. Behaviour of an oil-in-water emulsion stabilized by [beta]-lactoglobulin in an in vitro gastric model. Food Hydrocoll. 23, 1563−1569.

Singh, H., 2011. Aspects of milk-protein-stabilised emulsions. Food Hydrocoll. 25, 1938−1944.

Singh, H., Sarkar, A., 2011. Behaviour of protein-stabilised emulsions under various physiological conditions. Adv. Coll. Inter. Sci. 165, 47−57.

Singh, H., Ye, A.M., 2013. Structural and biochemical factors affecting the digestion of protein-stabilized emulsions. Curr. Opin. Coll. Inter. Sci. 18 (4), 360−370.

Singh, H., Ye, A., Horne, D., 2009. Structuring food emulsions in the gastrointestinal tract to modify lipid digestion. Prog. Lipid Res. 48, 92−100.

Singh, J., Dartois, A., Kaur, L., 2010. Starch digestibility in food matrix: a review. Trends Food Sci. Technol. 21, 168−180.

Thomas, A., 2006. Gut motility, sphincters and reflex control. Anaesth. Intens. Care Med. 7, 57−58.

Turgeon, S.L., Rioux, L.-E., 2011. Food matrix impact on macronutrients nutritional properties. Food Hydrocoll. 25, 1915−1924.

Tzen, J.T.C., Huang, A.H.C., 1992. Surface-structure and properties of plant seed oil bodies. J. Cell Biol. 117, 327−335.

van Aken, G.A., 2010. Relating food emulsion structure and composition to the way it is processed in the gastrointestinal tract and physiological responses: what are the opportunities? Food Biophys. 5, 258−283.

Vors, C., Pineau, G., Gabert, L., Drai, J., Louche-Pelissier, C., Defoort, C., et al., 2013. Modulating absorption and postprandial handling of dietary fatty acids by structuring fat in the meal: a randomized crossover clinical trial. Am. J. Clin. Nutr. 97, 23−36.

Walstra, P., 1995. Physical chemistry of milk fat globules. In: Fox, P.F. (Ed.), Advanced Dairy Chemistry, second ed. Chapman & Hall, London, UK, pp. 131−178.

Weintraub, M.S., Zechner, R., Brown, A., Eisenberg, S., Breslow, J., 1988. Dietary polyunsaturated fats of n-6 and n-3 series reduce post-prandial lipoprotein concentrations. J. Clin. Invest. 82, 1884−1893.

White, D.A., Fisk, I.D., Makkhun, S., Gray, D.A., 2009. In vitro assessment of the bioaccessibility of tocopherol and fatty acids from sunflower seed oil bodies. J. Agric. Food Chem. 57, 5720 5726.

Wong, J.M.W., Esfahani, A., Singh, N., Villa, C.R., Mirrahimi, A., Jenkins, D.J.A., Kendall, C.W.C., 2012. Gut microbiota, diet, and heart disease. J. AOAC Int. 95, 24−30.

Wu, N.-N., Huang, X., Yang, X.-Q., Guo, J., Yin, S.-W., He, X.-T., et al., 2012. In vitro assessment of the bioaccessibility of fatty acids and tocopherol from soybean oil body emulsions stabilized with iota-carrageenan. J. Agric. Food Chem. 60, 1567−1575.

Ye, A., Cui, J., Singh, H., 2010. Effect of the fat globule membrane on in vitro digestion of milk fat globules with pancreatic lipase. Int. Dairy J. 20, 822−829.

Ye, A., Cui, J., Singh, H., 2011. Proteolysis of milk fat globule membrane proteins during in vitro gastric digestion of milk. J. Dairy Sci. 94, 2762−2770.

Ye, A., Singh, H., Oldfield, D.J., Anema, S.G., 2004. Kinetics of heat-induced association of beta-lactoglobulin and alpha-lactalbumin with milk fat globule membrane in whole milk. Int. Dairy J. 14, 389–398.

Zeece, M., Huppertz, T., Kelly, A., 2008. Effect of high-pressure treatment on in-vitro digestibility of beta-lactoglobulin. Innov. Food Sci. Emerg. Technol. 9, 62–69.

Zúñiga, R.N., Troncoso, E., 2012. Improving nutrition through the design of food matrices. In: Valdez, B., Schorr, M., Zlatev, R. (Eds.), Scientific, Health and Social Aspects of the Food Industry. InTech, Rijeka, Croatia, p. 488.

Chapter 3

The Basis of Structure in Dairy-Based Foods: Casein Micelles and their Properties

Douglas G. Dalgleish

Department of Food Science, University of Guelph, Guelph, Ontario, Canada

CONTENTS

INTRODUCTION

The structures of many dairy foods depend almost exclusively upon the protein they contain. For example, the structures and properties of simple emulsions using whey proteins or caseinates as the emulsifiers depend only to a small extent on the oil within the lipid droplets, but extensively on the surfactant protein. At a higher level of structural complexity, for example in homogenized milks of different types, the fat droplets are stabilized not by monolayers of proteins but by fragments of casein micelles. Even in relatively simple products as these, the properties of the products (e.g., heat or

Food Structures, Digestion and Health. http://dx.doi.org/10.1016/B978-0-12-404610-8.00003-7

83

calcium sensitivity) depend on the protein. It is only in products such as ice-cream or whipped cream that the properties of the fat become important, because of the need for partial coalescence to create the structure of the product, but even here the initial fat droplets are stabilized by proteins.

The most significant influences of protein on structure are in those products created by the interaction of casein micelles, whose structures and reactivity have been modified by chemical, biochemical, or physical means. Cheeses or fermented milk products depend at least for their initial structures on the structures of the casein micelles and the manner in which they interact together. Although much research and thought has been expended in attempts to determine the internal structures of the casein micelles (Horne, 2006, 2009; Dalgleish, 2011, Dalgleish and Corredig, 2012; de Kruif et al., 2012), there is still room for uncertainty. Among the most important factors to be resolved is the way in which water is distributed throughout these particles, and how this distribution may change during processing.

This review seeks to define some of the influences on micellar structure and how this relates to the structures of dairy products. Recent thinking on micellar structure, and some of the problems arising from this in how micelles coagulate together to form products, will be explored.

THE STRUCTURE OF THE CASEIN MICELLE

The discussion that follows concerns bovine milk; there are differences between species in respect of the casein and whey protein contents and consequently the properties of the milk (Martin et al., 2003). Bovine milk contains approximately 32 g/l of protein, 80% of which is casein and 20% is composed of the whey proteins, mainly β-lactoglobulin and α-lactalbumin (Farrell et al., 2004). In contrast to the whey proteins, which are not aggregated, the caseins exist in particles containing many thousands of individual protein molecules, known as the casein micelles (hereinafter referred to simply as micelles) (de Kruif and Holt, 2003), and these colloidal particles give milk its unique processing properties. The particles are built up of the four casein proteins (α_{s1}-, α_{s2}-, β-, and κ-caseins), together with calcium phosphate (colloidal calcium phosphate, CCP). They possess average hydrodynamic diameters of about 200 nm, although any milk contains a wide distribution of sizes (de Kruif, 1998; Udabage et al., 2003). The internal structures of the micelles have been studied over the last 50 years, and several competing theories have been put forward; these have been reviewed by Fox and Brodkorb (2008) and Mezzenga and Fischer (2013).

The micelles are produced in the mammary gland. Their function is the nutrition of the neonate, and they have evolved as particles that sequester relatively large amounts of calcium and phosphate which would otherwise precipitate in the mammary gland (de Kruif and Holt, 2003). Functionally, they undergo clotting after consumption, either when the pH of the milk is decreased or by the action of proteolytic enzymes, or both combined. The micelles therefore form a very stable colloidal system that is easily destabilized under the appropriate conditions. Structurally, the question that must be answered is how can the micelles be assembled from such large numbers of protein molecules without undergoing infinite aggregation? Any model must take account of:

- the composition of the casein (approximately 35% α_{s1}-casein, 40% β-casein, 10% α_{s2}-casein, and 15% κ-casein) and the different functional properties of these proteins (Fox, 2003);
- the presence in the micelles of CCP as an integral part of the structure (de Kruif and Holt, 2003; Horne, 2006);
- the necessity for the κ-casein to be on the particle surface (Dalgleish *et al.*, 1989);
- the fact that micelles are not significantly disrupted either by heat or cold treatments (Anema and Li, 2003a);
- the physical properties of the particles, as revealed by light, small-angle, X-ray or small-angle neutron scattering (de Kruif *et al.*, 2012);
- the appearance of the micellar structure as revealed by different types of microscopy (Buchheim and Dejmek, 1990; Dalgleish *et al.*, 2004; McMahon and Oomen, 2008; Knudsen and Skibsted, 2010; Trejo *et al.*, 2011; Ouanezar *et al.*, 2012).

The average micelle contains approximately 20,000 individual casein molecules, held together by CCP and by non-covalent interactions (Fox, 2003; Horne, 2006). There is no strictly regular structure within the particles, probably because the individual casein molecules themselves do not appear to have fixed tertiary structures and can readily change conformation depending on their surroundings and the forces that act upon them (Holt and Sawyer, 1993; Gaspar *et al.*, 2008).

The α_s- and β-caseins contain groups of phosphorylated serine residues. These confer negative charge on the proteins, and cause them to interact strongly with calcium ions (Dalgleish and Parker, 1980; Parker and Dalgleish, 1981) and with CCP (Holt *et al.*, 1998), and they are precipitated in the presence of sufficient Ca^{2+}. The interaction with calcium and CCP is fundamental to the formation and structure of the micelle; the particles in calcium caseinate do not resemble native micelles in either their structures or properties (Srinivasan *et al.*, 1999). It is possible to produce artificial casein micelles in the laboratory by controlling the formation of CCP in the presence of the caseins, essentially as occurs in the secretory cells of the mammary gland (Schmidt, 1979). In contrast to the α_s- and β-caseins, κ-casein has only a single phosphoserine in its structure, and so does not interact strongly with Ca^{2+}. It does, however, interact with the other caseins, rather than with CCP, in the micelle. It is believed to be located on the surfaces of the micelles (Dalgleish *et al.*, 1989; Marchin *et al.*, 2007), providing a "hairy" layer of extended polypeptides formed from the glycomacropeptide moiety of the κ-casein (Walstra, 1979). The properties of this layer and its stabilizing effect have been extensively discussed (Horne, 1986, 2003; de Kruif, 1992, 1999; de Kruif and Zhulina, 1996). Thus, it is possible to define the micellar structure very approximately as having a "core" of the α_s- and β-caseins, together with CCP, and a "coat" of κ-casein (Waugh and van Hippel, 1956). However, there appears to be insufficient κ-casein to totally cover the micellar surfaces in a tight monolayer, so that it cannot provide a complete "coat" around the other caseins (Dalgleish, 1998).

Early electron microscope images, together with the observation that the caseins tend to self-aggregate even in the absence of calcium, led to the postulation of a "submicellar" structure for the particles. The micelles were viewed as composed of small aggregates of all of the caseins, held together by clumps of CCP (Schmidt, 1982). A later variant of this model suggested that the subunits contained the CCP within them (Walstra, 1990), although the details of the interactions between casein and CCP were not defined. None of the different submicellar proposals explained the manner in which caseins could interact with CCP. Nor could they explain why the caseins would segregate in such a way as to have all of the κ-casein on the micellar surface.

Consideration of the details of the interaction between the caseins and CCP leads to a more convincing model. The α_s- and β-caseins each contain clusters of phosphoserine. Through these clumps of phosphorylated residues the caseins can interact with CCP crystallites, the so-called nanoclusters (Holt *et al.*, 1996, 1998, 2003). On the other hand, the κ-casein, with its single phosphorylated serine residue, will interact weakly or not at all with CCP. Thus, as milk is synthesized in the mammary gland, calcium and inorganic phosphate interact with one another, and as saturation of CCP is reached, the nanoclusters begin to form. The α_s- and β-caseins interact with these growing domains of CCP via their phosphoserine groups until they have completely surrounded the CCP and created protein-stabilized CCP nanoclusters (de Kruif and Holt, 2003). The caseins most likely to be involved in this process are the α_{s1}- and α_{s2}-caseins, as being the most highly phosphorylated. However, the β-casein may also be partly involved. This process of particle formation will transiently create a mixture of casein/CCP nanoclusters, with some of the β-casein and all of the κ-casein free from the nanoclusters.

These nanoclusters are the building blocks of the micelles. The dual-binding model (Horne, 1998) considered the individual caseins as units with different numbers of reactive sites, because they contain phosphorylated domains (charged, hydrophilic) and hydrophobic domains (mainly uncharged, water repellent). By considering the possible interactions of the different domains with CCP and between the proteins, it is possible to explain the micelle structure as arising from the interaction of monofunctional (κ-casein), bifunctional (β-casein), and multifunctional (α_s-caseins) units with CCP and with each other (Horne, 2006). Bifunctional and multifunctional units can produce chains and networks respectively but monofunctional units can only terminate chain or network growth. Thus, κ-casein acts as a chain terminator and would be found only on the surface of the particle. This type of reasoning can be applied to the CCP nanoclusters. The phosphoserine regions of the caseins interact with the CCP, leaving the more hydrophobic parts of the caseins exposed on the surfaces of the nanoclusters. This would lead to aggregation of the nanoclusters via hydrophobic interactions, and so the cores of the micelles would be formed. Since the nanoclusters within the micelles are about 18 nm apart (Holt *et al.*, 2003), it is likely that there may be other caseins acting a spacers between the nanoparticles (Horne, 2006). As the hydrophobically-driven aggregation occurs, the sizes of the growing aggregates are limited by the binding of the "free" κ-casein to their surfaces, so that the sizes of the final micelles depend on the ratio of the κ-casein to the other caseins.

The dual-binding model cannot provide structural information about the organization of the material within the micelle. It also oversimplifies the forces involved. There must be other forces than hydrophobic holding the nanoclusters together (Dalgleish, 2011; de Kruif *et al.*, 2012). Cooling of micelles maintained purely by hydrophobic interactions would lead to their dissociation, but even prolonged cooling of milk leads only to the release of approximately half of the β-casein from the micelle and the other caseins remain in the structure (Creamer *et al.*, 1977). Thus, even though hydrophobic interactions may be responsible for the primary aggregation of the CCP/casein nanoclusters, other bonds (e.g., calcium bridges, salt bridges, hydrogen bonds, van der Waals interactions) must become important at some stage of the aggregation (Dalgleish, 2011). However, as the micelles are acidified, hydrophobic interactions may become more important (see below).

A schematic section through the micelle is shown in Figure 3.1. It appears to be now generally accepted that the micelle is a network of very small domains of CCP, surrounded by caseins, held together by a variety of bonds between the casein molecules. To try to resolve further details of the internal structure is difficult. Micelles can be dissociated in various ways, but none of them will necessarily provide the nanoclusters unaltered. Recent structural studies have depended heavily on two types of techniques, namely scattering (of light, neutrons, or X-rays) and electron microscopy.

Light scattering cannot give details of the interior of the micelle, because the wavelength is too large, although it is routinely used to determine the overall sizes of the micelles. Small-angle neutron (SANS) and X-ray (SAXS) scattering experiments are at much shorter wavelengths and can probe the internal structure. These two techniques give results that are consistent with the presence of the CCP nanoclusters, whose presence is denoted by a peak in the scattering plots, which disappears when the CCP is removed by acidification of the milk (Marchin *et al.*, 2007). It is known that the CCP can be removed in this way while maintaining the gross features of the micelles unaltered, so that the progressive removal of the peak in SANS or SAXS is consistent with the presence of unevenly distributed domains of CCP that are on average about 18 nm apart (Holt *et al.*, 2003; de Kruif

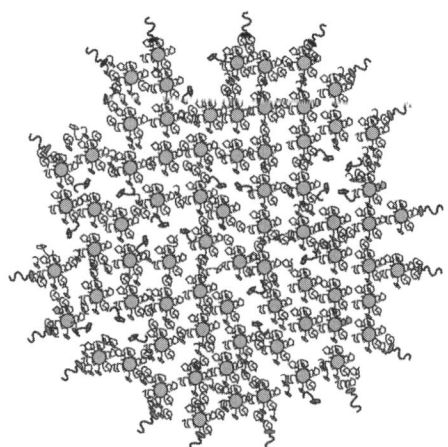

■ **FIGURE 3.1** Schematic representation of a cross-section through the casein micelle, showing the indented surface, with κ-casein macropeptide "hairs" protruding, and the porous interior (McMahon and Oomen, 2008; Bouchoux *et al.*, 2010; Trejo *et al.*, 2011), composed of the network of linked CCP/casein nanoclusters, and the presence of β-casein in the internal pores (Dalgleish, 2011). This is designed to show the different features of the putative structure, and no attempt has been made to draw it to scale.

et al., 2012). The caseins, however, appear to be much more evenly distributed. The interpretation of the SANS and SAXS data has recently been extensively discussed by de Kruif *et al.* (2012).

The scattering methods have the advantage that the micelles can be studied in conditions approximating to their native state. Electron microscopy requires greater or lesser preparation of the samples, and to that extent may be subject to error. Nevertheless, some significant results have been obtained, using a number of microscopic techniques attempting to minimize changes to the structure during sample preparation. Freeze–fracture techniques have shown an apparently non-uniform distribution of material within the micelle; this was originally taken to be evidence for "submicelles" but it may equally well demonstrate the presence of nanoclusters or nodes of caseins within the structure (Dalgleish, 2011). More recently, cryo-transmission electron microscopy (cryo-TEM), where disturbance of the sample is minimized, gives images that are consistent with reticulated structures containing small regions of high electron density, which can be interpreted as the CCP nanoclusters (Marchin *et al.*, 2007; Knudsen and Skibsted, 2010). Detailed scanning electron micrographs (FE-SEM) of the surfaces of the micelles suggest that they may be considered as being bicontinuous systems, with solvent channels running through the structures (Dalgleish *et al.*, 2004). This view is supported by recent work on TEM (McMahon and Oomen, 2008) and by cryo-TEM tomography (Trejo *et al.*, 2011) to produce three-dimensional structural images of the micelles. A recent study of immobilized micelles by AFM (Ouanezar *et al.*, 2012) shows a similar surface roughness of the micelles to that shown by FE-SEM of uncoated micelles (Dalgleish *et al.*, 2004). All of these appear to support the contention that there may be channels or domains of water within the micelles (Bouchoux *et al.*, 2010; Dalgleish, 2011), although this has been disputed by de Kruif *et al.* (2012) on the basis of detailed analysis of SANS and SAXS data. Thus, even though there may be agreement on the important factors governing the structures, there is still some uncertainty on the details, depending on what methods are used for its elucidation.

The micelles are highly hydrated. Native micelles contain as much as 3.5 grams of H_2O per gram of protein (Walstra, 1979). Measured values for hydration are technique dependent; estimates from dynamic physical measurements such as viscosity (Dewan *et al.*, 1973; de Kruif, 1998) or sedimentation velocity (Morris *et al.*, 2000) tend to give higher values of hydration than pelleting of the micelles by sedimentation and subsequent measurement of the water contained in the pellet (Snoeren *et al.*, 1984; Ahmad *et al.*, 2008). This can be explained by the presence of the hairy layer of the macropeptide chains (residues 106–169) of the surface κ-casein of the micelle

(Walstra, 1979). This layer contains little protein, but it is sufficient to affect the transport properties (i.e., viscosity, diffusion, and sedimentation rate) of the micelles. On the other hand, pelleting of the micelles by ultracentrifugation can possibly cause the collapse or squeezing of this layer, causing a decreased value of hydration (Dewan *et al.*, 1973). The pelleting process may also compress the micelles themselves to some extent, since they have been shown to be capable of being squeezed when dispersions of micelles are highly concentrated using osmotic pressure (Bouchoux *et al.*, 2010). Thus, part of the hydration of the particles is clearly associated with the hairy layer, but even taking account of this, the core hydration of the micelles is still high, at least $2-3$ g H_2O/g protein.

This water must be incorporated into the core of the micellar structure. It seems in principle unlikely that it will be evenly distributed throughout the structure because of the nature of the caseins. If the caseins interact with CCP via their phosphorylated serine residues, then their more hydrophobic parts will be on the exterior of the nanocluster particles. The micelle formed by the assembly of these particles would have a hydrophobic interior and would not provide a good matrix for water evenly distributed throughout the structure; there should be water-rich and water-poor regions. It has been proposed that the "free" β-casein (which is capable of being removed by cooling the micelles) stabilizes the water domains by interacting hydrophobically with the network of nanoclusters, in effect acting as surfactant to stabilize water within the system (Dalgleish, 2011). The sizes of the water domains are, however, not established. According to the interpretation of the SANS and SAXS scattering data, they cannot be large (de Kruif *et al.*, 2012), while according to the electron microscopic data they seem to be of significant size. Studies using NMR (Le Feunteun and Mariette, 2007) appeared to show that the pores can allow passage of molecules of molecular mass of several thousand, although later work (Salami *et al.*, 2013) has suggested that intra-micellar diffusion of the polymers may not be as important as was first proposed.

The micelles are stabilized by the hairy layer of κ-casein, which protrudes from the surface of the micelle and creates steric stabilization, that is, as the particles approach, the surface layers cannot interpenetrate, because they either compress each other (energetically unfavorable) or try to interpenetrate (entropically unfavorable). The charges on the macropeptide chains may also be of some importance. As long as this layer remains intact, the micelles remain as separate entities, and the "inner" portions of the micelles cannot interact. Removal of the hairy layer, however, allows interaction between the inner parts of the structures. Even collapse of the hairy layer allows close approach of the micelles, as a result of the decrease in steric stabilization. It should be pointed out that in the first stages of aggregation, only the micellar surfaces are of importance. Subsequent changes after aggregation, such as fusion of the micelles, can occur, but only after a primary aggregation stage.

MODIFICATION OF THE MICELLAR STRUCTURE BY ACIDIFICATION

Acidification of milk is important not only in yoghurt manufacture, but also in the production of many of the different varieties of cheese. A number of structural factors of the micelles are altered as milk is acidified. First and perhaps most importantly, the micellar CCP is progressively dissolved as the pH is decreased from the natural pH (≈ 6.7) of milk (Dalgleish and Law, 1989; Le Graët and Gaucheron, 1999). With slow acidification using lactic bacteria or small amounts of GDL, the release of Ca and P_i from the micelles is a pseudo-equilibrium process (Law and Leaver, 1998; Dalgleish *et al.*, 2005). The removal of the CCP is complete by a pH of approximately 5.1.

Perhaps unexpectedly in view of the importance of CCP to the micellar structure, its removal by slow acidification may not be accompanied by extensive changes in the micellar structure. Although it is not easy to determine whether the detailed internal structure alters, it can certainly be seen that the micelles do not dissociate extensively as long as the temperature is maintained above about 25°C (Dalgleish and Law, 1988); their particle size remains very similar throughout the acidification as long as the milk is not diluted (Moitzi *et al.*, 2011). Although the peak in SANS and SAXS measurements attributable to the CCP nanoclusters disappears during acidification, there is little other change in the scattering (Marchin *et al.*, 2007). The increasing importance of hydrophobic interactions in maintaining the structure is demonstrated, however, by the extensive dissociation of all of the casein types from the micelles when acidification is carried out at low temperature (Dalgleish and Law, 1988). Horne (2009) has suggested that in rapid acidification using high concentrations of GDL and/or at a temperature of 40°C, the dissolution of the CCP does not keep pace with the pH change on the exterior of the micelles, so that acid-induced aggregation precedes the final dissolution of the CCP, resulting in gel formation and then some loosening of the structure as the remaining CCP is dissociated; the elasticity of the gel increases and then decreases. This is also true for rennet gels that are subsequently acidified (Lucey *et al.*, 2000). Thus, the higher the pH at which gelation occurs, the greater is the possibility of subsequently weakening the gel by the dissolution of the remaining CCP nanoclusters.

Acidification causes collapse of the hairy surface layer. At neutral pH, this layer is charged, and charge repulsions maintain the hairs extended. As the charges are decreased during the drop in pH, the hairs become less extended and therefore offer less stabilization (de Kruif, 1997). This collapse of the hairy layer is suggested by the decrease in the hydrodynamic radius of the particles during acidification (Alexander and Dalgleish, 2005). Finally, at a pH of about 5.0, the hairs are fully collapsed, steric stabilization is minimized, and the charges on the proteins are at their minimum. The energy barrier to close approach is low and the micelles can then aggregate. It must be assumed that the collapse of the hairy layer would be homogeneous across the micellar surface, so that reactive "hot spots" would not be created.

The hydration of the micelles changes with pH. Because of the tendency of the micelles to aggregate, it is not possible to measure hydration using viscosity or diffusion and it is necessary to rely on the measurements of the hydration of sedimented micelles. There is a small decrease in the hydration between pH 6.7 and 6.0, followed by an increase until pH 5.4, below which there is a considerable loss of hydration until a pH of about 4.7 (Snoeren *et al.*, 1984; Ahmad *et al.*, 2008). The early change in hydration during acidification may be a consequence of the collapse of the hairy layer (Figure 3.2),

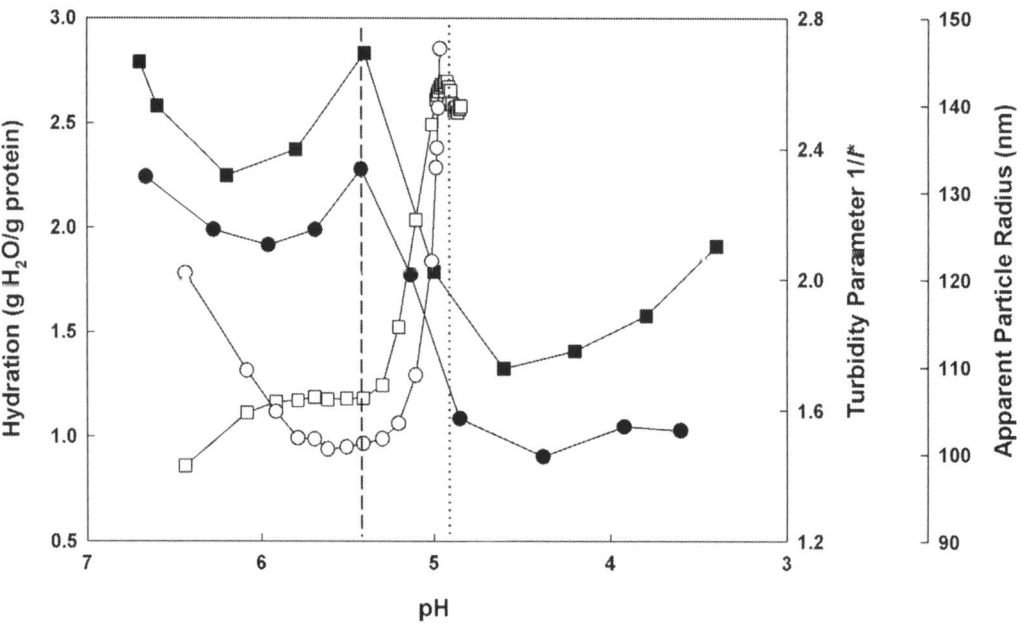

■ **FIGURE 3.2** Changes in the casein micelles during acidification. Hydration results (filled squares) of Snoeren *et al.* (1984) and (filled circles) of Ahmad *et al.* (2008), together with changes in particle radius (open circles) and turbidity parameter 1/*l** (open squares) measured by diffusing wave spectroscopy (Alexander and Dalgleish, 2005). The dotted line indicates the pH where extensive gelation occurs, and the broken line shows the correspondence of the changes in hydration, particle size, and optical properties.

since the apparent radii of the micelles decrease during this stage of the acid-ification (Alexander and Dalgleish, 2005, Moitzi *et al.*, 2011). However, (Figure 3.2) the light scattering properties of the particles change in this pH region (shown by the change of the turbidity ($1/l*$) of the milk). Thus, the micelles in the centrifugal pellet will be able to pack together more tightly. The considerable drop in hydration below pH 5.4 suggests that the micelles can now lose a considerable amount of water, possibly because they become more compressible because of reduced charge repulsions and the loss of CCP. The changes in hydration appear to occur before the main gelation of the micelles (Figure 3.2). Electron micrographs of particles from yoghurt or acid gels do not appear to contain collapsed micellar parti-cles. However, a recent study using AFM has suggested that there is signif-icant shrinkage of the micelles during acidification, and also that there is some restructuring of the particles (Ouanezar *et al.*, 2012).

The aggregation of the acidified micelles near their isoelectric point leads to the creation of a three-dimensional network of linked particles (Figure 3.3), and the particles do not lose their identity (i.e., there is no fusing of the mi-celles). This gel structure is the result of diffusion-limited cluster—cluster interaction, giving rise to a fractal type of aggregate: that is, there is no particular directionality in the formation of the gel structure. On the other hand, because the gel forms rather rapidly, and under kinetic control, the ag-gregates are not in their optimal free-energy configuration. Especially since the micellar particles within the gel are not held together by covalent bonds, it is possible for them to move within the gel. Thus, the gel may undergo

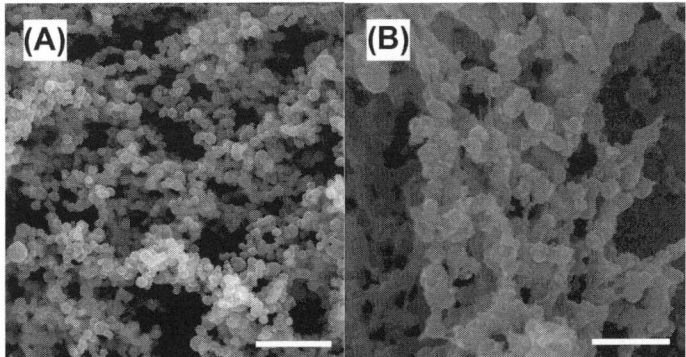

■ **FIGURE 3.3** Structures of acid gels from (A) unheated and (B) heated milks. The micrographs show the different types of contacts between the micelles in the two cases. In (A) the micelles remain distinct within the gel and in (B) they are drawn together by the whey protein attached to them and forming strands between them. Scale bar in (A) is 1 μm and in (B) 500 nm.

syneresis as its internal structure changes, with the expulsion of water as the constituent micelles make close contacts and form a tighter matrix.

RENNETING AND AGGREGATION OF CASEIN MICELLES

Removal of the κ-casein hairy layer by chymosin is the basis of many cheese-making procedures. Chymosin attacks the κ-casein at a specific point, so that the sterically stabilizing macropeptide is removed (Dalgleish, 1987). The decrease in hydrodynamic radius of the micelles as the hairy layer is removed is well documented (Horne, 1984; de Kruif, 1992; Sandra *et al.*, 2007). At the same time, the micelles remain stable (at least at pH 6.7) until a very large proportion (about 85%) of the κ-casein has been destroyed by the enzyme (Tunier and de Kruif, 2002; Sandra *et al.*, 2007). This appears to demonstrate the great stabilizing ability of the κ-casein macropeptide chains; the last few hairs of κ-casein remaining on the micelles are of disproportionate importance in affording stabilization to the particles.

As the pH of milk is decreased, the activity of chymosin increases, but importantly the extent of renneting required to permit aggregation to start is decreased (van Hooydonk *et al.*, 1986). This is in agreement with the postulate that the hairy layer governs the stability, because decreasing pH diminishes the charge and depth of the hairy layer and steric stabilization is decreased. Although the reaction may be described by an overall decrease in the interaction energy between micelles (Tunier and de Kruif, 2002), it is also possible to give a simple geometrical explanation of the interaction, where it is assumed that the micelles can stick together if gaps are formed in the hairy layer, and that the distribution of hairs remaining on the micellar surface defines the sizes of the gaps. If the action of chymosin on κ-casein is random, the aggregation potential of a micelle can be predicted simply by the average micellar radius and the depth of the hairy layer (Dalgleish and Holt, 1988). On this basis, the effects of acidification on the chymosin-induced aggregation can be explained. As the depth of the hairy layer decreases, the size of the gap required for interaction of the micelles becomes smaller, and hence the aggregation will start at a lower degree of proteolysis.

Although the preceding explanation depends on the formation of reactive "hot spots" on the micellar surface, it appears that these are produced randomly. This will not produce any favored directionality in the bonds between the micelles: the aggregation reaction depends on the random production of bare patches. Thus, the reaction can be described as a reaction-limited cluster–cluster aggregation, again giving rise to fractal-like flocs.

(It should be mentioned that in milk, the diffusion of the micelles is not limited. In more concentrated milks, the diffusion may be restricted and so the gel will not have the same structure as it has when the micelles diffuse freely.) However, the micelles beginning to aggregate will still have some of their κ-casein left, which is later removed by further proteolysis. The first gel that is formed by these particles will not have the same structure as one that would be formed if all of the κ-casein were destroyed before the aggregation was allowed to proceed. There will be a progressive increase of the reactivity of the micelles incorporated into the gel as their remaining κ-casein is removed, so that the final gel is not in its most stable state. Hence, disturbance of the gel will lead to extensive rearrangement of the now fully reactive micelles, i.e., syneresis will occur. This is important because in the cheese curd, in contrast to acid gels, the micelles begin to fuse and to lose their identity. Because in chymosin-treated milk the micelles have lost their stabilizing hairy layer, their "cores" are free to interact with one another. At least part of this interaction is hydrophobic, because it is well known that milk does not clot with rennet at low temperatures (Brinkhuis and Payens, 1984), but other forces are involved, because it is well established that addition of calcium can increase the efficiency of the clotting reaction. In acidified milk, the hairy layer on the micelles is still present, although it is collapsed. Its presence prevents the interior parts of the micelles from being in intimate contact, so that the micelles will be much less liable to fuse than in renneted milk.

MIXED COAGULATION WITH ACID AND CHYMOSIN

It is common practice to incorporate lactic acid bacteria as starter cultures into milk for cheese-making, and the properties of different cheeses depend on the activity of the culture used and the final pH that is achieved when the curd is collected. Thus, many cheeses are the result of mixed coagulation processes, where the micelles are destabilized both by acidification and by the loss of κ-casein. This process is more complex to describe than either simple acid gelation or rennet coagulation. Not only are the renneting properties of the micelles altered, but their internal properties are also changed by the loss of greater or smaller amounts of CCP, depending on the extent of pH change during the cheese-making process.

Even small extents of κ-casein destruction lead to a very much increased gel strength in the acid coagulum, combined with an increased pH at which the coagulation occurs (Li and Dalgleish, 2006). At relatively high concentrations of chymosin, coagulation occurs at high pH and the gel may be regarded as a typical rennet gel. After gel formation, further decrease of the pH causes a weakening (or loss of elasticity) (Li and Dalgleish, 2006; Lucey *et al.*, 2000) that may be attributed to the dissociation of the micellar CCP, weakening the internal structures of the micelles (Lucey *et al.*, 2000). Conversely, if only small amounts of chymosin are used, the gel is formed at lower pH and is strongly elastic (as strong as the gel from heated milk). This last observation is hard to explain in the way that has been used to discuss the other gelation processes. The extent of κ-casein breakdown that is required for a large increase in the strength of the acid gel is as low as 15% (Li and Dalgleish, 2006). At this extent of proteolysis, it would not be expected that the micellar surfaces could make good enough contact within the remains of the hairy layer, which must still cover most of the surface. If an acid gel from unheated milk is weak because the κ-casein is intact and prevents intimate contact between the "interiors" of the micelles, then removing a small proportion of the total κ-casein in principle would not be expected to make a great deal of difference. The strength of a gel formed from particles is determined partly by the strength of the bonds within the particle and partly by the inter-particle bonds. According to Horne (2009), the internal bonds decrease because of the dissolution of the CCP. In the case of mixed acidification, the internal bonds will be changed by pH irrespective of the degree of breakdown of the κ-casein. The increase in gel strength compared with an acid gel prepared from unrenneted milk must therefore be a consequence of the inter-particle interactions in the gel.

This mixed coagulation is important in the digestion of milk in the stomach. Milk will raise the pH of the stomach, which will then be decreased as more acid is produced. At the same time, the micelles are attacked by protease (pepsin) and clotting occurs. The properties of the clot and its manner of further digestion will depend on the relative rates of acidification and proteolysis.

MODIFICATION OF THE CASEIN MICELLES BY HEATING

Further structural modification of the micelles occurs when milk is heated. In the absence of whey proteins, heating of micelles to temperatures up to 100°C appears to have little effect (Anema and Li, 2003a). However, if whey proteins are present during the heating, they are denatured at temperatures above about 75°C, and interact with each other and with the micelles (Anema and Li, 2003b). The end product of these reactions is that some of the whey proteins become attached to the micelles, and some form complexes with κ-casein detached from the micelles, and remain in solution. The milk therefore contains a mixture of modified micelles and small whey protein/κ-casein complexes. The process is highly pH dependent; if heating is performed on milk adjusted to pH 6.2, nearly all of the whey protein is to be found on the surfaces of the micelles. Conversely, at pH 7.0, nearly all of the whey proteins are in the serum as free complexes (Anema and Li, 2003b). Adjustment of the pH after the heating has occurred does not alter the distribution of the complexes between the micelles and the serum.

It is not established whether there is a change in the internal organization of the micelles during heating, but, since the processing of milk depends mostly on the micellar surface, the reactivity and the structures that are formed are different from those in unheated milk. The most obvious of these is that the milk coagulates differently with chymosin; essentially, the heated milk does not clot efficiently (Singh and Waungana, 2001), although micelles heated in the absence of whey proteins coagulate normally (Park *et al.*, 1996). This does not appear to be the result of inaccessibility of κ-casein to the protease: experiments have shown that the breakdown of the κ-casein appears to be achieved to the same extent, and at a similar rate, to the process in unheated milk (Vasbinder *et al.*, 2003c; Anema *et al.*, 2007; Kethireddipalli *et al.*, 2011). It is the subsequent coagulation of the micelles that is inhibited. The whey protein complexes on the micellar surface may act as steric stabilizing agents and prevent the close approach of the micellar surfaces (Dalgleish, 1990; Vasbinder *et al.*, 2003c; Singh and Waungana, 2001). Observations of milk heated at high pH, where the whey protein complexes are largely detached from the micelles, show only small improvements in renneting (Kethireddipalli *et al.*, 2010), and it seems that the complexes may re-attach to the micelles during renneting (Renan *et al.*, 2007; Kethireddipalli *et al.*, 2011). It has also been established that the micelles from milk heated at any pH do not coagulate well with chymosin when dispersed in unheated milk serum, even those from milk heated at pH 7.1, where there are very few attached whey proteins (Kethireddipalli *et al.*, 2010).

Renneting at lower pH (6 or below) is more efficient, because of the increased activity of the chymosin, and the need for smaller extents of κ-casein proteolysis for clotting to occur. This has been used as a method of improving the coagulation of heated milks, and indeed workable curds can be produced in this way (Horne and Banks, 2004). However, the finished cheese has a structure very different from that made with unheated milk, so that the whey protein complexes have a significant effect on the behavior of the micelles in the curd after coagulation. Presumably, the fusion of the micelles is altered by the presence of surface-bound whey protein complexes; in addition, the micelles at the low pH values necessary for coagulation will have lost a considerable portion of their CCP and will be susceptible to structural change.

The effects of heat are very positive in yoghurt manufacture. Acid coagulation of unheated milk gives a fairly weak gel; heating the milk prior to acidification gives a much stronger gel, with lower tendency to syneresis (Lucey, 2002). This can be ascribed to the presence of the whey protein complexes in the serum and on the surfaces of the micelles. It has been demonstrated that the greater the proportion of soluble complexes, the stronger is the acid gel (Vasbinder *et al.*, 2003a; Anema *et al.*, 2004). This suggests that the complexes play a role by forming bridges between the micelles, and linking to the κ-casein on their surfaces (Donato *et al.*, 2007; Guyomarc'h *et al.*, 2009; Donato and Guyomarc'h, 2009). It has been suggested that these bridges are linked by the formation of disulfide bonds, despite the fact that these are unlikely to be formed at low pH (Vasbinder *et al.*, 2003b). We should therefore envisage the acid gel from heated milk as being formed by micelles whose surfaces are not in direct contact with one another. In gels from unheated milk, it is the collapsed layers of κ-casein that are in contact; in contrast, in heated milk the micellar surfaces are unlikely to be in direct contact, so that the strength of the gel is determined by the whey protein/κ-casein/micelle contacts. It is interesting that the gels from milk heated at lower pH values, where the bulk of the complexes are on the micellar surfaces, give weaker gels than those formed from milk heated at its natural pH, where most of the complexes are free (Anema *et al.*, 2004). Of course, addition of whey protein to the milk before heating will give rise to even more complexes, with consequent strengthening of texture, both because of increased bridging between micelles but also because of increased overall protein concentration.

ULTRA-HIGH PRESSURE TREATMENT AND THE STRUCTURES OF MICELLES

Ultra-high pressure (UHP) processing at pressures of above about 250 MPa causes the micelles to dissociate. The colloidal CCP is dissolved (Huppertz *et al.*, 2006), and there is generally a decrease in the size of the micelles and depletion of κ- and β-casein (Gaucheron *et al.*, 1997). These rearrangements, however, still do not cause full colloidal destabilization and precipitation, suggesting that proteins other than κ-casein can stabilize the particles. The effects of the pressure treatment appear to be quasi-reversible at the lower scale of pressure treatment. At moderate pressures (<200 MPa), the micelles reform to something similar to their original size distribution once the pressure is released (Anema *et al.*, 2005; Gebhart *et al.*, 2006). If this pressure is exceeded, the micelles do not reform, but exist as much smaller (≈ 30 nm) caseinate particles of unknown structure (Gebhart *et al.*, 2006). Results from different laboratories do not totally agree on sizes of the final particles (Anema *et al.*, 2005), perhaps because the original starting materials (skim milk powders) were different.

In the presence of whey proteins, and at pressure >100 MPa, the surface of the micelle is also modified by the association of whey proteins that have been denatured by pressure (Needs *et al.*, 2000b; Huppertz and de Kruif, 2007). The high-pressure treatment appears to cause changes in the way in which the milk coagulates with rennet. There seem to be a number of effects, both on the rate of action of the chymosin on the micelles and especially on the rate of coagulation of the renneted micelles (Lopez-Fandino *et al.*, 1996), depending on the pressure used, the time of pressurization, and the temperature; this has been reviewed by O'Reilly *et al.* (2001). The structure of the rennet gel is different from that of a control sample, especially if pressures above 450 MPa have been used (Needs *et al.*, 2000a); the gels are finer-stranded and have different rheological properties. The application of high-pressure treatment increases the yield of cheese, because of the incorporation of denatured whey protein into the curd (Voigt *et al.*, 2010), but the presence of whey protein, and perhaps the different structure of the curd, lead to differences in the properties of Cheddar cheese made from milk treated at 600 MPa before renneting (Voigt *et al.*, 2012).

As might be expected, the dissociation of the micelles at high pressures has an effect on yoghurt structure: the irreversible dissociation of the micelles leads to the formation of weaker gels than the standard (Udabage *et al.*, 2010). However, the pressurization of milk in the presence of added whey proteins leads to beneficial textural effects (Needs *et al.*, 2000b).

CONCLUSION

Once the final product is formed (cheese curd, yoghurt), the colloidal properties of the micelles become unimportant and the structure of the final product depends on other factors, such as the fusion of the micelles within the gel, or proteolytic degradation of the caseins. However, the initial structures of the products are defined by the method of destabilization of the micelles and the aggregation reactions that follow, coupled with the state of the interior of the micelle, for example the extent to which the calcium phosphate is intact. This may be especially important when the digestion of milk or dairy products in the stomach is being considered, when the material is subjected to acidification, proteolysis, and mechanical stress, all of which will affect the manner of digestion of the product.

NOTE ADDED IN PROOF

Since this article was written, a paper has been published which takes a new viewpoint on casein micellar structures (Holt, C., Carver, J.A., Ecroyd, H., Thorn, D.C., 2013. Caseins and the casein micelle: their biological functions, structures and behaviour in foods. J. Dairy Sci. 96, 6127–6146). The authors consider caseins and other phosphoproteins in the light of their molecular interactions, their tendency to form amyloid structures and the role of calcium phosphate in moderating their interactions.

REFERENCES

Ahmad, S., Gaucher, I., Rousseau, F., Beaucher, E., Piot, M., Grongnet, J.-F., Gaucheron, F., 2008. Effects of acidification on physico-chemical characteristics of buffalo milk: a comparison with cow's milk. Food Chem. 106, 11–17.

Alexander, M., Dalgleish, D.G., 2005. Interactions between denatured milk serum proteins and casein micelles studied by diffusing wave spectroscopy. Langmuir 21, 11380–11386.

Anema, S.G., Lee, S.K., Klostermeyer, H., 2007. Effect of pH at heat treatment on the hydrolysis of κ-casein and the gelation of skim milk by chymosin. LWT 40, 99–106.

Anema, S.G., Li, Y., 2003a. Association of denatured whey proteins with casein micelles in heated reconstituted skim milk and its effect on casein micelle size. J. Dairy Res. 70, 73–83.

Anema, S.G., Li, Y., 2003b. Effect of pH on the association of denatured whey proteins with casein micelles in heated reconstituted skim milk. J. Agric. Food Chem. 51, 1640–1646.

Anema, S.G., Lowe, E.K., Stockmann, R., 2005. Particle size changes and casein solubilisation in high-pressure-treated skim milk. Food Hydrocoll. 19, 257–267.

Anema, S.G., Lee, S.K., Lowe, E.K., Klostermeyer, H., 2004. Rheological properties of acid gels prepared from heated pH-adjusted skim milk. J. Agric. Food Chem. 52, 337–343.

Bouchoux, A., Gésan-Guiziou, G., Pérez, J., Cabane, B., 2010. How to squeeze a sponge: casein micelles under osmotic stress, a SAXS study. Biophys. J. 99, 3754–3762.

Brinkhuis, J., Payens, T.A., 1984. The influence of temperature on the flocculation rate of renneted casein micelles. Biophys. Chem. 19, 75–81.

Buchheim, W., Dejmek, P., 1990. Milk and dairy-type emulsions. In: Larsson, K., Friberg, S.E. (Eds.), Food Emulsions. Marcel Dekker, New York, pp. 203–246.

Creamer, L.K., Berry, G.P., Mills, O.E., 1977. A study of the dissociation of β-casein from the bovine casein micelle at low temperature. NZ J. Dairy Sci. Technol. 12, 58–66.

Dalgleish, D.G., 1987. The enzymatic coagulation of milk. In: Fox, P.F. (Ed.), Cheese—Chemistry, Physics and Microbiology. Applied Science Publishers, Barking, pp. 63–96.

Dalgleish, D.G., 1990. The effect of denaturation of β-lactoglobulin on renneting—a quantitative study. Milchwissenschaft 45, 491–494.

Dalgleish, D.G., 1998. Casein micelles as colloids: surface structures and stabilities. J. Dairy Sci. 81, 3013–3018.

Dalgleish, D.G., 2011. On the structural models of casein micelles—review and possible improvements. Soft Matter 7, 2265—2272.

Dalgleish, D.G., Corredig, M., 2012. The structure of the casein micelle in milk and its changes during processing. In: Doyle, M.P., Klaenhammer, T.R. (Eds.), Annual Review of Food Science and Technology, vol. 3. Annual Reviews, Palo Alto, pp. 449—467.

Dalgleish, D.G., Holt, C., 1988. A geometrical model to describe the initial aggregation of partly renneted casein micelles. J. Colloid Interface Sci. 123, 80—84.

Dalgleish, D.G., Law, A.J.R., 1988. pH-induced dissociation of bovine casein micelles. I. Analysis of liberated caseins. J. Dairy Res. 55, 529—538.

Dalgleish, D.G., Law, A.J.R., 1989. pH-induced dissociation of bovine casein micelles II. Mineral solubilization and its relation to casein release. J. Dairy Res. 56, 727—735.

Dalgleish, D.G., Horne, D.S., Law, A.J.R., 1989. Size-related differences in bovine casein micelles. Biochim. Biophys. Acta. 991, 383—387.

Dalgleish, D.G., Parker, T.G., 1980. Binding of calcium ions to bovine α_{s1}-casein and precipitability of the protein-Ca^{2+}complexes. J. Dairy Res. 47, 113—122.

Dalgleish, D.G., Spagnuolo, P.A., Goff, H.D., 2004. A possible structure of the casein micelle based on high-resolution field-emission scanning electron microscopy. Int. Dairy J. 14, 1025—1031.

Dalgleish, D.G., Verespej, E., Alexander, M., Corredig, M., 2005. The effect of acidification rate on the release of calcium from casein micelles and its effect on the ultrasonic properties of skim milk. Int. Dairy J. 15, 1105—1112.

de Kruif, C.G., 1992. Casein micelles: diffusivity as a function of renneting time. Langmuir 8, 2932—2937.

de Kruif, C.G., 1997. Skim milk acidification. J. Colloid Interface Sci. 185, 19—25.

de Kruif, C.G., 1998. Supra-aggregates of casein micelles as a prelude to coagulation. J. Dairy Sci. 81, 3019—3028.

de Kruif, C.G., 1999. Casein micelle interactions. Int. Dairy J. 9, 183—188.

de Kruif, C.G., Holt, C., 2003. Casein micelle structure, function and interactions. In: Fox, P.F., McSweeney, P.L.H. (Eds.), Advanced Dairy Chemistry, Proteins, vol. 1. Kluwer Academic/Plenum Publishers, New York, pp. 233—276.

de Kruif, C.G., Huppertz, T., Urban, V.S., Petukhov, A.V., 2012. Casein micelles and their internal structure. Adv. Colloid Interface Sci. 171—172, 36—52.

de Kruif, C.G., Zhulina, E.B., 1996. κ-Casein as a polyelectrolyte brush on the surface of casein micelles. Colloids Surf. A 117, 151—159.

Dewan, R.K., Bloomfield, V.A., Chudgar, A., Morr, C.V., 1973. Viscosity and voluminosity of bovine casein micelles. J. Dairy Sci. 56, 699—705.

Donato, L., Alexander, M., Dalgleish, D.G., 2007. Acid gelation in heated and unheated milks: interactions between serum proteins and the surfaces of casein micelles. J. Agric. Food Chem. 55, 4160—4168.

Donato, L., Guyomarc'h, F., 2009. Formation and properties of the whey protein/κ-casein complexes in heated skim milk—a review. Dairy Sci. Technol. 89, 3—29.

Farrell Jr., H.M., Jiménez-Flores, R., Black, G.T., Butler, J.E., Creamer, L.K., Hicks, C.L., et al., 2004. Nomenclature of the proteins of cows' milk—Sixth Revision. J. Dairy Sci. 87, 1641—1674.

Fox, P.F., 2003. Milk proteins: general and historical aspects. In: Fox, P.F., McSweeney, P.L.H. (Eds.), Advanced Dairy Chemistry, Proteins, vol. 1. Kluwer Academic/Plenum Publishers, New York, pp. 1—48.

Fox, P.F., Brodkorb, A., 2008. The casein micelle: historical aspects, current concepts and significance. Int. Dairy J. 18, 677–684.

Gaspar, A.M., Appavon, M.-S., Busch, S., Unruh, T., Doster, W., 2008. Dynamics of well-folded and natively-disordered proteins in solution: a time of flight neutron scattering study. Eur. Biophys. J. 37, 573–582.

Gaucheron, F., Famelart, M.-H., Mariette, F., Raulot, K., Michel, F., Le Graët, Y., 1997. Combined effects of temperature and high-pressure treatments on physicochemical characteristics of skim milk. Food Chem. 59, 439–447.

Gebhart, R., Doster, W., Friedrich, J., Kulozik, U., 2006. Size distribution of pressure-decomposed casein micelles studied by dynamic light scattering and AFM. Eur. Biophys. J. 35, 503–509.

Guyomarc'h, F., Jemin, M., Le Tilly, V., Madec, M.-N., Famelart, M.-H., 2009. Role of the heat-induced whey protein/κ-casein complexes in the formation of acid milk gels: a kinetic study using rheology and confocal microscopy. J. Agric. Food Chem. 57, 5910–5917.

Holt, C., 1998. Casein micelle substructure and calcium phosphate interactions studied by Sephacryl column chromatography. J. Dairy Sci. 81, 2994–3003.

Holt, C., de Kruif, C.G., Tunier, R., Timmins, P.A., 2003. Substructure of bovine casein micelles by small-angle X-ray and neutron scattering. Colloids Surf. A 213, 275–284.

Holt, C., Sawyer, L., 1993. Caseins as rheomorphic proteins: interpretation of primary and secondary structures of the α_s-, β- and κ-caseins. J. Chem. Soc. Farad. Trans. 89, 2683–2692.

Holt, C., Timmins, P.A., Errington, N., Leaver, J., 1998. A core-shell model of calcium phosphate nanoclusters stabilized by β-casein phosphopeptides, derived from sedimentation equilibrium and small-angle X-ray and neutron scattering experiments. Eur. J. Biochem. 252, 73–78.

Holt, C., Wahlgren, N.M., Drakenberg, T., 1996. Ability of a β-casein phosphopeptide to modulate the precipitation of calcium phosphate by forming amorphous dicalcium phosphate nanoclusters. Biochem. J. 314, 1035–1039.

Horne, D.S., 1984. Steric effects in the coagulation of casein micelles by ethanol. Biopolymers 23, 989–993.

Horne, D.S., 1986. Steric stabilization and casein micelle stability. J. Colloid Interface Sci. 111, 250–259.

Horne, D.S., 1998. Casein interaction—casting light on the Black Boxes, the structure in dairy products. Int. Dairy J. 8, 171–177.

Horne, D.S., 2003. Casein micelles as hard spheres: limitations of the model in acidified gel formation. Colloids Surf. A 213, 255–263.

Horne, D.S., 2006. Casein micelle structure: models and muddles. Curr. Opin. Coll. Interf. Sci. 11, 148–153.

Horne, D.S., 2009. Casein micelle structure and stability. In: Thompson, A., Boland, M., Singh, H. (Eds.), Milk Proteins: from Expression to Food. Elsevier, New York, pp. 133–162.

Horne, D.S., Banks, J.M., 2004. Rennet-induced coagulation of milk. In: Fox, P.F., McSweeney, P.L.H.M., Cogan, T.M., Guinee, T.P. (Eds.), Cheese—Chemistry, Physics and Microbiology. Elsevier Academic Press, San Diego, pp. 47–70.

Huppertz, T., de Kruif, C.G., 2007. Disruption and reassociation of casein micelles during high pressure treatment: influence of whey proteins. J. Dairy Res. 73, 194–197.

Huppertz, T., Kelly, A.L., de Kruif, C.G., 2006. Disruption and reassociation of casein micelles under high pressure. J. Dairy Res. 73, 294–298.

Kethireddipalli, P., Hill, A.R., Dalgleish, D.G., 2010. Protein interactions in heat-treated milk and effect on rennet coagulation. Int. Dairy J. 20, 838–843.

Kethireddipalli, P., Hill, A.R., Dalgleish, D.G., 2011. Interaction between casein micelles and whey protein/κ-casein complexes during renneting of heat-treated reconstituted skim milk powder and casein micelle/serum mixtures. J. Agric. Food Chem. 59, 1442–1448.

Knudsen, J.C., Skibsted, L.H., 2010. High pressure effects on the structure of casein micelles in milk as studied by cryo-transmission electron microscopy. Food Chem. 119, 202–208.

Law, A.J.R., Leaver, J., 1998. Effects of acidification and storage of milk on dissociation of bovine casein micelles. J. Agric. Food Chem. 46, 5008–5016.

Le Feunteun, S., Mariette, F., 2007. Impact of casein gel microstructure on self-diffusion coefficient of molecular probes measured by ^{1}H PFG-NMR. J. Agric. Food Chem. 55, 10764–10772.

Le Graët, Y., Gaucheron, F., 1999. pH-induced solubilization of minerals from casein micelles: influence of casein concentration and ionic strength. J. Dairy Res. 66, 215–224.

Li, J., Dalgleish, D.G., 2006. Mixed coagulation of milk—gel formation and mechanism. J. Agric. Food Chem. 54, 4687–4695.

Lopez-Fandino, R., Carrascosa, A.V., Olano, A., 1996. The effects of high pressure on whey protein denaturation and cheese making properties of raw milk. J. Dairy Sci. 79, 929–936.

Lucey, J.A., 2002. Formation and physical properties of milk protein gels. J. Dairy Sci. 85, 281–294.

Lucey, J.A., Tamehana, M., Singh, H., Munro, P.A., 2000. Rheological properties of milk gels formed by a combination of rennet and glucono-δ-lactone. J. Dairy Res. 67, 415–427.

Marchin, S., Putaux, J.-L., Pignon, F., Léonil, J., 2007. Effects of the environmental factors on the casein micelle structure studied by cryo transmission electron microscopy and small angle x-ray scattering/ultrasmall angle x-ray scattering. J. Chem. Phys. 126, 045101–1-10.

Martin, P., Ferranti, P., Leroux, C., Addeo, F., 2003. Non-bovine caseins: quantitative variability and molecular diversity. In: Fox, P.F., McSweeney, P.L.H. (Eds.), Advanced Dairy Chemistry, Proteins, vol. 1. Kluwer Academic/Plenum Publishers, New York, pp. 277–318.

McMahon, D., Oomen, B.S., 2008. Supramolecular structure of casein micelles. J. Dairy Sci. 91, 1709–1721.

Mezzenga, R., Fischer, P., 2013. The self-assembly, aggregation and phase transitions of food protein systems in one, two and three dimensions. Rep. Prog. Phys. 76, 046601.

Moitzi, C., Menzel, A., Schurtenberger, P., Stradner, A., 2011. The pH induced sol-gel transition in skim milk revisited. A detailed study using time-resolved light and X-ray scattering experiments. Langmuir 27, 2195–2203.

Morris, G.A., Foster, T.J., Harding, S.E., 2000. Further observations on the size, shape, and hydration of casein micelles from novel analytical ultracentrifuge and capillary viscometry approaches. Biomacromolecules 1, 764–767.

Needs, E.C., Stenning, R.A., Gill, A.L., Ferragut, V., Rich, G.T., 2000a. High-pressure treatment of milk: effects on casein micelle structure and on enzymic coagulation. J. Dairy Res. 67, 31–42.

Needs, E.C., Capellas, M., Bland, A.P., Manoj, P., MacDougal, D., Paul, G., 2000b. Comparison of heat and pressure treatments of skim milk, fortified with whey protein concentrate, for set yogurt preparation: effects on milk proteins and gel structure. J. Dairy Res. 67, 329–348.

O'Reilly, C., Kelly, A.L., Murphy, P.M., Beresford, T.P., 2001. High pressure treatment: applications in cheese manufacture and ripening. Trends Food Sci. Technol. 12, 51–59.

Ouanezar, M., Guyomarc'h, F., Bouchoux, A., 2012. AFM imaging of casein micelles: evidence for structural rearrangement upon acidification. Langmuir 28, 4915–4919.

Park, S.-Y., Nakamura, K., Niki, R., 1996. Effects of β-lactoglobulin on the rheological properties of casein micelle rennet gels. J. Dairy Sci. 79, 2137–2145.

Parker, T.G., Dalgleish, D.G., 1981. Binding of calcium ions to bovine β-casein. J. Dairy Res. 48, 71–76.

Renan, M., Guyomarc'h, F., Chatriot, M., Gamerre, V., Famelart, M.-H., 2007. Limited enzymatic treatment of skim milk using chymosin affects the micelle-serum distribution of the heat-induced whey protein/κ-casein aggregates. J. Agric. Food Chem. 55, 6736–6745.

Salami, S., Rondeau-Mouro, C., van Duynhoven, J., Mariette, F., 2013. PFG-NMR self-diffusion in casein dispersions: effects of probe size and protein aggregate size. Food Hydrocoll. 31, 248–255.

Sandra, S., Alexander, M., Dalgleish, D.G., 2007. The rennet coagulation mechanism of skim milk as observed by transmission diffusing wave spectroscopy. J. Colloid Interface Sci. 308, 364–373.

Schmidt, D.G., 1979. Properties of artificial casein micelles. J. Dairy Res. 46, 351–356.

Schmidt, D.G., 1982. Association of caseins and casein micelle structure. In: Fox, P.F. (Ed.), Developments in Dairy Chemistry. Applied Science Publishers, London, pp. 61–86.

Singh, H., Waungana, A., 2001. Influence of heat treatment of milk on cheese making properties. Int. Dairy J. 11, 543–551.

Snoeren, T.H.M., Klok, K.J., van Hooydonk, A.C.M., Damman, A.J., 1984. The voluminosity of casein micelles. Milchwissenschaft 39, 461–463.

Srinivasan, M., Singh, H., Munro, P.A., 1999. Adsorption behaviour of sodium and calcium caseinates in oil-in-water emulsions. Int. Dairy J. 9, 337–341.

Trejo, R., Dokland, T., Jurat-Fuentes, J., Harte, F., 2011. Cryo-transmission electron tomography of native casein micelles from bovine milk. J. Dairy Sci. 94, 5770–5775.

Tunier, R., de Kruif, C.G., 2002. Stability of casein micelles in milk. J. Chem. Phys. 117, 1290–1295.

Udabage, P., McKinnon, I.R., Augustin, M.A., 2003. The use of sedimentation field-flow fractionation and photon correlation spectroscopy in the characterization of casein micelles. J. Dairy Res. 70, 453–459.

Udabage, P., Augustin, M.A., Versteeg, C., Puvanethiran, A., Yoo, J.A., Allen, N., et al., 2010. Properties of low-fat yogurts made from high-pressure-processed skim milk. Innov. Food Sci. Emerg. Technol. 11, 32–38.

Van Hooydonk, A.C.M., Boerrigter, I.J., Hagedoorn, H.G., 1986. pH-induced physico-chemical changes of casein micelles in milk and their effect on renneting. 2. Effect of pH on renneting of milk. Neth. Milk Dairy J 40, 297–313.

Vasbinder, A.J., Alting, A.C., de Kruif, K.G., 2003a. Quantification of heat-induced casein—whey protein interactions in milk and its relation to gelation kinetics. Colloids Surf. B: Biointerfaces 31, 115—123.

Vasbinder, A.J., Alting, A.C., Visschers, R.W., de Kruif, C.G., 2003b. Texture of acid milk gels: formation of disulfide cross-links during acidification. Int. Dairy J. 13, 29—38.

Vasbinder, A.J., Rollema, H.S., de Kruif, C.G., 2003c. Impaired rennetability of heated milk: study of enzymatic hydrolysis and gelation kinetics. J. Dairy Sci. 86, 1548—1555.

Voigt, D.D., Donaghy, J.A., Patterson, M.F., Stephan, S., Kelly, A.L., 2010. Manufacture of Cheddar cheese from high-pressure-treated whole milk. Innov. Food Sci. Emerg. Technol. 11, 574—579.

Voigt, D.D., Chevalier, F., Donaghy, J.A., Patterson, M.F., Qian, M.C., Kelly, A.L., 2012. Effect of high-pressure treatment of milk for cheese manufacture on proteolysis, lipolysis, texture and functionality of Cheddar cheese during ripening. Innov. Food Sci. Emerg. Technol. 13, 23—30.

Walstra, P., 1979. The voluminosity of casein micelles and some of its implications. J. Dairy Res. 46, 317—324.

Walstra, P., 1990. On the stability of casein micelles. J. Dairy Sci. 73, 1965—1979.

Waugh, D.F., van Hippel, P.H., 1956. κ-Casein and the stabilization of casein micelles. J. Amer. Chem. Soc. 78, 4576—4582.

Chapter 4

The Milk Fat Globule Membrane: Structure, Methodology for its Study, and Functionality

Sophie Gallier,[1,3] Andrea Laubscher[2] and Rafael Jiménez-Flores[2]

[1]*Danone Nutricia Research, Uppsalalaan, Utrecht, The Netherlands,* [2]*Dairy Products Technology Center, California Polytechnic State University, San Luis Obispo, CA, USA,* [3]*Riddet Institute, Massey University, Palmerston North, New Zealand*

CONTENTS

Food Structures, Digestion and Health. http://dx.doi.org/10.1016/B978-0-12-404610-8.00004-9

INTRODUCTION

BIOLOGICAL ORIGIN AND CURRENT STATUS OF KNOWLEDGE

Milk lipids are mainly composed of triacylglycerols that originate within the bilayer of the rough endoplasmic reticulum (rER) of epithelial cells of the mammary glands where they start their journey towards the apical membrane (Nielsen *et al.*, 1999; Robenek *et al.*, 2006a, b). Freeze-fracture electron microscopy has shown that lipid droplets developed alongside the rER membrane, and is essential in the secretion of the fat from epithelial cells into milk. Regardless of the lipid droplet biogenesis, a membrane known as the milk fat globule membrane (MFGM) surrounds the lipid droplets in milk. This triple layer membrane acts as an emulsifier and prevents coalescence of fat droplets and lipase activity (Dewettinck *et al.*, 2008). The fat globule size depends on many factors, such as cow breed and lactation stage. For example, fat globules from Jersey cows are larger than those from Friesian cows, with average diameters of 4.5 and 3.5 µm, respectively (Singh, 2006). This fact, of course, has a profound effect in dairy products.

The MFGM is complex and 10 to 50 nm thick containing phospholipids, sphingolipids, and specific membrane proteins, as seen in Figure 4.1. Phospholipids and proteins account for over 90% of the membrane's dry weight (Singh, 2006). Some of the proteins are an integral part of the membrane, and others are peripheral or loosely attached within the triple-layer membrane (Gallier *et al.*, 2010a, b; Lopez *et al.*, 2010; Gallier *et al.*, 2011). The first membrane, derived from the rER, is a monolayer containing phospholipids and proteins. The second membrane is a bilayer containing glycosylated and non-glycosylated proteins, glycerol-phospholipids and sphingolipids, enzymes, cholesterol, and other minor components. Between the inner membrane and the bilayer, there is an electron-dense coat rich in proteins (Rombaut et al., 2006b; Lopez *et al.*, 2008). A glycocalyx forms a fourth layer that acts as a source of specific bacterial and viral ligands and varies during lactation (Spitsberg and Gorewit, 1998, 1999; Evers, 2004; Wilson *et al.*, 2008). A detailed description of milk fat synthesis during lactation is given by Bionaz and Loor (2008).

The gross composition of the MFGM, reported in literature, differs as a result of isolation, purification, and techniques used in analysis (Keenan *et al.*, 1971; Keenan and Mather, 2006; Dewettinck *et al.*, 2008). Furthermore, the composition can be altered by physiological, chemical/enzymatic, and physical/mechanical factors (Gallier *et al.*, 2010c, 2011). The latter

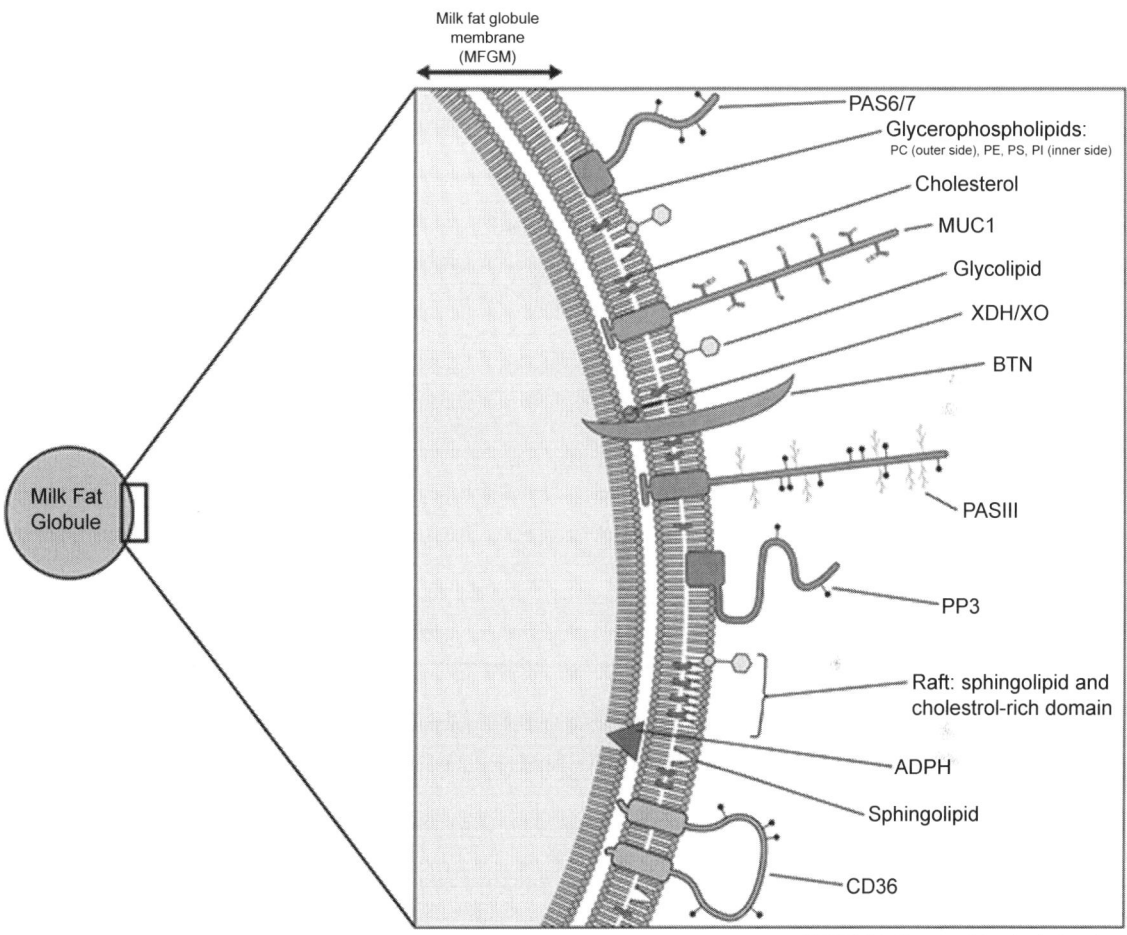

Milk fat globule
membrane
(MFGM)

Milk Fat
Globule

PAS6/7

Glycerophospholipids:
PC (outer side), PE, PS, PI (inner side)

Cholesterol

MUC1

Glycolipid

XDH/XO

BTN

PASIII

PP3

Raft: sphingolipid and
cholestrol-rich domain

ADPH

Sphingolipid

CD36

■ **FIGURE 4.1** Diagram of the distribution of the phospholipids and proteins that constitute the MFGM. *Elias, Laubscher and Jiménez-Flores (2013).* [A color version of this figure is available online at www.booksite.elsevier.com/9780124046108].

includes cooling, drying, separation, agitation, heating, and homogenization (Michalski and Januel, 2006; Dewettinck *et al.*, 2008; Brisson *et al.*, 2010; Elias-Argote and Jiménez-Flores, 2010; Gallier *et al.*, 2010b). With respect to physiological factors, Lopez *et al.* (2008) observed large differences in the fatty acid (FA) composition of the phospholipids in the milk from cows fed a regular diet and a diet rich in polyunsaturated FA. The latter resulted in a significant decrease in saturated FA content in milk, which enhances milk's nutrition quality (Jensen, 2002; Lopez *et al.*, 2008).

In an attempt to summarize the different views of the arrangement of the lipids and proteins in the MFGM, we analyzed the properties of MFGM using Langmuir trough isotherms, laser confocal microscopy, Raman spectroscopy, and atomic force microscopy, and we have developed a diagram of the structure of the MFGM presented in Figure 4.1. Our own work, along with that of others, helped in the graphic design (Brisson *et al.*, 2010; Gallier *et al.*, 2010a, c, 2011).

Lipids in the MFGM

Polar lipids, such as glycerophospholipids and sphingolipids, constitute less than 1% of the milk total lipids; nevertheless, they function as intracellular signaling molecules, provide a framework structure, and have nutritional and functional properties such as emulsification (Jiménez-Flores and Brisson, 2008; Lopez *et al.*, 2008). Indeed, research has also found a relationship between polar lipid consumption and enhanced health (Spitsberg, 2005). The phospholipids on the MFGM are distributed heterogeneously and present in different percentages (Dewettinck *et al.*, 2008; Jiménez-Flores and Brisson, 2008; Lopez *et al.*, 2008). These phospholipids contain high levels of long-chain FAs such as palmitic, stearic, and tricosanoic acids (Singh, 2006; Sánchez-Juanes *et al.*, 2009). No difference has been observed in the FA composition of MFGM phospholipids regardless of season (Fauquant *et al.*, 2005). Other lipids, such as triglycerides (TG), diglycerides, monoglycerides, sterols, and sterol esters, are mainly present in the milk fat globule core as seen in Table 4.1. Among these, triglycerides represent about 60% of the neutral lipids in the milk fat globules and cholesterol accounts for 90% of the total sterols in milk fat globules (Mather, 2000; Keenan and Mather, 2006; Rombaut *et al.*, 2006a; Dewettinck *et al.*, 2008).

Phospholipids and cholesterol are very interesting components of biological membranes, and they function as excellent emulsifiers, and accordingly their properties have been extensively studied (Krog, 1991; Xiong and Kinsella, 1991; McClements *et al.*, 1993; Goff, 1997; Singh and Dalgleish, 1998; Einhorn-Stoll *et al.*, 2002; Waninge *et al.*, 2005). In particular, sphingomyelin in milk has been studied as a bioactive lipid, and only animal sources of fat contain this particular class of polar lipid (Prostenik and Gospocic, 1966; Morrison, 1969; Schmelz *et al.*, 1996; Vesper *et al.*, 1999; Nyberg *et al.*, 2000; Motouri *et al.*, 2003; Waninge *et al.*, 2003b; Noh and Koo, 2004; Waninge *et al.*, 2005; Rombaut *et al.*, 2007). Sphingolipids constitute up to one-third of the MFGM polar lipid fraction (Dewettinck *et al.*, 2008; Jiménez-Flores and Brisson, 2008). They are characterized by a sphingoid base, which is a long-chain aliphatic amine (12−22 carbon atoms) (Rombaut *et al.*, 2007). Attachment of an FA via a hydroxyl group forms a ceramide that is the basic unit for the formation of sphingo-phospholipids, such as sphingomyelin. The latter has a high degree of saturation that facilitates

Table 4.1 Composition of Milk Lipids and their Distribution among Globule, MFGM and Serum

Lipid Class	Content in Total fat (g/kg)	Fraction in %		
		Globule Core	MFGM	Skim Phase
Neutral glycerides				
Triacylglycerol	958–983	100		
Diacylglycerol	2.8–22.5	≈90	≈10	?
Mono-acylglycerol	0.3–3.8	Traces	Traces	Traces
Free fatty acids	1.0–4.4	60	≈10	30
Phospholipids[a]	2.0–11.1	–	65	35
Small globule (3 μm)[b]			28.4[c]	
Large globule (6 μm)			36.9[c]	
Cerebrosides	1.0	–	70	30
Gangliosides	0.1	–	≈70	≈30
Stetols		80	10	10
Cholesterol	3.0–4.6			
Cholesteryl ester	≤0.2			
Careotenoids + Vit. A	0.02	≈95	≈5	Traces

[a]*PL in the MFGM pellet.*
[b]*Small globule tend to contain more sphingomyelin.*
[c]*PL% in total lipids of the MFGM from microfiltrated globules.*

complex formation with cholesterol, known as a lipid raft. These are rigid domains implicated in cellular processes, like signal transduction, endocytosis, and cholesterol trafficking (Dewettinck *et al.*, 2008; Lopez *et al.*, 2008, 2010). Sphingomyelin is primarily located in the outer membrane of the MFGM, and is present in higher percentages in milk derived from cows fed with unsaturated FAs due to the presence of smaller globules and consequently more surface area (Lopez *et al.*, 2008). An analysis of sphingolipids in the most commonly consumed dairy products found similar sphingolipid concentrations in non-fat dry milk and fermented products, indicating that starter cultures do not contribute to sphingomyelin in dairy products (Jiménez-Flores and Higuera-Ciapara, 2009); however, we can manipulate the composition of this essential polar lipid by changing the diet of cows (Lopez *et al.*, 2008). Furthermore, sphingomyelin has been shown to reduce aberrant crypt foci and the appearance of colon adenocarcinoma, to modulate immune responses, among other functions (Dewettinck *et al.*, 2008; Jiménez-Flores and Higuera-Ciapara, 2009). Sphingomyelin metabolites also provide health benefits (Schmelz *et al.*, 1996; Thompson and Singh, 2006) as seen in Table 4.2. Long-term feeding of sphingolipids

Table 4.2 Nutritional Aspects of Phospholipids Found in the MFGM

Component	PL %[a]	Nutritional Aspects
Sphingolipids and metabolites	18.0−34.1[b]	Reduction of the number of aberrant crypt foci and adenocarcinomas. Shift in tumor type (malignant → benign). Anticholesterolemic. Protection of the liver from fat- and cholesterol-induced steatosis. Suppression of gastrointestinal pathogens. Neonatal gut maturation. Myelination of the developing central nervous system. Endogenous modulators of vascular function. Associated with age-related diseases and the development of Alzheimer's
Sphingosine1-phosphate		Mitogenic
Phosphatidylcholine (PC)	19.2−37.3	Supports liver recovery from toxic chemical attack or viral damage. Protects the human GI mucosa against toxic attack. Reduction of necrotizing enterocolitis. Alleviates orotic acid-induced fatty liver
Lysophosphatidylcholine (lysoPC)	2	Bacteriostatic and bactericidal capacity. Strong gastroprotective role in the duodenal mucosa
Phosphatidylethanolamine (PE)	19.8−42.0	Maintains hemostasis
Phosphatidylinositol (PI)	5−11%	Substrate in cell signaling. Promotes plasma cholesterol transport and metabolism
Phosphatidylserine (PS)	1.9−10.5	Restores normal memory on a variety of tasks. Positive effects on Alzheimer's patients. Improves exercise capacity of exercising humans

[a]Relative phospholipid content (g per 100 g of polar lipid including sphingomyelin).
[b]Sphingomyelin content.
Adapted from Rombaut et al. (2006a, b); Jiménez-Flores and Higuera-Ciapara (2009); Wat et al. (2009).

(1%) to laboratory rats significantly decreases total blood cholesterol levels and elevates HDL, suggesting that sphingolipids represent a "functional" constituent of food.

Proteins in the MFGM

MFGM proteins constitute 1−2% of the total bovine milk proteins, and depending on the milk source and how it is processed, 25−70% of the MFGM may be polypeptides, ranging in molecular weight (MW) from 15,000 to 240,000 Da (Ye *et al.*, 2002, 2004; Dewettinck *et al.*, 2008; Jiménez-Flores and Brisson, 2008; Jiménez-Flores and Higuera-Ciapara, 2009). Indeed, Murgiano *et al.* (2009) detected differences between Chianina and Holstein cattle in the amount of proteins associated with mammary gland development, lipid droplets formation, and host defense mechanisms. Mather (2000) presents a detailed description and suggested nomenclature of the known and major MFGM proteins.

It is very important that some consideration has been given to MFGM protein composition from different sources. Affolter and his group have applied proteomic techniques to accurately quantify seven MFGM proteins from buttermilk or whey protein concentrate (Affolter *et al.*, 2010). They used sophisticated laboratory analytical techniques, based in isotope dilution mass spectrometry, that describe in detail the quantitative differences.

From our own research we have learned that in addition to the large number and diverse functions of the major MFGM proteins, they appear to have complex interactions with each other, other proteins, lipids, and other molecules such as hydrophobic and hydrophilic interactions, and the formation of disulfide bonds (Kim and Jiménez-Flores, 1995; Morin *et al.*, 2006, 2007a, 2008). It has recently been shown that XDH/XO and BTN form a high molecular weight aggregate through the formation of intermolecular disulfide bonds after a heat treatment at 60°C for 10 minutes (Ye *et al.*, 2002), while previous studies have shown that with higher temperature heat treatments MFGM proteins form similar high molecular weight complexes with milk proteins. Kim and Jiménez-Flores (1995) reported that with heat treatment of 87°C, β-lactoglobulin (BLG) and other milk serum proteins interact, forming high molecular weight complexes with the MFGM proteins PAS6/7. The mechanism of formation was undefined, but did not appear to be solely due to the formation of disulfide bonds (Kim and Jiménez-Flores, 1995).

MFGM and Health and Well-Being

Progress in the knowledge of the composition and role of milk fat globule membrane components has led to the realization that some possess biological properties beyond their nutritional significance, summarized in Table 4.3. Spitsberg (2005), in his review, refers to the MFGM as a nutraceutical with nutritional and pharmaceutical potential. In fact, the MFGM has antimicrobial proteins, illness suppressors, and micronutrient-binding proteins that bind compounds, such as iron, zinc, copper, folate, and vitamin B1, and other components, with anticarcinogenic properties (Snow *et al.*, 2010, 2011). Lonnerdal *et al.* (2006) complemented infant food with bovine MFGM and micronutrients to evaluate its effect on children's growth, nutrition, and morbidity. In a double-blind study with 6- to 11-month-old infants ($n = 550$), fortification of food was beneficial in the copper and vitamin B12 status among the subjects, which enhance growth and normal function of the brain and nervous system, respectively (Lonnerdal *et al.*, 2006). Wat *et al.* (2009) supplemented bovine PL to mice fed a high-fat diet resulting in a reduction in liver weight, total liver lipid, and serum lipid levels, which might be of therapeutic benefit in humans with non-alcoholic fatty liver disease, especially obese and diabetic

Table 4.3 Health Benefits Reported for Proteins and other Components of the MFGM

Proteins	MW (kg/mol)	Reported Health Benefits/Functions
Mucin 1 (MUC1)[a]	160	Antiviral action/anti-rotavirus, especially in neonates
Mucin 15 (MUC15 or PAS III)[a]	94—100	Antiviral action
Butyrophilin (BTN)[a]	66	Suppression of multiple sclerosis
Xanthine oxidase (XO)	150—155	Bactericidal agent
Cluster of differentiation (CD36 or PAS IV)[a]	78	Glycoproteins that act as receptors due to high sugar content
Fatty acid binding protein (FABP)	15	Cell growth inhibitor Anticancer factor (FABP as selenium carrier)
BRCA1 and BRCA2	210	Inhibition of breast cancer
Lactadherin (PAS 6/7)[a]	43—59	Role of epithelialization, cell polarization, cell movement and rearrangement, protection against viral infection in the gut
Adipophilin (ADPH)	52	Milk synthesis
Other components (glycosylated proteins)		
β-Glucoronidase inhibitor		Inhibition of colon cancer
Helicobacter pylori inhibitor		Prevention of gastric diseases
Cholesterolemia-lowering factor		Anticholesterol activity
Phosphoproteins		Source of organic phosphorus and calcium phosphate

[a]glycoproteins
Adapted from Dewettinck et al. (2008); Michalski and Januel (2006); Spitsberg (2005); Jiménez-Flores and Brisson (2008).

patients. Choline has recently been identified as an essential component of the human diet. Nutrition studies have revealed that choline-containing phospholipids play a central role in the process of signal transduction within the body and that alterations in this process may lead to abnormalities such as cancer and Alzheimer's disease (Zeisel et al., 1986; Fang and Dalgleish, 1995; Nyberg et al., 1998; Parodi, 2003, 2005; Miyaji et al., 2005). Based on this, we know firsthand that MFGM is now regarded as a valuable source of these essential components in our diet.

We now have some indication of the beneficial properties of MFGM as part of a dairy ingredient. Ward and coworkers have recently shown that an ingredient rich in bovine MFGM components from whey cream prevented intestinal cancer in rats (Snow et al., 2010). Using a similar component in the diet of rats, the same group demonstrated that an MFGM-rich ingredient inhibited gastrointestinal leakiness in mice treated with lipopolysaccharide

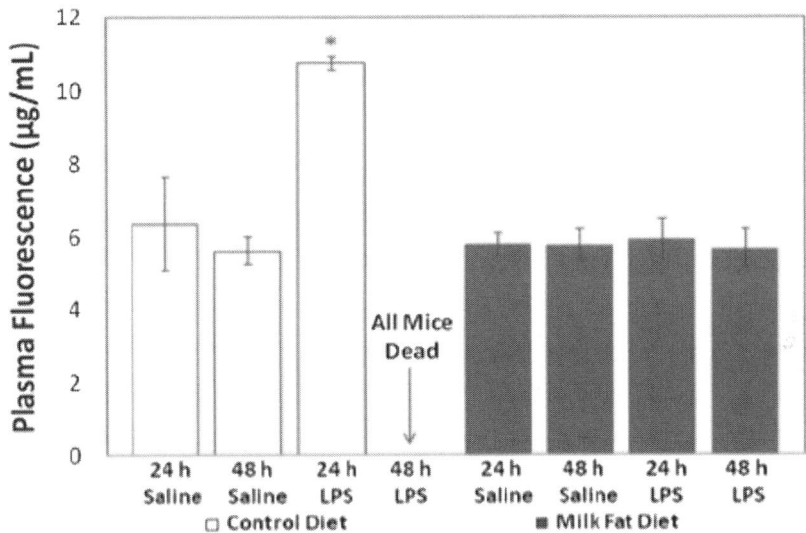

PROTECTION OF GUT BARRIER BY MILK FRACTION

■ **FIGURE 4.2** Reproduction of the Figure 3 in Snow et al. (2012) (our research group produced the MFGM-rich ingredient) demonstrating the protective effect of MFGM on the intestine of rats towards a challenge with lipopolysaccharide (LPS). *From Snow* et al. *(2011).*

(Snow *et al.*, 2011). Figure 4.2 presents the result of treating rats with MFGM and challenging them with inflammatory lipopolysaccharide. All the rats treated with MFGM survived, while all those in the control group died.

Physiological Function

The overall physiological function of the MFGM, as we understand it today, is to allow for secretion of the milk fat into the milk. The nature of lipid droplet synthesis and secretion follows a biochemical pathway that has been described recently by Bionaz and Loor (2006). In this work, they describe the action of 45 genes associated with lipid synthesis, and have postulated the interaction of these genes in the achievement of milk synthesis and secretion.

Proteomic analysis for the MFGM has been studied by several groups (Cavaletto *et al.*, 2002, 2004; Yamada *et al.*, 2002; Fortunato *et al.*, 2003; O'Donnell *et al.*, 2004; Hamdan and Righetti, 2005; Manso *et al.*, 2005; Wang *et al.*, 2006; Westermeier, 2006; Beddek *et al.*, 2008; Reinhardt and Lippolis, 2008; Wilson *et al.*, 2008; Gagnaire *et al.*, 2009; Murgiano *et al.*, 2009; Affolter *et al.*, 2010), and there are important and significant similarities among studies and across the studied species. The most abundant proteins in the MFGM, butyrophylin and xanthine oxido/reductase, indicate the importance of these proteins in the secretory process, and very likely in the mechanism of attachment of the MFGM to the surface of the fat globule. Other similarities are found on the high proportion of the proteins in the MFGM that form part of lipid metabolism, lipid transport and trafficking, immunity and defense, and membrane structure.

In summary, the genomic analysis of the origin of the MFGM indicates a highly conserved series of proteins that are essential for the synthesis and secretion of the fat into milk, and the group of proteins that communicate chemically important growth and immunological signals to the neonate, are the main functional proteins of the MFGM. This is organized and structured in a phospholipid matrix that allows their biological function in the secretory mammary cell, and one can only hypothesize at this point that it also has the elements for transmitting the nutrients and biologically active components and presenting them to the neonate in a functional manner.

Technological Relevance

The challenge issued to the food technologists of the future is a more efficient nutrition delivery system, which incorporates elements of function and efficiency of the natural food. Milk has an estimated 120 million year evolution history, and human technology has much to learn from biological systems. Our need for better foods is in correlation to the information we obtain to develop technologies. An example of this approach is what is known as biomimicry. This approach has been successful in other areas, such as materials design and efficient energy use. Our view is that food scientists will need information as to how Nature intends to deliver nutrients, so we can mimic and develop better foods.

Therefore, the information that we can generate on an efficient system such as milk, which is at the center of the survival strategy of mammals on Earth, is an attractive approach. To achieve part of this goal, elucidation of the structure and function of the MFGM is required.

Current View of the MFGM Structure

Enzymatic hydrolysis has been used to unravel the distribution of phospholipids and proteins within the three MFGM layers. Deeth (1997) showed that phosphatidylcholine and sphingomyelin were predominantly found on the external leaflet of the bilayer, and phosphatidylinositol, phosphatidylserine, and phosphatidylethanolamine were mainly located in the internal leaflet. Most carbohydrate moieties on glycoproteins and glycolipids are located on the outer MFGM surface forming the glycocalyx as revealed using lectins (Vanderghem *et al.*, 2011).

The location of proteins within the MFGM trilayer is still the subject of studies. Xanthine dehydrogenase/oxidase (XDH/XO) and lactadherin are recovered after washing cream with $MgCl_2$, suggesting that these proteins may be peripheral (Mather and Keenan, 1975). Using several washing mediums (Triton X-100, Cutscum, bile salts, and guanidine-HCl), Mather *et al.* (Mather *et al.*, 1977) showed that butyrophilin seemed anchored tightly in the MFGM. Also using guanidine-HCl, Shimizu *et al.* (Shimizu *et al.*, 1978) reported that some of XDH/XO and lactadherin were released from the globule surface indicating that they may be weakly bound to the MFGM. On the other hand, part of XDH/XO, MUC 1, PAS III, CD36 were more resistant to solubilization, and BTN and adipophilin (APDH) were fully resistant (Shimizu *et al.*, 1978). This indicates that the distribution of MFGM proteins is asymmetric, similar to the asymmetric organization of the MFGM trilayer of phospholipids. Using enzymatic proteolysis by pronase E and trypsin, Vanderghem *et al.* (Vanderghem *et al.*, 2011) attempted to locate MFGM proteins on the trilayer of phospholipids. BTN and APDH were easily digested unlike XDH/XO and lactadherin, which were partially resistant to proteolysis. Freeze-fracture immunocytochemistry analysis of MFGM showed that ADPH, XDH/XO, TIP47 are located in the monolayer surrounding the milk fat globules (Robenek *et al.*, 2006a). In addition, ADPH is also present in the inner leaflet of the bilayer. BTN is found in the monolayer and external leaflet of the bilayer, with more abundance in the former (Robenek *et al.*, 2006a). Only BTN presented a symmetrical distribution between layers. BTN and XDH/XO form a complex that binds to ADPH.

Methodologies for Characterization
Microscopy

Early views of the MFGM structure differ from the current view. It was proposed to be composed of a phospholipid monolayer including cholesterol and vitamin A and in contact with an inner layer of high-melting point triglycerides; the head groups of the phospholipids were thought to interact with a bilayer of proteins (King, 1955). Hayashi and Smith (1965) described the MFGM as a layer of water-insoluble lipids and proteins at the surface of the triglyceride core and onto which water-soluble proteins are adsorbed and can be released by sodium deoxycholate. Bargmann *et al.* (1961) studied secreted milk fat globules and showed the trilayer structure of the MFGM, made of an adsorbed layer of cytoplasmic molecules surrounded by a true cytoplasmic bilayer. The debate on the biological origin of the MFGM or its formation by adsorption of molecules post-secretion continued for a few years.

Confocal Microscopy

Confocal laser scanning microscopy (CLSM) is now widely available and used in food science. It allows the structural characterization of samples *in situ*. Confocal microscopy allows high spatial resolution imaging and blocks out-of-focus fluorescence to allow imaging of thin sections (Garcia-Saez and Schwille, 2010b). A large variety of fluorescent probes is available and fluorescent phospholipid analogues are commonly used to study biological membranes (Gallier *et al.*, 2010a). By scanning the sample in the *z*-axis, it is possible to obtain a three-dimensional view of the sample (Gallier *et al.*, 2010a). Dynamical features of membranes can also be followed using CLSM (Lopez *et al.*, 2010). CLSM also allows the simultaneous staining (and thus the location) of multiple compounds by wisely choosing the fluorescent dyes.

CLSM has been used to investigate the lateral distribution of MFGM compounds at the surface of milk fat globules (Evers *et al.*, 2008; Gallier *et al.*, 2010a; Lopez *et al.*, 2010; Ong *et al.*, 2010). Several fluorescent phospholipid analogues have been used to characterize the phospholipid organization at the surface of the MFGM. The bulky fluorescent dye molecule favors the distribution of the fluorescent phospholipid analogue in the liquid-disordered phase (Gallier *et al.*, 2010a; Lopez *et al.*, 2010). Indeed, liquid-ordered domains, whose tightly packed organization prevents access of the bulky fluorescent analogue, were observed at the surface of milk fat globules and are thought to be rich in sphingomyelin and cholesterol (Figure 4.3). Gallier *et al.* (2010a) used phospholipid analogues with the fluorescent dye molecule attached either to the polar headgroup or an FA tail of the phospholipid and did not observe any difference in the distribution of the MFGM phospholipids. Temperature affected the size, number and shape of the liquid-ordered domains (Gallier, 2010). Milk fat globule-associated proteins can be detected using Fast Green FCF which binds to charged groups of proteins. Using Fast Green FCF, proteins associated with the MFGM after pasteurization and homogenization were observed at the surface of the globules (Gallier *et al.*, 2010a). Lectins, mainly concanavalin A (con A) and wheat germ agglutinin (WGA), have also been used to detect the presence of glycoproteins and glycolipids at the surface of the MFGM (Lopez *et al.*, 2010; Gallier *et al.*, 2012b). Con A binds to α-mannopyranosyl and α-glycopyranosyl residues, whereas WGA binds to N-acetylglucosamine and N-acetylneuraminic acid (sialic acid) residues (Gallier *et al.*, 2012b). These glycoproteins were found only in the liquid-disordered phase and distributed as patches or a network.

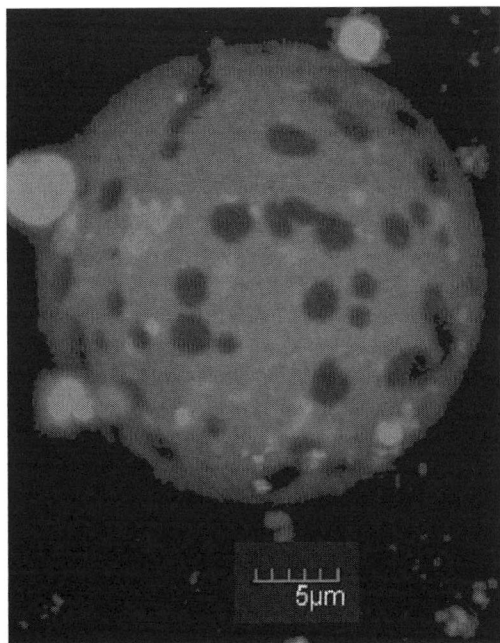

■ **FIGURE 4.3** CLSM three-dimensional image of a large milk fat globule from raw cream stained with the fluorescent head group-labeled phospholipid analogue 1,2-dioleoyl-sn-glycero-3-phosphoethanolamine-N-lissamine rhodamine B sulfonyl. *Reprinted with permission from Gallier et al. (2010a). Copyright 2010 American Chemical Society.* [A color version of this figure is available online at www.booksite.elsevier.com/9780124046108].

In human milk, cytoplasmic crescents from lactating cells have been observed between the triglyceride core of the globules and the MFGM using the fluorescent dye acridine orange and fluorescence microscopy (Huston and Patton, 1990). Only less than 1% of bovine milk fat globules contained crescents, unlike the milk fat globules of human, goat, guinea pig, rat, pig, sheep, and rabbit, with those of rabbit having the most crescents (Janssen and Walstra, 1982). Huston and Patton (1990) suggested that the amount, location, and acylation of two MFGM proteins, butyrophilin and xanthine oxidase, were the main factors regulating the formation of crescents. Crescents on the surface of human milk fat globules were also observed using fluorescence microscopy and the fluorescent dye $DilC_{18}(3)$-DS by Evers et al. (2008). Their role is currently unknown.

Spectroscopy Techniques

Phospholipids are commonly determined by chromatography and mass spectrometry techniques (Gallier *et al.*, 2010c). [31]Phosphorus nuclear magnetic resonance spectroscopy (^{31}P-NMR) has been used to quantify milk phospholipids (MacKenzie *et al.*, 2009). ^{31}P-NMR is a convenient method for routine analysis and requires less time and fewer chemicals than chromatography and mass spectrometry techniques. Similar to these techniques, ^{31}P-NMR analysis of phospholipids requires the samples to be highly rich in phospholipids. For phospholipids present in less than 1%, additional phospholipid extraction and a higher-field NMR instrument are required for their detection. The pH and temperature of the solution are also crucial as both influence the chemical shift of phospholipids, which can limit their detection (MacKenzie *et al.*, 2009). Most milk phospholipids are well resolved at pH 7–7.3, besides lactosylated phosphatidylethanolamine, which is resolved at pH 8.5.

It is possible to obtain the chemical fingerprint of phospholipids from their bending and stretching vibrational transitions using infrared (IR) spectroscopy or Raman spectroscopy (Li-Chan, 1996; Krafft *et al.*, 2005; Gallier *et al.*, 2011). IR and near-IR spectroscopy are commonly used vibrational spectroscopy techniques in food and dairy industries (Dufour, 2006; Li-Chan, 2007). Because of the weak Raman scattering of water, Raman spectroscopy is a better choice than IR spectroscopy for the analysis of aqueous and *in situ* samples (Li-Chan, 1996). Raman spectroscopy is based on the inelastic scattering of light whereby the wavelength of the light source (a laser) is altered through interaction with the irradiated molecules (Nafie, 2001). Raman spectroscopy has been used to study fat globules from bovine milk (Forrest, 1978; Gallier *et al.*, 2011) and human milk (Argov *et al.*, 2008). Small globules of less than 1 μm lacked a detectable triglyceride peak and could be structurally similar to phospholipid liposomes (Argov *et al.*, 2008; Gallier *et al.*, 2011). The Raman spectra of these very small globules contained the chemical fingerprints of MFGM lipids. These small globules could be the isolated MFGM vesicles reported by Waninge *et al.* (2003a) or the lipoprotein particles or "microsomes" studied by Mulder and Walstra (1974) or alternatively the small fat globules present at early stages of processing (pumping and agitation) detected by Michalski *et al.* (2002).

Differential scanning calorimetry (DSC) was used to study the structure of the MFGM (Appell *et al.*, 1982). DSC measurements involve placing of the sample in a cell and heating of the sample at a slow heating rate to obtain the endothermic transition temperatures of the sample. For MFGM, in phosphate-buffered saline solution at pH 7.4, six transition temperatures, 16°C, 28°C, 43°C, 58°C, 68°C, and 75°C, were recorded (Appell *et al.*, 1982). The first three transitions are reversible denoting the involvement of MFGM lipids melting at these temperatures. MFGM apolar lipids melted at 43°C. Butyrophilin unfolded and was denatured at 58°C. Trypsin-resistant proteins, likely MFGM glycoproteins, were responsible for the transition at 68°C and 75°C and, thus, likely denatured at these temperatures.

Scanning and Transmission Electron Microscopy

Transmission electron microscopy (TEM) and scanning electron microscopy (SEM) have the advantage of high-resolution imaging; however, they require harsh chemical treatment of the samples. TEM has been used for over 50 years to study the surface of the milk fat globules. Structural changes at the surface of the globules after thermal treatment have been observed using TEM after fixation with glutaraldehyde and osmium tetraoxide, embedding in epoxy resin and staining with lead citrate (Zamora *et al.*, 2012). The trilayer structure of the MFGM and the surrounding glycocalyx were observed at the surface of the raw milk fat globules. TEM has also been used to characterize the structure of the MFGM in buttermilk (Morin *et al.*, 2007b). The MFGM in buttermilk appeared as amorphous linear fragments. Lee and Morr (Lee and Morr, 1992) characterized dairy cream fat globules using three different fixation and staining techniques (imidazole osmium, ferricyanide osmium, or tricomplex osmium) and TEM and SEM. Osmium tetraoxide reacts with unsaturated FAs, making them visible by electron microscopy. With fixation and staining by imidazole-buffered osmium and tricomplex osmium, poorly-stained and irregular areas of the MFGM were associated with the presence of saturated FAs; these areas could be liquid-ordered domains. SEM allows a three-dimensional observation of the milk fat globules but gives little information on the MFGM structure. Lectin-labeled gold granules in combination with TEM and SEM allowed the location of human and bovine MFGM glycoproteins (Horisberger *et al.*, 1977). Clusters of wheat germ agglutinin-labeled gold granules were located all over the surface of milk fat globules using SEM, suggesting the presence of N-acetyl-D-glucosamine and sialic acid residues. Staining with concanavalin A was weak, related to a low amount of α-D-mannose and α-D-glucose.

Electron Microscopy

Electron microscopy with thin sectioning and freeze-etching provided images of cream and MFGM fragments (Henson, 1971). MFGM fragments formed vesicles or amorphous aggregates. The surface of the unwashed cream globules rarely showed a trilaminar structure using freeze-etching but a layer of crystallized triglycerides coated the triglyceride core. Thin sections of washed cream observed by electron microscopy showed an accumulation of electron-dense material included within the MFGM trilayer. The inclusions could be related to the cytoplasmic crescents observed by

Huston and Patton (1990). Electron microscopy, in combination with antibodies and colloidal gold labeling, was used to locate MUC1 at the surface of guinea-pig milk fat globules (Mather *et al.*, 2001). The presence of MUC1 in apical membranes and MFGM emphasizes the origin of the MFGM from the apical surface of the mammary secretory epithelial cells (Mather *et al.*, 2001). Electron microscopy after free-fracture or freeze-etching also allowed the observation of the membrane of milk and cream fat globules and MFGM fragments (Da Silva *et al.*, 1980). Using this technique, crescents were observed covered by the MFGM. MFGM fragments formed vesicles and disk-like aggregates.

Physical Analysis

The measurement of the ζ-potential was used as a mean to probe mechanical changes to the MFGM (Michalski *et al.*, 2002). Processing dramatically affects the MFGM and the ζ-potential measurement can determine changes in the surface charge of the globule surface induced by milk processing. The accumulation of casein micelles and whey proteins at the surface of the globules during pasteurization and homogenization increases the ζ-potential of the globules (Michalski *et al.*, 2002). The ζ-potential of a particle determines its stability and predisposition to flocculation or coalescence. For example, a low ζ-potential will indicate a tendency to aggregate because of low repulsion forces between particles. A high ζ-potential indicates high stability of the particles (Gallier *et al.*, 2012c). During gastric digestion, the presence of the glycocalyx surrounding the milk fat globules provided a surface charge high enough for the globules to remain stable (Gallier *et al.*, 2012c).

Langmuir Trough and AFM

Langmuir monolayers are monomolecular films spread at the air–water interface on a Langmuir trough. Amphiphilic molecules, such as FAs and phospholipids, are easily studied with a Langmuir trough with the hydrophilic part, i.e., the head, having an affinity for the water phase and the hydrophobic part, i.e., the tail, oriented towards the air and making the molecules insoluble (Kaganer *et al.*, 1999). Two-dimensional ordering of Langmuir monolayers can be followed as a function of temperature by changing the temperature of the aqueous subphase and pressure by moving the barriers along the aqueous surface. The pH and ionic strength of the subphase can also be varied and will affect the ionization of the polar head of the amphiphilic molecules.

Biological membranes are highly dynamic and organized two-dimensional complex structures and can be either composed of mono-, bi-, or trilayers of phospholipids with associated or intrinsic membrane proteins. There is an increased recognition that the self-assembly of phospholipids into membrane structures plays a role in cell functioning (Bagatolli *et al.*, 2010). Thus, Langmuir monolayer model systems are appropriate for understanding the physicochemical properties of biological membranes. Surface pressure–area compression and expansion isotherms provide information on structural phase transitions (from gas to liquid and to solid/crystalline). A dilute system, i.e., very few molecules in a large area, represents a two-dimensional gas phase. Upon compression of the monolayer, the molecules transition into a liquid-expanded phase where the conformation of the molecules is disordered. Upon further compression (i.e., increase of the surface pressure), liquid-expanded/liquid-condensed phase coexistence takes place. The molecules in the liquid-condensed phase are described as ordered and tightly packed. With further increase in surface pressure, the monolayer will transition into a solid-like phase until collapse. The isothermal compressibility of a monolayer gives information on how easily a film can be compressed, which is related to the fluidity (or rigidity) of a film. For example, the liquid-expanded phase is more compressible and more fluid than the rigid and less permeable liquid-condensed phase. Phase transitions occur at higher temperature for molecules with longer chain length (Kaganer et al., 1999). In systems with complex lipid composition, the liquid-expanded phase is defined as a liquid-disordered phase with similar high lateral diffusion coefficient in the plane of the monolayer and conformational freedom. The liquid-condensed phase is defined as a liquid-ordered phase with a low diffusion coefficient and parallel arrangement of the acyl lipid chains. The coexistence of liquid-ordered and liquid-disordered phases results from liquid–liquid immiscibility. The phase coexistence takes place under specific conditions of lipid composition, temperature, and surface pressure (Gallier et al., 2010b).

Each phase transition induces a kink in the compression isotherm. However, in some systems the detection of phase coexistence is not always obvious due to the absence of a kink in the compression isotherm. The combination of a Langmuir trough with fluorescence microscopy, Brewster-angle microscopy, or polarized fluorescence microscopy has unraveled more complex structural phase transitions than previously thought due to their higher sensitivity to structural changes within monolayers than isotherm measurements (Kaganer *et al.*, 1999). Langmuir–Blodgett films are obtained by the layer-by-layer transfer of Langmuir monolayers onto a solid support (e.g., glass or mica slides) (Kaganer *et al.*, 1999) and are easily studied with atomic force microscopy (AFM), a label-free technique to study surfaces in air or liquid. In AFM, a cantilever scans the surface and the interaction between the cantilever and the surface gives topographical nanoscale images of the interfacial structure of the scanned surface down to the molecular level

(Morris *et al.*, 2011). For example, AFM was used to show the orogenic displacement of proteins by surfactants at the oil–water and air–water interfaces and follow the digestion of interfacial proteins under simulated gastric conditions. In lipid monolayers in the phase coexistence, it is possible to detect liquid-ordered domains with their greater thickness caused by the extended conformation of the saturated acyl chains than the liquid-disordered phase (Gallier et al., 2010b). To minimize the line tension at the liquid-ordered–liquid-disordered phase boundaries, the domains will adopt circular shapes. In biological membranes, specific membrane proteins and lipids of intermediate acyl chain length accumulate at the phase boundaries to reduce the line tension (Garcia-Saez and Schwille, 2010a). Phase images of monolayers can be acquired when using the AFM in tapping mode and reveal the viscoelastic properties of the monolayers (Gallier et al., 2012a).

Biological membranes present lipid domains rich in cholesterol and sphingomyelin, called lipid rafts. The domains are in a liquid-ordered state and surrounded by the fluid liquid-disordered phase rich in unsaturated phospholipids. Lipid rafts are involved in various cellular processes such as signaling and membrane trafficking (Garcia-Saez and Schwille, 2010a). However, the formation and behavior of lipid rafts are still poorly understood. Lipid rafts were originally defined as detergent-resistant membranes, rich in cholesterol, sphingomyelin, phosphatidylcholine, and raft proteins, and were recovered after treatment of plasma membranes at 4°C with Triton X-100 (Bagatolli *et al.*, 2010). Patton *et al.* (1986) used a similar technique to recover MFGM fragments after treatment with Triton X-100 or conjugated bile salts. Model systems have provided most of the knowledge known so far on lipid rafts. Using binary and ternary systems of cholesterol and saturated and unsaturated phospholipids, lipid–lipid interactions were shown to play a crucial role in the formation of liquid-ordered domains; interactions between cholesterol and unsaturated phospholipids are repulsive unlike interactions between cholesterol and saturated phospholipids in the liquid-ordered state which are more favorable (Garcia-Saez and Schwille, 2010a). Cholesterol is a rigid, fused, four-ring molecule and possesses a polar head group too small to shield its hydrophobic tail oriented towards the hydrophobic side of the membrane (Ikonen, 2008). Thus, two models, the umbrella model and the condensed complex, have proposed that other lipids with larger head groups provide shielding. The former model suggests that large-head group phospholipids can shield one or two molecules, whereas the later model suggests the formation of hydrogen bonding between cholesterol and saturated acyl chains of phospholipids leads to condensed complexes. Thus, highly saturated sphingomyelin with large choline head group represents a more favorable candidate for association with cholesterol in biomembranes.

Phospholipids are the building blocks of biomembranes. The MFGM is a trilayer of phospholipids with proteins, cholesterol, and other compounds. Thus, phospholipid Langmuir monolayers provide a basic understanding of the biophysics of the MFGM (Gallier *et al.*, 2010b, 2012a). Most studies are limited to single phospholipid monolayers or binary or ternary systems due to the complexity of monolayer model systems (Kaganer *et al.*, 1999). However, the MFGM, like other biological membranes, is composed of various phospholipids and proteins. In an attempt to understand the lateral organization of the MFGM, Gallier *et al.* (Gallier *et al.*, 2010b, 2012a) extracted phospholipids from dairy products (raw milk, raw cream, pasteurized and homogenized milk, and buttermilk powder) and characterized the physical properties of these phospholipid monolayers with and without milk proteins at the air—water interface. A fluorescent phospholipid analogue was added to the phospholipid extracts at a concentration low enough to not affect the properties of the system and the phospholipid monolayers were studied on a Langmuir trough mounted on top of an epifluorescence microscope. Epifluorescence microscopy presents the advantage of following phase separation as a function of temperature and surface pressure. Processing of milk changed the phospholipid composition and the FA profile (Gallier *et al.*, 2010c). This led to phospholipid monolayers with different physical properties (Gallier *et al.*, 2010b). Raw milk, raw cream, and pasteurized and homogenized milk phospholipid monolayers presented a gas phase at surface pressure of less than 1 mN/m, a two-dimensional foam upon compression of the monolayers, followed by liquid-ordered–liquid-disordered phase coexistence. The area fraction of liquid-ordered domains grew with increased surface pressure. Buttermilk powder phospholipid monolayers did not present a gas phase or two-dimensional foam, possibly due to a lower concentration in phosphatidylethanolamine (Gallier et al., 2010b). All monolayers collapsed at about 50 mN/m. The phase separation was observed in the absence of cholesterol, denoting the importance of sphingomyelin in liquid-ordered domain formation (Gallier et al., 2010b). A three-phase coexistence was observed in monolayers at 15 mN/m after addition of cholesterol. Temperature did not affect the compression isotherms but had an effect on the size and shape of the liquid-ordered domains, with more stretched "floret" domains (Gallier et al., 2010b). As MFGM contains a large amount of proteins, Gallier et al. (2012a) studied MFGM phospholipid—milk protein monolayers to understand protein—lipid interactions at the surface of the globules. Phospholipids were spread at the air—water interface and β-casein was added to the subphase. Gallier et al. (2012a) showed that, in addition to lipid—lipid interactions, protein—protein and protein—lipid interactions played a crucial role in the lateral distribution of phospholipids in the MFGM. Indeed, as seen in Langmuir—Blodgett films using AFM, β-casein molecules formed a linear network, and interacted with and clustered small liquid-ordered domains (Figure 4.4). This agrees with the current definition of lipids rafts (Pike, 2006).

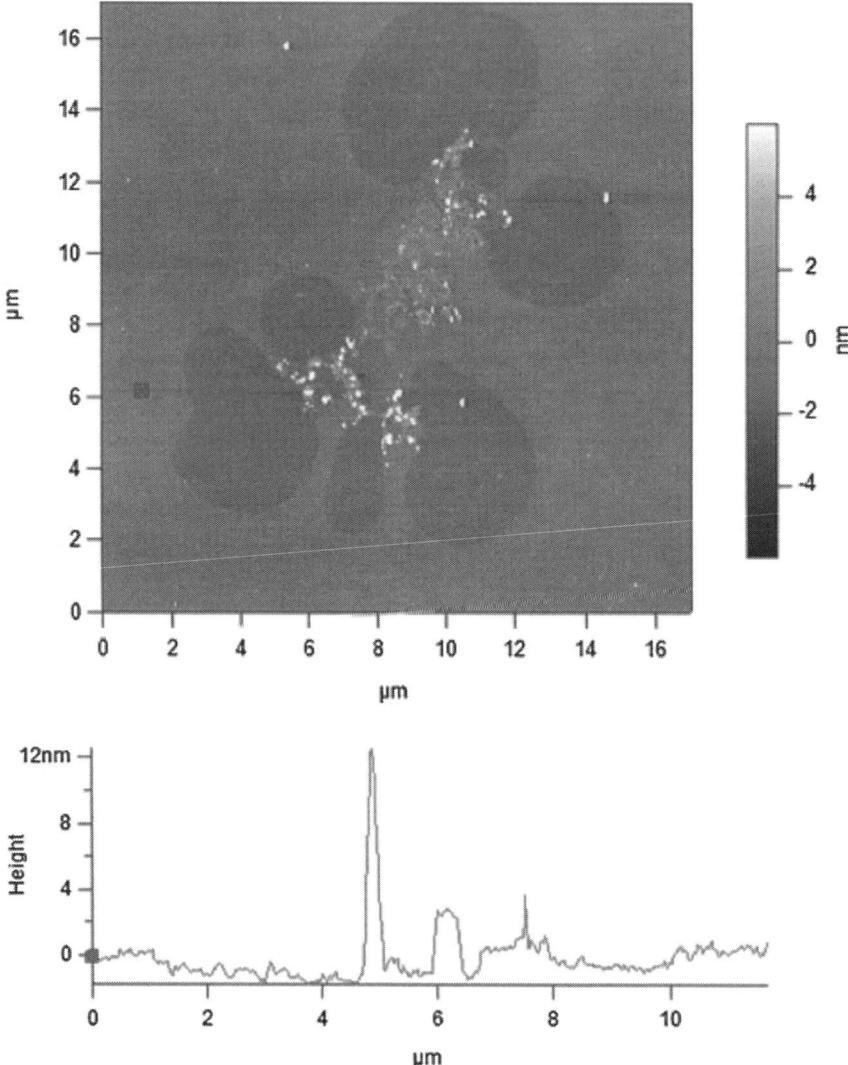

■ **FIGURE 4.4** Atomic force microscopy topographic image and respective height graphs of raw cream phospholipid-β-casein Langmuir—Blodgett films acquired using the alternating current mode. The height graph was obtained by measuring the height of the molecules along the line drawn on the topographic image. The liquid-ordered phase appeared in dark brown, the liquid-disordered phase in light brown, and the β-casein regions in white. *Adapted from Gallier et al. (2012a) with permission from Elsevier.* [A color version of this figure is available online at www.booksite.elsevier.com/9780124046108].

Liposomes and GUV

Membrane curvature plays an important role in the lateral organization of biomembranes (Garcia-Saez and Schwille, 2010b). Membrane curvature is not simulated in flat Langmuir monolayers. Phospholipids, previously dissolved in solvent, form a thin film after drying of the solvent using nitrogen or vacuum. Above the melting temperature of the phospholipids and after hydration of the film, multilamellar vesicles are formed with the help of freeze/thaw cycles or shaking. Small unilamellar vesicles or liposomes are formed by further sonication or extrusion through membranes of small pore size (Bacia and Schweizer, 2005). Liposomes are small in size, less than 100 nm, and have a very large membrane curvature compared with biological cells. They are commonly used as bilayer model systems of biological cells with an aqueous interior or encapsulation matrices.

Liposomes were prepared by microfluidization with MFGM phospholipids or soya phospholipids. MFGM phospholipid liposomes had a thicker membrane with lower permeability and higher phase transition temperature than soya phospholipid liposomes (Thompson *et al.*, 2006). They were also more stable under various pH, ionic strength, and storage temperature conditions (Thompson *et al.*, 2006). They showed greater encapsulation efficiency of hydrophilic and hydrophobic compounds (Bezelgues *et al.*, 2009; Thompson *et al.*, 2009). Tea polyphenols (Gulseren *et al.*, 2012), ascorbic acid (Farhang *et al.*, 2012), lactoferrin (Liu *et al.*, 2013), tocopherol, lycopene (Bezelgues *et al.*, 2009), β-carotene and potassium chromate (Thompson *et al.*, 2009) were successfully encapsulated with MFGM phospholipids. It is wise to think that the encapsulated material is not only located within the aqueous interior of the liposomes but may also partition within the phospholipid bilayer. MFGM phospholipid liposomes appeared as more efficient delivery systems than soya phospholipids due to their higher stability during *in vitro* gastrointestinal digestion (Liu *et al.*, 2012).

Giant unilamellar vesicles (GUVs) are model systems that closely resemble biological membranes in size and bilayer physical properties. Their size varies from a few microns to up to 300 μm. They have the advantage of being free-standing bilayers unlike supported-lipid bilayers (Garcia-Saez and Schwille, 2010b). However, their preparation method is still a hurdle in the use of this system to provide insight into the biophysics of biomembranes under physiological conditions. Electroformation is the most common method of preparation of GUVs (Angelova and Dimitrov, 1986). The addition of salts in the aqueous solution reduces the size and yield of the GUVs. GUVs can be observed with fluorescence microscopy and two-photon microscopy using fluorescent probes such as 6-dodecanoyl-2-dimethylaminonaphthalene (LAURDAN), a probe sensitive to the lipid state in membranes (Bagatolli and Gratton, 2001). This allows the study of translational motions of micrometer-sized liquid-ordered domains within the membrane plane. The fluorescent probe can partition preferentially into the liquid-ordered or liquid-disordered phase or emit light with a color dependent on the lipid packing as LAURDAN does.

Proteomic Techniques

The composition of MFGM can be altered by physical/mechanical factors (Evers, 2004), which include cooling, drying, separation, agitation, heating, and homogenization (Michalski and Januel, 2006; Dewettinck *et al.*, 2008; Lopez *et al.*, 2008; Jiménez-Flores and Brisson, 2008). Examining the effect of milk processing on the chemical changes of the MFGM is currently being performed through proteomic methodology. SDS-PAGE is an excellent technique to visualize proteins; however, identification of proteins is limited by protein resolution; therefore, to assess the effect of milk processing in more depth, liquid chromatography connected with tandem mass spectrometry (LC-MS/MS) can be used to detect more proteins from the SDS-PAGE gel.

MFGM proteins constitute 1−2% of the total bovine milk protein, and depending on the milk source and how it is processed, 25−70% of the MFGM may be polypeptides, ranging in molecular weight (MW) from 15,000 to 240,000 Da (Ye *et al.*, 2002; Riccio, 2004; Dewettinck *et al.*, 2008; Jiménez-Flores and Higuera-Ciapara, 2009). Using proteomic techniques, Reinhardt and Lippolis (2006) identified 120 proteins, and studies by Fong and Norris have demonstrated eight predominant proteins in the MFGM: butyrophilin, adipophllin (ADPH), xanthine oxidase, cluster of differentiation 36, fatty acid binding protein (FABP), lactadherin, mucin 1, and mucin 15 (Fong and Norris, 2009).

Interactions among milk constituents and other molecules occur mainly on the fat globule surface (Vanderghem *et al.*, 2011), and heat is well known to induce protein interactions and protein displacement as well as protein adsorption on the milk fat globule membrane (Evers, 2004; Singh, 2006). Indeed, heating milk above 70°C causes whey protein to denature and deposit on the surface of the fat globules causing interactions that determine the physicochemical and functional properties of MFGM isolates.

Washing the MFGM material during the extraction procedure can lower the yield (Kitchen, 1977; Sánchez-Juanes *et al.*, 2009) with corresponding literature extracting approximately 95 mg dry weight per liter of fresh milk and

171 to 315 mg of dry weight per liter of milk extracted without washing (Elias-Argote and Jiménez-Flores, 2010). In addition, MFGM isolates can differ depending on the sample's temperature as observed by Kim and Jiménez-Flores (1995) with higher temperatures resulting in less membrane and more insoluble product. Thermal treatments also resulted in higher protein content in the MFGM isolates compared to fresh milk, indicating interactions between MFGM proteins and milk proteins as the temperature increased (Elias-Argote and Jiménez-Flores, 2010).

The use of 2-DE gels coupled with MALDI-TOF allows for mapping of MFGM protein variation. Milk fat globule membrane proteins can be detected using peptide mass fingerprinting (PMF). There are limitations to successfully identifying spots via MALDI-TOF, such as incompleteness of bovine protein databases and the relatively high threshold parameters used to define a positive identification.

STRUCTURAL ANALYSIS AND ROLE IN DIGESTION

A recent study compared the hydrolysis of native milk fat globules with that of emulsified anhydrous milk fat by microbial lipases (Bourlieu *et al.*, 2012). This has some importance for the stability of some dairy products. The presence of the MFGM reduced the efficiency of the lipase activity. A lag time in the rate of hydrolysis was observed and reduced in the presence of phospholipase and protease able to hydrolyze MFGM components, thus facilitating access by the microbial lipases to the triglyceride core. The inhibition of lipolysis by pancreatic or microbial lipases was also demonstrated in the presence of MFGM, but was lost when MFGM was subjected to hydrolysis by trypsin (Shimizu *et al.*, 1982). This proves the role of the MFGM as a protective barrier against endogenous lipases such as lipoprotein lipase. This is also true for some exogenous lipases such as digestive lipases. Indeed *in vitro*, pancreatic lipase has little hydrolytic activity on native milk fat globules because of its inability to access the triglyceride core in the presence of the MFGM (Blackberg *et al.*, 1981). However, the *in vitro* digestion of milk fat globules by pancreatic lipase is enhanced in the presence of bile salts, colipase, and phospholipase A_2 (Bernback *et al.*, 1990). Phospholipase A_2 is able to hydrolyze MFGM phospholipids, colipase forms a complex with pancreatic lipase, bile salts adsorb onto the surface of the globules, and colipase interacts with bile salts, which activates pancreatic lipase. Patton *et al.* (1986) showed that physiologically relevant concentration of conjugated bile salts allowed the removal of the MFGM from the surface of the globules. But, *in vivo*, native milk fat globules are subjected first to gastric lipolysis which generates free FAs accumulating at the surface of the globules (Figure 4.5) and helping the binding of pancreatic lipase and the onset of pancreatic lipolysis upon reaching the small intestine (Bernback *et al.*, 1989, 1990). Gastric lipase is able to penetrate the MFGM and access the triglyceride core. However, gastric lipolysis is reduced in the presence of phosphatidyl-ethanolamine and sphingomyelin (Favé *et al.*, 2007), two major MFGM phospholipids.

Human milk is the first food that a newborn will receive. Dairy products are widely consumed in Western countries. Several studies have focused on the health benefits of human milk (Hamosh *et al.*, 1985, 1999; Armand *et al.*, 1996) and MFGM-rich diets (Snow *et al.*, 2010; Zavaleta *et al.*, 2011; Kuchta *et al.*, 2012). Human milk fat globules were found to be digested more efficiently than lipid droplets of infant formula despite their greater size, but this was explained by the presence of the MFGM on the surface of the human milk fat globules facilitating gastric lipolysis (Armand *et al.*, 1996). In addition, short-chain FAs and monoglycerides produced during digestive lipolysis of milk fat globules possess antibacterial and antiviral activity (Hamosh *et al.*, 1999).

In rats, a delay in the appearance of plasma TAG was observed after force-feeding native milk fat globules compared with non-emulsified milk fat (Michalski *et al.*, 2006). Dietary sphingolipids decreased total plasma cholesterol by 30% and increased hepatic cholesterol in rats (Kobayashi *et al.*, 1997).

MFGM glycoproteins showed resistance to proteolysis by Pronase and trypsin (Kobylka and Carraway, 1973). MFGM glycoproteins were responsible for the inhibition of lipase activity on MFGM-emulsified milk fat (Shimizu *et al.*, 1982). Mucin and lactadherin, two major MFGM glyco-proteins, are resistant to pepsin and retain their function at low gastric pH (Hamosh *et al.*, 1999). This resistance of glycoproteins was associated with the presence of carbohydrate moieties, forming the glycocalyx sur-rounding the milk fat globules (Figure 4.5). As well as providing a steric bar-rier against aggregation and coalescence, the glycocalyx presents a barrier to lipolysis (Gallier *et al.*, 2012c). Liquid-ordered domains were observed at the surface of milk fat globules during *in vitro* gastric digestion, suggesting that they may play a physiological role (Gallier *et al.*, 2012c). By subjecting MFGM fragments to *in vitro* gastric and intestinal digestion, Le *et al.* (Le *et al.*, 2012) showed that glycosylated MFGM proteins were more resis-tant to digestion by gastric and intestinal lipolysis than non-glycosylated pro-teins. Cluster of differentiation 36 (CD36) and PAS III and PAS 6/7 were detected intact or partially digested at the end of gastric digestion. MUC1

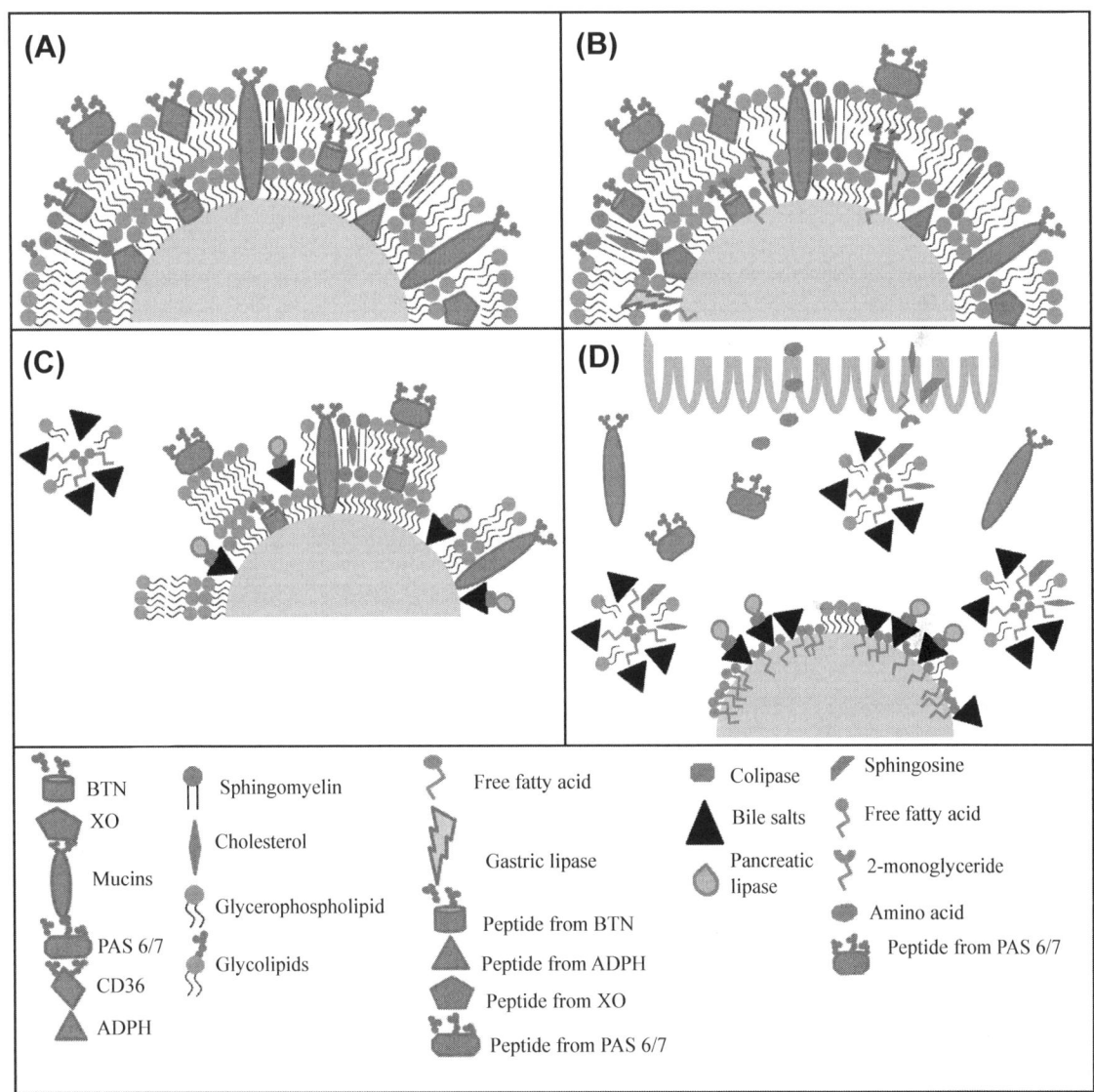

■ **FIGURE 4.5** Schematic diagram illustrating the interfacial changes occurring at the surface of the milk fat globule during gastric and intestinal digestion. (A) Native milk fat globule membrane (MFGM) structure. A phospholipid trilayer embedding proteins. (B) Milk fat globule in gastric environment. Pepsin hydrolyzes MFGM proteins at different rates and some MFGM glycoproteins are resistant to proteolysis; the resulting peptides are surface active enough to remain at the interface. Gastric lipase hydrolyzes the triglyceride core and the released free fatty acids accumulate at the surface of the globules. (C) Beginning of the intestinal digestion. The bile salts displace the phospholipids, proteins, and peptides and adsorb onto the interface. The colipase adsorbs on to the bile salt-rich interface and the pancreatic lipase forms a complex with the colipase. (D) Trypsin and chymotrypsin hydrolyze some of the MFGM proteins and peptides into amino acids, which further absorb through the intestinal wall; most of the mucins and part of periodic acid-Schiff (PAS) 6/7 are resistant to proteolysis. The pancreatic lipase hydrolyzes the triglyceride core, and lipolytic products accumulate at the oil—water interface and are then solubilized into mixed phospholipid—bile salt micelles for their transport through the intestinal wall. Cholesterol molecules are similarly transported. CD36 = cluster of differentiation 36; BTN = butyrophilin; ADPH = adipophilin; XO = xanthine oxidase. Not to scale. *Adapted from Gallier et al. (2012c) with permission from Elsevier.* [A color version of this figure is available online at www.booksite.elsevier.com/9780124046108].

was the most resistant with some intact fragments still present at the end of gastrointestinal digestion. The resistance of some MFGM glycoproteins contributes to their anti-adhesive properties against pathogens (Le *et al.*, 2012). MUC1 has anti-infection activity against rotavirus and common enteropathogenic bacteria and prevents HIV transmission from mother to child. PAS 6/7 and CD36 are also beneficial to the intestinal immune system by binding to pathogens. After feeding of human milk, large fragments of MUC1 from human MFGM were detected in stools of infants (Patton, 1994). After feeding mice with MFGM-rich diets, the β-glucoronidase activity of colonic enterobacteria was inhibited due to the presence of undigested MFGM glycoproteins. A randomized double-blind clinical trial on 499 infants fed either a MFGM-rich or skim milk-rich complementary food showed that the consumption of MFGM reduced the number of diarrhea episodes in infants (Zavaleta *et al.*, 2011). Snow *et al.* (Snow *et al.*, 2011) fed lipopolysaccharide-injected mice a diet rich in corn oil or MFGM-emulsified milk fat for 5 weeks. The MFGM group showed lower gut permeability and lower levels of pro-inflammatory cytokines.

MFGM proteins have proven anticancer properties. FABP was reported to inhibit *in vitro* the growth of breast cancer cells (Spitsberg, 2005). BRCA1 and BRCA2 proteins, present in human and bovine MFGM, act as an oncosuppressor. BTN is a suppressor of multiple sclerosis (Spitsberg, 2005) and is involved in lipid secretion (Robenek *et al.*, 2006a). XDH/XO is a potent bactericidal agent. As an enzyme, XDH/XO produces reactive oxygen and nitrogen species with antimicrobial activity (Ward *et al.*, 2006). The MFGM can act as a carrier of liposolubles, vitamins, organic phosphates, selenium, and anticancer agent through FABP and other bioactives (Couvreur and Hurtaud, 2007; Jiménez-Flores and Brisson, 2008; Bezelgues *et al.*, 2009).

Phospholipase A_2, an enzyme present in the small intestine, hydrolyzes glycerophospholipids, is sensitive to defects and lateral structure in the membrane, and is most active in the phase coexistence region (Bagatolli *et al.*, 2010). Thus, the lateral organization and presence of liquid-ordered domains in the MFGM, either at the surface of native milk fat globules or in liposomes, may facilitate the digestion of phospholipids by phospholipase A_2 in the intestine and thus the release of encapsulated material. The hydrolysis of phospholipids by phospholipase A_2 results in the release of FAs and lysophospholipids, the latter being efficient surfactants (Fujita and Suzuki, 1990) able to lyse Gram-positive bacteria (Sprong *et al.*, 2002, 2012). Sphingomyelin molecules are hydrolyzed by sphingomyelinase from the intestinal epithelium and bile in the intestines and in the colon (Kuchta *et al.*, 2012). An alkaline ceramidase from the intestinal epithelium converts ceramide into sphingosine in the intestinal lumen (Kuchta *et al.*, 2012). The interaction of sphingomyelin and cholesterol, for example in lipid rafts, slows down sphingomyelin digestion, as well as high concentration of bile salts in the duodenum. Sphingolipids are not all digested. Undigested sphingomyelin reaches the colon where it can be hydrolyzed and absorbed.

In addition to the antiviral, antibacterial, anti-inflammatory, and anticancer properties of MFGM proteins (Dewettinck *et al.*, 2008), MFGM phospholipids and, in particular, sphingomyelin have many health benefits, including gut health and a protective role against colon cancer (Spitsberg, 2005; Kuchta *et al.*, 2012). Phospholipids are involved in cell proliferation, molecular transport, memory, stress responses, Alzheimer's disease, myelination of the developing central nervous system, maturation of the neonatal gut, and protection against gastric ulceration (Ward *et al.*, 2006; Kuchta *et al.*, 2012). They are also antioxidants and antimicrobial and antiviral agents (Ward *et al.*, 2006). Gangliosides have been reported to inhibit foodborne pathogens (Ward *et al.*, 2006). MFGM lipids from buttermilk and whey cream may also provide protection against infection by rotavirus (Fuller *et al.*, 2013). Hydrolytic products of phosphatidylcholine and sphingomyelin are lipid secondary messengers such as ceramide, diacylglycerol, sphingosine, ceramide-1-phosphate, and sphingosine-1-phosphate and are involved in cellular processes of shut down in cancer such as apoptosis, cell growth, and differentiation (Ward *et al.*, 2006). Dietary bovine milk sphingomyelin decreased the amount of aberrant colonic crypt foci in mice. In addition, a change from adenocarcinomas to benign adenomas was reported after feeding milk sphingomyelin and glycosphingolipids to mice (Kuchta *et al.*, 2012). Sphingomyelin possesses chemotherapeutic and chemopreventive effects (Spitsberg, 2005). It has also an anticholesterolemic effect by binding to cholesterol and reducing its absorption (Spitsberg, 2005). Milk phospholipids have been shown to protect the skin from ultraviolet light damage using human skin models (Russell *et al.*, 2010; Dargitz *et al.*, 2012).

CONCLUDING REMARKS

Studies around the world on the MFGM are giving us an accurate idea of the complexity of its structure and its relationship to composition. The relative small variation in phospholipid content of the MFGM of mammals is indirect evidence of the importance that composition and structural relationships have for the correct secretory and nutrition delivery functions of this important milk component.

Whether there is a dependence of the proteins on phospholipids to be organized on the membrane or vice versa, it is an overall important issue of the food technologists of the future to design foods that work in the most biologically efficient manner.

Many questions remain on elucidation of the structure and function of the MFGM, and as technology advances, new and innovative techniques applied to understanding this system will undoubtedly yield beneficial data on which to build some foods of the future.

REFERENCES

Affolter, M., Grass, L., et al., 2010. Qualitative and quantitative profiling of the bovine milk fat globule membrane proteome. J. Proteom. 73, 1079–1088.

Angelova, M.I., Dimitrov, D.S., 1986. Liposome electroformation. Faraday Dis. 81, 303–311.

Appell, K.C., Keenan, T.W., et al., 1982. Differential scanning calorimetry of milk-fat globule membranes. Biochim. Biophys. Acta 690, 243–250.

Argov, N., Wachsmann-Hogiu, S., et al., 2008. Size-dependent lipid content in human milk fat globules. J. Agric. Food Chem. 56, 7446–7450.

Armand, M., Hamosh, M., et al., 1996. Effect of human milk or formula on gastric function and fat digestion in the premature infant. Pediatr. Res. 40, 429–437.

Bacia, K., Schweizer, J., 2005. Practical Course: Giant Unilamellar Vesicles. Technische Universität Dresden.

Bagatolli, L.A., Gratton, E., 2001. Direct observation of lipid domains in free-standing bilayers using two-photon excitation fluorescence microscopy. J. Fluoresc. 11, 141–160.

Bagatolli, L.A., Ipsen, J.H., et al., 2010. An outlook on organization of lipids in membranes: searching for a realistic connection with the organization of biological membranes. Prog. Lipid Res. 49, 378–389.

Bargmann, W., Fleischhauer, K., et al., 1961. Uber die morphologie der milchsekretion. 2. Zugleich eine kritik am schema der sekretionsmorphologie. Zeitschrift Fur Zellforschung Und Mikroskopische Anatomie 53, 545–568.

Beddek, A.J., Rawson, P., et al., 2008. Profiling the metabolic proteome of bovine mammary tissue. Proteomics 8, 1502–1515.

Bernback, S., Blackberg, L., et al., 1989. Fatty-acids generated by gastric lipase promote human-milk triacylglycerol digestion by pancreatic colipase-dependent lipase. Biochim. Biophys. Acta 1001, 286–293.

Bernback, S., Blackberg, L., et al., 1990. The complete digestion of human-milk triacylglycerol *in vitro* requires gastric lipase, pancreatic colipase-dependent lipase, and bile salt stimulated lipase. J. Clin. Invest. 85, 1221–1226.

Bezelgues, J.B., Morgan, F., et al., 2009. Short communication: milk fat globule membrane as a potential delivery system for liposoluble nutrients. J. Dairy Sci. 92, 2524–2528.

Bionaz, M., Loor, J.J., 2008. Gene networks driving bovine milk fat synthesis during the lactation cycle. BMC Genom. 9, 366–387.

Blackberg, L., Hernell, O., et al., 1981. Hydrolysis of human-milk fat globules by pancreatic lipase—role of colipase, phospholipase-A2, and bile salts. J. Clin. Invest. 67, 1748–1752.

Bourlieu, C., Rousseau, F., et al., 2012. Hydrolysis of native milk fat globules by microbial lipases: mechanisms and modulation of interfacial quality. Food Res. Int. 49, 533–544.

Brisson, G., Payken, H.F., et al., 2010. Characterization of Lactobacillus reuteri interaction with milk fat globule membrane components in dairy products. J. Agric. Food Chem. 58, 5612–5619.

Cavaletto, M., Giuffrida, M.G., et al., 2002. A proteomic approach to evaluate the butyrophilin gene family expression in human milk fat globule membrane. Proteomics 2, 850–856.

Cavaletto, M., Giuffrida, M.G., et al., 2004. The proteomic approach to analysis of human milk fat globule membrane. Clin. Chim. Acta 347, 41–48.

Couvreur, S., Hurtaud, C., 2007. Globule milk fat: secretion, composition, function and variation factors. INRA Productions Animales 20, 369–382.

Da Silva, P.P., De Menezes, A.P., et al., 1980. Structure and dynamics of the bovine milk fat globule membrane viewed by freeze fracture. Exp. Cell Res. 125, 127–140.

Dargitz, C., Russell, A., et al., 2012. Milk phospholipid's protective effects against UV damage in skin equivalent models. Proceedings of the SPIE—The International Society for Optical Engineering, 82251V.

Deeth, H.C., 1997. The role of phospholipids in the stability of milk fat globules. Aust. J. Dairy Technol. 52, 44–46.

Dewettinck, K., Rombaut, R., et al., 2008. Nutritional and technological aspects of milk fat globule membrane material. Int. Dairy J. 18, 436–457.

Dufour, E., 2006. Spectroscopic techniques (NMR, infrared and fluorescence) for the determination of lipid composiition and structure in dairy products. In: Fox, P.F., McSweeney, P.L.H. (Eds.), Advanced Dairy Chemistry: Lipids, Vol. 2. Springer, New York, pp. 697–707.

Einhorn-Stoll, U., Weiss, M., et al., 2002. Influence of the emulsion components and preparation method on the laboratory-scale preparation of o/w emulsions containing different types of dispersed phases and/or emulsifiers. Nahrung-Food 46, 294–301.

Elias-Argote, X., Jiménez-Flores, R., 2010. Effect of processing on the milk fat globule membrane constituents. J. Dairy Sci. 93, 606.

Evers, J.M., 2004. The milkfat globule membrane—compositional and structural changes post secretion by the mammary secretory cell. Int. Dairy J. 14, 661–674.

Evers, J.M., Haverkamp, R.G., et al., 2008. Heterogeneity of milk fat globule membrane structure and composition as observed using fluorescence microscopy techniques. Int. Dairy J. 18, 1081–1089.

Fang, Y., Dalgleish, D.G., 1995. Studies on interactions between phosphatidylcholine and casein. Langmuir 11, 75–79.

Farhang, B., Kakuda, Y., et al., 2012. Encapsulation of ascorbic acid in liposomes prepared with milk fat globule membrane-derived phospholipids. Dairy Sci. Technol. 92, 353–366.

Fauquant, C., Briard, V., et al., 2005. Differently sized native milk fat globules separated by microfiltration: fatty acid composition of the milk fat globule membrane and triglyceride core. Eur. J. Lipid Sci. Technol. 107, 80–86.

Favé, G., Lévêque, C., et al., 2007. Modulation of gastric lipolysis by the phospholipid specie: link to specific lipase-phospholipid interaction at the lipid/water interface? FASEB J. 21, A1010.

Fong, B.Y., Norris, C.S., 2009. Quantification of milk fat globule membrane proteins using selected reaction monitoring mass spectrometry. J. Agric. Food Chem. 57 (14), 6021−6028.

Forrest, G., 1978. Raman spectroscopy of milk globule membrane and triglycerides. Chem. Phys. Lipids 21, 237−252.

Fortunato, D., Giuffrida, M.G., et al., 2003. Structural proteome of human colostral fat globule membrane proteins. Proteomics 3, 897−905.

Fujita, S., Suzuki, K., 1990. Surface activity of the lipid products hydrolyzed with lipase and phospholipase A-2. J. Am. Oil Chem. Soc. 67, 1008−1014.

Fuller, K.L., Kuhlenschmidt, T.B., et al., 2013. Milk fat globule membrane isolated from buttermilk or whey cream and their lipid components inhibit infectivity of rotavirus in vitro. J. Dairy Sci. 96, 3488−3497.

Gagnaire, V., Jardin, J., et al., 2009. Invited review: proteomics of milk and bacteria used in fermented dairy products: from qualitative to quantitative advances. J. Dairy Sci. 92, 811−825.

Gallier, S., 2010. Understanding the structure of the bovine milk fat globule and its membrane by means of microscopic techniques and model systems. Doctor of Philosophy. University of Otago.

Gallier, S., Gragson, D., et al., 2012a. β-Casein−phospholipid monolayers as model systems to understand lipid−protein interactions in the milk fat globule membrane. International Dairy J 22 (1), 58−65.

Gallier, S., Gordon, K.C., et al., 2011. Composition of bovine milk fat globules by confocal Raman microscopy. Int. Dairy J. 21, 402−412.

Gallier, S., Gordon, K.C., et al., 2012b. Chemical and structural characterisation of almond oil bodies and bovine milk fat globules. Food Chem. 132, 1996−2006.

Gallier, S., Gragson, D., et al., 2010a. Using confocal laser scanning microscopy to probe the milk fat globule membrane and associated proteins. J. Agric. Food Chem. 58, 4250−4257.

Gallier, S., Gragson, D., et al., 2010b. Surface characterization of bovine milk phospholipid monolayers by langmuir isotherms and microscopic techniques. J. Agric. Food Chem. 58, 12275−12285.

Gallier, S., Gragson, D., et al., 2010c. Composition and fatty acid distribution of bovine milk phospholipids from processed milk products. J. Agric. Food Chem. 58, 10503−10511.

Gallier, S., Ye, A., et al., 2012c. Structural changes of bovine milk fat globules during in vitro digestion. J. Dairy Sci. 95, 3579−3592.

Garcia-Saez, A.J., Schwille, P., 2010a. Stability of lipid domains. FEBS Lett. 584, 1653−1658.

Garcia-Saez, A.J., Schwille, P., 2010b. Surface analysis of membrane dynamics. Biochim. Biophys. Acta−Biomembr. 1798, 766−776.

Goff, H.D., 1997. Colloidal aspects of ice cream—a review. Int. Dairy J. 7, 363−373.

Gulseren, I., Guri, A., et al., 2012. Encapsulation of tea polyphenols in nanoliposomes prepared with milk phospholipids and their effect on the viability of HT-29 human carcinoma cells. Food Dig. 3, 36−45.

Hamdan, M., Righetti, P.G., 2005. Proteomics Today: Protein Assessment and Biomarkers using Mass Spectrometry, 2D Electrophoresis, and Microarray Technology. John Wiley & Sons.

Hamosh, M., Bitman, J., et al., 1985. Lipids in milk and the first steps in their digestion. Pediatrics 75, 146–150.

Hamosh, M., Peterson, J.A., et al., 1999. Protective function of human milk: the milk fat globule. Sem. Perinatol. 23, 242–249.

Hayashi, S., Smith, L.M., 1965. Membranous material of bovine milk fat blobules. I. Comparison of membranous fractions released by deoxycholate and by churning. Biochemistry 4, 2550.

Henson, A.F., 1971. Physicochemical analyses of bovine milk fat globule membrane. 2. Electron microscopy. J. Dairy Sci. 54, 1752.

Horisberger, M., Rosset, J., et al., 1977. Location of glycoproteins on milk fat globule membrane by scanning and transmission electron microscopy, using lectin-labelled gold granules. Exp. Cell Res. 109, 361–369.

Huston, G.E., Patton, S., 1990. Factors related to the formation of cytoplasmic crescents on milk fat globules. J. Dairy Sci. 73, 2061–2066.

Ikonen, E., 2008. Cellular cholesterol trafficking and compartmentalization. Nat. Rev. Mol. Cell Biol. 9, 125–138.

Janssen, M.M.T., Walstra, P., 1982. Cytoplasmic remnants in milk of certain species. Netherlands Milk and Dairy J. 36, 365–368.

Jensen, R.G., 2002. The Composition of Bovine Milk Lipids: January 1995 to December 2000. J. Dairy Sci. 85, 295–350.

Jiménez-Flores, R., Brisson, G., 2008. The milk fat globule membrane as an ingredient: why, how, when? Dairy Sci. Technol. 88, 5–18.

Jiménez-Flores, R., Higuera-Ciapara, I., 2009. Beverages based on milk fat globule memembrane (MFGM) and other novel concepts for dairy-based functional beverages. In: Functional and speciality beverage technology. P. P. PhD, 1. CRC Press, Woodhead Publishing Limited, Boca Raton, New York, Washington DC, 281–296.

Kaganer, V.M., Möhwald, H., et al., 1999. Structure and phase transitions in Langmuir monolayers. Rev. Mod. Phys. 71, 779–819.

Keenan, T.W., Mather, I.H., 2006. Intracellular origin of milk fat globules and the nature of the milk fat globule membrane. In: Fox, P.F., McSweeney, P.L.H. (Eds.), Advanced Dairy Chemistry: Lipids, Vol. 2. Springer, New York, USA, pp. 137–171.

Keenan, T.W., Olson, D.E., et al., 1971. Origin of the milk fat globule membrane. J. Dairy Sci. 54, 295–299.

Kim, H.-H.Y., Jiménez-Flores, R., 1995. Heat-induced interactions between the proteins of milk fat globule membrane and skim milk. J. Dairy Sci. 78, 24–35.

King, N., 1955. The Milk Fat Globule Membrane and Some Associated Phenomena. Commonwealth Agricultural Bureaux, Farnham Royal, Bucks, Great Britain.

Kitchen, B.J., 1977. Fractionation and characterization of the membranes from bovine milk fat globules. J. Dairy Res. 44, 469–482.

Kobayashi, T., Shimizugawa, T., et al., 1997. A long-term feeding of sphingolipids affected the levels of plasma cholesterol and hepatic triacylglycerol but not tissue phospholipids and sphingolipids. Nutr. Res. 17, 111–114.

Kobylka, D., Carraway, K.L., 1973. Proteolytic digestion of proteins of the milk fat globule membrane. Biochim. Biophys. Acta (BBA)—Biomembr. 307, 133–140.

Krafft, C., Neudert, L., et al., 2005. Near infrared Raman spectra of human brain lipids. Spectrochim. Acta Part A—Mol. Biomol. Spectrosc. 61, 1529–1535.

Krog, N., 1991. Thermodynamics of interfacial films in food emulsions. ACS Symp. Ser. 448, 138–145.

Kuchta, A.M., Kelly, P.M., et al., 2012. Milk fat globule membrane—a source of polar lipids for colon health? A review. Int. J. Dairy Technol. 65, 315–333.

Le, T.T., Van de Wiele, T., et al., 2012. Stability of milk fat globule membrane proteins toward human enzymatic gastrointestinal digestion. J. Dairy Sci. 95, 2307–2318.

Lee, S.Y., Morr, C.V., 1992. Fixation and staining milk fat globules in cream for transmission and scanning electron microscopy. J. Food Sci. 57, 887–891.

Li-Chan, E.C.Y., 1996. The applications of Raman spectroscopy in food science. Trends Food Sci. Technol. 7, 361–370.

Li-Chan, E.C.Y., 2007. Vibrational spectroscopy applied to the study of milk proteins. Lait 87, 443–458.

Liu, W., Ye, A., et al., 2012. Structure and integrity of liposomes prepared from milk- or soybean-derived phospholipids during in vitro digestion. Food Res. Int. 48, 499–506.

Liu, W., Ye, Y., et al., 2013. Stability during in vitro digestion of lactoferrin-loaded liposomes prepared from milk fat globule membrane-derived phospholipids. J. Dairy Sci. 96, 2061–2070.

Lonnerdal, B., Valencia, N., et al., 2006. Effect of fortifying complementary food with a bioactive milk protein fraction with micronutrients on growth and micronutrient status of Peruvian infants. J. Pediatr. Gastroenterol. Nutr. 42, E88.

Lopez, C., Briard-Bion, V., et al., 2008. Phospholipid, sphingolipid, and fatty acid compositions of the milk fat globule membrane are modified by diet. J. Agric. Food Chem. 56, 5226–5236.

Lopez, C., Madec, M.N., et al., 2010. Lipid rafts in the bovine milk fat globule membrane revealed by the lateral segregation of phospholipids and heterogeneous distribution of glycoproteins. Food Chem. 120, 22–33.

MacKenzie, A., Vyssotski, M., et al., 2009. Quantitative analysis of dairy phospholipids by P-31 NMR. J. Am. Oil Chem. Soc. 86, 757–763.

Manso, M.A., LÈonil, J., et al., 2005. Application of proteomics to the characterisation of milk and dairy products. Int. Dairy J. 15, 845–855.

Mather, I.H., Keenan, T.W., 1975. Studies on structure of milk fat globule membrane. J. Membr. Biol. 21, 65–85.

Mather, I.H., 2000. A review and proposed nomenclature for major proteins of the milk-fat globule membrane. J. Dairy Sci. 83, 203–247.

Mather, I.H., Jack, L.J.W., et al., 2001. The distribution of MUC1, an apical membrane glycoprotein, in mammary epithelial cells at the resolution of the electron microscope: implications for the mechanism of milk secretion. Cell Tissue Res. 304, 91–101.

Mather, I.H., Weber, K., et al., 1977. Membranes of mammary gland. XII. Loosely associated proteins and compositional heterogenity of bovine milk fat globule membrane. J. Dairy Sci. 60, 394–402.

McClements, D.J., Monahan, F.J., et al., 1993. Effect of emulsion droplets on the rheology of whey-protein isolate gels. J. Text. Stud. 24, 411–422.

Michalski, M.-C., Januel, C., 2006. Does homogenization affect the human health properties of cow's milk? Trends Food Sci. Technol. 17, 423–437.

Michalski, M.-C., Michel, F., et al., 2002. Apparent zeta-potential as a tool to assess mechanical damages to the milk fat globule membrane. Coll. Surf. B—Biointerface 23, 23–30.

Michalski, M.C., Soares, A.F., et al., 2006. The supramolecular structure of milk fat influences plasma triacylglycerols and fatty acid profile in the rat. Eur. J. Nutr. 45, 215–224.

Miyaji, M., Jin, Z.X., et al., 2005. Role of membrane sphingomyelin and ceramide in platform formation for Fas-mediated apoptosis. J. Exp. Med. 202, 249–259.

Morin, P., Britten, M., et al., 2007. Microfiltration of buttermilk and washed cream buttermilk for concentration of milk fat globule membrane components. J. Dairy Sci. 90, 2132–2140.

Morin, P., Jiménez-Flores, R., et al., 2007b. Effect of processing on the composition and microstructure of buttermilk and its milk fat globule membranes. Int. Dairy J. 17, 1179–1187.

Morin, P., Pouliot, Y., et al., 2006. A comparative study of the fractionation of regular buttermilk and whey buttermilk by microfiltration. J. Food Eng. 77, 521–528.

Morin, P., Pouliot, Y., et al., 2008. Effect of buttermilk made from creams with different heat treatment histories on properties of rennet gels and model cheeses. J. Dairy Sci. 91, 871–882.

Morris, V.J., Woodward, N.C., et al., 2011. Atomic force microscopy as a nanoscience tool in rational food design. J. Sci. Food Agric. 91, 2117–2125.

Morrison, W.R., 1969. Polar lipids in bovine milk. I. Long-chain bases in sphingomyelin. Biochem. Biophys. Acta 176, 537–546.

Motouri, M., Matsuyama, H., et al., 2003. Milk sphingomyelin accelerates enzymatic and morphological maturation of the intestine in artificially reared rats. J. Pediatr. Gastroenterol. Nutr. 36, 241–247.

Mulder, H., Walstra, P., 1974. The Milk Fat Globule Emulsion Science as Applied to Milk Products and Comparable Foods. PUDOC, Wageningen, Netherlands.

Murgiano, L., Timperio, A.M., et al., 2009. Comparison of milk fat globule membrane (MFGM) proteins of Chianina and Holstein cattle breed milk samples through proteomics methods. Nutrients 1, 302–315.

Nafie, L.A., 2001. Theory of Raman scattering. In: Lewis, I.R., Edwards, H. (Eds.), Handbook of Raman Spectroscopy. From the Research Laboratory to the Process Line. Marcel Dekker Inc, New York, Basel, pp. 1–10.

Nielsen, R.L., Andersen, M.H., et al., 1999. Isolation of adipophilin and butyrophilin from bovine milk and characterization of a cDNA encoding adipophilin. J. Dairy Sci. 82, 2543–2549.

Noh, S.K., Koo, S.I., 2004. Milk sphingomyelin is more effective than egg sphingomyelin in inhibiting intestinal absorption of cholesterol and fat in rats. J. Nutr. 134, 2611–2616.

Nyberg, L., Duan, R.-D., et al., 2000. A mutual inhibitory effect on absorption of sphingomyelin and cholesterol. J. Nutr. Biochem. 11, 244–249.

Nyberg, L., Farooqi, A., et al., 1998. Digestion of ceramide by human milk bile salt-stimulated lipase. J. Pediatr. Gastroenterol. Nutr. 27, 560–567.

O'Donnell, R., Holland, J.W., et al., 2004. Milk proteomics. Int. Dairy J. 14, 1013–1023.

Ong, L., Dagastine, R.R., et al., 2010. The effect of milk processing on the microstructure of the milk fat globule and rennet induced gel observed using confocal laser scanning microscopy. J. Food Sci. 75, E135–E145.

Parodi, P.W., 2003. Anti-cancer agents in milkfat. Aust. J. Dairy Technol. 58, 114–118.

Parodi, P.W., 2005. Dairy product consumption and the risk of breast cancer. J. Am. Coll. Nutr. 24, 556S–568S.

Patton, S., 1994. Detection of large fragments of the human milk mucin MUC-1 in feces of breast-fed infants. J. Pediatr. Gastroenterol. Nutr. 18, 225–230.

Patton, S., Borgstrom, B., et al., 1986. Release of membrane from milk-fat globules by conjugated bile-salts. J. Pediatr. Gastroenterol. Nutr. 5, 262–267.

Pike, L.J., 2006. Rafts defined: a report on the Keystone Symposium on Lipid Rafts and Cell Function. J. Lipid Res. 47, 1597–1598.

Prostenik, M., Gospocic, L., 1966. Sphingomyelin from cow's milk. Naturwissenschaften 53, 407.

Reinhardt, T.A., Lippolis, J.D., 2006. Bovine milk fat globule membrane proteome. J. Dairy Res. 73 (4), 406–416.

Reinhardt, T.A., Lippolis, J.D., 2008. Developmental changes in the milk fat globule membrane proteome during the transition from colostrum to milk. J. Dairy Sci. 91, 2307–2318.

Riccio, P., 2004. The proteins of the milk fat globule in the balance. Trends Food Sci. Technol. 15, 458–461.

Robenek, H., Hofnagel, O., et al., 2006a. Butyrophilin controls milk fat globule secretion. Proc. Natl. Acad. Sci. USA 103, 10385–10390.

Robenek, H., Hofnagel, O., et al., 2006b. Adipophilin-enriched domains in the ER membrane are sites of lipid droplet biogenesis. J. Cell Sci. 119, 4215–4224.

Rombaut, R., Camp, J.V., et al., 2006a. Phospho- and sphingolipid distribution during processing of milk, butter and whey. Int. J. Food Sci. Technol. 41, 435–443.

Rombaut, R., Dejonckheere, V., et al., 2006b. Microfiltration of butter serum upon casein micelle destabilization. J. Dairy Sci. 89, 1915–1925.

Rombaut, R., Dewettinck, K., et al., 2007. Phospho- and sphingolipid content of selected dairy products as determined by HPLC coupled to an evaporative light scattering detector (HPLC-ELSD). J. Food Comp. Anal. 20, 308–312.

Russell, A., Laubscher, A., et al., 2010. Investigating the protective properties of milk phospholipids against ultraviolet light exposure in a skin equivalent model. Proc. SPIE 7569, Multiphoton Microscopy in the Biomedical Sciences X, 75692Z.

Sánchez-Juanes, F., Alonso, J.M., et al., 2009. Distribution and fatty acid content of phospholipids from bovine milk and bovine milk fat globule membranes. Int. Dairy J. 19, 273–278.

Schmelz, E.M., Dillehay, D.L., et al., 1996. Sphingomyelin consumption suppresses aberrant colonic crypt foci and increases the proportion of adenomas versus adenocarcinomas in CF1 mice treated with 1,2-dimethylhydrazine: implications for dietary sphingolipids and colon carcinogenesis. Cancer Res. 56, 4936–4941.

Shimizu, M., Kanno, C., et al., 1978. Selective solubilization of bovine milk fat globule membrane proteinswith guanidine hydrochloride and disposition of some of proteins. Agric. Biol. Chem. 42, 2309–2314.

Shimizu, M., Miyaji, H., et al., 1982. Inhibition of lipolysis by milk-fat globule-membrane materials in model milk-fat emulsion. Agric. Biol. Chem. 46, 795–799.

Singh, A.M., Dalgleish, D.G., 1998. The emulsifying properties of hydrolyzates of whey proteins. J. Dairy Sci. 81, 918–924.

Singh, H., 2006. The milk fat globule membrane—a biophysical system for food applications. Curr. Opin. Colloid Interface Sci. 11, 154–163.

Snow, D.R., Jiménez-Flores, R., et al., 2010. Dietary milk fat globule membrane reduces the incidence of aberrant crypt foci in Fischer-344 rats. J. Agric. Food Chem. 58, 2157–2163.

Snow, D.R., Ward, R.E., et al., 2011. Membrane-rich milk fat diet provides protection against gastrointestinal leakiness in mice treated with lipopolysaccharide. J. Dairy Sci. 94, 2201–2212.

Spitsberg, V.L., 2005. Bovine milk fat globule membrane as a potential nutraceutical. J. Dairy Sci. 88, 2289–2294.

Spitsberg, V.L., Gorewit, R.C., 1998. Solubilization and purification of xanthine oxidase from bovine milk fat globule membrane. Protein Expr. Purif. 13, 229–234.

Spitsberg, V.L., Gorewit, R.C., 1999. Method of detecting expression of and isolating the protein encoded by the BRCA1 gene. USPO.

Sprong, R.C., Hulstein, M.F.E., et al., 2002. Bovine milk fat components inhibit food-borne pathogens. Int. Dairy J. 12, 209–215.

Sprong, R.C., Hulstein, M.F.E., et al., 2012. Sweet buttermilk intake reduces colonisation and translocation of Listeria monocytogenes in rats by inhibiting mucosal pathogen adherence. Br. J. Nutr. 108, 2026–2033.

Thompson, A.K., Singh, H., 2006. Preparation of liposomes from milk fat globule membrane phospholipids using a microfluidizer. J. Dairy Sci. 89, 410–419.

Thompson, A.K., Couchoud, A., et al., 2009. Comparison of hydrophobic and hydrophilic encapsulation using liposomes prepared from milk fat globule-derived phospholipids and soya phospholipids. Dairy Sci. Technol. 89, 99–113.

Thompson, A.K., Haisman, D., et al., 2006. Physical stability of liposomes prepared from milk fat globule membrane and soya phospholipids. J. Agric. Food Chem. 54, 6390–6397.

Vanderghem, C., Francis, F., et al., 2011. Study on the susceptibility of the bovine milk fat globule membrane proteins to enzymatic hydrolysis and organization of some of the proteins. Int. Dairy J. 21, 312–318.

Vesper, H., Schmelz, E.-M., et al., 1999. Sphingolipids in food and the emerging importance of sphingolipids to nutrition. J. Nutr. 129, 1239–1250.

Wang, J.J., Li, D.F., et al., 2006. Proteomics and its role in nutrition research. J. Nutr. 136, 1759–1762.

Waninge, R., Kalda, E., et al., 2003a. Cryo-TEM of isolated milk fat globule membrane structures in cream, 17th Annual Meeting of the European Colloid and Interface Science Society. Florence, Italy.

Waninge, R., Nylander, T., et al., 2003b. Phase equilibria of model milk membrane lipid systems. Chem. Phys. Lipids 125, 59–68.

Waninge, R., Walstra, P., et al., 2005. Competitive adsorption between beta-casein or beta-lactoglobulin and model milk membrane lipids at oil-water interfaces. J. Agric. Food Chem. 53, 716–724.

Ward, R.E., German, J.B., et al., 2006. Composition, applications, fractionation, technological and nutritional significance of milk fat globule membrane material. In: Fox, P.F., McSweeney, P.L.H. (Eds.), Advanced Dairy Chemistry, Lipids, vol. 2. Springer, USA, pp. 213–244.

Wat, E., Tandy, S., et al., 2009. Dietary phospholipid-rich dairy milk extract reduces hepatomegaly, hepatic steatosis and hyperlipidemia in mice fed a high-fat diet. Atherosclerosis 205, 144–150.

Westermeier, R., 2006. Sensitive, quantitative, and fast modifications for Coomassie Blue staining of polyacrylamide gels. Proteomics 6, 61–64.

Wilson, N.L., Robinson, L.J., et al., 2008. Glycoproteomics of milk: differences in sugar epitopes on human and bovine milk fat globule membranes. J. Prot. Res. 7, 3687–3696.

Xiong, Y.L., Kinsella, J.E., 1991. Influence of fat globule-membrane composition and fat type on the rheological properties of milk based composite gels. 2. Results. Milchwissenschaft—Milk Sci. Int. 46, 207—212.

Yamada, M., Murakami, K., et al., 2002. Identification of low-abundance proteins of bovine colostral and mature milk using two-dimensional electrophoresis followed by microsequencing and mass spectrometry. Electrophoresis 23, 1153—1160.

Ye, A., Singh, H., et al., 2002. Characterization of protein components of natural and heat-treated milk fat globule membranes. Int. Dairy J. 12, 393—402.

Ye, A., Singh, H., et al., 2004. Interactions of fat globule surface proteins during concentration of whole milk in a pilot-scale multiple-effect evaporator. J. Dairy Res. 71, 471—479.

Zamora, A., Ferragut, V., et al., 2012. Changes in the surface protein of the fat globules during ultra-high pressure homogenisation and conventional treatments of milk. Food Hydrocoll. 29, 135—143.

Zavaleta, N., Kvistgaard, A.S., et al., 2011. Efficacy of an MFGM-enriched complementary food in diarrhea, anemia, and micronutrient status in infants. J. Pediatr. Gastroenterol. Nutr. 53, 561—568.

Zeisel, S.H., Char, D., et al., 1986. Choline, phosphatidylcholine and sphingomyelin in human and bovine milk and infant formulas. J. Nutr. 116, 50—58.

Impact of food structures and matrices on nutrient uptake and bioavailability

Chapter

5

Exploring the Relationship between Fat Surface Area and Lipid Digestibility

Matt Golding

Institute of Food, Nutrition & Human Health and Riddet Institute, Massey University,
Palmerston North, New Zealand

CONTENTS

INTRODUCTION

Fat forms an essential component of the human diet; providing a source of energy and facilitating in the delivery of lipophilic micronutrients. It also provides a hedonic contribution to many food products, imparting a range of desirable sensory attributes. As with most nutrients, there is an appropriate level for daily intake as part of maintaining a healthy lifestyle. Accordingly, both under- and overconsumption of fats can lead to malnutrition, and contribute to associated maladies. In westernized diets, it is primarily overconsumption of fat that is considered most detrimental to health, with high fat intake being considered a contributing factor to obesity, cardiovascular disease, and development of conditions associated with metabolic syndrome.

Food Structures, Digestion and Health. http://dx.doi.org/10.1016/B978-0-12-404610-8.00005-0

145

Fat intake can be managed through a number of approaches. A straightforward and widely applied approach provided by the food industry has been to simply reduce the fat content in manufactured foods. Understanding of the relationship between food structure and sensory properties is now increasingly enabling production of good quality foods with reduced fat content (Norton *et al.*, 2006, 2007; Lundin and Golding, 2009). For some products, for example yoghurts, there is now little perceived difference in quality between low fat and regular formats. However, for other foods, where fat may play a greater hedonic role in terms of texture and flavor, it can still be challenging to appreciably reduce fat content and still deliver acceptable quality to the consumer.

An alternative approach to managing fat intake is to inhibit or eliminate fat uptake during digestion, in order to reduce calorie intake derived from lipids (Lundin *et al.*, 2008). A number of examples are already commercially available, and have been widely investigated in their ability to limit the caloric contribution provided by fat. A well-known case in point is the use of Olestra as a fat replacer (Bimal and Zhang, 2006), particularly in the production of potato chips and snack foods. Olestra is a trade name for sucrose polyester, whereby fatty acids have been esterified with sucrose molecules. When incorporated into food products as a replacement for conventional fats and oils (usually on a weight-for-weight basis), Olestra is able to impart all the sensory attributes associated with triglycerides. However, the size and structure of the sucrose polyester molecule renders it resistant to the mechanisms responsible for lipid digestion. Accordingly, it remains undigested during gastrointestinal transit and is eliminated. While effective at essentially eliminating the energy contribution normally provided by fat in such products, the use of Olestra also demonstrates some of the challenges encountered when attempting to bypass normal digestive function. Lipophilic micronutrients can be solubilized in sucrose polyesters; however, the non-digestible nature of the synthetic fat significantly reduces uptake of these oil-soluble bioactives. The indigestibility of Olestra has been shown to contribute to the condition of steatorrhea. There are also reports that hormonal secretions associated with lipid digestion (e.g., cholecystokinin and peptide YY) are not activated by the presence of Olestra, and consumption of sucrose polyesters may not generate the same satiating responses as triglyceride fats and oils (Jandacek, 2012).

Similar issues have been observed for the use of Orlistat, a pharmaceutical drug that acts to block the action of the pancreatic lipase responsible for conversion of triglycerides to fatty acids (Filippatos and Mikhailidis, 2009). As unhydrolyzed triglycerides are non-absorbable by the human body, they are eliminated. As with Olestra, the triglyceride indigestibility negates any energy contribution from the ingested lipid; however, similar physiological side effects are likewise observed when Orlistat is used as a means of inhibiting fat digestion.

These two approaches show that energy intake from lipids can be reduced with no compromise in the quality (or fat content) of the food eaten. Research is now increasingly directed towards the question of whether the composition and structure of fat itself—before, during, and after ingestion—can influence lipid digestibility and uptake. A challenge to this approach is that humans are highly proficient at digesting fat, with healthy adults usually able to achieve 95% efficiency (as part of normal dietary behavior) (Lentle *et al.*, 2011). Remarkably, this consistency appears to be achieved regardless of the considerable diversity in fat type, triglyceride composition, and lipid microstructural arrangement of lipids consumed as part of daily fat intake.

While the body is highly effective at normalizing fat digestion, there remains considerable interest as to whether the biological processes responsible for fat digestion can be manipulated or circumvented in order to alter the uptake of consumed lipids (Singh *et al.*, 2009; Golding and Wooster, 2010; van Aken, 2010). Potential nutritional benefits from such an approach not only include the ability to reduce the energy input provided by fat, but may also extend into other areas such as controlling the delivery and release properties of lipophilic bioactive components (McClements *et al.*, 2009). Although there are numerous ways by which the structure and stability of dietary fats may be altered to achieve specific behaviors during digestion, this chapter specifically explores the hypothesis that lipid digestive efficiency is related to relative surface area of fat. Furthermore, if the surface area of the fat can be deliberately minimized during gastrointestinal transit, fatty acid uptake will likewise be reduced, leading to changes in the way in which lipids are ultimately metabolized. Consequences of these effects may be modified behaviors related to lipid digestion, such as satiation, but also additional intriguing aspects such as alterations to the degree of oxidation and inflammation associated with the digestion of lipids.

THE RELEVANCE OF THE COLLOIDAL STATE TO LIPID DIGESTION

Compositionally, triglycerides represent the dominant lipid fraction present in all fats and oils regularly found in the human diet. As triglycerides cannot be directly taken into the bloodstream as part of digestion, hydrolysis to component fatty acids is first required (Carey *et al.*, 1983; Mu and Hoy, 2004).

Triglycerides are apolar materials, and are therefore immiscible with the aqueous conditions in the gut. Accordingly, both fat digestion and uptake of fatty acids are colloidal processes. Triglyceride hydrolysis requires adsorption of lipase enzymes at the oil–water interface, with lipolysis occurring in both the stomach and the small intestine (Mukherjee, 2003). Gastric lipolysis is achieved through adsorption of human gastric lipase (HGL), an acid stable, bile salt-independent lipase. This hydrophobic lipase acts at the Sn-1 or Sn-3 position of fats and oils, such that one molecule of triglyceride will produce one molecule each of diglyceride and fatty acid. Gastric lipase has an optimal pH of ≈ 5.4 (Carriere *et al.*, 1993) and can account for 10–30% of total lipid hydrolysis, notably being effective at liberating short to medium chain fatty acids, which are more typically arranged at the Sn-3 position on triglyceride molecules (Hamosh, 1990). The remainder of lipid hydrolysis takes place in the small intestine by co-lipase-dependent human pancreatic lipase (HPL). Human pancreatic lipase has an optimal pH of 6.5, but is able to effectively hydrolyze triglycerides from pH 4.5 to 7.5 to produce two free fatty acids and one Sn-2 monoglyceride (Carriere *et al.*, 1993; Lowe, 1997; Armand, 2007). While HPL is able to adsorb to the oil-in-water interface to establish hydrolysis, adsorption can be inhibited by the presence of surface-active components already present at surface of oil droplets. This has been shown in the case of emulsion droplets stabilized by the polar lipid Tween 80, which acts as a barrier to HPL adsorption (in the absence of bile salts and co-lipase) (Gargouri *et al.*, 1983). However, in spite of the fact that the interfacial domain of ingested lipids may be highly varied in terms of surface composition, structure, and rheology, the human body has developed mechanisms to ensure that enzymatic adsorption is not adversely compromised, irrespective of the nature of the interfacial layer separating the oil and aqueous phases.

During gastric lipolysis, the resulting surface-active free fatty acids can serve to displace any pre-existing interfacial layer, providing a more favorable binding domain during lipid hydrolysis by HPL (Pafumi *et al.*, 2002). The small intestinal stage of lipolysis is likewise facilitated by the presence of bile salts (Lowe, 1997, 2002). Bile salts, which have pronounced amphiphilic characteristics, provide a significant physiological role in facilitating uptake of fatty acids into the bloodstream through the formation of mixed micelles. The amphiphilic structure of these molecules imparts a high propensity towards their adsorption at the oil–water interface. Where the interface is already occupied by an exogenous amphiphile, bile salt adsorption may take place either through a mechanism of orogenic displacement (where bile salt surface pressure is greater than that of the surface layer being displaced) (Maldonado-Valderrama *et al.*, 2008, 2011), or by co-adsorption (which can occur when the existing layer imparts a higher surface pressure relative to the bile salts) (Wickham *et al.*, 1998; Vinarov *et al.*, 2012).

In cases where bile salt is unable to adsorb at the interface, lipolysis can be inhibited (Vinarov *et al.*, 2012).

The adsorption of bile salts to the oil–water interface is important for two reasons. First, the presence of bile salts at the interface enables attachment of the co-lipase–HPL complex which provides access of HPL to the oil–water interface, ensuring that lipid hydrolysis can take place (Lowe, 2002). Second, free fatty acids formed during lipolysis are incorporated into so-called mixed micelles, which are primarily stabilized by the surface-active bile salts (as well as additional endogenous surfactants). The greater surface hydrophilicity of these entities allows their detachment from the lipid system, and subsequent diffusion and transport across the epithelium, prior to uptake into the bloodstream (Maldonado-Valderrama *et al.*, 2011). This complex biochemical process is further mediated by the dynamic pH conditions in the stomach and small intestine, the biomechanical shear forces generated in the mouth and gut, and a temperature maintained at $\approx 37°C$.

To reiterate, the delivery and development of the colloidal state during ingestion and transit through the gastrointestinal tract is essential to ensuring effective lipid digestion, providing binding sites for triglyceride lipolysis in the stomach and small intestine, and enabling the assembly of mixed micelles that enable uptake of the resultant fatty acids. From a colloidal perspective, a number of compositional and structural parameters can be considered of particular consequence in determining digestive efficiency (Golding *et al.*, 2010; van Aken, 2010). These include the relative size of the emulsion droplets, the surface composition at the oil–water interface, the type and extent of interactions present before or during digestion, and the triglyceride composition of the droplets (which includes secondary parameters such as solid fat content and positional specificity of fatty acids). In many respects, the dynamic evolution of the emulsified state during digestion is dependent on the interrelationship between all these aspects and their response to the biochemical and biomechanical conditions in the gut. Specifically, changes to surface composition and surface area of emulsified droplets can be expected to be highly significant in determining both the accessibility and availability of lipase binding to the interface.

THE RELATIONSHIP BETWEEN SURFACE AREA AND DIGESTIBILITY OF FATS AND OILS

The ability to control surface area of the dispersed lipid phase is considered to be a means by which the extent of lipase binding to the interface can be manipulated. A simple premise is that fine emulsion structures, with small droplets and correspondingly high surface areas, present more binding sites for lipase adsorption in the GI tract than coarse emulsions with low surface area, or in extreme cases, unhomogenized layers of free oil. Accordingly, low surface area emulsions will be hydrolyzed more slowly than high surface area systems.

An intriguing aspect of this hypothesis is that, as part of modern diets, dietary fat can be consumed in a diverse array of structural states, representing a broad cross-section of surface areas. For example, in homogenized foods, such as milk, ice-cream, or yoghurt, emulsion droplets are typically less than 1 μm. For other food systems, such as cheese and whipped cream, emulsion structuring may lead to significant increases in relative fat particle size (>10 μm). Furthermore, for many foods, the lipid component may initially be in a non-emulsified state (e.g., for fried foods). For such foods, the shear forces during oral processing have been shown to be sufficient to crudely homogenize any non-emulsified fat or oil, such that fats are delivered in a coarse emulsion state into the stomach. Studies on the effects of oral processing of corn and castor oils have shown that even oral rinsing is sufficient to homogenize the oil phase to produce large but metastable droplets, with droplets presumably being stabilized by amphiphilic salivary proteins (Adams et al., 2007).

Such variations in relative emulsion droplet size represent considerable differences in surface area, spanning several orders of magnitude. So a question arises as to whether the variability of surface area of ingested fat has any influence on its subsequent digestion. Given the high efficiency of fat digestion in the normal human diet, one would argue that relative surface area does not appear to affect lipid digestibility during the normal course of digestion. To explore this relationship, a number of studies have investigated and reported aspects of the lipolysis behavior of predominantly model emulsion systems (Armand et al., 1997).

Surface Area Effects Under Intestinal Conditions

Using relatively simple static *in vitro* models, it can be shown that under such conditions emulsion surface area does appear to have an effect on lipolysis behavior (Helbig et al., 2012). Figure 5.1 shows variation in the liberation of free fatty acids for emulsions of matching interfacial composition, but variable particle size, ranging from 0.16 up to 200 μm. Emulsions were exposed to simulated intestinal fluid comprising pancreatic lipase, co-lipase, bile salts, and calcium (at 37°C and pH 6.8) for 60 minutes, with changes in fatty acid concentration being determined through the use

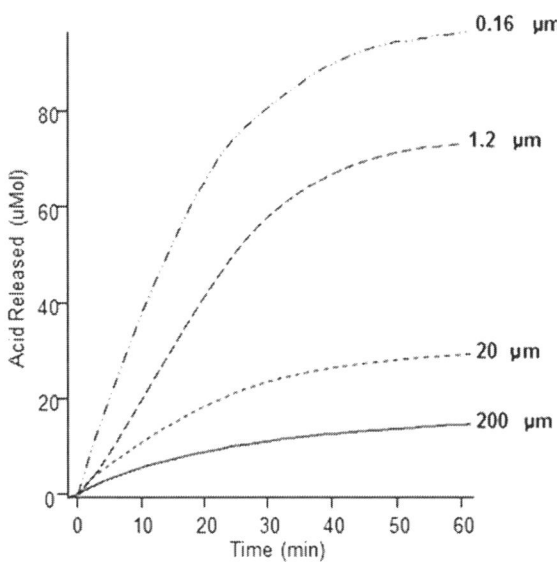

■ **FIGURE 5.1** Measurement of titratable acidity providing an indicator of fatty acid generation during *in vitro* pancreatic lipolysis of 15% peanut oil-in-water emulsions of varying particle size distribution. The pancreatic model comprised a lipase/co-lipase mixture, 12 mM mixed bile and 10 mM Ca^{2+}. *Wooster, unpublished data.*

of a pH-stat. The results in Figure 5.1 show that the rate and magnitude of lipolysis is maximized for the emulsion system with mean droplet size of 0.16 µm (and thus with the highest surface area). With increasing particle size, the rate and extent of acid release decreases, correlating with a reduction in emulsion surface area. Under these experimental conditions, it is entirely reasonable to surmise that the reduced availability of lipase binding sites, caused by decreasing surface area, is responsible for the lower extent of lipolysis taking place, since any other factors that might influence lipolysis, such as interfacial or lipid composition, are not varied within the emulsion system.

Digestion of emulsions of variable particle size, specifically under intestinal conditions, has been studied *in vivo* (Maljaars *et al.*, 2008). One recent human study (Seimon *et al.*, 2009), involving 10 participants, investigated model emulsion systems intranasally intubated into the small intestine. A manometric APD catheter was used to infuse emulsions of particle size 0.26, 30, and 170 µm into the duodenum, with saline solution as control. The catheter was also able to measure motility response, including antral and pyloric pressure waves, allowing the effects of surface area on intestinal biomechanical response to be studied. Digestive biomarkers, such as the secretion of peptide YY (PYY) and cholecystokinin (CCK) (hormones associated with lipid digestion, regulation of GI transit, and control of appetite), and uptake of plasma triglyceride were measured using blood sampling.

Findings from the study indicated that increasing droplet size was associated with a number of changes to gastrointestinal motility, including a reduced suppression of both antral and duodenal pressure waves, and demonstrating a stimulation of isolated and basal pyloric pressures (Table 5.1).

It has been established that the presence of fat in the small intestine serves to regulate intestinal motility, causing the rate of transit of digesta to be slowed down to ensure nutrient digestion (Gregory *et al.*, 1986; Pilichiewicz *et al.*, 2007). Here, delivering emulsions of droplets with higher surface area directly into the small intestine appeared to generate a more pronounced response, such that motility was further decreased when compared to the effect of droplets of lower surface area. The enhanced digestibility of high surface area fat was also apparent when considering associated biomarkers (Figure 5.2).

Relative to the saline control, fat infusion resulted in elevated response for both PYY and CCK. Hormonal response was seen to be markedly higher for droplets of 0.26 μm. Interestingly, while CCK and PYY profiles for droplets of 30 and 170 μm were greater than the control, there was no significant difference in response between these two droplet sizes, in spite of their different surface areas. Similar behavior was also observed for plasma triglyceride uptake. Plasma triglyceride concentration showed higher values over the 2 hours' assessment period for all emulsions relative to the saline control. Again, the response was greatest for the 0.26 μm emulsion, which was significantly different from that of the other emulsion samples. In comparison, plasma triglyceride concentrations for emulsions with droplet size

Table 5.1 Number and Amplitude of Antral and Duodenal Pressure Waves During 120-min Duodenal Infusions of Saline (Control) or Lipid Emulsions Varying Particle Size: 0.26, 30, and 170 μm

	Total No.		Mean Amplitude (mm Hg)	
	Antral PWs	Duodenal PWs	Antral PWs	Duodenal PWs
Control	205 ± 38	1134 ± 184	61 ± 9	26 ± 1
LE-0.26	21 ± 6	328 ± 39	22 ± 3	23 ± 3
LE-30	33 ± 15	540 ± 64	20 ± 3	23 ± 2
LE-170	116 ± 36	744 ± 97	3 ± 9	25 ± 1

(Taken with Permission from Seimon et al., 2009)

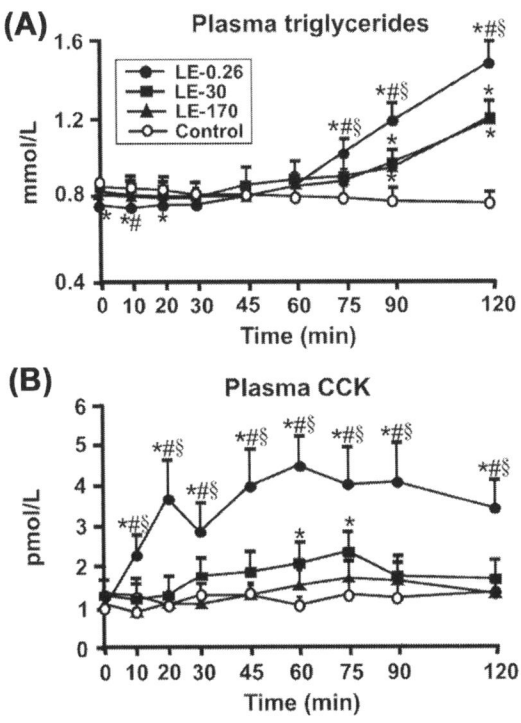

■ **FIGURE 5.2** Mean plasma concentrations of triglycerides (A) and cholecystokinin (CCK) (B) during 120 minutes of duodenal infusion of saline (control) or lipid emulsions of varying particle size. *Taken with permission from Seimon et al. (2009).*

30 and 170 μm were higher than those of the control, but showed no significant differences when compared to each other.

A study by Maljaars and coworkers (Maljaars *et al.*, 2012) compared duodenal and ileal digestive properties of emulsions of droplet size 0.65 and 5.40 μm (surface weighted mean diameter) that were infused into the small intestine as part of a 15 participant human study. Infusion of the 6 g fat emulsion took place after consumption of a fat-free milkshake preload, at $t = 30$ minutes for commencement of duodenal infusion and $t = 105$ minutes for ileal infusion. Findings showed that under the particular conditions of the study, when fat was infused to the duodenum, there was no significant observable difference in CCK or PYY hormonal response between the two emulsion systems (it was speculated that this was due to initial elevation of CCK and PYY by consumption of the preload). A comparison of gastric emptying time did, however, show that the finer emulsion resulted in a statistically longer gastric emptying half-time relative to that of the coarse emulsion, suggesting that the digestion of the emulsion with the greater surface area resulted in a reduction in the rate of meal transit.

Intriguingly, when the two emulsions were infused to the ileum, CCK and PYY hormonal responses were seen to be markedly different, with the fine emulsion with higher surface area promoting a higher hormonal response than the coarse emulsion. Paradoxically, in spite of these differences in hormonal response, no statistically significant differences in gastric emptying time were observed (here it was speculated that the stomach contents would have already been extensively emptied by the time ileal infusion commenced).

A summary of these findings indicates that direct intestinal intubation of very fine emulsions generates a differential response in terms of lipid digestion behavior when compared to coarse emulsions, most likely as a consequence of greater available surface. However, the relationship does appear dependent on comparative droplet size and specific location of the emulsion within the small intestine.

Arguably, these behaviors present something of an artificial scenario, given that daily consumption of fat does not usually involve intubation into the intestine. It does, however, provide an indication that fat surface area may have an influence on lipid digestive function in the event that such structures can be maintained to the point of entry to the small intestine. However, a particular challenge in extending this hypothesis further is that the diverse conditions in the mouth and stomach can dynamically influence droplet size during gastrointestinal transit. The initial particle size of an ingested emulsion may change considerably on mastication, exposure to the stomach, and on entering the small intestine. To explicitly determine that surface area influences digestion (particularly in relation to the efficiency of pancreatic lipolysis) is ultimately necessary to integrate changes to fat droplet size and distribution across the whole digestive cycle.

Dynamics of Surface Area Effects During Digestion

In a seminal and highly cited paper, Armand and coworkers (Armand *et al.*, 1999) investigated the effects of emulsion size on lipid digestion and absorption under *in vivo* conditions. For this human study, which involved eight participants, emulsions of particle size 0.7 and 10.1 μm (formulated identically) were intragastrically intubated. Intubation was carried out over 10 minutes, delivering 500 ml emulsion into the stomach, providing 48 g fat. At various time points, samples were aspirated from the stomach and duodenum in order to determine changes in particle size distribution during digestion. Additional measurements were taken to determine rate of gastric emptying, lipase activity, hormonal response, triglyceride levels in stomach, and intestinal contents, as well as serum triglycerides.

Findings showed a clear relationship between gastric lipid hydrolysis and initial droplet size, with the fine emulsion showing a markedly greater rate of triglyceride conversion relative to the coarse emulsion, consistent with the concept that higher surface area provides more domains for lipase adsorption (Figure 5.3). Interestingly, the fine emulsion was also seen to show a pronounced (four- to nine-fold) increase in mean particle diameter after infusion (Figure 5.3). In comparison, exposure to gastric conditions did not appear to affect the particle size distribution of the coarse emulsion. A number of explanations were presented, but the most likely was the adsorption of fatty acids at the oil−water interface as a consequence of gastric triglyceride lipolysis. While fatty acids are surface active, and capable of causing orogenic displacement of the original droplet interface, they are not particularly effective at imparting good surface mechanical stability. As a consequence, the presence of fatty acids at the interface may increase the propensity of emulsion droplets towards fusion and coalescence (Pafumi *et al.*, 2002). This effect is also dependent on droplet−droplet contact taking place. Consequently, it might be expected that the fine emulsion, with a greater number density of droplets and also more affected by Brownian motion, would be more susceptible to droplet collision and coalescence than the coarse emulsion. The propensity of fine emulsions to increase in particle size under gastric conditions has now been observed for a number of additional *in vitro* and *in vivo* studies.

A consequence of the increase in fine emulsion particle size distribution during gastric digestion was a reduction in the difference in surface area

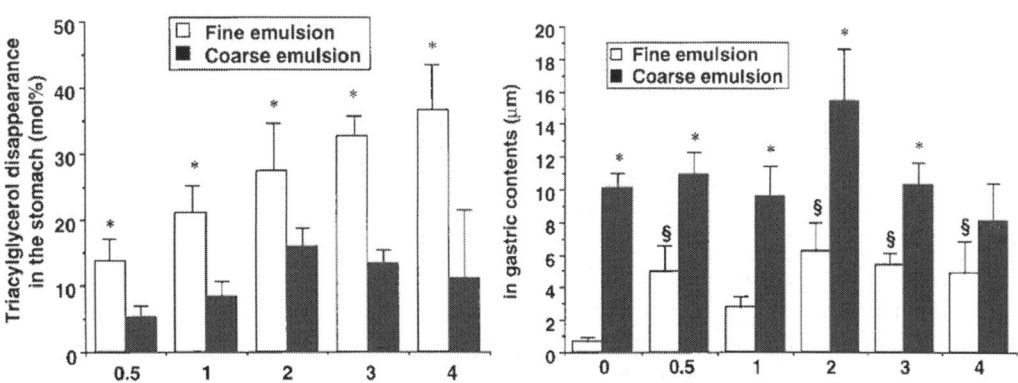

■ **FIGURE 5.3** Rate of triglyceride disappearance for a fine emulsion (0.7 μm) and a coarse emulsion (left-hand graph) and change in droplet size for fine and coarse emulsions (right-hand graph) within gastric contents during a 4-hour digestion period. *Taken with permission from Armand* et al. *(1999).*

between the fine and coarse emulsions on entry into the small intestine. Triglyceride consumption of both emulsions during intestinal lipolysis was observed to be higher than that of gastric lipolysis. However, comparison between the two emulsion types showed that while the fine emulsion still showed a greater reduction in triglyceride level compared to the coarse emulsion during intestinal lipolysis, the effect was less pronounced than the changes under gastric conditions. This is mostly likely due to the reduction in surface area difference between the two emulsions after the gastric stage of digestion.

In terms of whether these differences significantly alter the assimilation of fatty acids into the bloodstream, it is necessary to compare the plasma triglyceride profiles of the two emulsion systems. Interestingly, both emulsions showed remarkably similar bell-shaped profiles for plasma triglyceride, and although the fine emulsion exhibited peak response approximately 1 hour after that of the coarse emulsion, there was almost no difference in plasma triglyceride as measured by the area under the curve (Figure 5.4). The delayed response of the fine emulsion may be attributed to differences in the rate of gastric emptying: gastric partitioning of the coarse emulsion into oil-depleted and oil-rich domains would be expected to result in faster emptying of the oil-depleted region, leading to an earlier onset in plasma TAG response. However, the implication from the data is that while

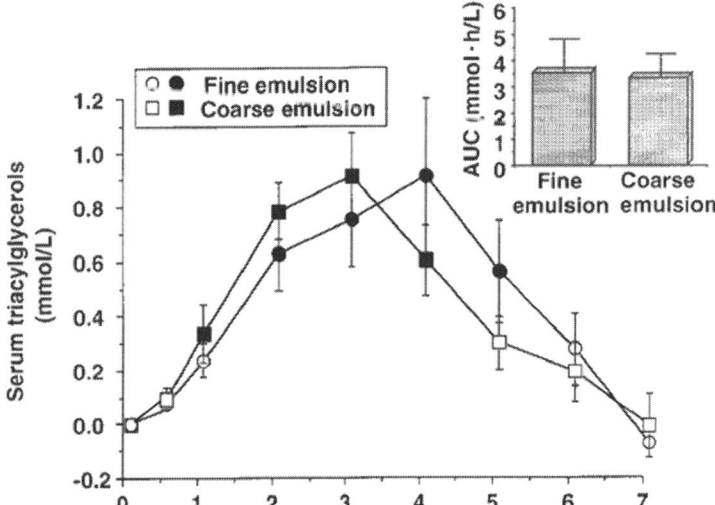

■ **FIGURE 5.4** Mean change in serum triacylglycerols during a 7-hour digestion period for intubated fine (0.7 μm) and coarse (10 μm) emulsions. Filled symbols indicate a significant difference from the fasting value. *Taken with permission from Armand* et al. *(1999).*

the time taken for lipid digestion may be affected by differences in particle size distribution, overall digestive efficiency is unaffected by particle size.

This observation was reinforced in a follow-up study which investigated the effects of emulsion droplet size in availability and uptake of oil-soluble vitamins A and E during digestion (Borel *et al.*, 2001). This human study, involving eight participants, retained the two earlier emulsion compositions with initial mean droplet size of 0.7 and 10.1 μm. The fat phase of the emulsion additionally contained 28 mg vitamin A and 440 mg vitamin E. As with the study carried out by Armand *et al.*, the fine emulsion displayed a significant increase in particle size during gastric digestion, while the coarse emulsion was more or less unaffected. Consequently, there was less difference in droplet size distribution between the two emulsions during intestinal digestion. Plasma profiles for lipophilic nutrients followed the same trends as for triglycerides, showing a 1 hour delay in peak value for the fine emulsion compared to the coarse emulsion, but showing no difference for area under the curve values between the two emulsions, indicating that relative particle size has no apparent impact on the overall absorption of the two lipophilic vitamins.

Approaches for Controlling Surface Area During Digestion

The observations from these and comparable studies indicate a degree of normalization of emulsion droplet size during digestion, which may account for the consistent plasma triglyceride profiles that are observed for emulsion digestion studies. A question therefore arises as to whether the structural state of the emulsion system, either before or during digestion, is able to have any role on overall lipid digestion, absorption, and uptake. To investigate this particular hypothesis, a recent study sought to examine how the conditions in the stomach could be employed to induce various emulsion structural states during the gastric stage of digestion, and as to whether the development of these diverse structures had any influence on overall lipid digestibility (Golding *et al.*, 2011). An intriguing aspect of the study was that each of the emulsions, irrespective of formulation, would be in a stable, unstructured state compared to before exposure to gastric conditions. Furthermore, all emulsions before being consumed would be of comparable particle sizes (droplets <1 μm, Figure 5.5A/B(i)) and would be isoviscous. Accordingly, structural and surface changes would be effected in the stomach, rather than being delivered in the food preload. While the primary purpose was to achieve specific gastric structuring pathways through emulsion design, a second motivation was to eliminate textural differences for the different emulsion systems during consumption, which might unduly influence aspects of the human study, such as satiety.

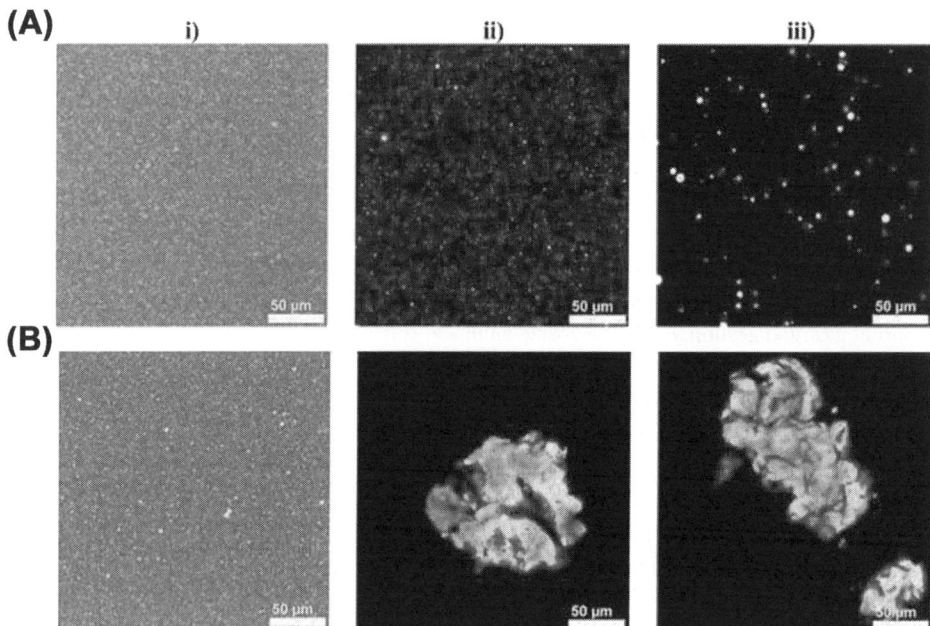

■ **FIGURE 5.5** Confocal images showing emulsion microstructural changes at various stages of *in vitro* digestion: (i) immediately after preparation; (ii) 30 minutes after incubation in simulated gastric fluid; (iii) 30 minutes after incubation in simulated intestinal fluid. (A) shows 20% oil-in-water emulsion stabilized by Tween 80 (liquid droplets); (B) shows 20% oil-in-water emulsion stabilized by sodium stearoyl lactylate (crystalline droplets). *Taken with permission from Golding* et al. *(2011). [A color version of this figure is available online at www.booksite.elsevier.com/9780124046108].*

An *in vitro* model comprising a gastric step and an intestinal step was initially used to demonstrate that, by controlling interfacial and lipid composition, various structural states with varying surface area could be generated in response to gastric pH, thermal conditions, biochemical and biomechanical environment. Structural states were imaged using confocal microscopy at various time points during the gastric incubation step. Emulsions stabilized by phospholipid or polysorbate emulsifiers were shown to be most structurally stable during gastric treatment, demonstrating a small increase in particle size (Figure 5.5A(ii)). Emulsions stabilized with milk protein (whey protein isolate) tended to undergo extensive flocculation, resulting in the formation of droplet networks under gastric conditions. This behavior was attributed to a loss of interfacial charge stabilization caused initially by transition through the isoelectric point of the protein on exposure to the acid conditions in the gastric model, coupled with detachment of hydrophilic protein domains as a consequence of interfacial proteolysis (Sarkar *et al.*, 2009).

Emulsions stabilized with the anionic emulsifier sodium stearoyl lactylate (SSL) were shown to undergo extensive coalescence during exposure to the gastric environment. Loss of stability was likewise attributed to negation of charge repulsion at low pH (with zeta potential approaching zero at pH 3). Under such conditions, the poor mechanical stability imparted by the SSL at the interface resulted in a markedly increased propensity towards coalescence. Two structural states were observed, depending on the oil composition. When the droplets were composed of liquid oil at 37°C, droplet coalescence resulted in the formation of larger droplets of $\approx 40-50 \, \mu m$. However, when the solid fat content (SFC) of the droplets was adjusted to 25% at 37°C by the blending of hydrogenated canola oil, the droplets effectively became partially crystalline at in-body temperatures. Consequently, solidification of the droplets resulted in the formation of large, partially coalesced agglomerates of irregular structure (Figure 5.5B(ii)). Both fully and partially coalescing emulsions caused a significant decrease in surface area on exposure to gastric conditions.

Partial coalescence phenomena were also observed for an emulsion system with SFC of 25% (at 37°C) with a mixed interface of sodium caseinate and saturated monoglyceride (Keogh *et al.*, 2011). Here, partial coalescence was seen to be more extensive than that of the emulsion stabilized by SSL, such that large particulates of fat were visibly observed in the *in vitro* cell during gastric incubation. The high extent of partial coalescence was believed to be a consequence of initial flocculation, arising from loss of protein charge and steric repulsion. Flocculated droplets in close contact were then highly susceptible to coalescence, due to further compromising of interfacial stability arising from competitive partial displacement of the protein by the monoglyceride. Removal of the hydrogenated fat from the emulsion, such that droplets were fully molten at 37°C, resulted in the emulsion breaking entirely, with a layer of free oil observed on the surface on the *in vitro* cell. This destabilization mechanism represented the most significant decrease in fat surface area on exposure to gastric conditions.

After gastric treatment, emulsion samples were exposed to simulated intestinal fluid. Lipolysis was measured during the incubation period using a pH-stat methodology to determine fatty acid release. In addition, confocal microscope observations were made to investigate the development of emulsion structure during intestinal digestion.

The *in vitro* intestinal model showed that the change in emulsion structure during exposure to gastric conditions had a significant impact on subsequent intestinal lipolysis. In the case of the polysorbate and phospholipid emulsions, there was an initial rapid release of fatty acids in the first 10 minutes of incubation, followed by a plateauing of fatty acid levels (Figure 5.6). These findings reflect the fact that these two emulsions were delivered to

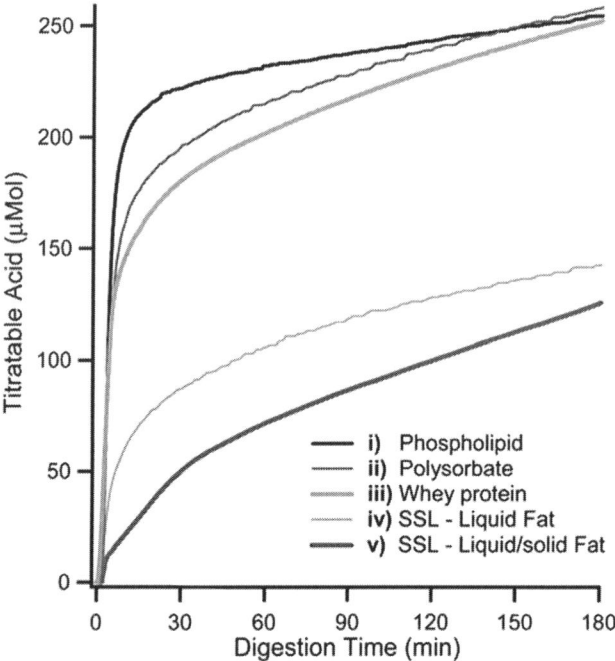

■ **FIGURE 5.6** Titratable acid release profiles of emulsions with different interfacial and lipid compositions during simulated intestinal digestion (following simulated gastric digestion). *Taken with permission from Golding et al. (2011).*

the intestinal model with high surface area retained after the gastric digestion step, and accordingly displayed very fast rates of lipolysis. Interestingly, the whey protein-stabilized emulsion displayed a similar lipolysis profile to the two gastric-stable emulsions (Figure 5.6). This was attributed to the fact that while the emulsion underwent flocculation under gastric conditions, this did not result in a change to the surface area of the emulsion, since there was no actual increase in particle size of the individual droplets associated with the floc structures.

For both emulsions stabilized by SSL, where coalescence and partial coalescence under gastric conditions did result in decreased surface area, the rate and extent of intestinal lipolysis was seen to be markedly reduced in comparison with the high surface area emulsion systems (Figure 5.6). This difference indicated that lipolysis efficiency was being compromised as the availability of sites for lipase binding was decreased. For the SSL-stabilized emulsions, the lipid composition provided further variance in lipolysis, with the partially crystalline emulsion showing a slower and less extensive degree of fatty acid generation, compared with the emulsion containing liquid oil. Although not presented in the Golding *et al.* study, lipolysis was shown to be likewise retarded for the mixed caseinate—monoglyceride emulsion, in which the surface area was also greatly reduced. (These findings were progressed and presented in the following Keogh *et al.* human study (Keogh *et al.*, 2011).)

The *in vitro* model showed that during the early stages of simulated intestinal lipolysis, surface area effects generated during gastric treatment had a significant impact on the relative lipolytic efficiency. However, considerable structural diversity was also observed for the various emulsion systems during exposure to simulated intestinal fluid. While the polysorbate emulsion showed no significant changes in particle size after 30 minutes of incubation (Figure 5.5A(iii)), the phospholipid and whey proteins were seen to have undergone significant coalescence. For the SSL emulsions, the droplet size and structure appeared less affected, with both emulsions still appearing highly coalesced. In the case of the SSL partially crystalline emulsion, irregularly shaped partially coalesced agglomerates were retained during intestinal incubation (Figure 5.5B(iii)).

To demonstrate whether these behaviors had consequences for human lipid digestion, a number of samples were progressed to a clinical study (Keogh *et al.*, 2011). The 20 participant study involved the consumption of a 350 ml milkshake preload, delivering 30 g fat. The control emulsion was based on the phospholipid emulsion composition that was shown under *in vitro* conditions to be gastric stable, thereby retaining a high surface area with a fast lipolysis rate as observed in the Golding study. The additional emulsions in the study, which were intended to display decreasing surface area during gastric digestion, were also based on observations from the Golding study. (Both SSL-stabilized emulsions and the partially crystalline mixed caseinate—monoglyceride interface emulsion were used.) Measurements included hormonal response (CCK and PYY), plasma triglyceride, carbon 13 breath test, and plasma paracetamol (a surrogate for the gastric emptying rate).

Findings from the study showed the caseinate—monoglyceride emulsion, which was seen to undergo partial coalescence in the *in vitro* gastric model, had markedly different digestion behaviors in comparison with the control emulsion. In the case of the control, plasma triglyceride measurements showed a steady increase in the 30—180 minute period after ingestion. This behavior is consistent with previous studies using similar compositions. In the case of the partially coalescing caseinate—monoglyceride emulsion, plasma triglyceride levels were significantly suppressed relative to the control emulsion, showing no observable elevation in levels during the timescale of

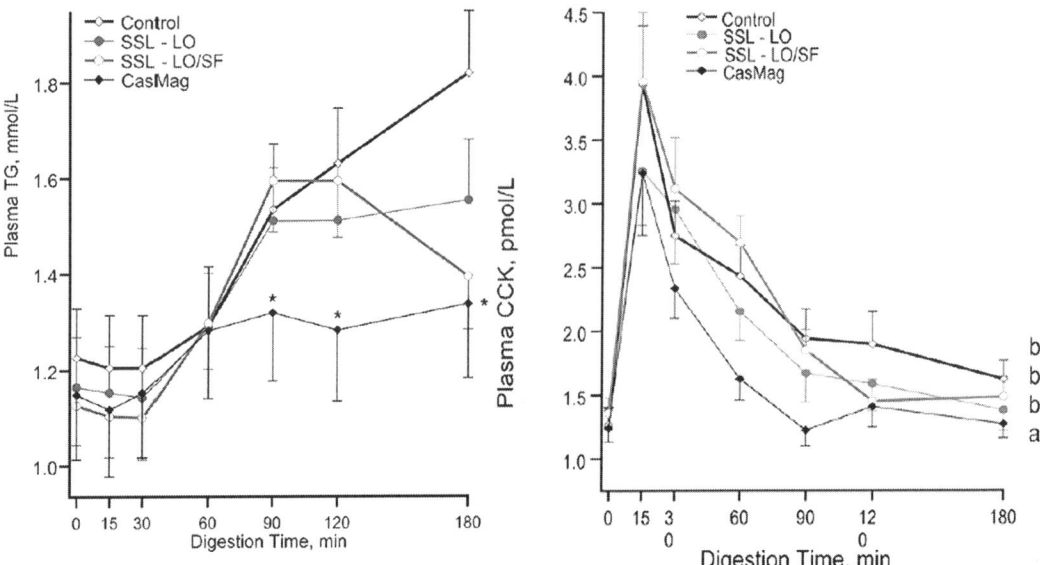

■ **FIGURE 5.7** Plasma triglyceride and plasma CCK concentration in men and women (*n* = 20) before and after consumption of emulsion meals delivering 30 g fat. *Taken with permission from Keogh et al. (2011).*

the analysis (Figure 5.7). A second experiment, carried out over a 6-hour digestion period, comparing the partially coalescing emulsion to a tween stabilized emulsion (also shown to be gastric stable), showed similar behavior for both emulsions, with no significant change in whole blood triglyceride levels observed for the agglomerated emulsion over the 6 hours of the study.

The differences were also represented in other aspects of the study. Carbon 13 breath testing showed a significantly lower cumulative dose response for the partially coalesced emulsion in comparison compared to gastric-stable control. Likewise, plasma CCK (Figure 5.7), PYY, and GLP-1 measurements all showed a significantly different hormonal response compared to gastric-stable controls, with notably lower levels being detected. These findings support the hypothesis that partial coalescence during the gastric stage of digestion causes a pronounced reduction in surface area, greatly reducing lipolytic efficiency. This effect is greatly compounded by the assembly of crystalline droplets into solid agglomerates. These are mechanically resistant to the shear forces under in-body conditions, and can therefore be only slowly digested by incremental lipolysis at the surface of agglomerates. The reduction in lipolysis of the agglomerates is further reflected by reduced hormonal response arising from the lower levels of fatty acids liberated and detected during the digestion process.

The human study also provided indirect evidence that emulsion agglomeration was taking place. Plasma paracetamol experiments (Figure 5.8) showed that for the first 75 minutes of digestion, the caseinate–monoglyceride emulsion displayed a statistically faster rate of gastric emptying relative to the control emulsion. This behavior is attributed to partitioning of the less dense fat phase in the stomach, such that large lipid agglomerates are able to rapidly form a phase separated lipid layer on the surface of the stomach contents (as observed in the *in vitro* model). Under such conditions, the lipid-depleted lower layer has a tendency to empty more quickly, with the rate of emptying being regulated by a hormonal response arising from the detection of fat entering the small intestine. In comparison, for the gastric stable emulsion, uniform distribution of droplets throughout the stomach contents results in higher fat levels initially being released into the small intestine, leading to a reduced rate of emptying. This behavior has been consistently observed in previous studies, most notably where the use of magnetic resonance imaging has highlighted that partitioned or phase-separating emulsions emptied more rapidly in the early stages of emptying than stable emulsions with homogeneous fat distributions (Schwizer *et al.*, 2006; Marciani *et al.*, 2007, 2009).

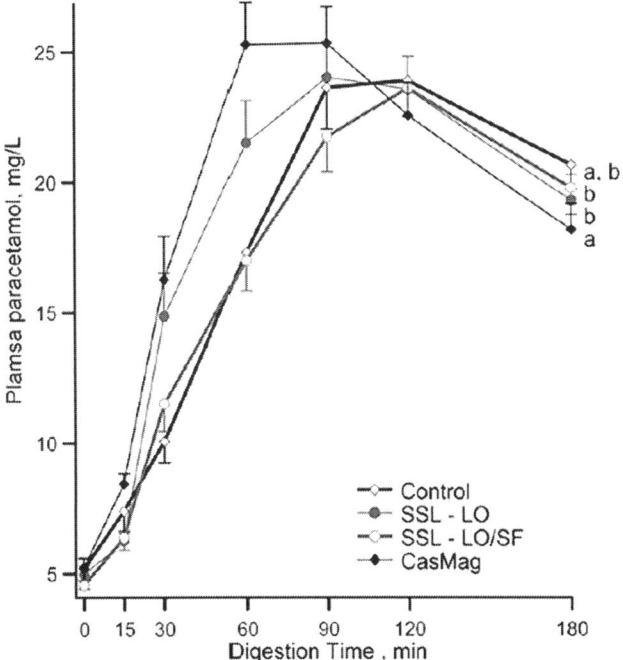

■ **FIGURE 5.8** Plasma paracetamol concentrations in men and women ($n = 20$) before and after consumption of emulsion meals delivering 30 g fat. *Taken with permission from Keogh* et al. *(2011).*

One unanswered aspect of the Keogh study was the ultimate fate of the lipid in the partially coalescing emulsion system. The fact that there was little change in plasma triglyceride levels for this system over the digestion period may imply that not all of the lipid fraction is able to be hydrolyzed during transit through the gastrointestinal tract. Given that non-hydrolyzed triglycerides are non-digestible, there is a possibility that the agglomeration of the emulsion into solid particulates ultimately results in a portion of the lipid content being defecated along with other non-digestible matter. If this is so, then there is an implication that this form of in-body emulsion structuring may enable reduction in fat uptake without necessarily reducing the fat content. Intriguingly, no physiological side effects were observed as a consequence of the emulsion preload consumption. It is, however, recognized that chronic exposure of the colon to lipids may lead to adverse health effects, and therefore it would be of relevance to determine whether lipid from structured emulsions was in fact transiting into the colon.

Evidence from other human trials does seem to indicate that consumption of (non-emulsified) high melting point fats does not appear to lead to significant malabsorption (Denke and Grundy, 1991; Dougherty *et al.*, 1995). In one particular study (Berry *et al.*, 2007), Berry and coworkers showed that fecal fat levels remained within expected ranges after consumption of 50 g blended shea butter (delivered as a muffin preload, and with a solid fat content of 22% at 37°C). However, this study is not able provide any indication as to the structural state of the fat after consumption.

CONCLUSIONS

The human body is able to digest fats and oils with remarkable efficiency, regardless of the fact that their material state in food systems can be highly diverse. For normal dietary lipid intake, aspects such as emulsified state, structure, and relative surface area appear to have little impact on the overall digestibility of lipids when integrated over the entire GI tract, although differences in aspects such as gastric emptying may be observed. The apparent normalization of digestion of fatty foods may be in part due to the fact that almost all dietary lipids are fully molten at in-body temperatures, and that the biochemical and surfactant conditions in the stomach and small intestine are biologically evolved to be effective across a broad range of colloidal states and structures.

However, when the colloidal state is altered beyond the normal parameters required for effective digestion, the digestive pathway can be manipulated. This has been demonstrated for model emulsions in which a combination of elevated droplet solid fat content and propensity for flocculation at low pH causes emulsion partial coalescence. The subsequent decrease in surface area, coupled with the solid nature of the agglomerates has been shown to greatly reduce lipolysis and rate of uptake. The ability to manipulate fat structures and their digestion has implications for the management of fat uptake and may provide new pathways for achieving more effective regulation of energy intake, or improved bio-delivery of lipophilic nutrients. This remains a challenge, in that while novel digestive pathways are now being established, the physiological benefits associated by these pathways are not necessarily as well defined. A final consideration is that even where new structuring pathways can provide improvements to dietary-related health and well-being, such structures have to be constructed into food systems that are able to deliver appropriate in-body functionality as well as meeting consumer expectations in terms of quality. Given the complexity and diversity of food materials this is certainly far from trivial.

REFERENCES

Adams, S., Singleton, S., Juskaitis, R., Wilson, T., 2007. In-vivo visualisation of mouth-material interactions by video rate endoscopy. Food Hydrocoll. 21, 986–995.

Armand, M., 2007. Lipases and lipolysis in the human digestive tract: where do we stand? Curr. Opin. Clin. Nutr. Metab. Care 10, 156–164.

Armand, M., Pasquier, B., Andre, M., Borel, P., Senft, M., Peyrot, J., et al., 1999. Digestion and absorption of 2 fat emulsions with different droplet sizes in the human digestive tract. Am. J. Clin. Nutr. 70, 1096–1106.

Armand, M., Pasquier, B., Borel, P., Andre, M., Senft, M., Peyrot, J., et al., 1997. Emulsion and absorption of lipids: the importance of physicochemical properties. Ocl-Oleagineux Corps Gras Lipides 4, 178–185.

Berry, S.E.E., Miller, G.J., Sanders, T.A.B., 2007. The solid fat content of stearic acid-rich fats determines their postprandial effects. Am. J. Clin. Nutr. 85, 1486–1494.

Bimal, C., Zhang, G.N., 2006. Olestra: a solution to food fat? Food Rev. Int. 22, 245–258.

Borel, P., Pasquier, B., Armand, M., Tyssandier, V., Grolier, P., Alexandre-Gouabau, M.C., et al., 2001. Processing of vitamin A and E in the human gastrointestinal tract. Am. J. Physiol. Gastr. Liver Physiol. 280, G95–G103.

Carey, M.C., Small, D.M., Bliss, C.M., 1983. Lipid digestion and absorption. Annu. Rev. Physiol. 45, 651–677.

Carriere, F., Laugier, R., Barrowman, J.A., Douchet, I., Priymenko, N., Verger, R., 1993. Gastric and pancreatic lipase levels during a test meal in dogs. Scand. J. Gastroenterol. 28, 443–454.

Denke, M.A., Grundy, S.M., 1991. Effects of fats high in stearic-acid on lipid and lipoprotein concentrations in men. Am. J. Clin. Nutr. 54, 1036–1040.

Dougherty, R.M., Allman, M.A., Iacono, J.M., 1995. Effects of diets containing high or low amounts of stearic-acid on plasma-lipoprotein fractions and fecal fatty-acid excretion of men. Am. J. Clin. Nutr. 61, 1120–1128.

Filippatos, T.D., Mikhailidis, D.P., 2009. Lipid-lowering drugs acting at the level of the gastrointestinal tract. Curr. Pharm. Des. 15, 490–516.

Gargouri, Y., Julien, R., Bois, A.G., Verger, R., Sarda, L., 1983. Studies on the detergent inhibition of pancreatic lipase activity. J. Lipid Res. 24, 1336—1342.

Golding, M., Wooster, T., 2010. The influence of emulsion structure and stability on lipid digestion. Curr. Opin. Colloid Interface Sci. 15, 90—101.

Golding, M., Wooster, T.J., Day, L., Xu, M., Lundin, L., Keogh, J., Clifton, P., 2011. Impact of gastric structuring on the lipolysis of emulsified lipids. Soft Matter 7, 3513—3523.

Gregory, P.C., Rayner, V., Wenham, G., 1986. The influence of intestinal infusion of fats on small intestinal motility and digesta transit in pigs. J. Physiol.—London 379, 27—37.

Hamosh, M., 1990. Lingual and gastric lipases. Nutrition 6, 421—428.

Helbig, A., Silletti, E., Timmerman, E., Hamer, R.J., Gruppen, H., 2012. In vitro study of intestinal lipolysis using pH-stat and gas chromatography. Food Hydrocoll. 28, 10—19.

Jandacek, R.J., 2012. Review of the effects of dilution of dietary energy with olestra on energy intake. Physiol. Behav. 105, 1124—1131.

Keogh, J.B., Wooster, T.J., Golding, M., Day, L., Otto, B., Clifton, P.M., 2011. Slowly and rapidly digested fat emulsions are equally satiating but their triglycerides are differentially absorbed and metabolized in humans. J. Nutr. 141, 809—815.

Lentle, R., Janssen, P., Golding, M., Wooster, T., 2011. Colloidal dynamics and lipid digestive efficiency. In: Lentle, R., Janssen, P. (Eds.), The Physical Processes of Digestion. Springer, New York, pp. 63—90.

Lowe, M.E., 1997. Structure and function of pancreatic lipase and colipase. Annu. Rev. Nutr. 17, 141—158.

Lowe, M.E., 2002. The triglyceride lipases of the pancreas. J. Lipid Res. 43, 2007—2016.

Lundin, L., Golding, M., 2009. Structure design for healthy food. Aust. J. Dairy Technol 64, 68—74.

Lundin, L., Golding, M., Wooster, T.J., 2008. Understanding food structure and function in developing food for appetite control. Nutr. Dietet. 65, S79—S85.

Maldonado-Valderrama, J., Wilde, P., Macierzanka, A., Mackie, A., 2011. The role of bile salts in digestion. Adv. Colloid Interface Sci. 165, 36—46.

Maldonado-Valderrama, J., Woodward, N.C., Gunning, A.P., Ridout, M.J., Husband, F.A., Mackie, A.R., et al., 2008. Interfacial characterization of beta-lactoglobulin networks: displacement by bile salts. Langmuir 24, 6759—6767.

Maljaars, P.W.J., van der Wal, R.J.P., Wiersma, T., Peters, H.P.F., Haddeman, E., Masclee, A.A.M., 2012. The effect of lipid droplet size on satiety and peptide secretion is intestinal site-specific. Clin. Nutr. 31, 535—542.

Maljaars, P.W.J., van der Wal, R.J.R., Haddeman, E.A., Peters, H.P.F., Beindorff, C., Masclee, A.A.M., 2008. Effect of droplet size of a fat emulsion delivered in the small intestine on satiety and food intake in healthy volunteers. Eur. J. Gastroenterol. Hepatol. 20, A54.

Marciani, L., Faulks, R., Wickham, M., Bush, D., Pick, B., Wright, J., et al., 2009. Effect of intragastric acid stability of fat emulsions on gastric emptying, plasma lipid profile and postprandial satiety. Br. J. Nutr. 101, 919—928.

Marciani, L., Wickham, M., Singh, G., Bush, D., Pick, B., Cox, E., et al., 2007. Enhancement of intragastric acid stability of a fat emulsion meal delays gastric emptying and increases cholecystokinin release and gallbladder contraction. Am. J. Physiol.—Gastr. Liver Physiol. 292, G1607—G1613.

McClements, D.J., Decker, E.A., Park, Y., 2009. Controlling lipid bioavailability through physicochemical and structural approaches. Crit. Rev. Food Sci. Nutr. 49, 48−67.

Mu, H.L., Hoy, C.E., 2004. The digestion of dietary triacylglycerols. Prog. Lipid Res. 43, 105−133.

Mukherjee, M., 2003. Human digestive and metabolic lipases—a brief review. J. Mol. Cat. B—Enzym. 22, 369−376.

Norton, I., Fryer, P., Moore, S., 2006. Product/process integration in food manufacture: engineering sustained health. AICHE J. 52, 1632−1640.

Norton, I., Moore, S., Fryer, P., 2007. Understanding food structuring and breakdown: engineering approaches to obesity. Obestiy Rev. 8, 83−88.

Pafumi, Y., Lairon, D., de la Porte, P.L., Juhel, C., Storch, J., Hamosh, M., Armand, M., 2002. Mechanisms of inhibition of triacylglycerol by human gastric lipase. J. Biol. Chem. 277, 28070−28079.

Pilichiewicz, A.N., Papadopoulos, P., Brennan, I.M., Little, T.J., Meyer, J.H., Wishart, J.M., et al., 2007. Load-dependent effects of duodenal lipid on antropyloroduodenal motility, plasma CCK and PYY, and energy intake in healthy men. Am. J. Physiol.—Reg. Integr. Comp. Physiol. 293, R2170−R2178.

Sarkar, A., Goh, K.K.T., Singh, R.P., Singh, H., 2009. Behaviour of an oil-in-water emulsion stabilised by beta-lactoglobulin in an in vitro gastric model. Food Hydrocoll. 23, 1563−1569.

Schwizer, W., Steingoetter, A., Fox, M., 2006. Magnetic resonance imaging for the assessment of gastrointestinal function. Scand. J. Gastroenterol. 41, 1245−1260.

Seimon, R.V., Wooster, T., Otto, B., Golding, M., Day, L., Little, T.J., et al., 2009. The droplet size of intraduodenal fat emulsions influences antropyloroduodenal motility, hormone release and appetite in healthy males. Am. J. Clin. Nutr. 89, 1729−1736.

Singh, H., Ye, A.Q., Horne, D., 2009. Structuring food emulsions in the gastrointestinal tract to modify lipid digestion. Prog. Lipid Res. 48, 92−100.

van Aken, G.A., 2010. relating food emulsion structure and composition to the way it is processed in the gastrointestinal tract and physiological responses: what are the opportunities? Food Biophys. 5, 258−283.

Vinarov, Z., Tcholakova, S., Damyanova, B., Atanasov, Y., Denkov, N.D., Stoyanov, S.D., et al., 2012. Effects of emulsifier charge and concentration on pancreatic lipolysis: 2. Interplay of emulsifiers and biles. Langmuir 28, 12140−12150.

Wickham, M., Garrood, M., Leney, J., Wilson, P.D.G., Fillery-Travis, A., 1998. Modification of a phospholipid stabilized emulsion interface by bile salt: effect on pancreatic lipase activity. J. Lipid Res. 39, 623−632.

Chapter **6**

Protein–Polysaccharide Interactions and Digestion of the Complex Particles

M.G. Semenova, D.V. Moiseenko, N.V. Grigorovich, M.S. Anokhina, A.S. Antipova, L.E. Belyakova, Yu.N. Polikarpov, and E.N. Tsapkina

N.M. Emanuel Institute of Biochemical Physics of Russian Academy of Sciences, Moscow, Russian Federation

CONTENTS

INTRODUCTION

It is well established that the protein–polysaccharide interactions are of great importance in determining the structure-forming properties of these biopolymers in the bulk and at interfaces of multicomponent food systems, controlling principally the structure, physical stability, and appearance of the latter (Semenova and Dickinson, 2010). These interactions may occur via physical bonding such as van der Waals, electrostatic, hydrophobic, hydrogen bonding, and excluded volume effects, or by chemical bonding as in the case of Maillard-type protein–polysaccharide conjugates. The strength and character (net attractive or net repulsive) of protein–polysaccharide non-covalent physical interactions may vary substantially, depending primarily on such environmental conditions as pH, ionic strength,

Food Structures, Digestion and Health. http://dx.doi.org/10.1016/B978-0-12-404610-8.00006-2

and temperature (Semenova *et al.*, 2009; Semenova and Dickinson, 2010). This diversity and sensitivity to such external influences makes protein—polysaccharide complexes very versatile as building blocks for the development of stimuli-sensitive carriers for controlled delivery, such as the oral delivery of biologically active compounds, containing prophylactic and therapeutic agents for the improvement of human health (or so-called nutraceuticals), to specific sites of the gastrointestinal (GI) tract and at a specified rate of their release (Benichou *et al.*, 2004; Shaw *et al.*, 2008; Livney, 2008; Semenova and Dickinson, 2010).

There is a considerable current interest in protecting bioactive molecules against release in the stomach or intestine (Singh and Horne, 2009). At the same time it is well known that a large number of polysaccharides (dextran, chitosan, chemically modified starch, alginates, pectins, carrageenan, xanthan, etc.) are resistant to the action of gastric and intestinal enzymes and bacteria, while they are specifically hydrolyzed only by colonic bacteria. Hence, these properties of polysaccharides could be useful, for example, for the elaboration of coating and matrix materials for specific delivery of bioactive molecules to the colon (Macleod *et al.*, 1999; Vandamme *et al.*, 2002).

From the literature data, two main approaches to using protein—polysaccharide interactions for the elaboration of stimuli-sensitive "switchable" carriers can be recognized. One of them, which has been widely investigated, is the use of hydrophilic or hydrophobic monolayer or multi-layer coatings at droplet surfaces in (macro)emulsions (size $>0.5\,\mu m$), which can be used to microencapsulate hydrophobic and/or hydrophilic nutraceuticals (Benichou *et al.*, 2004; McClements *et al.*, 2008, 2009; Grigoriev and Miller, 2009). For example, Benichou *et al.* (2004) have demonstrated that electrostatic complexes of whey protein isolate (WPI) with xanthan gum can be successfully used for both the effective multilayered coating of multiple (double) emulsion droplets and the controlled release of vitamin B1 entrapped within the inner aqueous phase of such droplets under the variation of the pH of the external aqueous phase.

Another promising approach is the use of protein—polysaccharide nanocomplexes to entrap bioactive compounds. Zimet and Livney (2009) used this approach successfully to entrap docosahexaenoic acid (DHA) by using electrostatic nanocomplexes of β-lactoglobulin with low-methoxy pectin. The nanocomplexes were found to possess outstanding colloidal stability (zeta potential more negative than $-50\,mV$, mean particle size $\approx 100\,nm$), and protection ability against oxidation of DHA during an accelerated shelf-life stress tests: only $\approx 5-10\%$ was lost during 100 hours at $40°C$, as compared to $\approx 80\%$ when the unprotected DHA was monitored (Zimet and Livney, 2009).

The promising abilities of nanosized protein—polysaccharide covalent conjugates as delivery vehicles for hydrophobic nutraceuticals were clearly demonstrated by the work of Grigorovich *et al.* (2011) and Markman and Livney (2012). Grigorovich *et al.* (2011) revealed that complex formation between covalent conjugates of sodium caseinate (SCN) + maltodextrin (MD) with polyunsaturated phosphatidylcholine (PC) led to the marked protection of PC against oxidation over a wide range of pH of the aqueous medium. This included the protein isoelectric point (pI), i.e., the pH point where SCN lost its solubility and functionality. Moreover, it was established that the molar protein—polysaccharide ratio could control both the protection ability of the covalent conjugates and the release of PC from the complex particles as a result of the successive action of the digestive enzymes (α-amylase; pepsin; α-amylase + trypsin + alpha-chymotrypsin) under *in vitro* conditions of the stomach and intestine. In turn, Markman and Livney (2012) have shown that the co-assembly of the casein—maltodextrin covalent conjugates with such hydrophobic nutraceuticals as vitamin D and epigallocatechin gallate (EGCG) conferred better protection against oxidation to both these nutraceuticals than the SCN + MD simple mixture. They suggested using such nanovehicles for the enrichment of clear beverages even at the pI of SCN. During simulated gastric digestion, Nile red (a fluorescent model for a hydrophobic nutraceutical) was not released from the conjugates, suggesting their potential application in enteric delivery.

All these data show clearly that protein—polysaccharide interactions can be used as a promising tool to regulate the protein—polysaccharide ability to behave like a nanovehicle for the controlled delivery of different kinds of nutraceuticals at a particular site of action, at a determined rate, and/or in response to a specific environmental trigger (pH, ionic strength, temperature, enzymatic action).

The nanoscale dimensions of such nanovehicles, which are typical of biopolymer molecules/associates, can offer additional advantages, because, as was suggested by Acosta (2009), the size of the delivery systems, especially below 500 nm, has enabled the promise of tackling problems of low oral bioavailability or inefficient delivery of poorly water-soluble nutraceuticals/drugs. This is a consequence of the enhancements based on the following factors (Acosta, 2009; Semenova and Dickinson, 2010): (1) the apparent solubility of the active ingredients; (2) the rate of mass transfer; (3) the gastrointestinal retention time in the mucus covering the intestinal epithelium; (4) the rate of release (due to large surface area); and (5) the direct uptake of particles by the intestinal epithelium (Horn and Rieger, 2001; Chen *et al.*, 2006; Medina *et al.*, 2007).

Research activities in the area of the elaboration of the nanoparticle delivery systems for micronutrients and nutraceuticals, which can be incorporated into food products, have increased almost exponentially during the past decade (Ransley *et al.*, 2001; Velikov and Pelan, 2008; McClements *et al.*, 2008, 2009; Faulks and Southon, 2008; Augustin and Hemar, 2009; Semenova and Dickinson, 2010; Semenova *et al.*, 2012). However, the physicochemical factors influencing both the susceptibility of the biopolymer complex nanocarriers to the enzymatic action under the simulated GI conditions and the controlled release (bioaccessibility) of the entrapped nutraceuticals from them are still not clearly understood.

The present work attempts to elucidate the structural basis underlying the digestibility (under the action of the gastric and intestinal enzymes in the simulated conditions of the GI tract *in vitro*) of protein–polysaccharide nanocomplexes involving soy phosphatidylcholine (PC). In parallel, a more penetrating insight is sought into the generality and the differences in behavior of the nanocomplexes in the GI tract *in vitro*, based on the different kinds of the protein–polysaccharide interactions, in particular electrostatic (SCN + dextran sulfate) and covalent bonding (SCN + MD).

It is noted that the choice of the polyunsaturated PC (Lipoid S100 (lecithin)) as a nutraceutical was due to its properties as an anti-aging agent, a superior

Table 6.1 Chemical Composition of Phosphatidylcholine (PC) (Lipoid S 100, Lipoid GmbH, Germany)

Constituent	g/100 g
phosphatidylcholine (by anhydrous weight)	94
N-acyl-phosphatidylethanolamine	0.5
phosphatidylethanolamine	0.1
phosphatidylinositol	0.1
lysophosphatidylcholine	3.0
triglycerides	2.0
free fatty acids	0.5
Typical fatty acid composition in % to the total fatty acids:	
palmitic acid	12 ÷ 17
stearic acid	2 ÷ 5
oleic acid	11 ÷ 15
linoleic acid	59 ÷ 70
linolenic acid	3 ÷ 7
DL-α-tocopherol	0.15 ÷ 0.25

protectant against liver damage (Kidd, 1996, 2000) and a supplier of essential polyunsaturated fatty acids (more than 80% of the easily oxidized unsaturated hydrocarbon chains of such essential fatty acids as oleic, linoleic, and linolenic, Table 6.1). Moreover, PC was also of interest because it can form environment-responsive liposomes in an aqueous medium, which could be promising as additional carriers for both hydrophobic and hydrophilic healthy food ingredients and drugs (Gennis, 1989).

PECULIARITIES OF THE STRUCTURAL AND THERMODYNAMIC PARAMETERS OF THE INITIAL ("BEFORE DIGESTION") TERNARY (PC + SCN + POLYSACCHARIDE) COMPLEX PARTICLES, FORMED BY THE DIFFERENT KINDS OF PROTEIN–POLYSACCHARIDE INTERACTIONS

The initial ternary (PC/DPPC $(1 \times 10^{-3} \text{ M})$ + SCN + polysaccharide) complex particles based on the covalent protein–polysaccharide bonding (SCN + MD-SA2) and on their electrostatic interactions (SCN + DS) are prepared in an aqueous medium as described previously (Grigorovich *et al.*, 2012; Semenova *et al.*, 2012).

The complex particles are analyzed using a combination of multiangle static and dynamic laser light scattering, which is a particularly useful tool for exploring the structure and interactions of both individual biopolymer molecules and their self-assembled or co-assembled particles in dilute solutions on length scales of the order of ≈ 1 µm and below (Burchard, 1994; Semenova and Dickinson, 2010). Using this combination, the following structural (the weight–average molar mass M_w, the radius of gyration R_G, the hydrodynamic radius R_h) and thermodynamic (the second virial coefficient A_2) parameters of the ternary complexes were measured.

It is noted that SCN is a commercial food ingredient of variable composition and aggregation state, depending on the origin, manufacturing, and storage conditions (Semenova *et al.*, 2009; Semenova and Dickinson, 2010). For the pure SCN sample in aqueous medium at neutral pH (6 and 7) and 37°C it was found that the size distribution measured by dynamic light scattering was bimodal (Figure 6.1A). This bimodal distribution seems to be a characteristic of caseinate solutions, both in the absence and presence of calcium ions (Chu *et al.*, 1995; Nash *et al.*, 2002; Müller-Buschbaum *et al.*, 2007; HadjiSadok *et al.*, 2008; Semenova *et al.*, 2009), with a variable relationship between peak areas depending on the origin of the caseinate sample and the environmental conditions. Similar bimodal size distributions were found for samples of pure conjugates (SCN + MD-SA2

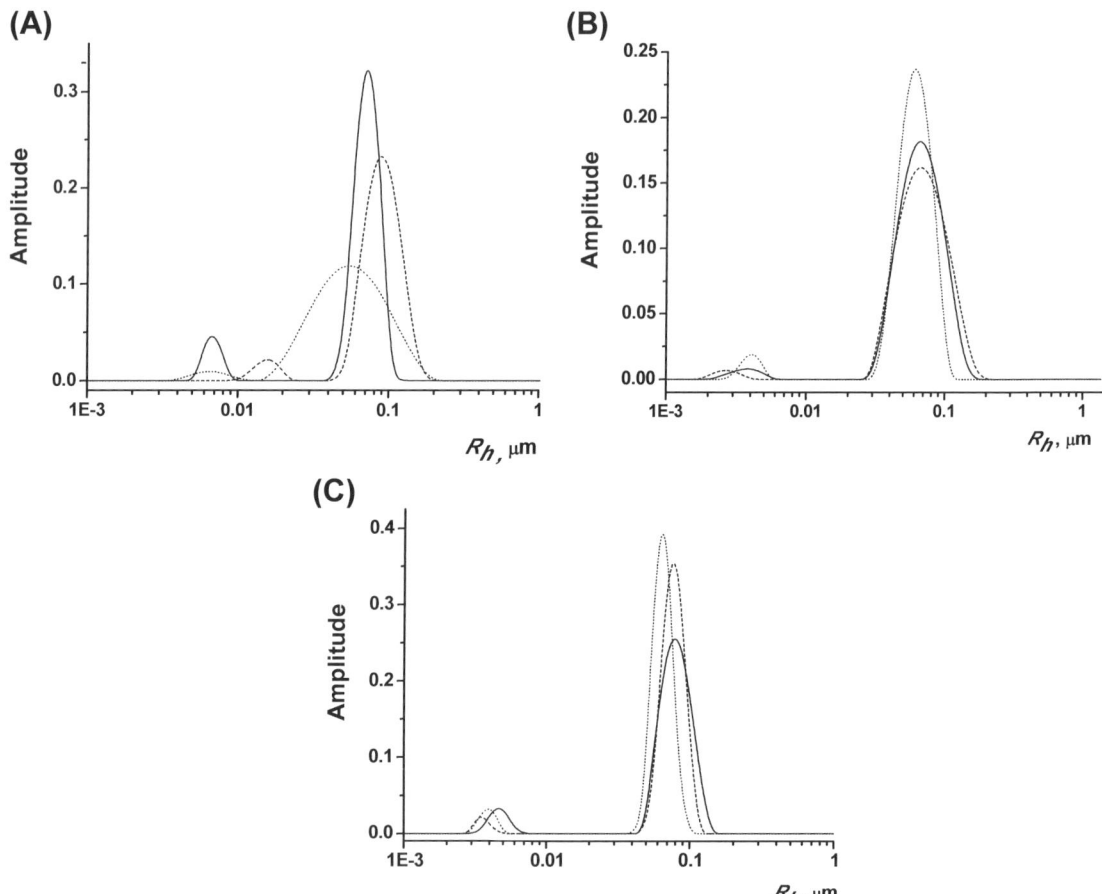

■ **FIGURE 6.1** The size distributions of the studied particles fractionated by filtration through membranes with different pore sizes ($\varnothing$) in aqueous medium (phosphate buffer, pH = 7.0, ionic strength = 0.001 M, $t = 37°$C): (A) SCN alone, (B) (SCN + MD-SA2) at $R_w = 0.4$, (C) (SCN + MD-SA2) at $R_w = 2.0$: $\varnothing = 0.8\ \mu$m (Dash); $\varnothing = 0.22\ \mu$m (Solid); $\varnothing = 0.03\ \mu$m (Dot).

at the two weight ratios of the polysaccharide to the protein: $R_w = 0.4$ and 2.0) in the aqueous medium at neutral pH (pH = 7) and 37°C (Figure 6.1B, C). In order to determine the weight fraction of the sample material that had been assembled initially into large particles, which dominate generally in the light scattering signal, samples were consecutively filtered through the membrane filters of different pore sizes (0.80 and 0.22 μm (millipore) and 0.03 μm (the membranes were made from lavsan (Russian equivalent of Darcon)). The filtrates were characterized by UV spectrophotometry ($\lambda = 280$ nm) to determine the residual protein content in the ultrafiltrates and the size distributions of the samples in the filtrates were measured using dynamic light scattering. The results are presented in

Table 6.2 The Residual Protein Content in the Filtrates as the Percentage of the Initial Protein Content (0.50 wt/v% of SCN) in the Sample Solutions (of the Pure SCN and the Pure Conjugates (SCN + MD-SA2) at $R_w = 0.4$ and 2.0) and the Total Average Hydrodynamic Radius of the Studied Particles (pH 7.0, Ionic Strength = 0.001 M, $t = 37°C$)

Sample	The residual protein content and the total average size of the sample particles in the aqueous solutions after consecutive filtration through the membrane filters having different pore size (Ø):					
	Ø = 0.80 µm		Ø = 0.22 µm		Ø = 0.03 µm	
SCN	100.0% ($R_h^{total} = 81.4$ nm)		98.4% ($R_h^{total} = 71.9$ nm)		90.5% ($R_h^{total} = 66.1$ nm)	
	The minor peak, R_h (nm)	The basic peak, R_h (nm)	The minor peak, R_h (nm)	The basic peak, R_h (nm)	The minor peak, R_h (nm)	The basic peak, R_h (nm)
	14.3	89.5	7.5	78.7	7.4	68.4
Conjugate SCN + MD-SA2, $R_w = 0.4$	100.0% ($R_h^{total} = 78.4$ nm)		98.3% ($R_h^{total} = 74.3$ nm)		96.2% ($R_h^{total} = 70.0$ nm)	
	The minor peak, R_h (nm)	The basic peak, R_h (nm)	The minor peak, R_h (nm)	The basic peak, R_h (nm)	The minor peak, R_h (nm)	The basic peak, R_h (nm)
	3.0	80.1	4.1	76.4	12.1	80.8
Conjugate SCN + MD-SA2, $R_w = 2.0$	100.0% ($R_h^{total} = 81.5$ nm)		98.9% ($R_h^{total} = 82.8$ nm)		98.1% ($R_h^{total} = 66.8$ nm)	
	The minor peak, R_h (nm)	The basic peak, R_h (nm)	The minor peak, R_h (nm)	The basic peak, R_h (nm)	The minor peak, R_h (nm)	The basic peak, R_h (nm)
	3.9	84.7	5.1	89.4	4.3	71.2

Table 6.2 and in Figure 6.1, respectively. Table 6.2 presents an ultimate protein weight loss during filtration of 9.5% in the SCN sample and only about 2—4% for the conjugates. As ultrafiltration takes place, the size distributions (Figure 6.1) show only small-scale shifts towards smaller sizes of light-scattering biopolymer particles in the samples. In the case of pure SCN, this shift was most pronounced and it was concomitant with a lowering of the intensity of light scattering that, in turn, could be attributable to the maximal protein weight lost. Thus, there seems to be the total contribution from all particles, having different sizes determined by their polydispersity, into the average values of the structural parameters measured by light scattering. The parameters that have been measured, in all cases, are structural (M_w, R_G, R_h) and thermodynamic (A_2) parameters of the ternary complex particles formed by unfractionated samples of pure SCN and the conjugates (SCN + MD-SA2, at $R_w = 0.4$ and 2.0).

Under the experimental conditions of formation of the initial ternary complexes ($C_{PC} = 10^{-3}$ M, pH = 7.0/6.0, ionic strength = 0.001 M, 37°C) the PC was present as liposomes, having a total average R_h of 60 nm and a rather narrow size distribution (Figure 6.2). It is noteworthy that the range of sizes of PC liposomes (diameter <200 nm) corresponds generally to the unilamellar vesicles (Bai *et al.*, 2010). Moreover, the R_h was slightly smaller than that of both pure SCN (Figure 6.1A) and the covalent conjugates of SCN + MD-SA2 (Figure 6.1B, C) independent of the R_w (Table 6.2), which can facilitate somehow the encapsulation ability of both the protein and the conjugates in relation to the PC liposomes (Semenova *et al.*, 2008, 2012).

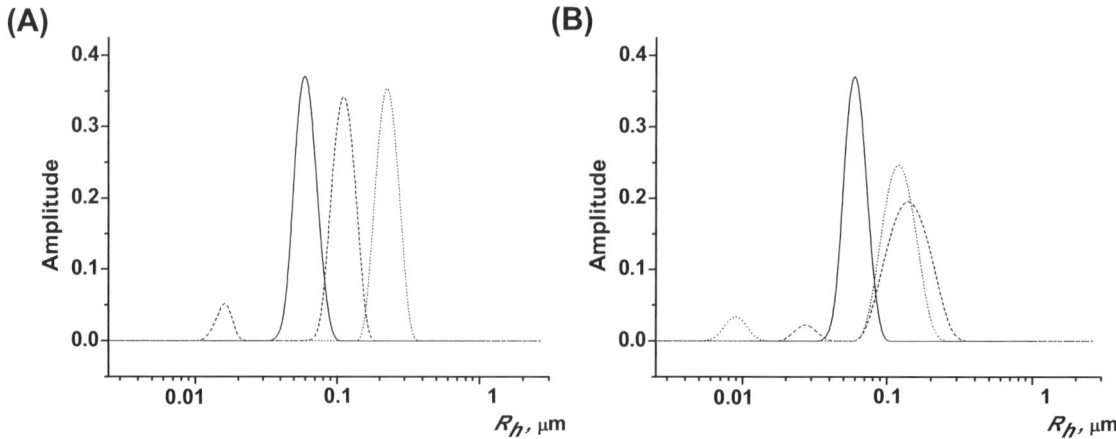

■ **FIGURE 6.2** The size distributions of both the PC liposomes in their pure form (Solid) and the ternary complex particles (A) ((SCN + PC) + DS) and (B) ((SCN + MD-SA2) + PC) in aqueous medium (ionic strength = 0.001 M; 37°C) at pH = 6.0 and pH = 7.0, respectively. R_w = 0.2/0.4 (Dash) and R_w = 2.0 (Dot).

In turn, the size distributions (Figure 6.2A, B) of the unfractionated ternary complex particles ((SCN + PC) + DS) and ((SCN + MD-SA2) + PC) show bimodal distributions with the main intensity of the light scattering peaks having rather narrow size distributions and showing a total average R_h in the range from 126 nm ($R_w = 0.4$) to 240 nm ($R_w = 2.0$) for the ((SCN + PC) + DS) particles and from 136 nm ($R_w = 2.0$) to 146 nm ($R_w = 0.4$) for the ((SCN + MD-SA2) + PC) particles (Table 6.3). There are also small-intensity peaks in the range of particle sizes of about 10 nm in each case. These characteristics of size distribution are similar to those inherent to both pure SCN and conjugates, which are determined by their original polydispersity (Figure 6.1A–C), which is apparently retained under the ternary complex formation. In addition, these distributions indicate the evident involvement of PC liposomes in the complex formation. As a result of the complex formation, such particles have larger sizes compared to the original components (compare Tables 6.2 and 6.3), as if zwitterionic PC liposomes behave as active internal cross-linking agents between the protein and polysaccharide particles. Data on the extraction of free PC by diethyl ether from aqueous solutions of the complex particles show about 90%, within experimental error, binding of the PC liposomes with both the ((SCN + PC) + DS) and ((SCN + MD-SA2) + PC) particles, independent of the R_w.

In order to gain more insight into the state of PC liposomes in such ternary complex particles, DSC measurements were made of the thermodynamic parameters of the inherent phase transition for the phospholipid bilayers using the example of a model phosphatidylcholine, namely dipalmitoyl phosphatidylcholine (DPPC). It is well known that alteration of the phase state of the bilayers of phospholipid liposomes as a result of interactions with biopolymers can be characterized by dramatic changes in the thermodynamic parameters and functions of the temperature-induced phase transition of the phospholipid bilayers from a solid-like gel state to a fluid liquid–crystalline state (Bai *et al.*, 2010; de Oliveira Tiera *et al.*, 2010).

Figure 6.3A and B show the thermograms for both the ((SCN + DPPC) + DS) and the ((SCN + MD-SA2) + DPPC) particles, respectively. They indicate that the DPPC bilayers are not destroyed by the formation of the different kinds of complex particles. Moreover, the intermolecular interactions of the DPPC molecules inside the bilayers seem to become stronger for both the ((SCN + MD-SA2) + DPPC) and (SCN + DPPC + DS) complex particles (Table 6.4) (Stenekes *et al.*, 2001; Bai *et al.*, 2010; de Oliveira Tiera *et al.*, 2010). This manifests itself as a larger area under the endothermic peak of the transition on the thermogram giving a greater value of the molar enthalpy of the transition, ΔH_c, in comparison with the value of

Table 6.3 The Initial ("Before Digestion") Structural (M_w, R_G, R_h, ρ, d) and Thermodynamic (A_2) Parameters of the Ternary Complex Particles (((SCN + PC) + DS) and ((SCN + MD-SA2) + PC)) in the Aqueous Medium (Ionic Strength = 0.001 M; 37°C) at pH = 6.0 and pH = 7.0, Respectively. The Polysaccharide Contents Refer to the Two Polysaccharide—Protein Weight Ratios, R_w

	Experiment		Theory		Ternary complex particles				
R_w	M_w (kDa)	A_2^* (m³mol⁻¹)	A_2^{exc} (m³mol⁻¹)[a]	A_2^{el} (m³mol⁻¹)[b]	R_G (nm)	R_h (nm)	$\rho = R_G/R_h$	d (mg/ml)[c]	the % of PC oxidation[d]
(SCN + PC) + DS									
0.2	9450	4.5	41.3	−36.8	127	126	1.01	1.8	14.7
2.0	6500	2.1	278.7	−276.6	132	240	0.55	0.2	21.1
(SCN + MD-SA2) + PC									
0.4	7400	3.7	62.7	−59.0	152	146	1.04	0.8	3.7
2.0	2300	0.2	50.7	−50.5	131	136	0.96	0.4	11.1

[a] $A_2^{exc} = 10^{-3} 4\pi N_A/3 (2R)^3$, where R is the radius of the equivalent hard sphere representing the macroion (Tanford, 1961). In our case R = R_h;

[b] $A_2^{el} = A_2^* - A_2^{exc}$;

[c] $d = (M_w/N_A) V^{-1}$, where N_A is the Avogadro number, $V = 4/3\pi R_h^3$ Tanford, 1961);

[d] after 1 day of the storage of the tested solutions at t = 37°C and in the presence of 10^{-5} M CuSO₄. The level of the oxidation of pure PC was taken as 100%.

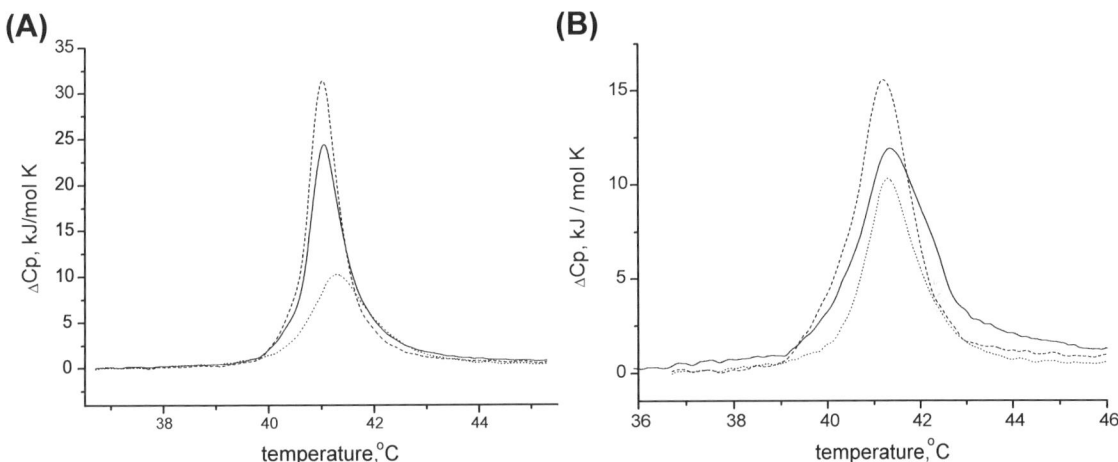

■ **FIGURE 6.3** Thermograms of the phase transition of the DPPC bilayers from the solid-like gel state to the fluid liquid—crystalline state for DPPC liposomes in a pure form (Dot) and in the ternary complex particles (A) ((SCN + DPPC) + DS) and (B) ((SCN + MD-SA2) + DPPC) in aqueous medium (ionic strength = 0.001 M; 37°C) at pH = 6.0 and pH = 7.0, respectively. The polysaccharide contents refer to the two polysaccharide—protein weight ratios: R_w = 0.2/0.4 (Solid) and R_w = 2.0 (Dash). DSC measurements were performed by the high sensitivity differential scanning calorimetry using a DASM-4M model, differential scanning calorimeter (Pushino, Russia). The heating scan rate was maintained at 0.5°C/min for all the experiments.

Table 6.4 Thermodynamic Parameters of the Phase Transition from the Solid-like Gel State to the Fluid Liquid—Crystalline State for DPPC Bilayers in a Pure Form and in the Ternary Complex Particles (((SCN + DPPC) + DS) and ((SCN + MD-SA2) + DPPC)) in the Aqueous Medium (Ionic Strength = 0.001 M; 37°C) at pH = 6.0 and pH = 7.0, Respectively. The Polysaccharide Contents Refer to the Two Polysaccharide—Protein Weight Ratios, R_w

Samples	t, °C	ΔH_c, kJ/mol$_{DPPC}$
DPPC	41.3	17.4
((SCN + DPPC) + DS)		
R_w = 0.2	41.0	26.4
R_w = 2.0	41.1	29.6
(SCN + MD-SA2) + DPPC		
R_w = 0.4	41.4	33.37
R_w = 2.0	41.2	35.47

ΔH_c for the pure DPPC (Figure 6.3A, B; Table 6.4). The greatest increase in the value of ΔH_c is revealed for the ((SCN + MD-SA2) + DPPC) particles at $R_w = 2.0$. In turn, the lowest increase in the value of the ΔH_c is revealed for the ((SCN + DPPC) + DS) particles at $R_w = 0.2$. The observed positive impact of complex formation on the stability of the DPPC bilayer could be attributable to the DPPC charge neutralization under its interactions with the complex particles, in particular with the complex-combined protein, and the compensating negative effect to the addition of negative charges from the DS molecules attached to the initial (SCN + DPPC) complex particles. Shoemaker and Vanderlick (2003) suggested that an increase in the surface charge of lipid membranes decreased their mechanical stability. In addition, some destabilizing effect of the highly charged DS molecules on the DPPC membranes could be attributable to the partial transformation of the vesicles into mixed micelle-like aggregates (Bai *et al.*, 2010) with the formation of the additional cross-links in the complex particles. It is interesting to note that the phase transition of the DPPC bilayer seems to become more cooperative in the ternary complexes ((SCN + DPPC) + DS), which is more pronounced at the weight excess of DS molecules, i.e., at $R_w = 2.0$.

In order to develop a deeper insight into the peculiarities of the initial ternary complex particles formed, it is instructive to consider their structural and thermodynamic parameters as measured by a combination of dynamic and static laser light scattering. Table 6.3 shows the structural (M_w, R_G, R_h, the structural factor ρ, d) and thermodynamic (A_2) parameters of the ternary complexes formed.

The lower value of the M_w of the ternary complexes at the excess of the polysaccharides is noteworthy. In the case of DS this result could be mainly attributable to the partial dissociation of the original binary (SCN + PC) particles in response to the interactions with the excess concentration of highly negatively charged DS molecules. The expected strong electrostatic repulsion between the negatively charged sulfate groups on the attached DS molecule is the most likely driving force for such a dissociation (Semenova *et al.*, 2009). In turn, for the ternary particles containing the covalent conjugates (SCN + MD-SA2), the observed result could be attributable to the partial destruction of the internal contacts that initially stabilized the sodium caseinate particles, in response to the covalent attachment of the excess of the maltodextrin molecules (Grigorovich *et al.*, 2012).

The architecture of the ternary complex particles has been further characterized by the values of the ratio R_g/R_h (the structural factor, ρ) (Table 6.3), the utility of which in distinguishing among different architectures of both individual macromolecules and their aggregates has been proven in many light scattering experiments (Kunz *et al.*, 1983; Kajiwara and Burchard, 1984; Burchard, 1994; Ioan et al., 2000; Tuteja *et al.*, 2006). Relying on these literature data, we can suggest that the all studied particles have the architectures of homogeneous impenetrable spheres: $0.788 < \rho < 1.1$ except for the ternary $((SCN + PC) + DS)$ particles at $R_w = 2$, for which $\rho \approx 0.55$ and the architecture of a gel-like particle with many dangling chains at the surface can be suggested (Burchard, 1994). The variations observed in the values of the structural factor ρ in the case of the homogeneous impenetrable spheres seem to be governed mainly by the differences in both the thermodynamic affinity of the particles for the solvent (A_2) and the level of their polydispersity (Figure 6.2). It was established that an increase in both polydispersity and affinity for the solvent of the polymer particles led to an increase in the value of ρ, but this increase was not pronounced and was about 14% as shown by the examples given by Burchard in his review (1994).

As a result of the observed destruction of the ternary complex particles at an excess of both DS and MD-SA2, the values of the second virial coefficients, A_2, which characterize the thermodynamics (both the character and strength) of the pairwise interactions between the complex particles, become less positive (Table 6.3). A positive value of A_2 indicates thermodynamically unfavorable biopolymer—biopolymer interactions in a solution (an increase in the magnitude of the excess chemical potential of the biopolymer in the solution), in other words, mutual biopolymer repulsions, whereas a negative value indicates mutual biopolymer attractions (Semenova and Dickinson, 2010). The current theories of the second virial coefficient indicate that, in the case of polyelectrolytes (proteins and anionic/cationic polysaccharides), a positive value of the second virial coefficient is mainly determined by a contribution from the thermodynamically excluded volume (A_2^{exc}) of the macroions, as well as by a contribution from the electrostatic forces acting between macroions (A_2^{el}) (Tanford, 1961; Nagasawa and Takahashi, 1972; Semenova and Dickinson, 2010). In this analysis, it is implied that the second virial coefficient may be written as a sum of two terms $A_2^* = A_2^{exc} + A_2^{el}$, where we use the subscript 2 for the case of interactions between pairs of identical macroion species. To a first approximation, it can be assumed that the excluded volume term (A_2^{exc}) is determined only by the physical volume occupied by the biopolymer molecule/particle. For the spherical particles found in our experiments, the simplest case of the interacting solid spheres can be used. Table 6.3 shows a comparison of the values of the second virial coefficients obtained experimentally by static laser light scattering, A_2^*, with their excluded volume, A_2^{exc}, and electrostatic, A_2^{el}, terms calculated theoretically. On the basis of this calculation, it can be suggested that the excluded volume effects determine predominantly the observed positive values of A_2^*, whereas despite the expected electrostatic repulsions, hydrophobic attraction likely occurs between the ternary complex particles in the aqueous medium, which appear to be stronger in the case of the gel-like particles formed in the presence of the excess of DS at $R_w = 2$. This result could be attributable, on the one hand, to the marked neutralization of the charged functional groups of SCN, DS, and PC as a result of ternary complex formation and, on the other hand, to the simultaneous exposure of hydrophobic patches on the surface of the protein molecules in the complex particles.

It is notable that for each kind of ternary complex particle, the excess of polysaccharides in the system at $R_w = 2$ leads to a decrease in the density of the particles of an order of the magnitude for the gel-like ((SCN + PC) + DS) particle and two-fold for the hard sphere-like ((SCN + MD-SA2) + PC) particle (Table 6.3).

RELATIONSHIPS BETWEEN STRUCTURAL PARAMETERS OF THE TERNARY COMPLEX PARTICLES AND THEIR FUNCTIONALITY AS DELIVERY VEHICLES FOR POLYUNSATURATED PC

The requirements for efficient delivery systems for nutraceuticals have been suggested by McClements and coworkers (2009). The basic among them are the following: (1) the delivery system should efficiently encapsulate an appreciable amount of the functional component in a form that is easily incorporated into food systems; (2) depending on the application, the delivery system may have the capability to protect the functional component from chemical degradation (e.g., oxidation, hydrolysis) keeping it in its active state; (3) depending on the application, the delivery system may have the capability to release the functional component at a particular site of action, at a controlled rate and/or in response to a specific environmental trigger (e.g., pH, ionic strength or temperature); (4) the delivery system should have the capability of maintaining the bioactivity of the functional component within the human body prior to its being delivered to the desired site of action, e.g., resisting the high acidity and enzyme activity of the stomach.

We have focused on *in vitro* studies to elucidate the relationships between the structural parameters of the ternary complex particles that are formed and their abilities both to protect the polyunsaturated PC against oxidation and to release the PC under enzymatic action in the targeted part of the gastrointestinal tract (i.e., either stomach or intestine).

Protection of the Polyunsaturated PC Against Oxidation

Quantitative assessment of the ability of the ternary complex particles to protect the polyunsaturated PC against oxidation was carried out by the estimation of the percentage of PC oxidation in the ternary complex particles as compared to the level of PC oxidation in its pure form, which was taken as 100% as described in our previous paper (Semenova *et al.*, 2012).

Direct correlations were observed between the density of the ternary complex particles and the extent of PC oxidation within their specific type, namely either ((SCN + MD-SA2) + PC) or ((SCN + PC) + DS) (Table 6.3): the lower the density of the ternary particles the higher the extent of the PC oxidation. The protective ability of the ternary complex particles could be attributed to the probability that the relatively high values of the density of the particles could hinder the diffusion of small molecules such as oxygen to the unsaturated hydrocarbon chains of PC, which are in the interior of the complex particles. The general importance of such parameters as density of the complex particles, for their protective ability against oxidation for the unsaturated PC, could also be supported by data obtained for the complex particles formed between both sodium caseinate and β-casein associates with PC liposomes in an aqueous medium (Semenova *et al.*, 2008, 2012). Moreover, the architecture of the particles seems to have an effect on the level of oxidation of the PC. Thus, the highest level of the oxidation was obtained for the architecture of the gel-like particles inherent to the ((SCN + PC) + DS) particles at $R_w = 2$.

Targeted Release of PC Under Enzymatic Action in the Gastrointestinal Tract (Either Stomach or Intestine) *In Vitro*

Enzymatic hydrolysis of the ternary complex particles ((SCN + PC) + DS) and ((SCN + MD-SA2) + PC)) was performed under simulated conditions of the gastrointestinal digestion *in vitro*. Considering that the molecular containers for PC consist of both SCN and MD-SA2, the complex particles were treated with the main enzymes (E) responsible for the hydrolysis of food proteins and starch polysaccharide (substrate, S) in the GI tract. The *in vitro* simulated digestion included three stages. In the first stage of the hydrolysis "in the mouth," the test sample solutions were treated with α-amylase for 5 minutes at pH 7.0/6.0, ionic strength = 0.001 M and 37°C. In the second stage of the proteolysis "in the stomach," the test sample solutions were treated with pepsin for 2 hours at pH 2.0, ionic strength = 0.1 M and 37°C. In the third stage of the proteolysis "in the small intestine," the test sample solutions were treated by an equimass mixture of α-amylase + trypsin + α-chymotrypsin for 2 hours at pH 8.0, ionic strength = 0.15 M and 37°C. The values of the ionic strength on the second and third stages were chosen in accordance with the recommendations of McClements and Li (2010). The ratio of enzyme to substrate was 1:1000 for each type of enzyme.

Determination of the initial rate of hydrolysis, V_0, and the estimation of the extent of PC, both the initial binding and the release from the ternary complex particles under the simulated GI conditions, were conducted using the procedures described previously for β-casein and SCN complexes with PC (Semenova *et al.*, 2012).

Let us consider the transformation of the molecular parameters of the ternary complex particles during their passage through the model GI tract, using the example of the ternary complex particles, having an excess of polysaccharide, i.e., at $R_w = 2$.

Figure 6.4A shows the changes in the size distribution for the ternary complex particles ((SCN + PC) + DS). There is little change during the first step of the digestion in the "mouth." By contrast, marked changes occur in the second stage, the "stomach," where, along with enzymatic action, the total charge of the protein changes from negative to positive when the pH is reduced from above (pH 7.0) to below (pH 2.0) the protein isoelectric point (pI ≈ 4.6). It is noted that dextran sulfate evidently has little or no negative charge. As a result of these changes, new particles with both smaller and larger sizes are formed, increasing the polydispersity of the

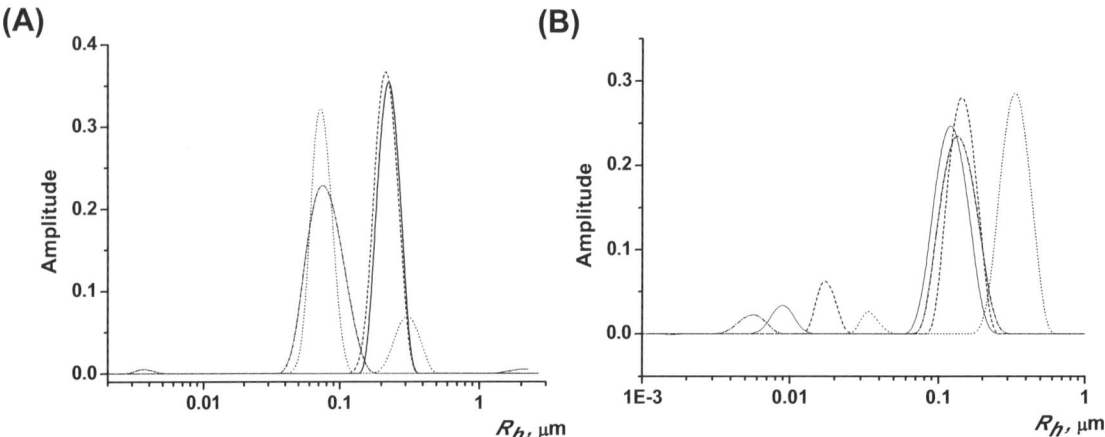

■ **FIGURE 6.4** The size distributions of the ternary complex particles at $R_w = 2.0$ (A) ((SCN + PC) + DS) and (B) ((SCN + MD-SA2) + PC) during their consecutive passage through the model GI tract *in vitro* (the experimental conditions are given in the text): without enzymes (Solid); stage I is "a mouth" (Dash); stage II is "a stomach" (Dot); stage III is "an intestine" (Dash Dot).

samples. In the subsequent step of digestion in the "small intestine," the size distribution shows greater polydispersity, with smaller average sizes of the particles. This result could be attributable to further proteolysis of the complex particles and to the simultaneous disruption of the electrostatic contacts between sodium caseinate and dextran sulfate, mostly due to the pH increase up to 8.0 and the increase in the ionic strength up to 0.15 M (Semenova *et al.*, 2009).

In turn, Figure 6.4B shows the change in size distribution for the ternary complex of the covalent conjugate (SCN + MD-SA2) with PC under the consecutive digestion steps in the model GI tract *in vitro*. There are only slight changes during the first step of digestion in the "mouth," which is manifest in a small shift of the peaks towards larger particle sizes. In contrast, marked changes occur during the second stage of digestion in the "stomach," where markedly larger particles were formed. This result could be attributable to either the aggregation of the proteolyzed particles or the changes in their architecture, accordingly to well-known direct relationships between R_h and the structural factor, ρ. In the next step of the digestion, in the "small intestine," the size distribution shows smaller average sizes of the particles. This result indicates further proteolysis of the complex particles. In line with the observed differences in the changes of the size distributions for the different types of the ternary complex particles during their passage through the *in vitro* GI tract, Figure 6.5 shows the marked contrast between

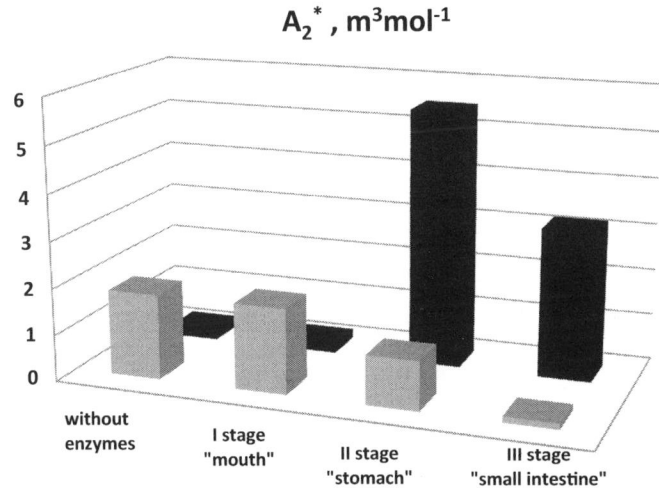

$$A_2^{\,*}\,,\ m^3mol^{-1}$$

■ **FIGURE 6.5** Alteration of the thermodynamic affinity of the ternary complex particles ($R_w = 2$) for aqueous medium (the second virial coefficient, A_2) during their consecutive passage through the model GI tract *in vitro*: ((SCN + PC) + DS) (gray columns); ((SCN + MD-SA2) + PC) (black columns).

the alterations of their thermodynamic affinity for the aqueous medium. A pronounced decrease in the thermodynamic affinity for the aqueous medium (a decrease in the positive value of A_2) was observed at both the second and third stages of the digestion in the case of the ((SCN + PC) + DS) ternary complexes. In contrast, a marked increase in the thermodynamic affinity for the solvent (an increase in the positive value of A_2) was observed at the same stages for the complex particles formed by the covalent conjugates (SCN + MD-SA2) with PC, in particular at the stage of the "stomach." This contrast could be attributable either to the loss of some hydrophilic DS molecules from the complex particles due to the increase in both pH to 8.0 and ionic strength to 0.15 M in the "small intestine" or to the evident decrease in the total negative charge of DS in the stage of the "stomach" at pH 2.0. In an alternative explanation, this contrast could be attributable to the pronounced aggregation, hindering access to the hydrophobic patches of the protein and PC molecules in the interior of the large aggregates, and to architecture changes of the complex particles formed by the covalent conjugates.

Dramatic differences in the alterations of the architecture of the initial ternary complex particles formed by the ((SCN + PC) + DS) ternary complexes as compared with the particles formed by the covalent conjugates (SCN + MD-SA2) with PC have been revealed (Figure 6.6). In the case of the complex particles based on electrostatic protein–polysaccharide interactions, the particles changed from gel-like particles in the first stage of digestion, in the "mouth," to hard spheres during the further stages. By contrast, the ternary particles based on the covalent conjugates changed from hard spheres in the first stage of the digestion, in the "mouth," to the gel-like

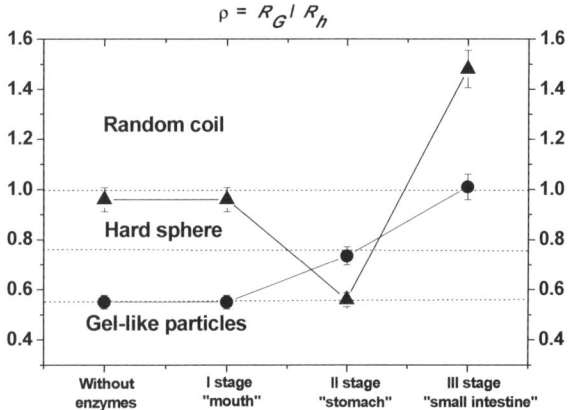

■ **FIGURE 6.6** Alteration of the structural parameter ρ of the ternary complex particles ($R_w = 2$) during their consecutive passage through the model GI tract *in vitro*: ((SCN + PC) + DS) (●); ((SCN + MD-SA2) + PC) (▲).

particles in the stage of the "stomach" and further to random coils in the third stage of the "small intestine."

The transformations found for the particles based on the electrostatic SCN−DS interactions ((SCN + PC) + DS) can be attributed to the partial loss of DS molecules from the complex particles due to the screening of opposite charges on DS, SCN, and PC molecules by the added electrolyte (up to 0.1 M in the "stomach" and further to 0.15 M in the "small intestine") presumably resulting in a great reduction in the strength of any attractive electrostatic interactions between these molecules. As a result of this loss, the architecture of the particles seems to be mainly dictated by the hydrolyzed protein. In the case of the complex particles based on the covalent conjugates (SCN + MD-SA2), the observed transformation from compact particles to ones with either entangled or more open architecture could be attributable to intensive modifications of the particles by proteolysis, most likely followed by hydrophobic aggregation.

The observed transformations in the molar masses (k_{Mw}) (Table 6.5) and the architecture (Figure 6.6) of the ternary complex particles led to marked changes in their density (Figure 6.7). There was one order of the magnitude increase in the density of the ternary complex particles formed on the basis of the SCN−DS electrostatic interactions ((SCN + PC) + DS) and, by contrast, one order of the magnitude decrease in the density of the ternary complex particles formed by the covalent conjugates (SCN + MD-SA2) with PC.

Figure 6.8 shows similar, rather low, levels of release of PC from the ternary complex particles specifically under the conditions of the "small intestine." This is dramatically different from the release of PC from the pure protein

Table 6.5 The Distinctive Characteristics of the Behavior of the Ternary Complex Particles in the Simulated Conditions of the GI Tract *In Vitro*

Sample	Release of PC, %	$V_0 \times 10^2$ (I_{90}/min)	d, mg/ml	$\rho = R_G/R_h$	k_{Mw}[a]
((SCN + MD-SA2) + PC) "stomach"	0	7.3	0.41	0.96	2.6
((SCN + MD-SA2) + PC) "small intestine"	22.4	7.3	0.06	0.56	2.7
((SCN + PC) + DS) "stomach"	0	0.6	0.21	0.55	1.2
((SCN + PC) + DS) "small intestine"	21.2	2.1	1.23	0.74	0.7

[a]k_{Mw} *is the extent of the aggregation of the complex particles, which is defined by the ratio of the weight-average molar masse of the ternary complex particle under the simulated GI tract conditions to the initial one.*

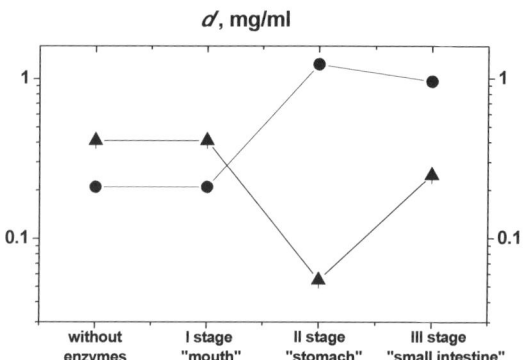

■ **FIGURE 6.7** Alteration of the density of the ternary complex particles ($R_w = 2$) during their consecutive passage through the model GI tract *in vitro*: ((SCN + PC) + DS) (●); ((SCN + MD-SA2) + PC) (▲).

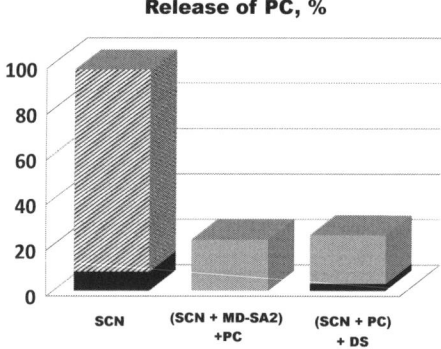

■ **FIGURE 6.8** Release of PC from the ternary complex particles ($R_w = 2$) during their consecutive passage through the model GI tract *in vitro*: without enzymes (the black part of the columns); stage II in the "stomach" (the striped part of the columns); stage III in the "small intestine" (the gray part of the columns).

particles that occurs completely under the conditions of the "stomach." Such result could be attributed to the steric hindrance of direct access of the protease to the protein combined in the complexes with the bulky polysaccharide molecules.

On the basis of the molecular parameters found for each stage of the proteolysis, we have tried to reveal the relationships between them and the initial velocity of the digestion, V_0, of the ternary complex particles (Table 6.5). Thus, the lower density of the particles and their more open architecture seem favourable for their increased susceptibility to proteolysis. At the same time, the gel-like architecture of the particles along with their efficient density apparently hinders the formation of the contacts between the enzyme and the substrate that leads to the very slow velocity of the hydrolysis. It seems that, under all other factors being equal, as in the case of the particles based on the conjugates, the lower density determines the higher extent of the PC release (Table 6.5). In turn, in the case of the particles based on the electrostatic SCN−DS interactions both the higher initial velocity of the hydrolysis and the lower extent of the aggregation, k_{Mw}, of the biopolymer particles can underline the higher extent of the PC release.

CONCLUSIONS

For the ternary complexes formed on the basis of different kinds of the interactions between proteins and polysaccharides, their behavior in general is mainly determined by the steric effects of the polysaccharides attached to the protein, while the differences are dictated by the responsiveness of the electrostatic interactions to the environmental conditions. Protein−polysaccharide electrostatic and covalent interactions have much to offer in the formulation of "intelligent" nanoscale encapsulation systems for food applications.

ACKNOWLEDGMENTS

The authors would like to express their gratitude to the manufacturers AVEBE (Netherlands) and Lipoid GmbH (Germany) for the supply of free samples for this work.

REFERENCES

Acosta, E., 2009. Bioavailability of nanoparticles in nutrient and nutraceutical delivery. Curr. Opin. Colloid Interface Sci. 14, 3−5.

Augustin, M.A., Hemar, Y., 2009. Nano- and micro-structured assemblies for encapsulation of food ingredients. Chem. Soc. Rev. 38, 902−912.

Bai, G., Nichifor, M., Bastos, M., 2010. Association and phase behavior of cholic acid-modified dextran and phosphatidylcholine liposomes. J. Phys. Chem. Lett. 1, 932−936.

Benichou, A., Aserin, A., Garti, N., 2004. Double emulsions stabilized with hybrids of natural polymers for entrapment and slow release of active matters. Adv. Colloid Interface Sci. 108−109, 29−41.

Burchard, W., 1994. Light scattering. In: Ross-Murphy, S.B. (Ed.), Physical Techniques for the Study of Food Biopolymers. Blackie, Glasgow, pp. 151−214.

Chen, H., Weiss, J., Shahidi, F., 2006. Nanotechnology in nutraceuticals and functional foods. Food Technol. 60, 30−36.

Chu, B., Zhou, Z., Wu, G., Farrell Jr., H.M., 1995. Laser light scattering of model casein solutions: effects of high temperature. J. Colloid Interface Sci. 170, 102−112.

de Oliveira Tiera, V.A., Winnik, F.M., Tiera, M.J., 2010. Interaction of amphiphilic derivatives of chitosan with DPPC (1,2-dipalmitoyl-sn-glycero-3-phosphocholine). J. Therm. Anal. Calor. 100, 309—313.

Faulks, R.M., Southon, S., 2008. Assessing the bioavailability of nutraceuticals. In: Garti, N. (Ed.), Delivery and Controlled Release of Bioactives in Foods and Nutraceuticals, Ch. 1. Woodhead Publishing Limited, CRS Press, Cambridge, England, pp. 3—25.

Gennis, R.B., 1989. Biomembranes Molecular structure and function. In: Cantor, C.R. (Ed.), Advanced Text in Chemistry. Springer, Berlin.

Grigoriev, D.O., Miller, R., 2009. Mono- and multilayer covered drops as carriers. Curr. Opin. Coll. Inter. Sci. 14, 48—59.

Grigorovich, N.V., Moiseenko, D.V., Antipova, A.S., Anokhina, M.S., Belyakova, L.E., Polikarpov, Y.N., et al., 2011. Structural and thermodynamic features of covalent conjugates of sodium caseinate with maltodextrins underlying formation, functionality and digestibility of their complexes with phosphatidylcholine. In the book of Abstracts of the 4th International Symposium "Delivery of Functionality in Complex Food Systems Physically-Inspired Approaches from the Nanoscale to the Microscale,". The University of Guelph, Canada, pp. 77—79.

Grigorovich, N.V., Moiseenko, D.V., Antipova, A.S., Anokhina, M.S., Belyakova, L.E., Polikarpov, et al., 2012. Structural and thermodynamic features of covalent conjugates of sodium caseinate with maltodextrins underlying their functionality. Food Funct. 3, 283—289.

HadjiSadok, A., Pitkowski, A., Nicolai, T., Benyahia, L., Moulai-Mostefa, N., 2008. Characterization of sodium caseinate as a function of ionic strength, pH and temperature using static and dynamic light scattering. Food Hydrocoll. 22, 1460—1466.

Horn, D., Rieger, J., 2001. Organic nanoparticles in the aqueous phase—theory, experiment, and use. Angewandte Chemie, International Edition 40, 4330—4361.

Ioan, C.E., Aberle, T., Burchard, W., 2000. Structure properties of dextran. 2. Dilute solution. Macromolecules 33, 5730—5739.

Kajiwara, K., Burchard, W., 1984. Rotational isomeric state calculations of the dynamic structure factor and related properties of some linear chains. 1. The $\rho = (S^2)^{1/2}(R_H^{-1})$ parameter. Macromolecules 17, 2669—2673.

Kidd, P.M., 1996. Phosphatidylcholine: a superior protectant against liver damage. Alter. Med. Rev. 1, 258—274.

Kidd, P.M., 2000. Dietary phospholipids as anti-aging nutraceuticals. In: Klatz, R.A., Goldman, R. (Eds.), Anti-Aging Medical Therapeutics. Health Quest Publications, Chicago, IL, pp. 283—301.

Kunz, D., Thurn, A., Burchard, W., 1983. Dynamic light scattering from spherical particles. Coll. Polymer Sci. 261, 635—644.

Livney, Y.D., 2008. Complexes and conjugates of biopolymers for delivery of bioactive ingredients via food. Ch. 7. In: Garti, N. (Ed.), Delivery and Controlled Release of Bioactives in Foods and Nutraceuticals. Woodhead Publishing Limited, CRS Press, Cambridge, England, pp. 234—250.

Macleod, G.S., Collett, J.H., Fell, J.T., 1999. The potential use of mixed films of pectin, chitosan and HPMC for bimodal drug release. J. Contr. Rel. 58, 303—310.

Markman, G., Livney, Y.D., 2012. Maillard-conjugate based core—shell co-assemblies for nanoencapsulation of hydrophobic nutraceuticals in clear beverages. Food Funct. 3, 262—270.

McClements, D.J., Li, Y., 2010. Review of in vitro digestion models for rapid screening of emulsion-based systems. Food Func 1, 32—59.

McClements, D.J., Decker, E.A., Park, Y., Weiss, J., 2008. Designing food structure to control stability, digestion, release and absorption of lipophilic food components. Food Biophys. 3, 219—228.

McClements, D.J., Decker, E.A., Park, Y., Weiss, J., 2009. Structural design principles for delivery of bioactive components in nutraceuticals and functional foods. Crit. Rev. Food Sci. Nutr. 49, 577—606.

Medina, C., Santos-Martinez, M.J., Radomski, A., Corrigan, O.I., Radomski, M.W., 2007. Nanoparticles: pharmacological and toxicological significance. Br. J. Pharmacol. 150, 552—558.

Müller-Buschbaum, P., Gebhardt, R., Roth, S.V., Metwalli, E., Doster, W., 2007. Effect of calcium concentration on the structure of casein micelles in thin films. Biophys. J. 93, 960—968.

Nagasawa, M., Takahashi, A., 1972. Light scattering from polyelectrolyte solutions. In: Huglin, M.B. (Ed.), Light Scattering from Polymer Solutions. Academic Press, London, pp. 671—723.

Nash, W., Pinder, D.N., Hemar, Y., Singh, H., 2002. Dynamic light scattering investigation of sodium caseinate and xanthan mixtures. Int. J. Biol. Macromol. 30, 269—271.

Ransley, J.K., Donnelly, J.K., Read, N.W. (Eds.), 2001. Food and Nutritional Supplements: Their Role in Health and Disease. Springer-Verlag, Berlin.

Semenova, M.G., Dickinson, E., 2010. Biopolymers in Food Colloids: Thermodynamics and Molecular Interactions. In: Burlakova, E.B., Zaikov, G.E. (Eds.). Brill, Leiden, p. 350.

Semenova, M.G., Belyakova, L.E., Polikarpov, Yu, N., Antipova, A.S., Anokhina, M.S., 2008. Utilization of sodium caseinate nanoparticles as molecular nanocontainers for delivery of bioactive lipids to food systems: relationship to the retention and controlled release of phospholipids in the simulated digestion conditions. In: Williams, P.A., Phillips, G.O. (Eds.), Gums and Stabilisers for the Food Industry 14. Royal Society of Chemistry, Cambridge, UK, pp. 326—333.

Semenova, M.G., Belyakova, L.E., Polikarpov, Yu.N., Antipova, A.S., Dickinson, E., 2009. Light scattering study of sodium caseinate+dextran sulfate in aqueous solution: relationship to emulsion stability. Food Hydrocoll. 23, 629—639.

Semenova, M.G., Antipova, A.S., Anokhina, M.S., Belyakova, L.E., Polikarpov, Yu.N., Grigorovich, N.V., Tsapkina, E.N., 2012. Thermodynamic and structural insight into the underlying mechanisms of the phosphatidylcholine liposomes—casein associates co-assembly and functionality. Food Funct. 3, 271—282.

Shaw, L.A., Faraji, H., Aoki, T., Djordjevic, D., McClements, D., Decker, E.A., 2008. Emulsion droplet interfacial engineering to deliver bioactive lipids into functional foods. Ch. 7. In: Garti, N. (Ed.), Delivery and Controlled Release of Bioactives in Foods and Nutraceuticals. Woodhead Publishing Limited, CRS Press, Cambridge, England, pp. 184—206.

Shoemaker, S.D., Vanderlick, T.K., 2003. Calcium modulates the mechanical properties of anionic phospholipid membranes. J. Colloid Interface Sci. 266, 314—321.

Singh, H., Ye, A., Horne, D.S., 2009. Structuring food emulsions in the gastrointestinal tract to modify lipid digestion. Prog. Lipid Res. 48, 92—100.

Stenekes, R.J.H., Loebis, A.E., Fernandes, C.M., Crommelin, D.J.A., Hennink, W.E., 2001. Degradable dextran microspheres for the controlled release of liposomes. Int. J. Pharm. 214, 17–20.

Tanford, C., 1961. Physical Chemistry of Macromolecules. Wiley, New York.

Tuteja, A., Mackay, M.E., Hawker, C.J., Van Horn, B., Ho, D.L., 2006. Molecular architecture and rheological characterization of novel intramolecularly crosslinked polystyrene nanoparticles. J. Polymer Sci.: Part B: Polymer Phys. 44, 1930–1947.

Vandamme, Th.F., Lenourry, A., Charrueau, C., Chaumeil, J.–C., 2002. The use of polysaccharides to target drugs to the colon. Carb. Polymers 48, 219–231.

Velikov, K.P., Pelan, E., 2008. Colloidal delivery systems for micronutrients and nutraceuticals: tools and resources. Soft Matter 4, 1964–1980.

Zimet, P., Livney, Y.D., 2009. Beta-lactoglobulin and its nanocomplexes with pectin as vehicles for ω-3 polyunsaturated fatty acids. Food Hydrocoll. 23, 1120–1126.

Chapter 7

Muscle Structure and Digestive Enzyme Bioaccessibility to Intracellular Compartments

Thierry Astruc

INRA Clermont-Ferrand Theix, "Quality of Animal Products" Research Unit, Saint Genès Champanelle, France

CONTENTS

Food Structures, Digestion and Health. http://dx.doi.org/10.1016/B978-0-12-404610-8.00007-4

INTRODUCTION

The body draws its energy from carbohydrates, fats, and proteins. These three energy nutrients form the class of macronutrients from which oxidative degradation provides energy to the body. Energy is extracted by the digestive system, which converts sugars into glucose (or galactose), proteins into amino acids, and fats into fatty acids.

Meat, whose origin is skeletal muscle of farm animals, is composed mainly of proteins. The muscle intracellular proteins contain 40 to 45% of essential amino acids, which confer their high nutritional value. In addition to high protein content, meat is a carrier of other nutrients such as vitamins and minerals.

The muscle is an organ that is highly structured at several scales. This structure is likely to hinder the accessibility of digestive enzymes to their major substrates, which are the myofibrillar proteins within cells. Technological treatments applied to meats modify their structures but do not completely upset the general organization of muscle, except in highly fragmented products such as protein emulsions. Chewing partially deconstructs meat but at the microscopic level this breakdown is very incomplete and different levels of barriers to the ingress of digestive hydrolases are likely to be retained.

Understanding the effect of processing on the meat structure in connection with the bioaccessibility of digestion juices is a prerequisite for improving the nutritional quality of meat products through modulation of technological processes.

PHYSIOLOGY OF DIGESTION

Digestion is the mechanical and chemical breakdown of food into smaller components that are more easily absorbed into the bloodstream.

The first stage of digestion is mastication, which consists of converting food into a bolus ready for swallowing. After chewing, the particle size is reduced to a few millimeters. The average particle size of cooked meat and cooked ham is close to 1.5 mm, but some particles exceed 4 mm (Jalabert-Malbos *et al.*, 2007). Thus, even if chewing breaks down the structure of the food eaten, the microscopic structure of the food is partially preserved, which may affect the accessibility of digestive enzymes in the following steps of digestion. Saliva makes the food easier to swallow. It goes down the throat and esophagus and reaches the stomach. Stomach gastric juice, which mainly contains hydrochloric acid and pepsin, starts protein digestion. Peristaltic contractions promote mixing of the food with gastric juice and its progression in the stomach. The resulting thick liquid, which is called chyme, enters the duodenum where it is mixed with digestive enzymes from the pancreas, and then passes through the small intestine, in which digestion continues. When the chyme is fully digested, it is absorbed into the blood; 95% of absorption of nutrients occurs in the small intestine. Water and minerals are reabsorbed back into the blood in the colon (large intestine) where the pH is about 5.6 to 6.9. Waste material is eliminated from the rectum during defecation.

TOOLS FOR DIGESTION STUDIES

Two approaches are used to assess the digestibility of foods: *in vivo* using animals or humans, and *in vitro* using systems that simulate the process of digestion.

In Vivo Systems

In vivo systems generally provide the most relevant results because they include all stages of digestion and their interactions. Studies are often conducted on rats, pigs, and humans. This approach is regulated and ethically restrictive, expensive, and time consuming. The number of individuals employed (mainly when it comes to humans) is often reduced to limit the experimental cost. Finally, a large variability can be observed in the efficiency of mastication and digestion, which sometimes makes it difficult to interpret the results.

In Vitro Systems

In contrast, *in vitro* digestion systems are less expensive and can be used in the laboratory without special regulatory constraints. They allow the separate study of each stage of digestion. There are, for example, artificial masticators (Salles *et al.*, 2007; Mielle *et al.*, 2010; Woda *et al.*, 2010) and artificial systems where gastric digestion can be disconnected from the intestinal digestion (Savoie and Gauthier, 1986; Gatellier and Santé-Lhoutellier, 2009; Kaur *et al.*, 2010a, b).

Hur *et al.* (2011) published an overview of the different *in vitro* digestion systems that have been used to simulate human digestion. If the enzymes used are generally common to the various scientists, the authors highlight the disparity of incubation periods. It is also likely that the activities of enzymes are variable from one study to another. Unlike *in vivo* digestion, the peristaltic contractions of the digestive organs are difficult to reproduce, as are the phenomena of compensation and adaptation of digestion in response to sensory stimuli and to the parasympathetic system which modulates digestive secretions and mobility of the gastrointestinal tract.

Methods for Assessing the Bioaccessibility of Digestive Enzymes

Indirect and direct imaging approaches are the two major ways to characterize the accessibility of digestive enzymes in food.

The indirect approach is to characterize and interpret the structural changes resulting from the hydrolytic action of enzymes. Thus, it is not the digestive enzymes of interest that are highlighted, but rather the result of their action in the food. These structural changes can be characterized at different scales by light and electron microscopy or by spectral imaging. The advantage of this indirect approach is that methods of sample preparation are adapted from known techniques. The difficulties lie in the fact that the textural changes consecutive to processing and enzymatic digestion sometimes make it difficult to prepare the sample, which disintegrates or no longer adheres to its support for imaging observation. A high degree of technical ability of the operator is then required to overcome these problems and get usable preparations for structural analysis.

The second difficulty is related to the interpretation of images. During digestion, changes in food morphology result from the hydrolytic action of enzymes but also from all physicochemical conditions such as the pH of the digestion juices and peristaltic contractions of the digestive tract (or their simulations). The observed structural changes are most often due to the interaction of physicochemical factors and the hydrolytic action of enzymes. To better understand the mechanisms of digestion, it is sometimes necessary to decouple physicochemical conditions from enzymatic actions, which can be done with *in vitro* systems.

The direct approach consists in marking the digestive enzymes of interest to highlight them in the food by immunohistological methods (optical microscopy) or immunocytology (electron microscopy) according to the required level of scale. This is theoretically the most relevant way because it allows location of the digestive enzyme in the food and local characterization of the morphological changes generated by digestion at the same time.

Antigen–antibody labeling is generally used. A second antibody coupled to an identifiable marker allows location of the antigen–antibody complex in the matrix by microscopy. The main challenge is to find an antibody specific for the target enzyme and to preserve the enzyme target antigenicity during sample preparation.

MUSCLE COMPOSITION AND STRUCTURE

Meat comes from the skeletal muscles of farm animals. Skeletal muscle consists of roughly 75% water, 19% protein, 2.5% intramuscular fat, 1.2% carbohydrates, and 2.3% other soluble non-protein substances.

Skeletal muscle is essentially composed of cells called "muscle fibers," due to their elongated shape, that account for up to 90% of muscle volume.

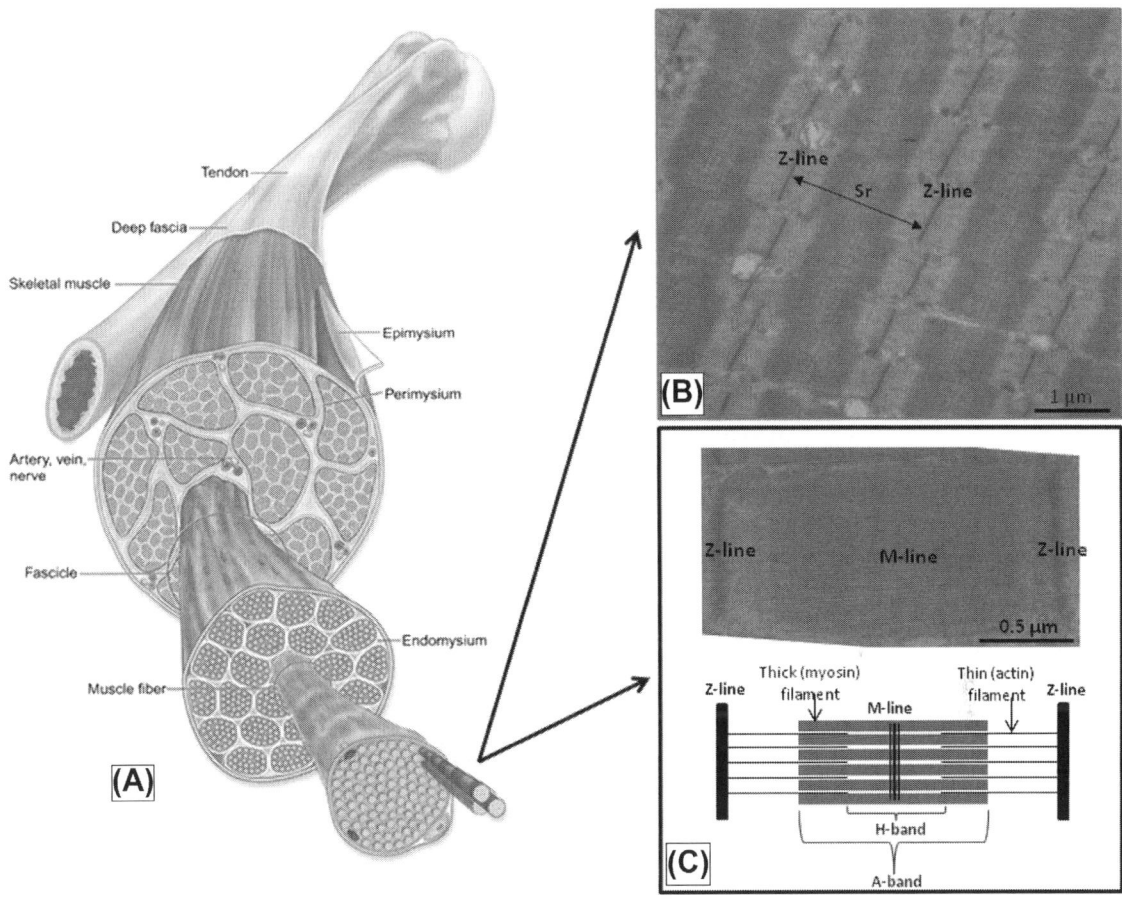

■ **FIGURE 7.1** Skeletal muscle organization.
A: General muscle organization. B: longitudinal
section of a muscle fiber showing the
myofibrils structure and organization. Myofibrils
are composed of a succession of sarcomeres
(Sr) located between the Z lines.
C: Ultrastructure of a sarcomere composed of
myosin and actin myofilaments. [A color
version of this figure is available online at
www.booksite.elsevier.com/9780124046108].

These muscle fibers are ensheathed in connective tissue (endomysium) and packed into fascicle groups of several cells ensheathed in another layer of connective tissue (perimysium) to form fiber bundles. Muscle structure organization is presented in Figure 7.1.

The muscle fiber is a spindle-shaped cylindrical filament measuring 10 to 100 μm in diameter but up to several centimeters long. Each muscle fiber is wrapped in its plasma membrane called the sarcolemma. The intracellular compartment contains the cytoplasm (called sarcoplasm), myofibrils, organelles, and nuclei. Adult mammalian muscle has four types of muscle fiber, called type I, type IIA, type IIX, and type IIB, which vary in terms of muscle metabolism and contractile speed (Pette and Staron, 1990, 2000). Muscles generally comprise various proportions of each muscle fiber type depending on the muscle's function.

Optical microscopy reveals that muscle fibers contain alternating light iso-trope bands (I bands) and dark anisotrope bands (A bands) that give skeletal muscle its striated appearance. The ultrastructural level reveals that the sarcoplasm contains mainly myofibrils that are arranged longitudinally (along the long axis of the cell) and that measure about 1 to 2 μm in diameter and fill almost all of the intracellular volume (Figure 7.1).

Electron microscopy allows access to a more complex structure: the A band is in fact composed of an outer part, dark and homogeneous, and of a central, clearer part called the H band. The central part of the H band is a little darker, and is called the M band. Band I is separated into two parts by the Z line (darker). Between two Z lines is the sarcomere which is the contractile unit of the myofibril. The sarcomere is composed of at least 28 different proteins (Craig and Padron, 2003) but consists mainly of two types of filaments: the thin myofilaments, essentially composed of actin, troponin, and tropomyosin proteins, and the thick myofilaments, mainly composed of myosin molecules, which account for 50% of the myofibrillar proteins. Other fibrillar proteins, among others titin, myomesin, M protein, and nebulin, maintain the internal structure and cohesion of myofibrils. It is assumed that the intracellular proteins are distributed one-third in sarcoplasmic proteins and two-thirds in myofibrillar proteins.

Meat is also composed of connective tissue, from 1.5 to about 10% of muscle dry weight depending on muscle function. Intramuscular connective tissue is principally composed of collagen and, to a lesser extent, of elastin proteins within a proteoglycan matrix (Bailey and Light, 1989). Among the 28 types of collagen, types I and III are predominant in muscle.

Intramuscular connective tissue is divided into three main structures (Figure 7.1): the epimysium, which surrounds the muscle, the perimysium, which surrounds bundles of muscle fibers, and the endomysium, which surrounds the muscle fibers (Bailey and Light, 1989; Gillies and Lieber, 2011).

Perimysium represents about 90% of total muscle connective tissue, where it is organized as a network of interconnected segments that vary extensively according to muscle type, species, age, and region in the muscle.

Endomysium surrounds each muscle fiber. It is composed of basal lamina proteins, proteoglycans, and laminin, plus collagens I, III, and IV, and is very similar in all muscle types and even across species.

Connective tissue content and composition significantly affect the organoleptic and technological qualities of meat and meat products, particularly texture and water-holding capacity (Bailey and Light, 1989; Purslow, 2005).

Meat connective tissue is at least 90% collagen, which has very poor nutritional value because it contains no tryptophan and low levels of the other dietary essential amino acids. Contrary to what was believed for a long time, digestibility of collagen seems to be as good as that of other meat proteins (Leibovich and Weiss, 1970; Bailey and Light, 1989; Hooda *et al.*, 2012).

Elastin also has poor nutritional value because of a low content of several dietary essential amino acids. Moreover, this protein generally represents about 5% of the protein connective tissue except in *semitendinosus* and *latissimus dorsi* where its proportion reaches about 40% of connective tissue (Rowe, 1986; Bailey and Light, 1989). The other components of connective tissue, such as proteoglycans, glycoproteins, and components of the basement membrane, would be expected to have high nutritional value, but they represent a minor part of the intramuscular connective tissue. Thus, it can be assumed that overall connective tissues have a poor nutritional value.

On the other hand, the structural organization of muscle could impede the access of digestive proteases to intracellular muscle proteins. Although collagen is known to be degradable by digestive proteases, the extracellular matrix is made up of a complex of molecules that could lead to a decrease in connective tissue degradability by the digestive juices. Meat is rarely eaten fresh, but undergoes certain transformations, the most common being cooking. Processing changes the composition and structure of muscle components and so can affect the bioaccessibility of digestive hydrolases and consequently the digestibility of meat. However, few studies have been conducted to characterize and understand the role of the structure of meat products on the access of digestive enzymes to the intracellular proteins that have the highest nutritional value.

EFFECTS OF PROCESSING ON THE MICROSTRUCTURE OF MEAT

Postmortem Changes in Muscle

Meat is the result of skeletal muscle changes after the slaughter of meat animals (Greaser, 1986). After bleeding, muscle metabolism is modified due to the cessation of blood flow. Muscles are thus in anoxia, and ATP synthesis is then based on anaerobic glycolysis. As the level of ATP decreases and glycogen is degraded, protons and lactate molecules are formed and accumulate in muscle tissue, resulting in a decrease of muscle pH. When muscle glycogen stores are depleted, the pH stabilizes at a value called ultimate pH (pHu), typically between 5.4 and 5.9, depending on the muscle.

Bovine meat is stored in a cold room for several days before its consumption in order to improve the texture and develop flavors (Ouali, 1990). This storage leads changes in the tissue structure as shown in Figure 7.2. Usually, meat storage leads to a decrease in muscle fiber cross-sectional area and an increase in the size of the extracellular space (Figures 7.2B) associated with a loss of water through exudation (Offer and Knight, 1988a).

During this storage, endogenous proteases are in a favorable environment for the degradation of muscle proteins. Their action causes cracks perpendicular to the muscle fibers (Figure 7.2D) and results in a more or less significant change in some myofibrillar structures such as Z lines, M bands, and cytoskeletal structures composed of titin, desmin, and other proteins. The alteration of these structures results in a cross-fragmentation of myofibrils most often observed in the I bands and near the Z lines (Davey and

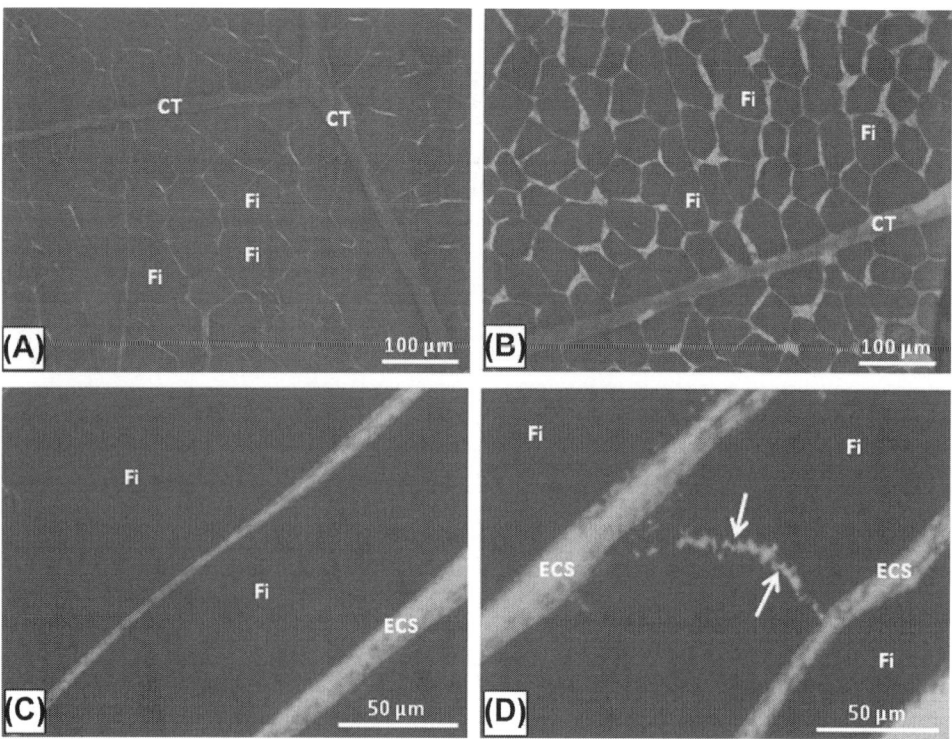

■ **FIGURE 7.2** Structural changes during meat storage. Digital images of histological skeletal muscle sections (optical microscopy). Transversal and longitudinal sections of 1-day postmortem (A; C) and 12 days postmortem (B; D) bovine *semitendinosus* muscle. Increasing the postmortem time leads to lateral fiber shrinkage, an increase in the area of extracellular space, changes in the connective tissue structure (B) and cracks in the fibers (D; arrows). Fi: muscle fiber; CT: connective tissue; ECS: extracellular spaces.

Dickson, 1970; Ouali, 1990; Taylor *et al.*, 2002). An example of these changes is given in Figure 7.3B.

Indirect measurements suggest that the plasma membrane deteriorates rapidly after the death of the animal (Heffron and Hegarty, 1974; Bertram *et al.*, 2004). Ultrastructural study shows perforations of the mouse sarcolemma in the hours following its death (Astruc, 2008a; Figure 7.3C).

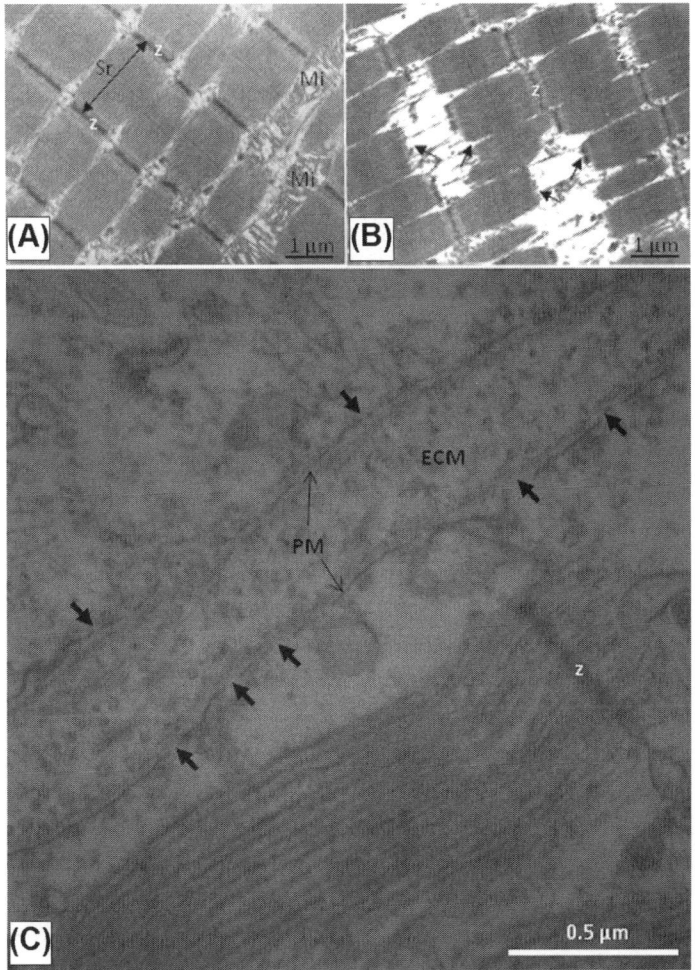

■ **FIGURE 7.3** Ultrastructural changes during meat storage. Digital images of skeletal muscle (transmission electron microscopy). A: 1-day postmortem *semitendinosus* bovine muscle. B: 14 days postmortem *semitendinosus* bovine muscle. C: 1-day postmortem *longissimus dorsi* mouse muscle. Postmortem proteolysis results in myofibrils breaks down along the Z line (B). Plasma membrane is altered (C; arrows). PM: plasma membrane; ECM: extracellular matrix.

Mechanisms are still poorly understood but may be a consequence of an increase in the osmotic pressure and to enzymatic hydrolysis.

The connective tissue is much less altered during maturation, but still undergoes structural changes in both the perimysium (Figure 7.2B) and the endomysium (Stanton and Light, 1987, 1988, 1990; Liu *et al.*, 1994; Nishimura *et al.*, 1995; Nishimura, 2010).

All these structural changes are likely to facilitate the bioaccessibility of digestive enzymes to the myofibrillar proteins and consequently enhance their degradation. Thus, the deterioration of the sarcolemma suggests an easier transfer of digestive juice to the intracellular compartment. Proteolysis of contractile proteins is also foreseen to facilitate the access of digestive enzymes to myofilaments.

Marination

Before cooking, the meat can be subjected to marinating. Acidic marinating is usually applied to improve the tenderness of meat rich in connective tissue. Acidity causes swelling of the collagen fibers (Figure 7.4) as a result of the rupture of weak bonds between collagen fibrils, which decreases the intramuscular collagen strength and improves meat tenderness (Offer and Knight, 1988b; Bailey and Light, 1989; Rao *et al.*, 1989; Lewis and Purslow, 1991; Aktas and Kaya, 2001a, b; Berge *et al.*, 2001; Chang *et al.*, 2010).

Acidic solutions increase the water-binding capacity of meat which increases the volume of meat pieces (Offer and Knight, 1988b) and muscle fibers (Wilding *et al.*, 1986; Offer and Knight, 1988b; Rao *et al.*, 1989). We recently observed that 1-day postmortem muscle, incubated 24 hours in acetic acid at pH 4.5, led to an increase in the cross-sectional area of the cell and in the area of the extracellular spaces of 16% ($p < 0.01$) and 33% ($p < 0.01$), respectively, which implies a gain in the volume of meat. However, no difference was observed when the muscle was incubated after 12 days' postmortem storage.

Meat acidification also causes an alteration of the structure of the myofibrils such as solubilization of the M lines and Z lines (Figure 7.4). We can expect that these acidic treatments, which lead to a partial breakdown of meat, will facilitate subsequent transfers of aqueous solutions such as digestive juices.

Salting

Salting or brining meat is a common practice, especially for pork and poultry. Meat transformation processes add sodium chloride as a flavor enhancer, a preservative, and a texture-control agent. Moreover, in

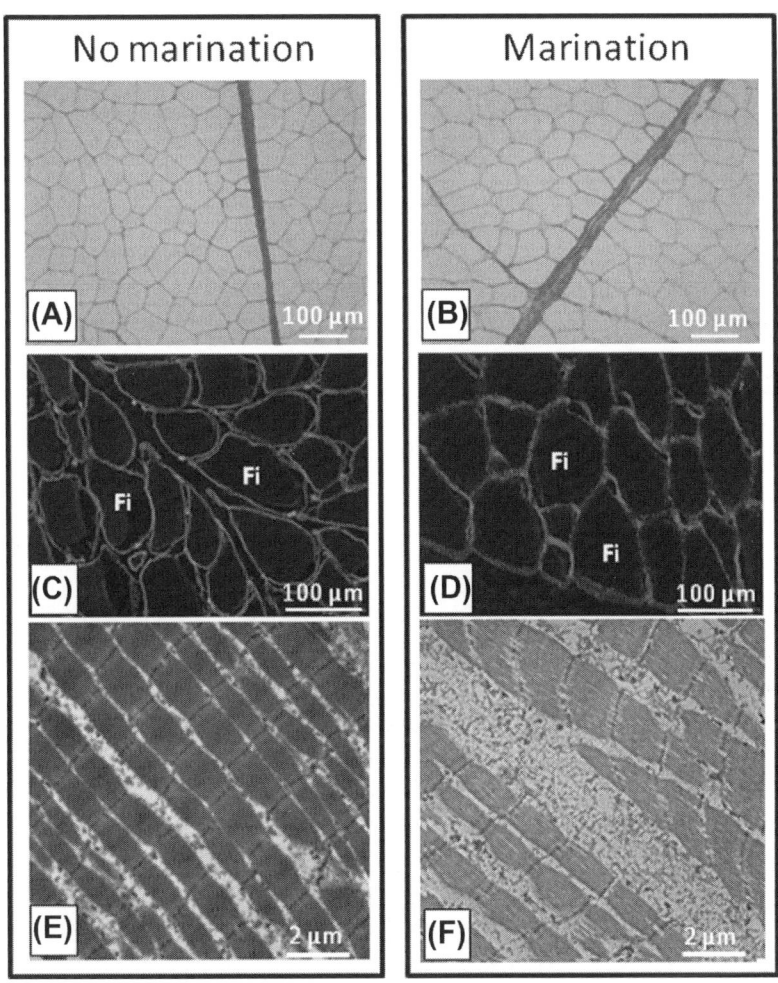

■ **FIGURE 7.4** Effect of acidic marination on meat structure. Bovine *semitendinosus* muscle samples before (A, C, E) and after (B, D, F) incubation in acetic acid at pH 4.75. A and B: Sirius red staining of muscle fibers cross-section. Collagen appears in dark gray around muscle fibers stained light gray (optical microscopy, bright field). C and D: Immunostaining of laminin, a component of the extracellular matrix (optical microscopy, epifluorescence). E and F: Longitudinal ultrathin section of muscle fibers observed in transmission electron microscopy. After acidic marination, collagen (B) and laminin (D) are more diffuse, reflecting damaged tissue. Marination causes a swelling of cells (B, D) and a significant change in the structure of myofibrils (F). The M lines and Z lines have a weakness, and extracted proteins are seen between the myofibrils. [A color version of this figure is available online at www.booksite.elsevier.com/9780124046108].

processed meats, salt improves technological quality by increasing the water-holding capacity of meat products. Salting is usually followed by drying or cooking. However, the simple fact of salting causes substantial structural modifications (Figure 7.5) that are able to affect the transfer of digestion juices into the intracellular compartments of muscle fibers.

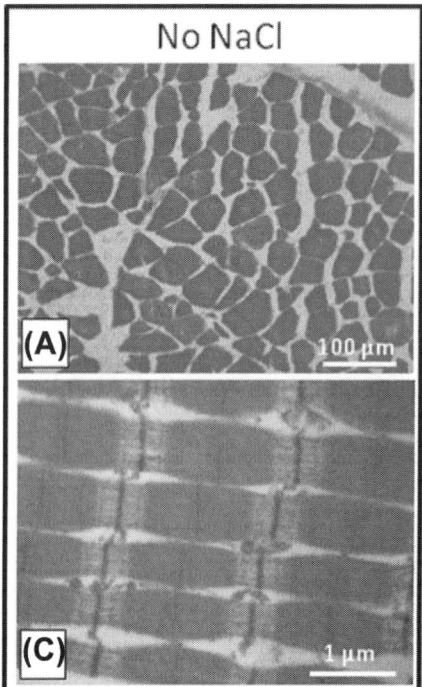
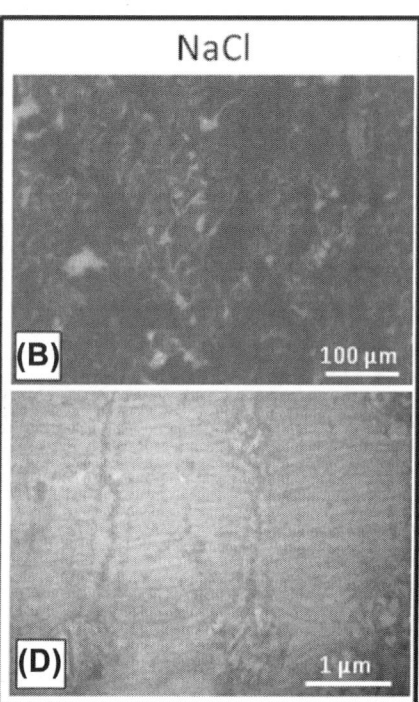

■ **FIGURE 7.5** Effect of sodium chloride on meat structure. Bovine *semitendinosus* muscle samples were incubated in a 5.8% NaCl bath. A and B: Cross-sections of not brined (A) and brined (B) muscle (optical microscopy, hematoxylin, eosin, safran staining). C and D: Longitudinal sections of not brined (A) and brined (B) muscle (transmission electron microscopy). Brining causes a profound change in the structure at the tissue, cellular, and ultrastructural scales. Salting leads to an increase in the volume of muscle fibers and myofibrils, and to a partial extraction of myofibrillar proteins. [A color version of this figure is available online at www.booksite.elsevier.com/9780124046108].

Adding sodium chloride to meat increases the ionic strength of muscle tissue and denatures the proteins, changing their molecular structure (Böcker *et al.*, 2006; Graiver *et al.*, 2006; Wu *et al.*, 2006; Chang *et al.*, 2010). Differential scanning calorimetry measurements showed that salting led to a large decrease in the denaturation temperature of collagen (Aktas and Kaya, 2001a; Chang *et al.*, 2010). Imaging techniques have also demonstrated that connective tissue undergoes structural changes during the salting of meat (Graiver *et al.*, 2006; Astruc *et al.*, 2008b; Chang *et al.*, 2010). However, the connective tissue was still visible and perimysium seemed to be a barrier limiting the salt diffusion (Astruc *et al.*, 2008b; Filgueras *et al.*, 2011a).

The addition of salt, most often in the form of brine, causes swelling of pieces of meat that accumulate water (Offer and Knight, 1988b). Several studies have shown that salt causes a swelling of muscle fibers (Offer and Knight, 1988b) and myofibrils (Offer and Trinick, 1983; Wilding *et al.*, 1986; Knight and Parsons, 1988; Parsons and Knight, 1990). These authors clearly demonstrated extraction of Z lines and A bands gradually as the NaCl concentration increased from 0 to 1 M. Wilding *et al.* (1986) also showed that the degree of swelling of muscle fibers was greater in the vicinity of areas where the endomysium was injured. In addition, the action of collagenase significantly increased swelling of the muscle fibers. Thus, the envelopes of the endomysium seem to slow the diffusion of ions, and probably small molecules, from the extracellular space to the intracellular space.

Cooking

In addition to the destruction of microorganisms and parasites, cooking causes important structural and chemical changes that affect the sensory attributes (texture, color, flavor, and juiciness) and nutritional qualities of meat. Examples of structural changes in meat induced by heating are shown in Figures 7.6 and 7.7.

Heating causes shrinkage of meat pieces in all three dimensions when fried or roasted, partly due to extrusion and evaporation of water from the meat, decreasing meat tenderness and juiciness (Bouton and Harris, 1972; Davey and Gilbert, 1974; Tornberg, 2005). Conversely, slow cooking in moist

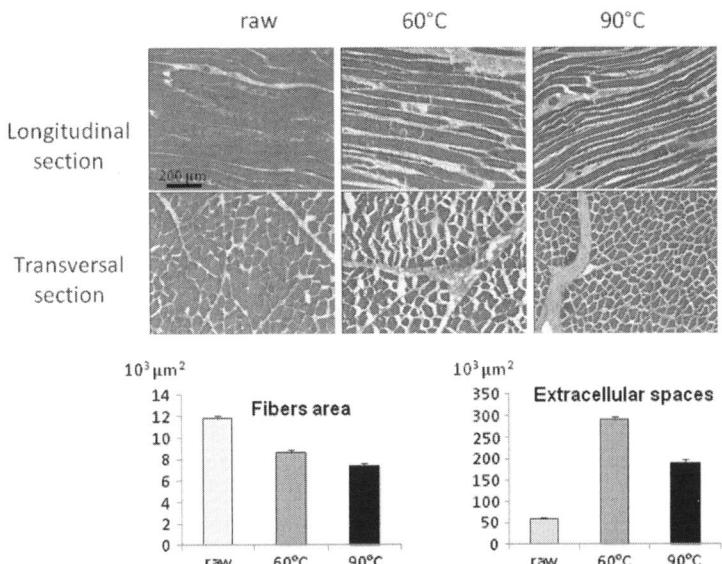

■ **FIGURE 7.6** Heating effect on fiber structure. Bovine *semitendinosus* muscle samples heated at 60 and 90°C for 45 minutes. The heating caused a reduction of the fibers' cross-sectional area and an increase in the extracellular spaces. Quantification by image analysis showed a less pronounced decrease of the extracellular space area at 90°C than at 60°C. [A color version of this figure is available online at www.booksite.elsevier.com/9780124046108].

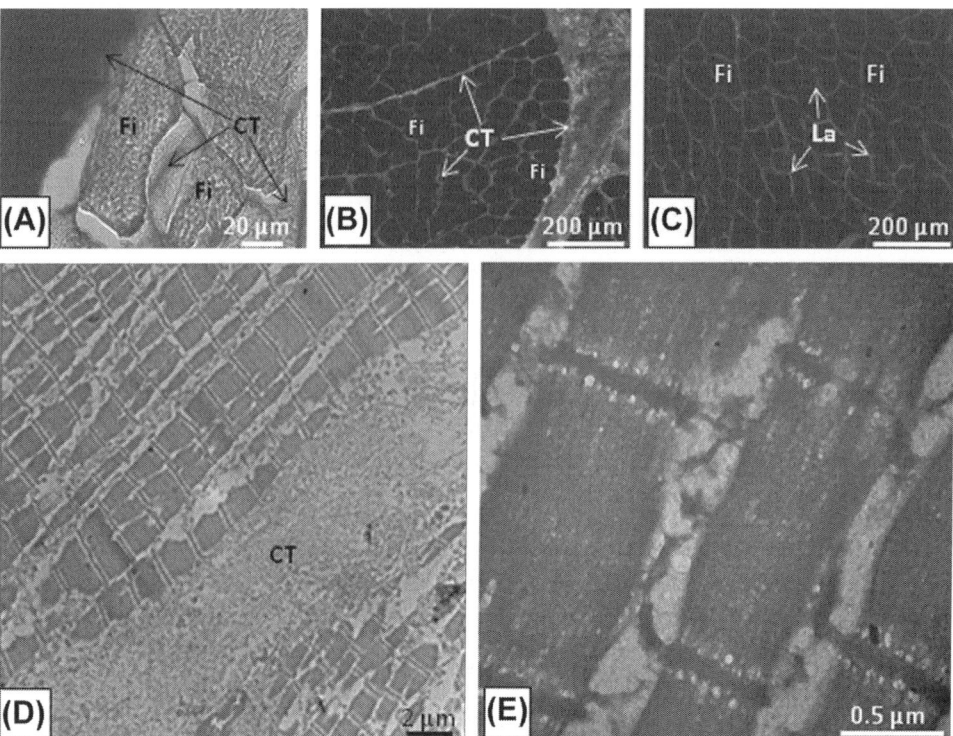

■ **FIGURE 7.7** Heating effects on muscle connective tissue and myofibrils. Bovine *semitendinosus* muscle heated at 60°C for 45 minutes. A: Sirius red staining of a cross-section, connective tissue (CT) appears in dark gray between the muscle fibers (Fi) (optical microscopy). B and C: Immunolabeling of collagen (B) and laminin (C) on muscle fiber cross-sections (optical microscopy). These extracellular proteins are stained light gray and surround the fibers. D and E: Longitudinal sections of muscle fibers observed in transmission electron microscopy. Collagen is altered but still recognizable (CT, D), and the myofilaments and Z lines (E) are coagulated. Fi. muscle fiber; CT: connective tissue; La: laminin.

conditions improves the texture of muscles rich in connective tissue. Evolution in meat quality is accompanied by structural changes in the tissue and muscle fibers, depending on heating parameters and muscle composition. The myofibrillar mass of cells shrinks laterally and longitudinally as a function of temperature (Hostetler and Landman, 1968; Schmidt and Parrish, 1971; Bendall and Restall, 1983; Tornberg, 2005; Astruc *et al.*, 2010). The collagen of intramuscular connective tissue, i.e., endomysial and perimysial collagen, shrinks and solubilizes at temperatures from 60 to 90°C, depending on the number of cross-links in the collagen (Schmidt and Parrish, 1971; Davey and Gilbert, 1974; Hamm, 1977, Light *et al.*, 1985). In recent work we saw that the structure of the perimysium and endomysium of bovine muscles was largely modified by heating the sample at 60°C for 45 min, but the collagen, which was partially solubilized, remained localized around the fibers and fiber bundles (Figure 7.7). Variation of this structural modification can be expected according to the heating conditions. It is likely that cooking of very long duration in wet conditions leads to a more pronounced disruption of the collagen layer structure.

At the ultrastructural level, increasing the heating temperature of fibers generally results in disintegration of the Z line structure, an increase in inter-myofibrillar spaces, shortening of the sarcomeres, cracks and breaks in the myofibrils at the Z lines, fragmentation or granulation of the myofibrils, destruction of cell membranes, and shrinkage and solubilization of connective tissue (Schmidt and Parrish, 1971; Jones *et al.*, 1977; Hearne *et al.*, 1978; Leander *et al.*, 1980; Wu *et al.*, 1985; Rowe, 1989; Palka and Daun, 1999; Wattanachant *et al.*, 2005; Astruc *et al.*, 2010). These structural changes result from the thermal denaturation of meat proteins (Hamm, 1977; Tornberg, 2005; Astruc *et al.*, 2012a).

At the molecular scale, protein denaturation leads to exposure of hydrophobic residues and their interactions lead to the formation of protein–protein aggregates (Hamm, 1977; Tornberg, 2005; Santé-Lhou-tellier *et al.*, 2008; Promeyrat *et al.*, 2010; Astruc *et al.*, 2010). Denaturation of myosin, in particular, leads to the formation of a gel (Xiong, 1994) which could affect the penetration of digestive solutions into the meat matrix. As explained above, collagen denatures at about 60°C and dissolves above about 80°C, depending on the level of intermolecular cross-links and the collagen type. However, using infrared microspectroscopy, we found that, *in situ*, bovine *semitendinosus* collagen denatures slightly below 80°C (Astruc *et al.*, 2012a). Furthermore, other components of intramuscular connective tissue probably play a role in the structure evolution of endo- and perimysium during heating meat: elastin, for example, is very thermo stable (Bailey and Light, 1989; Astruc *et al.*, 2012a) and can help to maintain the muscle organization during heating.

CONSEQUENCES OF MEAT PROCESSING ON BIOACCESSIBILITY TO DIGESTION JUICES AND DIGESTIBILITY EFFICIENCY

Bibliographic data on this subject are almost nonexistent. The majority of results and illustrations that are presented in this chapter are derived from a study whose preliminary results were presented at the conference "Food Structure, Digestion and Health" in Palmerston North (New Zealand) from March 7 to 9, 2012 (Astruc *et al.*, 2012b).

In this study, two bovine *semitendinosus* muscles were collected and stored in a cold room during, respectively, 1 day and 12 days. From each muscle, samples of $1 \times 1 \times 1.5$ cm were collected and subjected to different techno-logical treatments: a 24-hour immersion in a solution of acetic acid at pH 4.5 to simulate an acid marination; a heating of samples at, respectively, 60 and 90°C for 45 minutes; and various combinations of these treatments. Samples were then subjected to *in vitro* gastrointestinal digestion. Gastric digestion was simulated by immersing the sample in a pepsin solution (125 U/mg pro-teins in pH 1.8 glycine buffer) for 2 hours. Intestinal digestion was simulated by a subsequent immersion of the sample for 2 hours in a mixture of trypsin (150 U/mg protein) and chymotrypsin (0.1 U/mg protein) in glycine buffer titrated to pH 8. Digestions were carried out under agitation.

Sampling and analyses were performed at each step of the protocol to char-acterize the separate and interactive effects of treatments.

Fresh Meat

Although meat is rarely eaten raw, it was necessary to characterize and un-derstand how the digestive proteases diffuse into samples of raw muscle. These analyses are justified by the fact that certain meats, such as steaks or roasts, are often still raw inside. Furthermore, the effect of digestive pro-teases in raw muscle was used as a control to characterize the digestibility of processed meats.

Samples of 24-hour postmortem bovine *semitendinosus* muscle of $1 \times 1 \times 1.5$ cm immersed for 2 hours in a solution of pepsin were prepared for observation by optical microscopy. A cross-section of the entire sample is presented in Figure 7.8. The penetration of the digestive juice is represented by a halo of about 2 mm depth on the periphery of the sample (Figure 7.8A and B). In this halo, the cell size had increased by 65% (Figure 7.8A, B and D) compared to the cells located in the central part of the sample, where the digestive juice had not penetrated (Figure 7.8C). The connective tissue is partially degraded in the halo of digestion in contrast to the rest of the sample. Immersion of the sample in the same solution but without pepsin showed an increase in the cell size in the halo too (results not shown), indicating that cell swelling was due to the acidic pH of the digestion solution (pH 1.8), rather than to the action of pepsin. This swelling of the cells is consistent with studies on the effect of acid incubation on the structure of muscle (Wilding *et al.*, 1986; Offer and Knight, 1988b; Rao *et al.*, 1989). In addition, the degra-dation of connective tissue (Figure 7.8D) is likely to result from an interaction between the acidic environment, which causes swelling of collagen fibers (Offer and Knight, 1988b; Bailey and Light; 1989; Rao *et al.*, 1989; Aktas and Kaya, 2001a), and the action of pepsin, which degrades connective tissue proteins (Leibovich and Weiss, 1970; Bailey and Light, 1989).

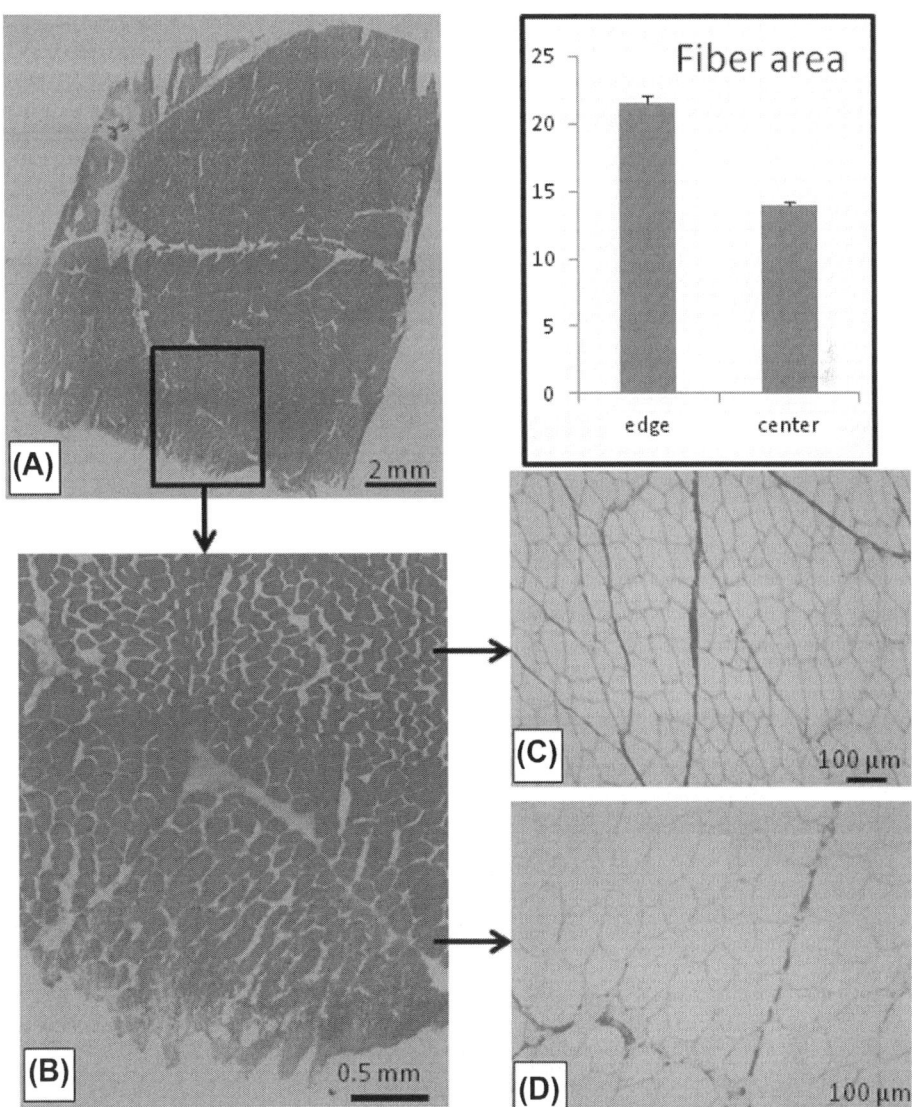

■ **FIGURE 7.8** Pepsin digestion of raw ST muscle. Cross-section of a raw muscle sample (section 1 × 1 cm) incubated for 2 hours in a pepsin solution. The incubation caused the appearance of a darker halo around the sample corresponding to the depth of penetration of the enzyme solution (A). A zoom shows the tissue digestion at the sample periphery and an increase in fiber size in the halo (B, histogram) due to the acidity of the digestion solution (pH 1.8). The perimysium is partially degraded in the halo (D) in contrast to the central part which shows no signs of degradation (C). [A color version of this figure is available online at www.booksite.elsevier.com/9780124046108].

These results indicate that enzymatic activity is not solely responsible for the degradation of the meat product and that the physicochemical environment of the food during digestion is to be taken into account.

It is likely that the acidic pH, by partially denaturing muscle protein, facilitates the access of digestive enzymes to their cleavage sites. However, cell swelling caused by the acidity also could limit access of the digestive solution to the center of the sample.

Meat is usually consumed after a period of storage called "maturation" (see the section on postmortem proteolysis). Breaks transverse to the muscle fibers, characteristic of postmortem proteolysis, are very thin on the undigested sample (Figure 7.9A). In contrast, successive enzymatic digestions greatly expand these fracture zones (Figure 7.9B and C). Areas weakened by postmortem proteolysis undergo deeper changes in the enzymatic digestions.

Pepsin digestion leads to degradation of the Z line (Figure 7.9E), but the rest of the sarcomere remains well preserved. The subsequent digestion with trypsin and chymotrypsin alters the whole structure of the sarcomere. Changes in the pH of the digestive juices, pH 1.8 for gastric juice (pepsin solution) and pH 8 for intestinal juice (trypsin and chymotrypsin), are

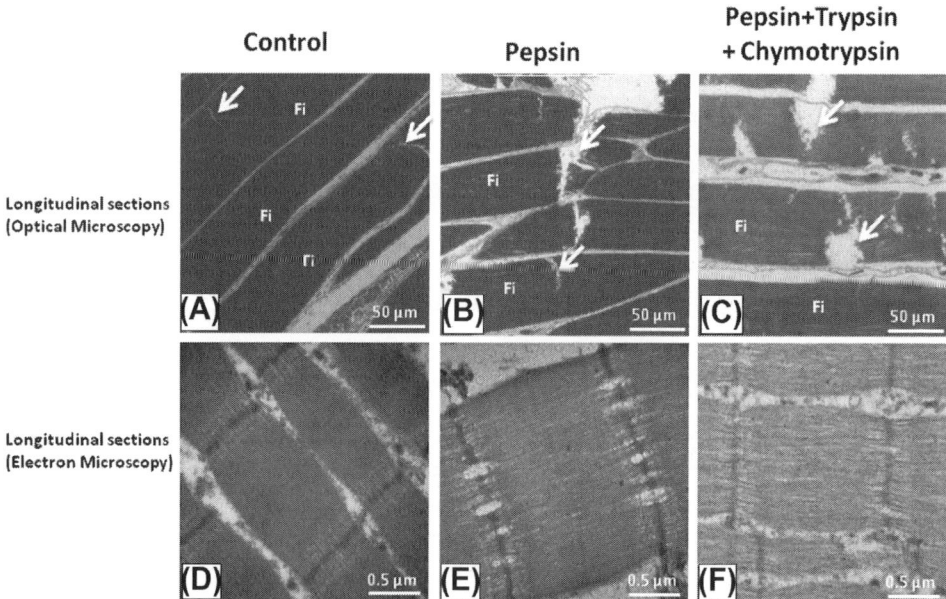

■ **FIGURE 7.9** Effect of enzymatic digestion on muscle fibers' structure and ultrastructure. Twelve days postmortem bovine *semitendinosus* muscle (A, D) subjected to gastric digestion (B, E) and intestinal digestion (C, F). Postmortem proteolysis has generated some breaks perpendicular to the fiber axis (A; arrows). The majority of sarcomeres show little alteration (D). After incubation in a solution of pepsin, alterations are predominant at the fiber breaks (B; arrows) and Z lines have perforations (E). Additional digestion with trypsin and chymotrypsin enhanced degradation along fiber fractures (C; arrows). The sarcomere degraded more uniformly and the myofilaments are undistinguishable. [A color version of this figure is available online at www.booksite.elsevier.com/9780124046108].

probably also responsible for changes in the microstructure. Extracted proteins appear in the extramyofibrillar spaces located near the proteolysis breaks (results not shown), but the sarcomeres have the same morphology as in the non-ruptured zones.

Postmortem proteolysis appears to facilitate the access of digestion juices into the intracellular space of the muscle fibers through damaged areas, but digestion does not lead to a dramatic change of sarcomere ultrastructure.

The kinetics of digestion of raw meat have often been evaluated *in vitro* by quantifying the release of muscle peptides into the digestion solution. These studies have highlighted the main degradation of extracellular matrix proteins (collagen) and intracellular proteins (sarcoplasmic and myofibrillar) systems by gastric and intestinal digestion (Kamin-Belsky *et al.*, 1996; Escudero *et al.*, 2010; Kaur *et al.*, 2010a, b). The results are in agreement with those obtained from *in vivo* digestion in extracts of duodenum and jejunum digesta of pigs having consumed beef meat (Bauchart *et al.*, 2007). Some myofibrillar proteins of crustaceans, such as tropomyosin, appear to be resistant to digestion (Liu *et al.*, 2010, 2011; Huang *et al.*, 2010).

Kaur *et al.* (2010, 2011) found that the addition of kiwifruit extracts to gastric juices improved the digestion of beef meat. The actinidin protease, naturally present in the kiwifruit juice, is assumed to play a role in improving meat disruption.

Marinated Meat

Analyses were performed to characterize accurately the effect of *in vitro* digestion on meat previously marinated in acetic acid (pH 4.5) for 24 hours. Preliminary data showed a swelling of the cells that had been penetrated by the digestion solutions and a significant degradation of connective tissue (Figure 7.10A).

However, no significant difference was seen compared to non-marinated meat that had undergone the same *in vitro* digestion.

In contrast, ultrastructural data showed a very significant change in the structure of myofibrils (Figure 7.10B and C). After pepsin digestion, extramyofibrillar spaces had disappeared, indicating a swelling of the myofibrils. Sarcomere structure was substantially modified: I bands and A bands were no longer visible, and only the Z lines remained visible (Figure 7.10B). Surprisingly, although modified, the ultrastructure of the sarcomere seemed less altered following the (trypsin + chymotrypsin) digestion. Ultrastructural differences depending on the type of digestion (gastric or intestinal) are probably more related to the physicochemical environment of the samples than

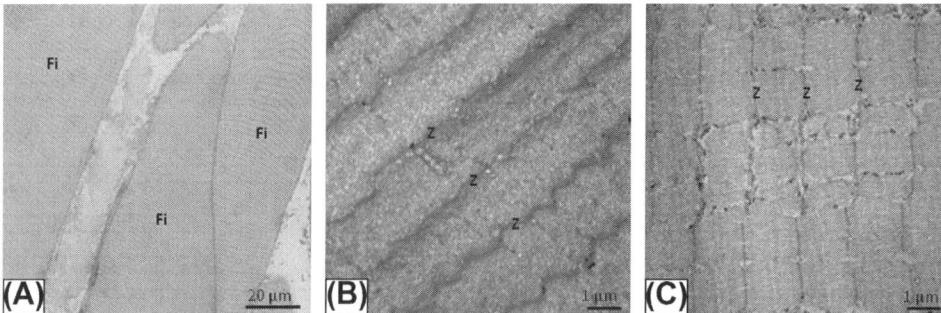

■ **FIGURE 7.10** Effect of enzymatic digestion on acid-marinated muscle fibers' structure. Twelve days postmortem bovine *semitendinosus* muscle marinated in acetic acid pH 4.5 for 24 hours and subjected to pepsin digestion (A, B) and pepsin + trypsin + chymotrypsin digestion (C). Pepsin digestion has led to the swelling of the cells and to the connective tissue disruption in the areas that had been penetrated by the digestion solutions (A). These structural changes are due to the low pH of the pepsin solution (pH 1.8) and to the proteolytic action of pepsin. After pepsin digestion, extramyofibrillar spaces have disappeared, indicating a swelling of myofibrils (B). I bands and A bands have disappeared, only the Z lines remain visible. The ultrastructure of the sarcomere seems less altered following the trypsin and chymotrypsin digestion, perhaps because of the less aggressive pH of the simulated intestinal juice (pH 8). [A color version of this figure is available online at www.booksite.elsevier.com/9780124046108].

the activity of digestive proteases. In particular, the pH, which changed from 1.8 in the pepsin solution to 8 in the trypsin and chymotrypsin solution, could have caused changes in molecular interactions that resulted in the ultrastructural changes observed in Figure 7.10.

Salted Meat

Although the process is commonly applied to meat and meat products, no study appears to have yet been published on the effect of salt on the digestion of cured meat products. Given the large structural changes caused by this process, it is obvious that the salt has an effect on the accessibility of digestion solutions to myofibrillar proteins. In addition, salt increases the ionic strength in meat, which favors the denaturation and solubilization of certain muscle proteins. The accessibility of digestive proteases to their specific cleavage sites could be affected. Furthermore, the activity of digestive proteases may also suffer from this high ionic strength environment. Further studies on this topic are needed to characterize and understand the effect of salting on the digestibility of meat products.

Cooked Meat

Heating is by far the most widely used process to prepare meat for consumption. As in the case of raw meat, incubation of cooked meat in pepsin solution caused a swelling of the cells in the periphery of the sample (Figure 7.11A and B). The central part of the sample, where the digestion juice had not accessed, had cells laterally contracted (small cross-section area) and large extracellular

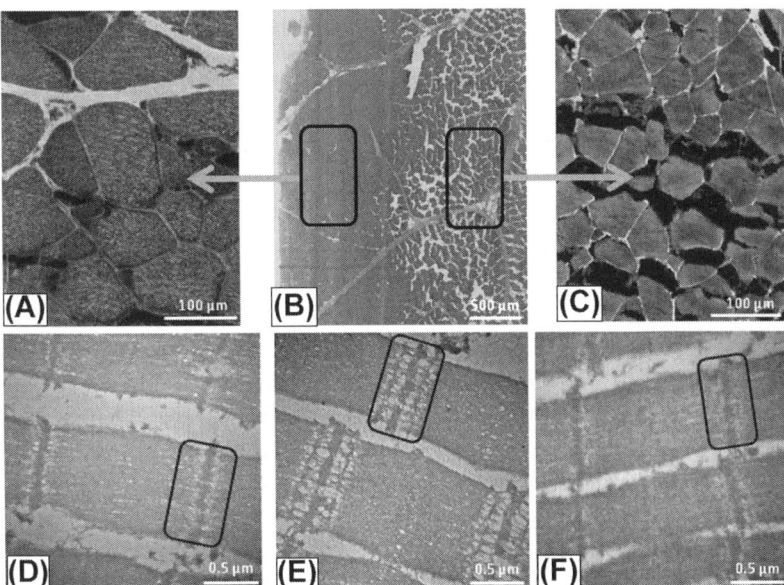

■ **FIGURE 7.11** Enzymatic digestion of cooked meat. As with raw meat, incubation of cooked meat (60°C for 45 min) in the pepsin solution caused swelling of the muscle fibers (located around the sample B). A and B: Double-labeling of myosin (intracellular, dark gray) and collagen (extracellular, light gray) of the part in contact with the pepsin solution, pH 1.8 (A) and the central part where the pepsin solution had not accessed (C). Myosin and collagen have bound antibodies, even in the area penetrated by the solution of pepsin, suggesting that these proteins are partially digested. C, D, E: Ultrastructure of a muscle successively cooked to 60°C (D), digested with pepsin (E) and a subsequent mixture of trypsin and chymotrypsin (F). Cooking has resulted in the coagulation of myofibrillar proteins (C). The digestion with pepsin substantially degraded the I band on either side of the Z line. The additional action of trypsin + chymotrypsin had a lesser effect on the sarcomere. [A color version of this figure is available online at www.booksite.elsevier.com/9780124046108].

spaces (Figure 7.11C). As in the case of raw meat, cell swelling is mostly a result of the acidity of the solution rather than of the action of pepsin. Despite the heat denaturation, myosin and collagen retained their antigenic properties as they bound their respective antibodies. Perimysium networks are also clearly visible (Figure 7.11A and B). The fluorescence intensity was lower in the cells of the halo of pepsin penetration than in the rest of the sample. Cell swelling, which dilutes the signal, and degradation of myosin are probably the cause of this decrease in fluorescence intensity.

In the edge of the sample, the large increase in cell volume has resulted in an almost complete disappearance of extracellular spaces. However, it is assumed that the transfer of aqueous solutions takes place preferentially in the extracellular spaces (Offer and Knight, 1988b). This swelling of cells in the edge of the sample could lead to clogging and slow down the penetration of the digestion solution.

Ultrastructural analyses of the halo of digestion are presented in Figure 7.11D, E and F. The ultrastructural changes produced by heating (Figure 7.11D) are consistent with that of Figure 7.6 and bibliography on the subject, namely, a coagulation of contractile protein and altered Z lines. The A bands have been little modified by pepsin digestion while the I bands and Z lines have been highly altered, showing numerous holes in this area (Figure 7.11E). Paradoxically, after intestinal digestion, these holes had disappeared, probably as a result of shrinkage of myofibrils, whose width had decreased.

Structural variations of myofibrils during *in vitro* digestion were probably the result of the pH—enzyme action interaction.

Recent studies on muscle protein extracts revealed the effect of heating on their digestibility (Santé-Lhoutellier *et al.*, 2008; Gatellier and Santé-Lhoutellier, 2009; Filgueras *et al.*, 2011b Bax *et al.*, 2012, 2013). Heating at 100°C reduced the rate of digestion of the myofibrillar proteins by pepsin, while the results were less clear for trypsin—chymotrypsin digestion (Santé-Lhoutellier *et al.*, 2008; Gatellier and Santé-Lhoutellier, 2009; Filgueras *et al.*, 2011b).

Compared to native muscle proteins, the lower digestibility of heated muscle protein was explained by thermal denaturation and oxidation that led to the formation of protein aggregates in which the cleavage sites of proteases were less accessible.

Postmortem time has little effect on the digestibility of heated myofibrillar proteins (Filgueras *et al.*, 2011b; Bax *et al.*, 2012, 2013). These results contradict the results reported by Astruc *et al.* (2012b). Previous authors worked on protein extracts while Astruc *et al.* worked on a piece of muscle. Postmortem proteolysis favors the bioaccessibility of digestive juices within the solid and structured meat, but not necessarily the digestibility of proteins in contact with digestive proteases.

However, recent studies suggest that cooking conditions modulate significantly the digestibility of meat protein (Bax *et al.*, 2012, 2013). These authors have shown that cooking meat at 70°C optimized the *in vitro* digestion rate compared to raw samples or samples heated at 100 and 140°C. Heating at 70°C denatures muscle proteins and exposes cleavage sites to digestive enzymes. At higher temperatures, the proteins form aggregates that decrease the accessibility of digestive proteases to their cleavage sites.

Process Interactions

The efficiency of digestion was assessed by spectrophotometric determination at 280 nm of peptides released in digestion solutions. At the same time, the thickness of the halo of digestion was measured on the meat sample to evaluate the penetration of the digestive solutions.

Our preliminary results (not shown) indicate that the thicker the halo, the greater is the amount of peptide released. Under our experimental conditions, results indicate that the more aggressive the process, the less effective is the *in vitro* digestion. During the *in vitro* gastrointestinal digestions, the fresh muscle released almost twice the amount of peptides as the muscle that had been marinated and heated to 90°C. The efficiency of digestion varied in the following order: fresh > marinated > cooked 60°C > marinated and cooked 60°C > cooked 90°C > marinated cooked 90°C. In addition, in all processes, 12-day postmortem muscles released twice the amount of peptides in the digestion process compared with 1-day postmortem muscles treated under the same conditions.

These results highlight the role of processes and structure on the quality of the digestion of meat products. However, it is important to be aware that the experimental protocol does not reflect the reality of *in vivo* digestion, where the swallowed particle size is much smaller and where saliva has begun some damage to the structure of the food.

PEPSIN BIOACCESSIBILITY AND LOCALIZATION

The foregoing paragraphs deal with the effect of digestive solutions on the structure of the meat. A more accurate method for assessing the bioaccessibility of digestive solutions in food is to locate a specific component of the digestive solutions. We located pepsin by immunohistofluorescence in digested meat. A pepsin antibody (primary antibody) is deposited on a digested tissue section. After several washes, a second antibody directed against the first antibody and carrying a fluorescent marker (secondary antibody) is deposited on the tissue section (Figure 7.12). The Ag/Ab complex is observed using a fluorescence microscope. Pepsin commonly used for *in vitro* digestion comes from the digestive tracts of pigs. The pig pepsin antibody is commercially available. We tested two different antibodies, produced by rabbits and goats inoculated with pig pepsin. The secondary antibodies, anti-rabbit and anti-goat labeled with fluorescent markers, are widely used in immunofluorescence studies and are commercially available.

The observation of the location of pepsin was performed on digested raw muscle samples (Figure 7.12). The fluorescent labeling was observed in the halo of penetration of the digestion juice, and more precisely in the perimysium and endomysium. It seems that pepsin is essentially bound in the connective tissue, suggesting that the digestive solution passes through the extracellular matrix. By increasing the magnification of the microscope, we observed high intensity fluorescence points in the cells, located in the halo of digestion. This labeling corresponds to the presence of pepsin. By matching

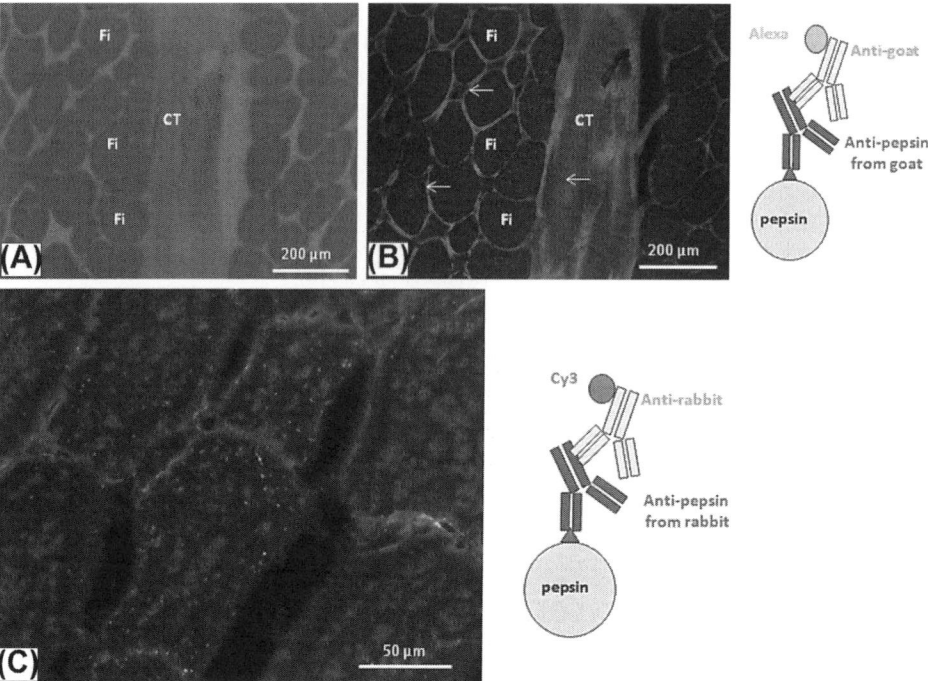

■ **FIGURE 7.12** Pepsin localization in raw *semitendinosus* muscle after *in vitro* gastric digestion. A and B: Serial cross-sections of muscle digested by pepsin showing the cells and perimysium morphology (A) and pepsin location evidenced by immunohistofluorescence (B). C: Immunolabeling of pepsin using another set of primary and secondary antibodies than on picture B. Pepsin is selectively identified in the connective tissue (B; arrows), and in muscle cells evidenced by the bright spots (C). Paler fluorescence indicated by a light gray, visible in and around the cells, is non-specific staining.

this information with the loss of antigenicity of myosin in the halo of digestion (Figure 7.11) we conclude that the pepsin first enters the tissue through the extracellular matrix and subsequently enters the intracellular space.

However, these results are not clear-cut, because the controls without the first antibody (replaced with normal serum) showed background fluorescence. This background may have several origins. Antibodies are tested by manufacturers to ensure specificity. However, the low pH of the pepsin solution probably alters the molecular structure of muscle proteins significantly. These structural changes can promote non-specific interactions with primary and secondary antibodies and create more noise than usual. In addition, some endogenous molecules such as collagen are known to be autofluorescent and contaminate the signal relative to the fluorescence of the second antibody.

Finally, the signal of pepsin is limited by its low concentration in the muscle. This low concentration may be "normal" but it can also be linked to implementation of the immunohistofluorescence technique, which can lead to the loss of pepsin in different incubation baths during preparation of the histological section.

Despite our best efforts, we have not been able to locate trypsin in the muscle after digestion, probably due to inadequate anti trypsin antibody.

Methodological developments are needed to accurately identify the digestive enzymes and better understand their modes of action in structured solid foods.

CONCLUSIONS

The digestibility of meat has been little studied and very little work has been done to understand the effect of the structure of muscle on its digestion. Digestion is a complex process that involves, at the same time, physicochemical, enzymatic, and mechanical actions. To better understand the mechanisms involved, it is necessary to decouple the various stages of digestion, often by using *in vitro* digesters. The product "meat" is meanwhile also complex. Meat comes from the transformation of muscle after slaughter, but muscles exhibit highly variable biochemical and structural characteristics, according to their anatomical position and function, and also depending on the species from which they come.

Various technological transformations that are applied to the meat before consumption result in changes in the structure and composition of meat products that change the transfer of digestion juices in food and the accessibility of digestive enzymes to myofibrillar proteins. Most technological processes lead to denaturing environment for proteins. The acidity of the marinades, the increase in ionic strength caused by salting, or thermal denaturation as a consequence of cooking, all cause changes in the secondary and tertiary structure of meat proteins, which in some cases can aggregate.

These protein aggregates are generally less rapidly degraded by digestive proteases than native proteins. The development of artificial masticators and digesters, multiscale and multimodal imaging technologies, and sophisticated biochemical techniques will, in the future, allow a better understanding of the mechanisms that govern the digestion of meat products. The information gained will enable adaptation of processing to improve the digestibility of meat and meat products.

REFERENCES

Aktas, N., Kaya, M., 2001a. Influence of weak organic acids and salts on the denaturation characteristics of intramuscular connective tissue. A differential scanning calorimetry study. Meat Sci. 58, 413–419.

Aktas, N., Kaya, M., 2001b. The influence of marinating with weak organic acids and salts on the intramuscular connective tissue and sensory properties of beef. Eur. Food Res. Technol. 213, 88–94.

Astruc, T., 2008a. Morphologie du sarcolemme de fibres musculaires anoxiques: résultats préliminaires. 12emes journées des Sciences du muscle et technologies de la Viande, Tours (France), 8 et 9 Octobre 2008. Hors série Viande et Produits Carnés, 161–162.

Astruc, T., Labas, R., Vendeuvre, J.L., Martin, J.L., Taylor, R.G., 2008b. Beef sausage structure affected by sodium chloride and potassium lactate. Meat Sci. 80, 1092–1099.

Astruc, T., Gatellier, P., Labas, R., Sante-Lhoutellier, V., Marinova, P., 2010. Microstructural changes in m. rectus abdominis bovine muscle after heating. Meat Sci. 85, 743–751.

Astruc, T., Peyrin, F., Venien, A., Labas, R., Abrantes, M., Dumas, P., Jamme, F., 2012a. In situ thermal denaturation of myofiber sub-type proteins studied by immunohistofluorescence and synchrotron radiation FT-IR microspectroscopy. Food Chem. 134, 1044–1051.

Astruc, T., Lucas, J., Peyrin, F., Venien, A., Sante-Lhoutellier, V., 2012b. Muscle structure affects digestive enzyme bio-accessibility to intracellular compartment. Food Structure Digestion & Health congress, 7–9 march, Palmerston North, New Zealand. Proceeding p. 39.

Bailey, A.J., Light, N.D., 1989. Connective Tissue in Meat and Meat Products. Elsevier Applied Science, London.

Bauchart, C., Morzel, M., Chambon, C., Patureau Mirand, P., Reynès, C., Buffière, C., Rémond, D., 2007. Peptides reproducibly released by *in vivo* digestion of beef meat and trout flesh in pigs. Br. J. Nutr. 98, 1187–1195.

Bax, M.-L., Aubry, L., Ferreira, C., Daudin, J.-D., Gatellier, P., Rémond, D., Santé-Lhoutellier, V., 2012. Cooking temperature is a key determinant of *in vitro* meat protein digestion rate: investigation of underlying mechanisms. J. Agric. Food Chem. 60, 2569–2576.

Bax, M.-L., Sayd, T., Aubry, L., Ferreira, C., Viala, D., Chambon, C., et al., 2013. Muscle composition slightly affects *in vitro* digestion of aged and cooked meat: identification of associated proteomic markers. Food Chem. 136, 1249–1262.

Bendall, J.R., Restall, D.J., 1983. The cooking of single myofibres, small myofibre bundles and muscle strips from beef M. psoas and M. sternomandibularis muscles at varying heating rates and temperatures. Meat Sci. 8, 93–117.

Berge, P., Ertbjerg, P., Larsen, L.M., Astruc, T., Vignon, X., Møller, A.J., 2001. Tenderization of beef by lactic acid injected at different times post mortem. Meat Sci. 57, 347–357.

Bertram, H.C., Stagsted, J., Young, J.F., Andersen, H., 2004. Elucidation of membrane destabilization in post-mortem muscles using an extracellular paramagnetic agent (Gd-DTPA): an NMR study. J. Agric. Food Chem. 52, 6320–6325.

Bocker, U., Ofstad, R., Bertram, H.C., Egelandsdal, B., Kohler, A., 2006. Salt-induced changes in pork myofibrillar tissue investigated by FT-IR microspectroscopy and light microscopy. J. Agric. Food Chem. 54, 6733–6740.

Bouton, P.E., Harris, P.V., 1972. The effect of cooking temperature and time on some mechanical properties of meat. J. Food Sci. 37, 140–144.

Chang, H.-J., Wang, Q., Zhou, G.-H., Xu, X.-L., Li, C.-B., 2010. Influence of weak organic acids and sodium chloride marination on characteristics of connective tissue collagen and textural properties of beef *semitendinosus* muscle. J. Text. Stud. 41, 279—301.

Craig, R.W., Padron, R., 2003. Molecular structure of the sarcomere. In: Engel, A.G., Franzini-Armstrong, C. (Eds.), Myology. McGraw-Hill, New York, pp. 129—166.

Davey, C.L., Dickson, D.R., 1970. Studies in meat tenderness and ultrastructural changes in meat during aging. J. Food Sci. 35, 56—60.

Davey, C.L., Gilbert, K.V., 1974. Temperature-dependent toughness in beef. J. Sci. Food Agric. 25, 931—938.

Escudero, E., Sentandreu, M.A., Toldra, F., 2010. Characterization of peptides released by *in vitro* digestion of pork meat. J. Agric. Food Chem. 58, 5160—5165.

Filgueras, R.S., Peyrin, F., Henot, J.M., Venien, A., Labas, R., Astruc, T., 2011a. Salt (NaCl) diffusion and distribution in rat skeletal muscle. 57th International Congress in Meat Science and Technology, Ghent, Belgium. August 7—12.

Filgueras, R.S., Gatellier, P., Ferreira, C., Zambiazi, R.C., Santé-Lhoutellier, V., 2011b. Nutritional value and digestion rate of rhea meat proteins in association with storage and cooking processes. Meat Sci. 89, 6—12.

Gatellier, P., Santé-Lhoutellier, V., 2009. Digestion study of proteins from cooked meat using an enzymatic microreactor. Meat Sci. 81, 405—409.

Graiver, N., Pinotti, A., Califano, A., Zaritzky, N., 2006. Diffusion of sodium chloride in pork tissue. J. Food Eng. 77, 910—918.

Greaser, M.L., 1986. Conversion of muscle to meat. In: Bechtel, P.J. (Ed.), "Muscle as Food", Edition. Academic Press, Orlando, Florida, pp. 37—102.

Gillies, A.R., Lieber, R.L., 2011. Structure and function of the skeletal muscle extracellular matrix. Muscle Nerve 44, 318—331.

Hamm, R., 1977. Changes of muscle proteins during the heating of meat. In: Höyem, T., Kvåle, O. (Eds.), Physical, chemical and biological changes in food caused by thermal processing. Applied Science Publishing, pp. 101—134.

Hearne, L.E., Penfield, M.P., Goertz, G.E., 1978. Heating effects of bovine *semitendinosus*: phase contrast microscopy and scanning electron microscopy. J. Food Sci. 43, 13—16.

Heffron, J.J., Hegarty, P.V., 1974. Evidence for a relationship between ATP hydrolysis and changes in extracellular space and fibre diameter during rigor development in skeletal muscle. Comp. Biochem. Physiol., A. Comp. Physiol 49, 43—56.

Hooda, S., Ferreira, S.G., Latour, M.A., Bauer, L.L., Fahey, G.C., Swanson Jr., K.S., 2012. *In vitro* digestibility of expanded pork skin and rawhide chew, and digestion and metabolic characteristics of expanded pork skin chews in healthy adult dogs. J. Anim. Sci. 90, 4355—4361.

Hostetler, R.L., Landman, W.A., 1968. Photomicrographic studies of dynamic changes in muscle fiber fragments. 1. Effects of various heat treatments on length, width and birefringence. J. Food Sci. 33, 468—470.

Huang, Y.-Y., Liu, G.-M., Cai, Q.-F., Weng, W.-Y., Maleki, S.J., Su, W.-J., Cao, M.-J., 2010. Stability of major allergen tropomyosin and other food proteins of mud crab (Scylla serrata) by *in vitro* gastrointestinal digestion. Food Chem. Toxicol. 48, 1196—1201.

Hur, S.J., Lim, B.O., Decker, E.A., McClements, D.J., 2011. *In vitro* human digestion models for food applications. Food Chem. 125, 1—12.

Jalabert-Malbos, M.-L., Mishellany-Dutour, A., Woda, A., Peyron, M.-A., 2007. Particle size distribution in the food bolus after mastication of natural foods. Food Qual. Pref. 18, 803—812.

Jones, S.B., Carroll, R.J., Cavanaugh, J.R., 1977. Structural changes in heated bovine muscle. A scanning electron microscope study. J. Food Sci. 42, 125.

Kamin-Belsky, N., Brillon, A.A., Arav, R., Shaklai, N., 1996. Degradation of myosin by enzymes of the digestive system: comparison between native and oxidatively cross-linked protein. J. Agric. Food Chem. 44, 1641—1646.

Kaur, L., Rutherfurd, S.M., Moughan, P.J., Drummond, L., Boland, M.J., 2010a. Actinidin enhances gastric protein digestion as assessed using an *in vitro* gastric digestion model. J. Agric. Food Chem. 58, 5068—5073.

Kaur, L., Rutherfurd, S.M., Moughan, P.J., Drummond, L., Boland, M.J., 2010b. Actinidin enhances protein digestion in the small intestine as assessed using an *in vitro* digestion model. J. Agric. Food Chem. 58, 5074—5080.

Knight, P., Parson, N., 1988. Action of NaCl and polyphosphates in meat processing: responses of myofibrils to concentrated salt solutions. Meat Sci. 24, 275—300.

Leander, R.C., Hedrick, H.B., Brown, M.F., White, J.A., 1980. Comparison of structural changes in bovine longissimus and *semitendinosus* muscles during cooking. J. Food Sci. 45, 1—12.

Leibovich, S.J., Weiss, J.B., 1970. Electron microscope studies of the effects of endo- and exopeptidase digestion on tropocollagen. A novel concept of the role of terminal regions in fibrillogenesis. Biochem. Biophys. Acta 214, 445—454.

Lewis, G.J., Purslow, P.P., 1991. The effect of marination and cooking on the mechanical properties of intramuscular connective tissue. J. Muscle Food 2, 177—195.

Light, N., Champion, A.E., Voyle, C., Bailey, A.J., 1985. The role of epimysial, perimysial and endomysial collagen in determining texture in six bovine muscles. Meat Sci. 13, 137—149.

Liu, A., Nishimura, T., Takahashi, K., 1994. Structural changes in endomysium and perimysium during post-mortem ageing of chicken *semitendinosus* muscle—contribution of structural weakening of intramuscular connective tissue to meat tenderization. Meat Sci. 38, 315—328.

Liu, G.-M., Cao, M.-J., Huang, Y.-Y., Cai, Q.-F., Weng, W.-Y., Su, W.-J., 2010. Comparative study of *in vitro* digestibility of major allergen tropomyosin and other food proteins of Chinese mitten crab (Eriocheir sinensis). J. Sci. Food Agric. 90, 1614—1620.

Liu, G.-M., Huang, Y.-Y., Cai, Q.-F., Weng, W.-Y., Su, W.-J., Cao, M.-J., 2011. Comparative study of *in vitro* digestibility of major allergen, tropomyosin and other proteins between Grass prawn (Penaeus monodon) and Pacific white shrimp (Litopenaeus vannamei). J. Sci. Food Agric. 91, 163—170.

Mielle, P., Tarrega, A., Sémon, E., Maratray, J., Gorria, P., Liodenot, J.J., et al., 2010. From human to artificial mouth, from basics to results. Sensors and Actuators B 146, 440—445.

Nishimura, T., Hattori, A., Takahashi, K., 1995. Structural weakening of intramuscular connective tissue during conditioning of beef. Meat Sci. 39, 127—133.

Nishimura, T., 2010. The role of intramuscular connective tissue in meat texture. Anim. Sci. J. 81, 21—27.

Offer, G., Knight, P., 1988a. The structural basis of water holding in meat. Part 2: Drip losses. In: Lawrie, R. (Ed.), Developments in Meat Science, fourth ed. Elsevier Applied Science Publishers, London, pp. 172—243.

Offer, G., Knight, P., 1988b. The structural basis of water holding in meat. Part 1: General principles and water uptake in meat processing. In: Lawrie, R. (Ed.), Developments in Meat Science, fourth ed. Elsevier Applied Science Publishers, London, pp. 63—172.

Offer, G., Trinick, J., 1983. On the mechanism of water holding in meat: the swelling and shrinking of myofibrils. Meat Sci. 8, 245—281.

Ouali, A., 1990. Meat tenderization: possible causes and mechanisms. A review. J. Muscle Foods 1, 129—165.

Palka, K., Daun, H., 1999. Changes in texture, cooking losses, and myofibrillar structure of bovine m. *semitendinosus* during heating. Meat Sci. 51, 237—243.

Parsons, N., Knight, P., 1990. Origin of variable extraction of myosin from myofibrils treated with salt and pyrophosphate. J. Sci. Food Agric. 51, 71—90.

Pette, D., Staron, R.S., 1990. Cellular and molecular diversities of mammalian skeletal muscle fibers. Rev. Physiol., Biochem. Pharmacol. 116, 1—76.

Pette, D., Staron, R.S., 2000. Myosin isoforms, muscle fiber types, and transitions. Microsc. Res. Tech. 50, 500—509.

Promeyrat, A., Gatellier, P., Lebret, B., Kajak-Siemaszko, K., Aubry, L., Santé-Lhoutellier, V., 2010. Evaluation of protein aggregation in cooked meat. Food Chem. 121, 412—417.

Purslow, P.P., 2005. Intramuscular connective tissue and its role in meat quality. Meat Sci. 70, 435—447.

Rao, M.V., Gault, N.F.S., Kennedy, S., 1989. Changes in the ultrastructure of beef muscles as influenced by acidic conditions below the ultimate pH. Food Microstr. 8, 115—124.

Rowe, R.W.D., 1986. Elastin in bovine *semitendinosus* and longissimus dorsi muscles. Meat Sci. 17, 293—312.

Rowe, R.W.D., 1989. Electron microscopy of bovine muscles: II—The effects of heat denaturation on post rigor sarcolemma and endomysium. Meat Sci. 26, 281—294.

Salles, C., Tarrega, A., Mielle, P., Maratray, J., Gorria, P., Liaboeuf, J., Liodenot, J.J., 2007. Development of a chewing simulator for food breakdown and the analysis of *in vitro* flavor compound release in a mouth environment. J. Food Eng. 82, 189—198.

Santé-Lhoutellier, V., Astruc, T., Marinova, P., Greve, E., Gatellier, P., 2008. Effect of meat cooking on physicochemical state and *in vitro* digestibility of myofibrillar proteins. J. Agric. Food Chem. 56, 1488—1494.

Savoie, L., Gauthier, S.F., 1986. Dialysis cell for the *in vitro* measurement of protein digestibility. J. Food Sci. 51, 494—498.

Stanton, C., Light, N., 1987. The effects of conditioning on meat collagen: part I—evidence for gross in situ proteolysis. Meat Sci. 21, 249—265.

Stanton, C., Light, N., 1988. The effects of conditioning on meat collagen: part 2—direct biochemical evidence for proteolytic damage in insoluble perimysial collagen after conditioning. Meat Sci. 23, 179—199.

Stanton, C., Light, N., 1990. The effects of conditioning on meat collagen: part 3—evidence for proteolytic damage to endomysial collagen after conditioning. Meat Sci. 27, 41—54.

Schmidt, J.G., Parrish, F.C., 1971. Molecular properties of *postmortem* muscle. 10. Effect of internal temperature and carcass maturity on structure of bovine longissimus. J. Food Sci. 36, 110−119.

Taylor, R.G., Labas, R., Smulders, F.J.M., Wiklund, E., 2002. Ultrastructural changes during aging in *M. Longissimus thoracis* from moose and reindeer. Meat Sci. 60, 321−326.

Tornberg, E., 2005. Effects of heat on meat proteins—implications on structure and quality of meat products. Meat Sci. 70, 493−508.

Wattanachant, S., Benjakul, S., Ledward, D.A., 2005. Effect of heat treatment on changes in texture, structure and properties of Thai indigenous chicken muscle. Food Chem. 93, 337−348.

Wilding, P., Hedges, N., Lillford, P.J., 1986. Salt-induced swelling of meat: the effect of storage time, pH, ion-type and concentration. Meat Sci. 18, 55−75.

Woda, A., Mishellany-Dutour, A., Batier, L., Francois, O., Meunier, J.-P., Reynaud, B., et al., 2010. Development and validation of a mastication simulator. J. Biomech. 43, 1667−1673.

Wu, F.Y., Dutson, T.R., Smith, S.B., 1985. A scanning electron microscopic study of heat-induced alterations in bovine connective tissue. J. Food Sci. 50, 1041−1044.

Wu, Z., Bertram, H.C., Kohler, A., Bocker, U., Ofstad, R., Andersen, H.J., 2006. Influence of aging and salting on protein secondary structures and water distribution in uncooked and cooked pork. A combined FT-IR microspectroscopy and 1H NMR relaxometry study. J. Agric. Food Chem. 54, 8589−8597.

Xiong, Y.L., 1994. Myofibrillar protein from different muscle fiber types: implications of biochemical and functional properties in meat processing. Crit. Rev. Food Sci. Nutr. 34, 293−320.

Cotyledon Cell Structure and *In Vitro* Starch Digestion in Navy Beans

Jaspreet Singh[1], Thilo Berg[1], Allan Hardacre[2] and Mike J. Boland[1]

[1]*Riddet Institute, Massey University, Palmerston North, New Zealand,*
[2]*Institute of Food Nutrition and Human Health, Massey University, Palmerston North, New Zealand*

CONTENTS

INTRODUCTION

Beans (*Phaseolus vulgaris*), a good source of protein and carbohydrates in human diets, are widely grown and consumed in developed as well as developing nations of the world. Apart from proteins (20−38%) and complex carbohydrates (50−60%), beans are rich in minerals, vitamins, and polyunsaturated free fatty acids (Rehman and Shah, 2005). The glycemic index of beans is generally low and postprandial glucose response is moderate after ingestion, which makes them a preferred source of energy (Jenkins *et al.*, 1981). Furthermore, they contain high levels of starch that escapes

Food Structures, Digestion and Health. http://dx.doi.org/10.1016/B978-0-12-404610-8.00008-6

hydrolysis in the small intestine (resistant starch) and is also known for its prebiotic properties (Rehman *et al.*, 2001; Vargas-Torres *et al.*, 2004). Starch in legumes is naturally situated inside the living cotyledon cells (Hahn *et al.*, 1977). Primary cell walls of growing and fleshy tissues have a conserved general composition of cellulose, hemicelluloses, and pectin (Chanda, 2005). The non-cellulosic material acts as a "glue" that holds the microfibrils of cellulose together, which in turn is responsible for the stability of cell walls (Carpita and Gibeaut, 1993). The starch granules in beans are present in the cotyledon cells and are embedded in the protein matrix of the cellular contents (Daussant *et al.*, 1983). This situation might restrict the complete swelling of the bean starch during gelatinization due to steric hindrance and other limiting effects including restricted water availability. Hahn *et al.* (1977) and Kon *et al.* (1971) observed birefringence of intracellular starch granules when microscopically examining cooked beans using plain polarized light. Wursch *et al.* (1986) pointed out that the thick and mechanically resistant nature of the cotyledon cell walls in legumes can prevent complete swelling of starch granules during gelatinization which may restrict their interaction with digestive enzymes.

Alpha amylase inhibitors present in raw beans are known to inhibit the activity of porcine pancreatic amylase; however, these inhibitors are generally inactivated at or above 100°C (Singh *et al.*, 2010). Processing leads to an alteration in the food structure and also influences the nutritional characteristics of the food including starch digestibility. The physical characteristics of food; the presence of other food components, such as proteins, lipids, and non-starch polysaccharides; and the changes and interactions occurring in them during food processing affect the enzymatic digestibility of starch to a considerable extent (Kaur *et al.*, 2007; Dartois *et al.*, 2010; Singh *et al.*, 2010). There are many reports available in the literature on the digestibility of starch in beans in relation to soaking time, microwave, and conventional cooking (Oliveira *et al.*, 2001; Salgado *et al.*, 2005; Ramirez-Cardenas *et al.*, 2008). However, the role of cotyledon cell structure during digestion of cooked beans remained unclear until Berg *et al.* (2012) reported the influence of cotyledon cell integrity, effect of storage and reheating on the extent and rate of starch digestibility in navy beans, using an *in vitro* model simulating the human digestive system. This study is discussed in detail below. The microstructural characteristics of cooked navy beans are also reviewed and discussed to understand the changes that occur to the surfaces and particle size distribution of intact cotyledon cells during the *in vitro* digestion. These insights might further lead to the possibility of changing processes and plant structures in order to enhance the positive properties of carbohydrates present in beans.

MICROSTRUCTURE OF RAW AND COOKED WHOLE NAVY BEANS, BEAN FLOUR, AND STARCH

Microstructural characteristics of natural foods provide important information on structural organization such as cell wall characteristics, starch granule arrangement, and presence of proteins, etc. Several authors have studied the microstructural characteristics of natural foods such as potatoes, rice, and beans using microscopic techniques (Ogawa *et al.*, 2000; Singh *et al.*, 2008; Berg *et al.*, 2012; Bordoloi *et al.*, 2012). Berg *et al.* (2012) studied the microstructure of raw and cooked whole navy beans using electron microscopy. For cooking, beans were first soaked in water for 1 h followed by autoclaving. The autoclaving procedure comprised heating to 121°C, holding of temperature for 15 min, and rapid cooling to 100°C. The autoclave was opened upon reaching this temperature and the bottles containing beans and water were put in cold water at 18–20°C to cool the contents. Thin bean cotyledon sections were cut with a razor blade, which were fixed with glutaraldehyde in a phosphate buffer and post-fixed in osmium tetroxide before viewing under a scanning electron microscope.

Within the raw bean cotyledon cells, the starch granules were observed to be embedded in, and surrounded by, a thick proteinaceous matrix derived from the cell contents (Berg *et al.*, 2012; Figure 8.1A). A similar morphology for the cellular contents of legumes has been described by Daussant *et al.* (1983). The diameter of the bean cotyledon cells ranged between 50 and 100 µm and they were hexagonal or angular in shape. The walls of the cotyledon cells appeared as thick robust structures surrounding the starch granules embedded within the proteinaceous cellular contents (Berg *et al.*, 2012).

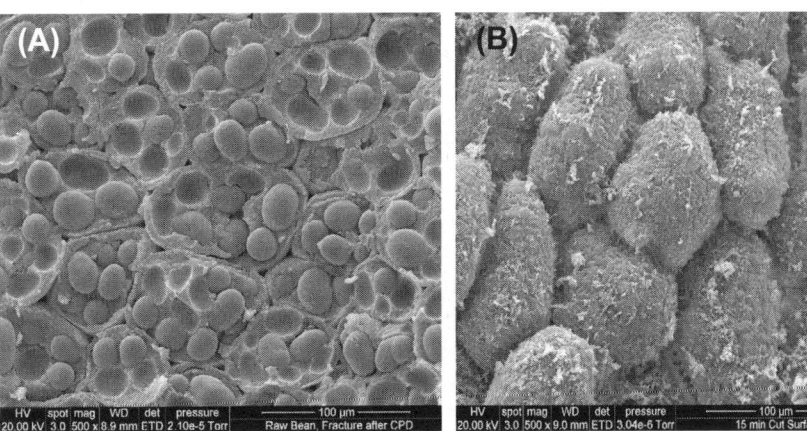

■ **FIGURE 8.1** Scanning electron micrographs showing sections of cotyledon cells: (A) raw navy beans, (B) cooked navy beans. *Reproduced from Berg* et al. *(2012) with permission from Elsevier.*

The interaction of the cell wall materials such as cellulose and non-cellulosic polysaccharides has been reported to be responsible for their stability and resistance towards turgor pressure (Carpita and Gibeaut, 1993).

The cotyledon cells were observed to swell due to hydration during cooking, but generally stayed intact during the cooking process (Figure 8.1B). Strands of dried soluble material were observed on the top surface of cotyledon cell walls. Berg *et al.* (2012) reported that this soluble material might consist of soluble starch containing mainly amylose, soluble sugars, and non-starch polysaccharides which oozed out of the cotyledon cells during cooking. Cell wall xyloglucans are composed of β-(1-4) linked D-glucose molecules, an identical primary structure to cellulose but with additional xylosyl units attached to the O-6 position of the glucosyl units. These xyloglucans occur at relatively high levels in legumes, which might be the reason for the extremely high rigidity of cotyledon cells of navy beans when cooked (Carpita and Gibeaut, 1993). In the dry state, the cell walls are very brittle and are easily disrupted during milling (Berg *et al.*, 2012).

Light and scanning electron microscopy of bean flour presented detailed information on the microstructure of the material. Bean flour was mainly composed of disrupted cotyledon cells, free starch granules, protein bodies, and cell wall fragments (Figure 8.2A). Morphological features captured through scanning electron microscopy are shown in Figure 8.2B and granule size distribution of the bean starch in terms of percentages of small (0−20 μm), medium (25−45 μm), and large (>45 μm) granules is presented in Figure 8.2C. The bean starch granules varied from round for the small granules to oval or irregular for the larger granules. The biochemistry of the chloroplast or amyloplast, as well as the physiology of the plant, has been reported to mainly dictate the morphology of starch granules (Badenhuizen, 1969). The membranes and the physical characteristics of the plastids have also been reported to be responsible for providing a particular shape or morphology to starch granules during granule development (Jane *et al.*, 1994; Lindeboom *et al.*, 2004). The bean starch contained a fairly high percentage (≈75 %) of medium granules whereas it had a very low percentage (7%) of large and small (17%) granules (Figure 8.2C). The majority of the starch granules ranged between 10 and 20 μm. The granule size distribution of starch has been reported to change during the development of the storage organs of plants (Chojecki *et al*, 1986). The total starch content and starch amylose measured in the navy beans have been reported to be around 36 and 39%, respectively, on a dry weight basis (Kim *et al.*, 1996; Chung *et al.*, 2008; Berg *et al.*, 2012). Starch granules in wet extracted starch, and in dry milled bean flour, have a similar appearance and it is assumed that dry milling does not cause major changes to the

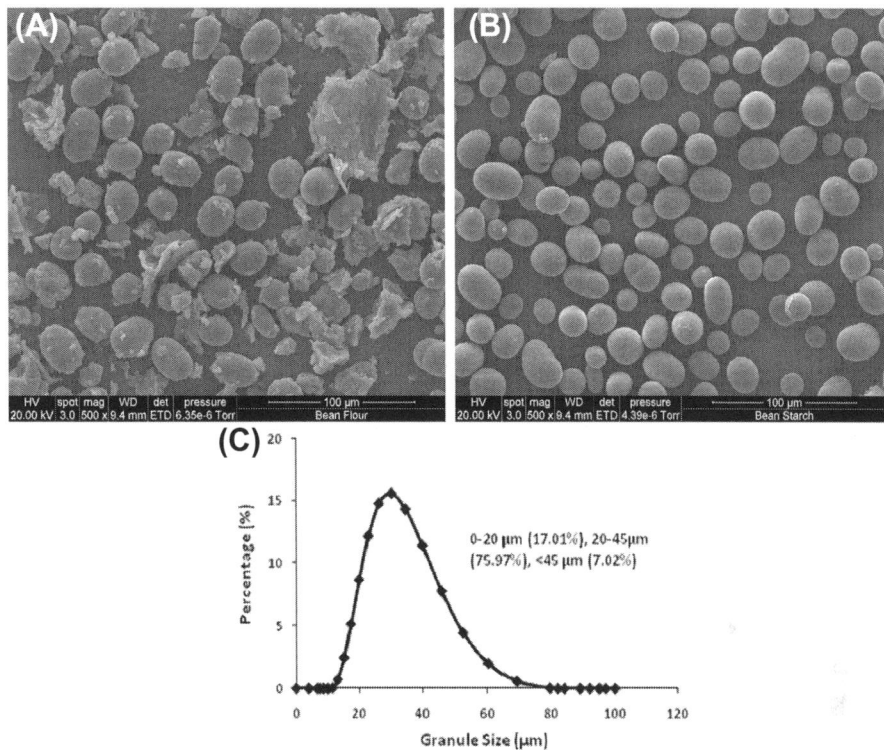

■ **FIGURE 8.2** Scanning electron micrograph of (A) milled navy bean flour, (B) navy bean starch, and (C) granule size distribution of navy bean starch. *Reproduced from Berg et al. (2012) with permission from Elsevier.*

granules. Legume and bean starches have been reported to have higher amylose content than the other cereal starches (Gujska *et al.*, 1994; Betancur-Ancona *et al.*, 2001). The differences among the amylose content of the starches could be attributed to differences in the activities of the enzymes involved in the biosynthesis of linear and branched components within the starch granules (Krossmann and Lloyd, 2000). The amylose content of the starch granules has also been reported to be affected by climatic conditions and soil type during growth, and granule size distribution (Singh *et al.*, 2004; 2007).

IN VITRO DIGESTION OF STARCH
Cooked Whole Navy Beans, Bean Flour, and Starch

In vitro digestion attempts to simulate the digestive process in human beings. Since digestion of food in humans is a very complex process, perfect simulations are not yet possible. Enzymes are present in digestive fluids as well as in the brush border of the small enzyme (Smith and Morton, 2001). The enzymes present in the human body are difficult to extract or expensive to buy, therefore enzymes from other mammals (e.g., pepsin and pancreatin from pigs) or from microorganisms are usually used in *in vitro* systems (Dartois *et al.*, 2010; Kaur *et al.*, 2010a, b). Mammalian enzymes show similarities to human enzymes, whereas the enzymes from microorganisms may work differently even though they may be similarly classified (e.g., glucoamylase) (BeMiller and Whistler, 2009). Furthermore, diffusion processes, which play an important role in the final uptake of monosaccharides into the human body, are difficult to simulate in *in vitro* systems.

The *in vitro* system generally used to study starch digestion imitates stomach conditions using a simulated stomach digestive fluid (Dartois *et al.*, 2010; Berg *et al.*, 2012; Bordoloi *et al.*, 2012). The simulated gastric fluid (SGF) is composed of pepsin dissolved in buffer at pH 1.2. The *in vitro* assay continues with conditions simulating the small intestine. The simulated small intestinal fluid (SIF) is composed of pancreatin dissolved in a potassium phosphate buffer. Pancreatin is an extract from porcine pancreas. It is composed of different enzymes, which can be classified as proteolytic, lipolytic, amylolytic, and nucleic acid splitting enzymes. α-Amylase (EC 3.2.1.1), the major amylolytic enzyme in pancreatin, is an endohydrolase specific for α-$(1 \rightarrow 4)$ glycosidic bonds (Sim *et al.*, 2008). It hydrolyzes amylose, amylopectin and related oligosaccharides to smaller subunits. The final products of hydrolysis are maltose and maltotriose for amylose and maltose, maltotriose and α-limit dextrins for amylopectin (Wursch, 1989). Additionally, to simulate the brush border enzymes in the human small intestine, amyloglucosidase (EC 3.2.1.3) and invertase (EC 3.2.1.26) are added (Woolnough *et al.*, 2007). Amyloglucosidase (EC 3.2.1.3) is an exohydrolase acting on both α-$(1 \rightarrow 4)$ and α-$(1 \rightarrow 6)$ glycosidic bonds, hydrolyzing from the non-reducing ends (Sim *et al.*, 2008). Action on α-$(1 \rightarrow 6)$ bonds is 30 times slower than on α-$(1 \rightarrow 4)$ linkages (Belitz *et al.*, 2009). The end product is D-glucose. Invertase (EC 3.2.1.26) hydrolyzes terminal non-reducing α-D-fructofuranoside residues in α-D-fructofuranosides such as that in sucrose. A colorimetric D-glucose assay kit (Megazyme) was used to measure glucose contents of samples. It is likely that some dextrins and oligosaccharides in the human body escape the brush border enzymes and reach the colon. A slight overestimation of glucose production might therefore occur in *in vitro* models.

A two-stage *in vitro* model as described above was used by Berg *et al.* (2012) to study digestion of starch in cooked navy beans, bean flour, and bean starch. The SGF and SIF were prepared in accordance with the US Pharmacopeia (2000). Samples were taken for glucose analysis after 0, 15,

30, and 0, 5, 10, 15, 30, 45, 60, 90, and 120 min of simulated gastric and in-testinal digestion, respectively (Figure 8.3). For studying microstructural characteristics through scanning electron microscopy, individual samples were frozen by immersion in liquid nitrogen and stored at -80°C before freeze drying. Samples were taken from the digestion reactor, immediately before adding the gastric juice, at the end of the *in vitro* gastric digestion, after 10 min, and at the end of *in vitro* intestinal digestion as per the flow chart (Figure 8.3).

The results of the study carried out by Berg *et al.* (2012) are discussed in detail below in order to understand the relationship between the rate and amount of hydrolysis of starch and the microstructural changes in cotyledon cells during the various stages of simulated gastric and small intestinal diges-tion. Cells observed under the light microscope in a sample of freshly pre-pared bean paste were generally intact, with little evidence of disrupted cells. The starch granules present in the cotyledon cells also showed birefrin-gence when viewed in polarized light suggesting incomplete gelatinization in cooked whole navy beans. This was reported to be presumably due to re-strictions to water uptake imposed by the thick cell walls (Figure 8.4). Figure 8.5 gives an overview of the kinetics of starch hydrolysis during the *in vitro* digestion process for cooked navy beans. The first 30 minutes of hydrolysis represented simulated gastric conditions having low pH while the next 120 minutes represented the simulation of small intestinal condi-tions at neutral pH. In the gastric step, no glucose has been reported to be released from bean starch, whereas in bean flour and autoclaved beans con-stant levels of glucose in a very low range (2.89 ± 0.54 %) were detected. This glucose was reported to be derived from sugars or dextrins initially pre-sent in navy beans. Similar observations were recorded during simulated gastric digestion for pure starch and other starch-based foods such as po-tatoes (Dartois *et al.*, 2010; Bordoloi *et al.*, 2012). In the case of bean starch, all soluble sugars would have been washed away during the purification pro-cess. The constant level of glucose throughout the *in vitro* gastric simulation period showed that negligible starch hydrolysis occurred under these condi-tions. Under simulated intestinal conditions, the amount of starch hydrolysis of freshly cooked bean paste, bean flour, and starch paste increased progres-sively over the 120 min period of digestion. However, the rate of starch hy-drolysis was lower in the digestion of freshly prepared bean paste compared to the rate seen with samples of cooked bean starch and cooked bean flour (Figure 8.5). A very low percentage ($\approx 20\%$) of starch hydrolysis was observed during the first 15 minutes of hydrolysis for the cooked beans whereas it was significantly higher ($>80\%$) for cooked bean flour and starch paste. The starch component of cooked bean flour showed a very similar rate

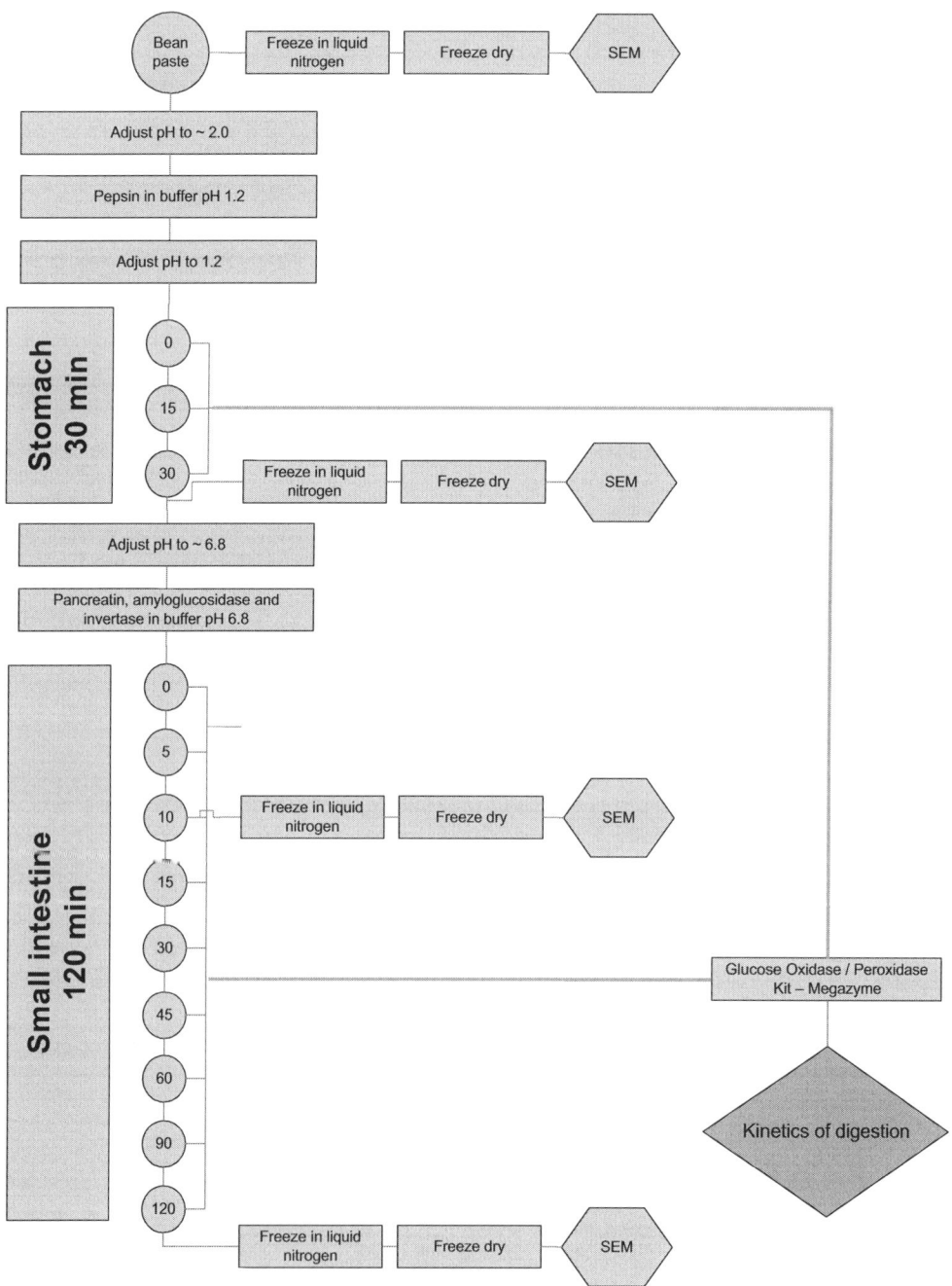

■ **FIGURE 8.3** Flow chart illustrating the *in vitro* digestion of starch in navy beans. *Reproduced from Berg et al. (2012) with permission from Elsevier.* [A color version of this figure is available online at www.booksite.elsevier.com/9780124046108].

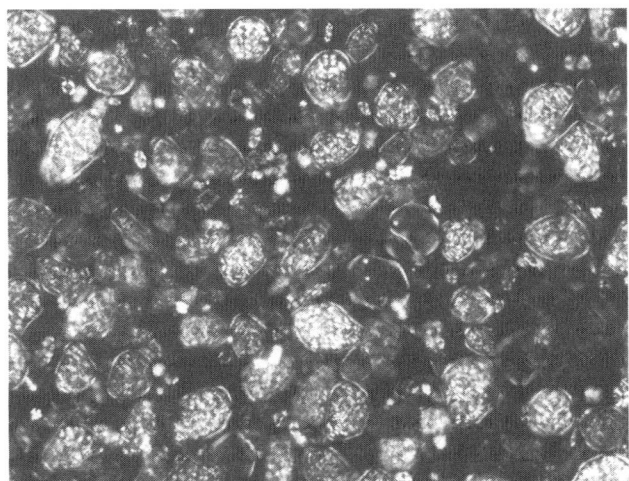

■ **FIGURE 8.4** Light microscopy picture (polarized light) showing cotyledon cells in navy bean paste before *in vitro* digestion. *Reproduced from Berg* et al. *(2012) with permission from Elsevier.*

of hydrolysis to that of cooked bean starch. During cooking, the starch granules in both cases were completely surrounded by water and unconstrained, allowing them to swell and gelatinize to full extent. Slightly lower levels of hydrolysis of starch in flour compared to extracted starch was reported to be

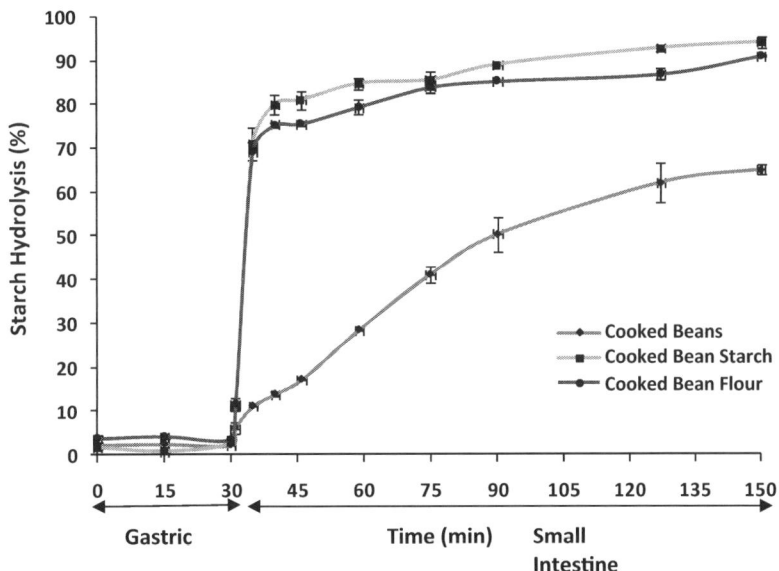

■ **FIGURE 8.5** Starch hydrolysis (%) during *in vitro* digestion of cooked navy beans, navy bean starch, and navy bean flour. *Reproduced from Berg* et al. *(2012) with permission from Elsevier.* [A color version of this figure is available online at www.booksite.elsevier.com/9780124046108].

possibly due to the cell clusters which were sieved out in the sample preparation, resulting in an overestimation of theoretical available starch. The presence of proteins and other components in the digestion matrix has been reported to decrease the amount of starch hydrolysis (Singh *et al.*, 2010). However, the levels of starch hydrolysis for bean flour and pure bean starch did not show an appreciable difference from each other. The total hydrolysis of starch in cooked bean paste at the end of *in vitro* digestion period was ≈65% of the total, while it reached more than 90% for the cooked bean flour and starch paste (Berg *et al.*, 2012; Figure 8.5). The enzymatic digestibility of different starches is influenced by starch source, granule size, crystallinity, and amylose to amylopectin ratio (Singh *et al.*, 2010). Starches with higher amylose content are more resistant towards digestion (Wolf *et al.*, 1999); however, Zhou *et al.* (2010) reported that molecular weight of amylose also significantly affects the digestion of bean starch in the human small intestine with high molecular weight amylose preferentially digested.

Effects of Storage and Reheating of Cooked Navy Beans

The influence of storage and reheating on the *in vitro* digestibility of starch in beans is shown in Figure 8.6. Starch digestibility significantly decreased during storage mainly due to retrogradation of the cooked starch. Navy bean

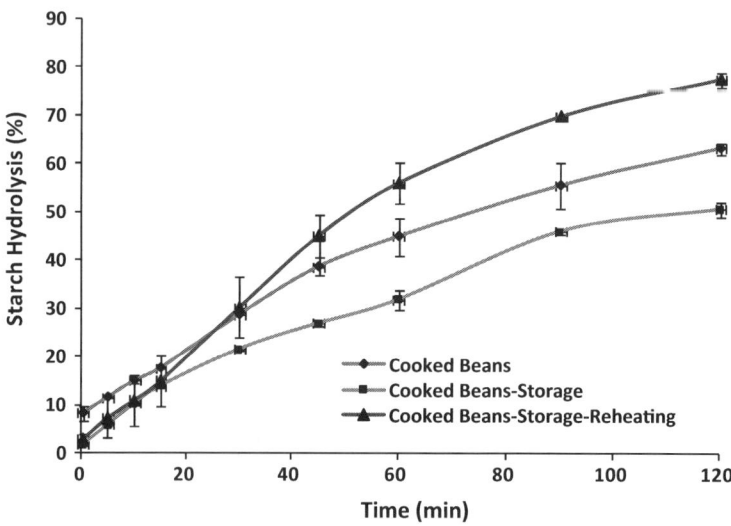

■ **FIGURE 8.6** Starch hydrolysis (%) during *in vitro* intestinal digestion of cooked, cooked—stored, and cooked—stored—reheated navy beans. *Reproduced from Berg* et al. *(2012) with permission from Elsevier.* [A color version of this figure is available online at www.booksite.elsevier.com/9780124046108].

starch contained high amylose content and it has been reported that the higher amylose content of starch lowers the starch digestibility because of a positive correlation between amylose content and resistant starch formation after cooking (Singh *et al.*, 2010). When Berg *et al.* (2012) studied the effect of storage at room temperature, a clear reduction in final values of hydrolysis was observed. Reheating rendered the starch more susceptible towards enzymatic hydrolysis, resulting in final values higher than the ones obtained from freshly cooked beans. During the reheating of the stored cooked beans, partially cooked starch granules may have been further gelatinized. Another possibility mentioned by Berg *et al.* (2012) was that the cell structure might have loosened during the initial cooking and subsequent holding period resulting in a more permeable structure after the second heating. A combination of these factors might have led to the increased rates of hydrolysis.

Effects of High Pressure (French Press) Treatment of Cooked Navy Beans

Micrographs of the cooked and high pressure treated bean samples under normal and polarized light are shown in Figure 8.7A−B (Berg *et al.*, 2012). Complete disruption of cells in the cooked and ground bean paste was carried out using a French press. The French press comprises a cylinder filled with the material to be processed and a piston that is forced down the cylinder with a hydraulic ram. The material flows out of the cylinder through a needle valve which is used to control the pressure drop (38 to 42 MPa) during processing. The French press treatment resulted in the disruption of cotyledon cells, which led to the release of starch granules. Most of these granules showed birefringence patterns under polarized light, which confirms the partial gelatinization of starch (Figure 8.7B). Figure 8.8 compares

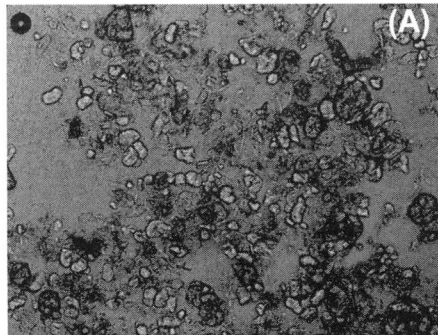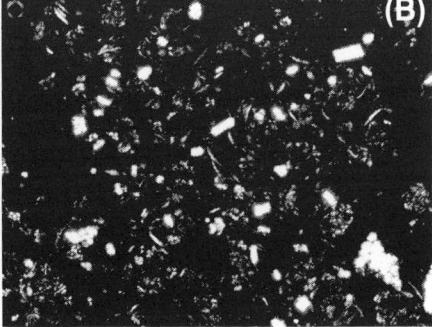

■ **FIGURE 8.7** Light microscopy picture of high pressure (French press) treated navy bean paste showing disrupted cotyledon cells. (A) Normal light and (B) polarized light. *Reproduced from Berg* et al. *(2012) with permission from Elsevier.* [A color version of this figure is available online at www.booksite.elsevier.com/9780124046108].

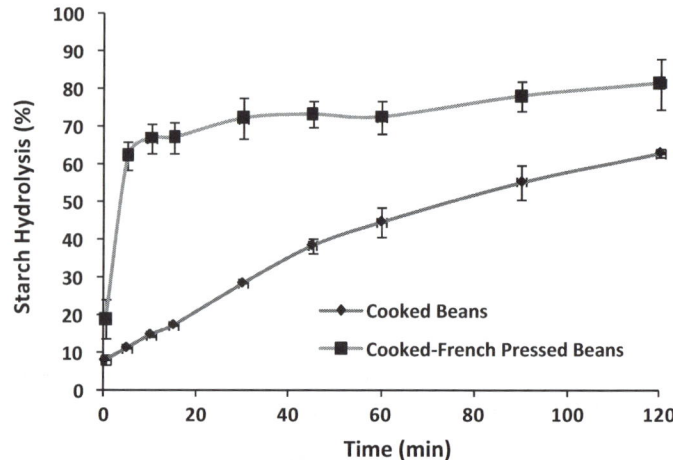

■ **FIGURE 8.8** Starch hydrolysis (%) during *in vitro* intestinal digestion of cooked and cooked—high pressure (French press) treated navy beans. *Reproduced from Berg* et al. *(2012) with permission from Elsevier.* [A color version of this figure is available online at www.booksite.elsevier.com/9780124046108].

in vitro starch hydrolysis of French press treated beans to that of freshly cooked beans. Compared to bean paste samples with intact cotyledon cells, the high pressure treated samples with disrupted cotyledon cells showed significantly higher final values of starch hydrolysis. Similar to starch hydrolysis in bean flour and pure bean starch, a sharp increase in the rate of hydrolysis during the first 10 minutes of *in vitro* simulated small intestinal digestion was observed for French press treated bean paste. However, the overall and end point *in vitro* intestinal digestibility values of French press treated samples were still lower than those obtained for bean starch or bean flour. This lower digestibility was attributed to the low degree of gelatinization in the cooked bean starch (Berg *et al.*, 2012). The French press treatment disrupted the cells, exposing the partially gelatinized starch granules to the intestinal enzymes. The gelatinized parts of the exposed starch granules were hydrolyzed quickly whereas the non-gelatinized parts were not hydrolyzed completely by the enzymes. The cell walls of bean cotyledons appeared to prevent the complete swelling and gelatinization during cooking as indicated by the presence of birefringence remaining in the cooked granules under polarized light (Figure 8.4). Kon *et al.* (1971) reported a similar result. The reduced gelatinization leads to lower starch hydrolysis values as observed in the comparison between French press treated and cooked beans (Figure 8.8).

A second reason for reduced hydrolysis of starch was reported to be the intactness of cells throughout the *in vitro* digestion process which had a retarding effect on the enzymatic action (Berg *et al.*, 2012). Comparative particle size measurements revealed that the particle size of bean cells does not change during *in vitro* digestion of beans (Figure 8.10). Scanning electron microscopy (Figure 8.9) reveals a high level of intact cells throughout and after the process (Berg *et al.*, 2012). These two findings

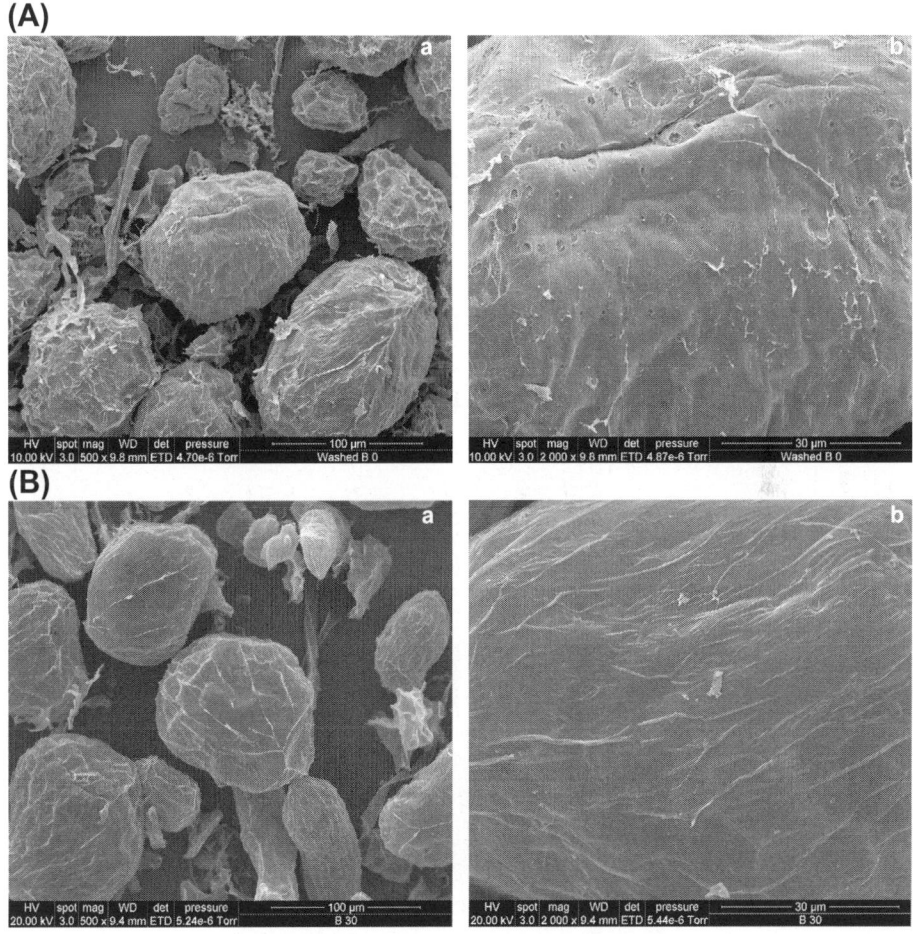

■ **FIGURE 8.9** (A) Scanning electron micrographs collected during *in vitro* gastric and intestinal digestion of starch in cooked navy beans: (a) undigested sample taken at 0 min showing cotyledon cells, (b) magnified view of cotyledon cell wall. (B) Scanning electron micrographs collected during *in vitro* gastric and intestinal digestion of starch in cooked navy beans: sample taken after 30 min of gastric digestion: (a) cotyledon cells in navy bean digesta, (b) magnified view of cotyledon cell wall. (C) Scanning electron micrographs collected during *in vitro* gastric and intestinal digestion of starch in cooked navy beans: sample taken after 10 min of intestinal digestion: (a) cotyledon cells in navy bean digesta, (b) magnified view of cotyledon cell wall. (D) Scanning electron micrographs collected during *in vitro* gastric and intestinal digestion of starch in cooked navy beans: sample taken after 120 min of intestinal digestion: (a) cotyledon cells in navy bean digesta, (b) magnified view of cotyledon cell wall, (c) broken cotyledon cells (picture not representative, see text). *Reproduced from Berg* et al. *(2012) with permission from Elsevier.*

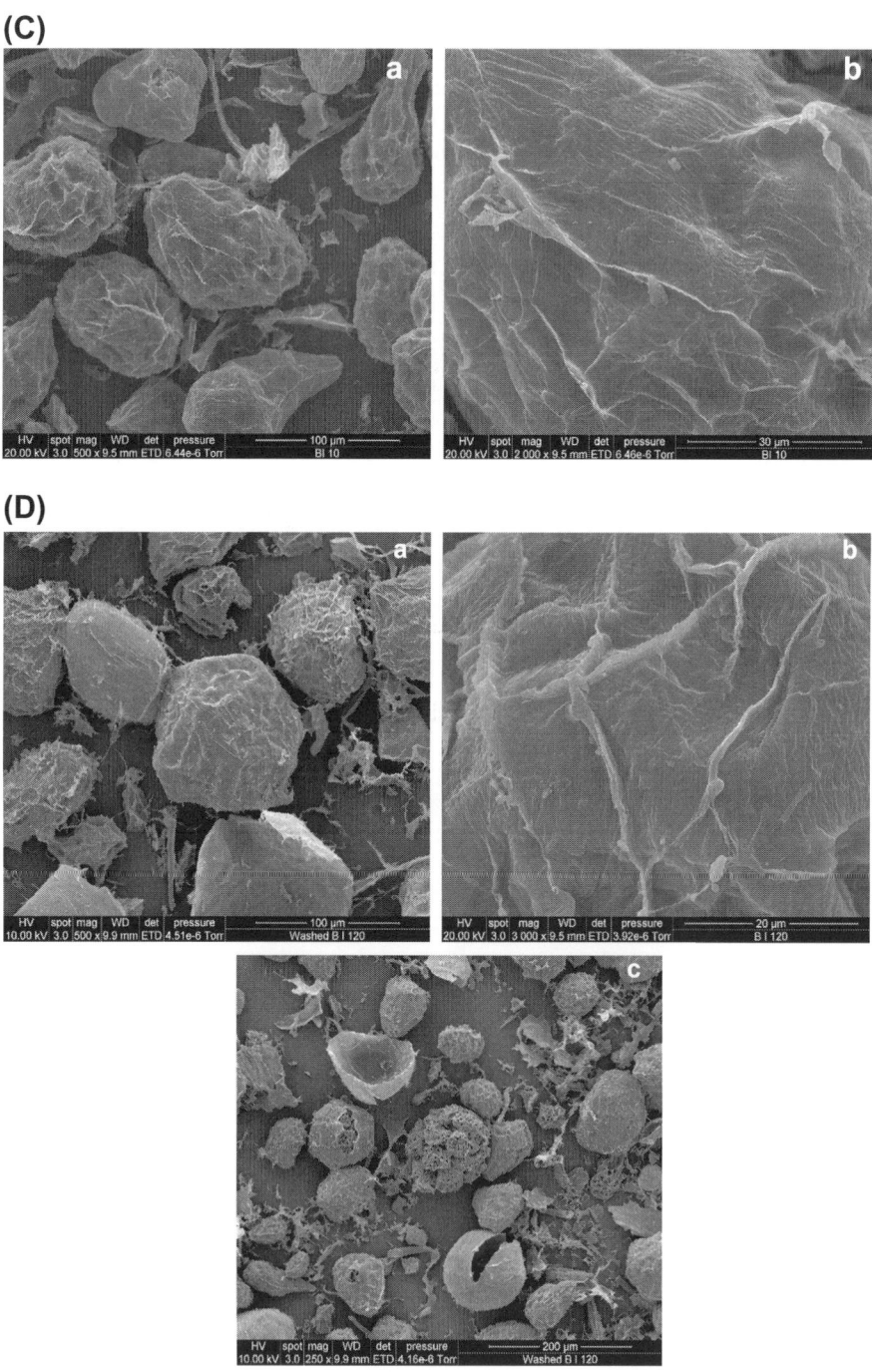

■ FIGURE 8.9 cont'd.

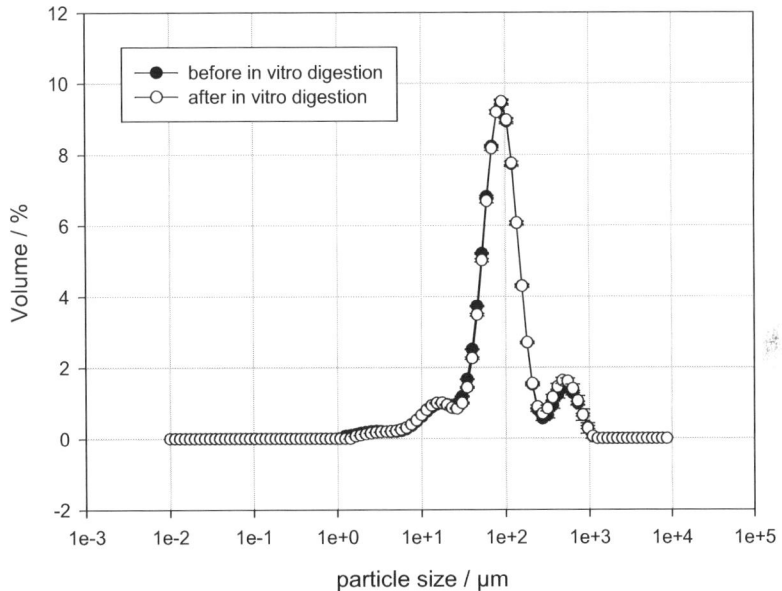

■ **FIGURE 8.10** Particle size distribution of freshly prepared cooked bean paste before and after *in vitro* gastric and intestinal digestion. *Reproduced from Berg et al. (2012) with permission from Elsevier.*

suggested that the hydrolysis of starch takes place inside the cotyledon cells. The observed intactness of cells correlated with a major decrease in the rate of hydrolysis. The most reasonable cause for this was that the partly gelatinized starch granules were tightly packed inside the cotyledon cells. This situation would have restricted free access of enzymes to the starch granules and therefore retarded their hydrolysis. In the case of disrupted cells, the partly gelatinized starch granules were separated from each other and from non-starch components. The free surface area of the starch granules was therefore increased and the enzymatic access was facilitated. Another reason reported for the increase in the rate of hydrolysis was that the relatively high pressure and high shear that occurred during the French press treatment had changed the structure of the partly gelatinized starch granules rendering them easier to hydrolyze. Furthermore, it cannot be ruled out that the cell wall acts as a barrier for the products of starch hydrolysis. In this case, the dextrins resulting from the initial hydrolytic action of α-amylase would have to be hydrolyzed to a certain degree, decreasing their size before the resulting oligosaccharides can diffuse out of the cells.

Storage and reheating of stored beans clearly affected the final values of hydrolysis but the rate of hydrolysis did not change to a major extent (Figure 8.6). Stored and aged starch gels showed a characteristic melting endotherm of recrystallized amylopectin around 55–60°C, which was absent in fresh starch gels and breads (Eliasson, 2006). This suggested that reheating of pre-cooked and stored (e.g., canned) beans to low but acceptable temperatures might preserve the lower final value of starch hydrolysis, which in turn would lead to a higher content of resistant starch. The stability of cotyledon cell walls in navy beans prevented the complete gelatinization of starch and was therefore responsible for the resistant nature of bean starch and that the intactness of cells throughout the simulated digestive process somehow retarded the hydrolysis of starch and thereby the availability of glucose.

MICROSTRUCTURE OF NAVY BEAN DIGESTA

Figure 8.9A–D show freeze-dried samples taken throughout the *in vitro* digestion process of freshly cooked beans (Berg *et al.*, 2012). The regular cellular structure of cotyledons was maintained to a good extent during and after cooking. A high level of cell integrity was observed in all samples. The cotyledon cells appeared to have shrunk slightly and showed indentation and wrinkles on their surface. The swelling of starch granules during auto-claving resulted in the enlargement of cotyledon cells due to absorption of water during cooking. This was evident in the case of undigested samples (Figure 8.9A). A magnified view of the cotyledon outer cell wall of the un-digested sample showed less wrinkling compared to the samples subjected to intestinal digestion. It might be the case that hydrolysis of starch during *in vitro* digestion and removal of water left a space in which the cell wall could be folded in during freeze drying, causing an even more wrinkled sur-face and indentation (Berg *et al.*, 2012). This phenomenon was distinctive in cells that underwent the whole simulated digestive process (120 min). How-ever, the possibility of artifacts produced by the freeze drying process could not be ruled out. The development of holes in the cotyledon cell walls during the *in vitro* digestion was considered to be unlikely because the cell wall is composed of β-(1-4) linked D-glucose molecules, and enzymes capable of hydrolyzing these linkages are not present in the simulated digestive fluids (Berg *et al.*, 2012). The cells stayed mainly intact during the enzymatic ac-tion. The *in vitro* digestion of starch therefore has to at least partially take place inside the bean cells, implying permeability of the cell wall to digestive enzymes. Figure 8.9D(c) is not representative for the sample, but showed a phenomenon that occurred only in a minority of cells which might have been damaged during sample preparation or mixing in the digestion reactor. The cell wall was completely or partly removed perhaps during preparation, and the cellular contents were exposed (Figure 8.9D(d)). The sponge-like open structure showed empty cavities from which starch might have been removed during digestion.

INFLUENCE OF PARTICLE SIZE ON PARTICLE SIZE OF NAVY BEAN PASTES

The change of particle size of bean paste samples during *in vitro* digestion was determined with a laser diffraction particle size analyzer (Malvern Mastersizer 2000, Malvern Instruments Ltd., Worcestershire, UK) (Berg *et al.*, 2012). Bean pastes were measured before starting the *in vitro* digestion and immediately afterwards. Sample material (≈ 10 ml) was added into the 800 ml water reservoir of the dispersion unit (Hydro 2000 MU) until an obscuration level of $15 \pm 2\%$ was obtained. The pump speed was set to 2000 rpm. Refractive indices of 1.50 and 1.33 were used for the bean paste and the water phase, respectively. The particle absorbance of the bean paste was set to 0.1. The samples taken before and after the *in vitro* digestion for particle size analysis showed almost identical particle size patterns (Figure 8.10). It can therefore be concluded that no major changes in the size of particles in bean paste occur. The peak of the actual particle sizes of $\approx 100\,\mu m$ is in good accordance with the size of bean cells in scanning electron microscopy pictures as seen in Figure 8.9, whereas the peak around $500\,\mu m$ might represent fragments of the hull of bean cotyledons (Berg *et al.*, 2012).

CONCLUSIONS

Cotyledon cell walls of navy beans are very stable and they impose restrictions on the swelling and gelatinization of bean starch during cooking. The pectinaceous middle lamella, which firmly connects cells in the raw state, is dissolved during cooking. Therefore, the cells are rather separated than disrupted during blending processes and presumably during chewing as well. The incomplete gelatinization of starch granules ultimately reduces the rate and extent of starch hydrolysis measured as glucose release during 120 minutes of *in vitro* digestion with simulated gastric and small intestinal fluids. The stability of cotyledon cells and the small surface area of the starch granules, which are tightly packed inside the cells, also appear to restrict the free access of amylolytic enzymes during *in vitro* digestion. Experiments with stored and reheated beans suggest that the extent of gelatinization and availability of water during cooking could be a major factor influencing starch hydrolysis during *in vitro* digestion.

ACKNOWLEDGMENTS

Permission from Elsevier to reproduce parts of the article (Berg *et al.*, 2012) is gratefully acknowledged.

REFERENCES

Badenhuizen, N.P., 1969. The Biogenesis of Starch Granules in Higher Plants. Appleton Crofts, New York.

Belitz, H.D., Grosch, W., Schieberle, P., 2009. Food Chemistry. Springer-Verlag, Berlin, Heidelberg.

BeMiller, J.N., Whistler, R.L., 2009. Starch: Chemistry and Technology. Academic Press, London.

Berg, T., Singh, J., Hardacre, A., Boland, M.J., 2012. The role of cotyledon cell structure during in vitro digestion of starch in navy beans. Carbohydr. Polym. 87, 1678—1688.

Betancur-Ancona, D.A., Guerrero, L.A.C., Matos, R.C., Ortiz, G.D., 2001. Physicochemical and functional characterization of Baby Lima bean (Phaseolus lunatus) starch. Starch 53, 219—226.

Bordoloi, A., Singh, J., Kaur, L., Singh, H., 2012. In vitro digestibility of starch in cooked potatoes as affected by guar gum: microstructural and rheological characteristics. Food Chem. 133, 1206—1213.

Carpita, N.C., Gibeaut, D.M., 1993. Structural models of primary-cell walls in flowering plants—consistency of molecular structure with the physical properties of the walls during growth. Plant J. 3, 1—30.

Chanda, S.V., 2005. Evaluation of effectiveness of the methods for isolation of cell wall polysaccharides during cell elongation in *Phaseolus vulgaris* seedlings. Acta. Physiol. Plant 27, 371—378.

Chojecki, A.J.S., Gale, M.D., Bayliss, M.W., 1986. The number and size classes of starch granules in the wheat endosperm, and their association with grain weight. Ann. Bot. 58, 819—831.

Chung, H., Liu, Q., Pauls, K.P., Fan, M.Z., Yada, R., 2008. *In vitro* starch digestibility, expected glycemic index and some physicochemical properties of starch and flour from common bean (Phaseolus vulgaris L.) varieties grown in Canada. Food Res. Int. 41, 869—875.

Dartois, A., Singh, J., Kaur, L., Singh, H., 2010. The influence of guar gum on the *in vitro* starch digestibility—rheological and microstructural characteristics. Food Biophys. 5, 149—160.

Daussant, J., Mosse, J., Vaughan, J.G., 1983. Seed Proteins. Academic Press, London.

Eliasson, A.-C., 2006. Carbohydrates in Food. CRC Press, Taylor and Francis, Florida.

Gujska, E., Reinhard, W.D., Khan, K., 1994. Physicochemical properties of field pea, pinto and navy bean starches. J. Food Sci. 59, 634—636.

Hahn, D.M., Jones, F.T., Akhavan, I., Rockland, L.B., 1977. Light and scanning electron-microscope studies on dry beans—intracellular gelatinization of starch in cotyledons of large lima beans (*Phaseolus lunatus*). J. Food Sci. 42, 1208—1212.

Jane, J.L., Kasemsuwan, T., Leas, S., Ia, A., Zobel, H., Il, D., Robyt, J.F., 1994. Anthology of starch granule morphology by scanning electron microscopy. Starch 46, 121—129.

Jenkins, D.J.A., Wolever, T.M.S., Taylor, R.H., Barker, H., Fielden, H., Baldwin, J.M., et al., 1981. Glycemic index of foods—a physiological basis for carbohydrate exchange. Am. J. Clin. Nutr. 34, 362—366.

Kaur, L., Rutherfurd, S.M., Moughan, P.J., Drummond, L.N., Boland, M.J., 2010b. Actinidin enhances protein digestion in the small intestine as assessed using an in vitro digestion model. J. Agric. Food Chem. 58, 5074—5080.

Kaur, L., Rutherfurd, S.M., Moughan, P.J., Drummond, L.N., Boland, M.J., 2010a. Actinidin enhances gastric protein digestion as assessed using an in vitro gastric digestion model. J. Agric. Food Chem. 58, 5068—5073.

Kaur, L., Singh, J., McCarthy, O.J., Singh, H., 2007. Physico-chemical, rheological and structural properties of fractionated potato starches. J. Food Eng. 82, 383—394.

Kim, Y.S., Wiesenborn, D.P., Lorenzen, J.H., Berglund, P., 1996. Suitability of edible bean and potato starch noodles. Cereal Chem. 73, 302—308.

Kon, S., Wagner, J.R., Becker, R., Booth, A.N., Robbins, D.J., 1971. Optimizing nutrient availability of legume food products. J. Food. Sci. 36, 635—638.

Krossmann, J., Lloyd, J., 2000. Understanding and influencing starch biochemistry. Crit. Rev. Biochem. Mol. Biol. 35, 141—196.

Lindeboom, N., Chang, P.R., Tylera, R.T., 2004. Analytical, biochemical and physicochemical aspects of starch granule size, with emphasis on small granule starches: a review. Starch 56, 89—99.

Ogawa, Y., Sugiyama, J., Kuensting, H., Ohtani, T., Hagiwara, S., Kokubo, M., et al., 2000. Development of visualization technique for three-dimensional distribution of protein and starch in a brown rice grain using sequentially stained sections. Food Sci. Technol. Res. 3, 176—178.

Oliveira, A.C., Queiroz, K.S., Helbig, E., Reis, S.M.P.M., Carraro, F., 2001. The domestic processing of the common bean resulted in a reduction in the phytates and tannins antinutritional factors, in the starch content and in raffinose, stachiose and verbascose flatulence factors. Arch. Latino Amer. Nutr. 51, 276—283.

Ramirez-Cardenas, L., Leonel, A.J., Costa, N.M.B., 2008. Effect of domestic processing on nutrient and antinutritional factor content in different cultivars of common beans. Cienc. Tecnol. Aliment. 28, 200—213.

Rehman, Z.U., Shah, W.H., 2005. Thermal heat processing effects on antinutrients, protein and starch digestibility of food legumes. Food Chem. 91, 327—331.

Rehman, Z., Salariya, A.M., Zafar, S.I., 2001. Effect of processing on available carbohydrate content and starch digestibility of kidney beans (*Phaseolus vulgaris* L.). Food Chem. 73, 351—355.

Salgado, S.M., Melo Filho, A.B., Andrade, S.A.C., Maciel, G.R., Livera, A.V.S., Guerra, N.B., 2005. Modification of the concentration of resistant starch in macassar bean (*Vigna unguiculata* L. Walp) hydrothermal process and freezing. Cienc. Tecnol. Aliment. 25, 259—264.

Sim, L., Quezada-Calvillo, R., Sterchi, E.E., Nichois, B.L., Rose, D.R., 2008. Human intestinal maltase-glucoamylase: crystal structure of the N-terminal catalytic subunit and basis of inhibition and substrate specificity. J. Mol. Biol. 375, 782—792.

Singh, J., Dartois, A., Kaur, L., 2010. Starch digestibility in food matrix: a review. Trends Food Sci. Tech. 21, 168—180.

Singh, J., Kaur, L., McCarthy, O.J., 2007. Factors influencing the physico-chemical, morphological, thermal and rheological properties of some chemically modified starches for food applications—a review. Food Hydrocoll. 21, 1—22.

Singh, J., McCarthy, O.J., Singh, H., Moughan, P.J., 2008. Low temperature post-harvest storage of New Zealand Taewa (Maori potato): effects on starch physico-chemical and functional characteristics. Food Chem. 106, 583—596.

Singh, N., Kaur, L., Singh, J., 2004. Relationships between various physicochemical, thermal and rheological properties of starches separated from different potato cultivars. J. Sci. Food. Agric. 84, 714−720.

Smith, M.E.P.D.D., Morton, D.G., 2001. The Digestive System. Churchill Livingstone, Edinburgh.

US Pharmacopeia, 2000. Pharmacopeia, simulated gastric fluid, TS, simulated intestinal fluid, TS, United States Pharmacopeial Convention, vol. 24. The National Formulary 9 (US Pharmacopeia Board of Trustees), Rockville, MD, 2235.

Vargas-Torres, A., Osorio-Díaz, P., Islas-Hernández, J.J., Tovar, J., Paredes-López, O., Bello-Pérez, L.A., 2004. Starch digestibility of five cooked black beans (*Phaseolus vulgaris* L.) varieties. J. Food Compos. Anal. 17, 605−612.

Wolf, W.B., Bauer, L.L., Fahey Jr, G.C., 1999. Effects of chemical modification in vitro rate and extent of food starch digestion: an attempt to discover a slowly digested starch. J. Food Agric. Chem. 47, 4178−4183.

Woolnough, J.W., Monro, J.A., Brennan, C.S., Bird, A.R., 2007. Glycemic carbohydrates: standardisation of in vitro methods. Proc. Nutr. Soc. New Zealand 32, 38−42.

Wursch, P., 1989. Starch in human nutrition. World Rev. Nutr. Diet. 60, 199−256.

Wursch, P., Delvedovo, S., Koellreutter, B., 1986. Cell structure and starch nature as key determinants of the digestion rate of starch in legume. Amer. J. Clin. Nutr. 43, 25−29.

Zhou, Z., Topping, D.L., Morell, M.K., Bird, A.R., 2010. Changes in starch physical characteristics following digestion of foods in the human small intestine. Br. J. Nutr. 104, 573−581.

Modelling the gastrointestinal tract

Chapter **9**

Mathematical Models of Food Degradation in the Human Stomach

A.S. Van Wey[1,2] and P.R. Shorten[1,2]

[1]*AgResearch, Ruakura Research Centre, Hamilton, New Zealand*, [2]*Riddet Institute, Massey University, Palmerston North, New Zealand*

CONTENTS

INTRODUCTION

The human stomach is a dynamic and complex system, where the degradation of food particles is the result of mechanical forces and chemical reactions. Peristaltic movements of the stomach compress the bolus which may

Food Structures, Digestion and Health. http://dx.doi.org/10.1016/B978-0-12-404610-8.00009-8

lead to fracturing of particular food matrices as well as move the viscous food bolus in a retropulsive flow pattern. This leads to shearing and erosion of the individual food surfaces through contact with other food components in the bolus. In addition to mechanical forces, acidic hydrolysis and enzymatic reactions cause leaching of nutrient from the food matrix to be absorbed in the small intestine. Food digestion kinetics are not limited to mechanical forces, but also depend on food structure, particle size, meal volume and composition, viscosity, pH, temperature, and enzymatic reaction.

Food particles break down by the processes of erosion, fragmentation, and tenderization (Kong and Singh, 2009a, b, 2011; Kong *et al.*, 2011). These processes are shown in Figure 9.1. Erosion describes the process of the gradual wearing away of the food surface when the applied stresses are less than that required to fracture or break the food particle. Fragmentation describes the breaking of food particles into two or more similar sized particles. The process of tenderization occurs as a result of the transport of gastric fluid into the food particle and the resulting softening of the food matrix. In general, the process of tenderization increases the rate of erosion and fragmentation. The transport of gastric fluid into food particles also facilitates leaching of solids into the gastric medium.

Here we review the mathematical modeling work currently in the literature which addresses digestion in the human stomach. There are remarkably few mathematical models in the literature that describe food degradation

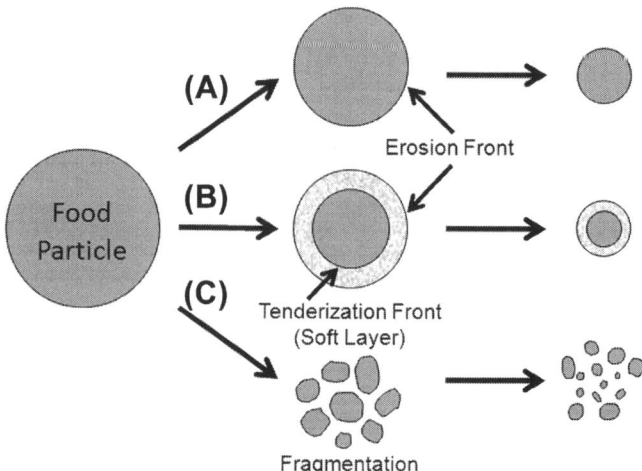

■ **FIGURE 9.1** The three processes of food particle degradation: erosion (A), tenderization (B), and fragmentation (C).

in the human stomach. We also outline some new modeling approaches to describe the degradation of food particles in the stomach.

MODELS OF FLUID AND FOOD PARTICLE FLOW IN THE STOMACH

The area that has received the most attention is that of modeling the motility and geometry of the human stomach (Ferrua *et al.*, 2011). These models are based on computation fluid dynamics (CFD). Many of these models have, for simplicity, reduced the geometry and motility to two dimensions; however, Ferrua and Singh (2010) using a three-dimensional (3D) model have shown that the flow of gastric contents is three dimensional and that the two-dimensional (2D) models cannot account for the differences in velocity and flow patterns as described with a 3D model. In particular, the 2D model cannot reproduce the retropulsive flow pattern and eddy structures predicted by the 3D model. The 3D model has yet to be expanded to describe the shearing or fracturing of food components in the human stomach. CFD has also been used in industrial/engineering contexts to model particle−fluid flow, particle−fluid interactions, particle−particle interactions, particle−wall interactions, and particle erosion (which has been described as a function of the particle impact velocities) (Li *et al.*, 2009).

EMPIRICAL MODELS OF WET MASS RETENTION DURING DIGESTION

The first mathematical models to describe food digestibility have primarily focused on the wet mass retention ratio of the studied food (wet sample weight after digestion time, *t*, divided by the initial wet sample weight). The wet mass retention ratio has been observed to follow three distinct profiles (delayed sigmoidal, sigmoidal, and exponential; Figure 9.2), yet all three profiles can be described using one equation:

$$y(t) = (1 + k\beta t)\exp(-\beta t) \qquad (9.1)$$

where k and β are constants found using regression (Kong and Singh, 2008, 2009b). The sigmoidal profile can be described by an initial slow disintegration stage, followed by an exponential disintegration profile. The delayed sigmoidal profile occurred in foods with low moisture content. In these foods, the absorption of gastric fluid increased the wet weight of the food to such an extent that it exceeded the degradation rate, which resulted in a wet mass retention ratio greater than 1. The delayed sigmoidal profile can be seen with studies of raw and roasted almonds; in this case the wet mass retention ratio increased, as a result of gastric fluid absorption, even

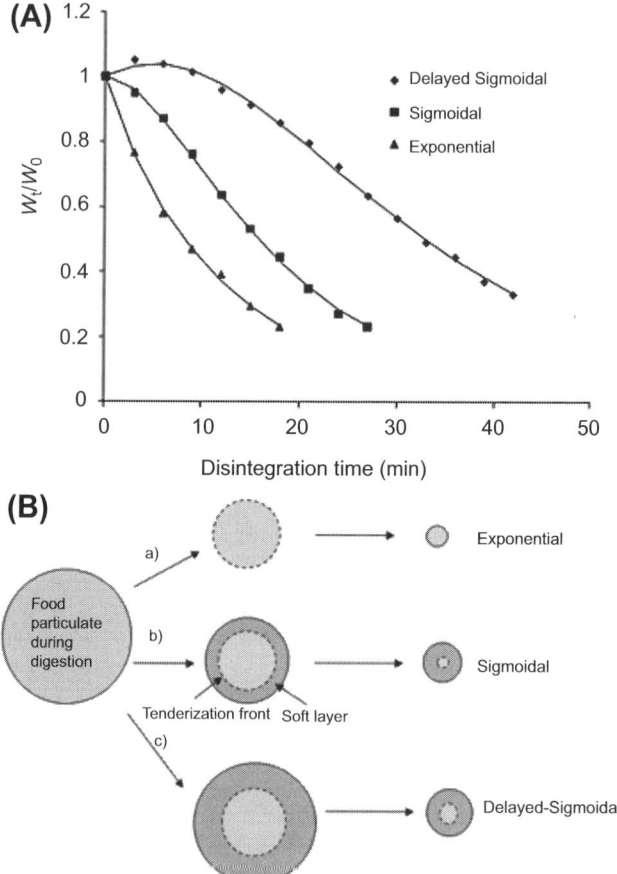

■ **FIGURE 9.2** (A) Three different profiles describing the wet mass retention rate during gastric digestion. (B) Diagram depicting how gastric fluid absorption influences the wet mass retention rate profiles and subsequent erosion of the food matrix during gastric digestion (Ferrua *et al.*, 2011). *Reprinted with permission from Trends in Food Science & Technology.*

though the dry solid mass decreased over the same digestion time (Kong and Singh, 2009a).

While it is useful to have a simple model that can be fit to the data using regression methods, this model does not provide additional understanding of the underlying processes facilitating degradation as it is an empirical model, rather than a mechanistic model. The use of the wet mass ratio may more accurately describe some of the satiation properties of a food matrix by revealing insights into how meal volume changes during the digestive process, rather than depicting the bioaccessibility of nutrients within a food matrix (the fraction of a nutrient released from a food matrix).

EMPIRICAL MODELING OF DRY SOLID LOSS DURING DIGESTION

In order to ascertain the bioaccessibility of nutrients from a food matrix, it is important to be able to determine the dry solid loss during the digestive process. The loss of the dry solid would constitute the soluble and insoluble nutrients that have leached into the gastric medium and thus are available for absorption in the small intestine. There have been only two mathematical models in the literature to describe solid loss, both of which have used raw carrot as the model substrate. The first was an empirical model proposed by Kong and Singh (2011), who used a Weibull function to describe the ratio of dry solid loss at time, t, hours of digestion:

$$\frac{S_0 - S(t)}{S_0} = 1 - \exp(-at^b) \tag{9.2}$$

where S_0 is the initial dry solid weight, $S(t)$ is the solid weight after t hours of digestion, and a and b are constants fit by regression (Kong and Singh, 2011). However, this description of the dry solid loss does not adequately describe the solid loss after an extended time (e.g., 36 hours or more) nor does it provide any insight into the mechanisms responsible for the solid loss. The second model is a mechanistic model that takes into account the role of acidic hydrolysis in solid loss and is described over the next four sections.

MODELING OF THE DYNAMICS OF STOMACH pH DURING DIGESTION

The pH of the contents within the human stomach is dynamic and dependent on meal frequency, amount, and composition. The stomach pH after the ingestion of a meal composed of coarsely ground tenderloin steak (cooked and seasoned with salt), white bread with butter, vanilla ice-cream topped with chocolate syrup, and a glass of water is shown in Figure 9.3 (Malagelada *et al.*, 1976). The pH rapidly increased after meal ingestion due to dilution and thereafter equilibrated to the pre-meal pH. The pH dynamics can be described by:

$$\frac{dpH}{dt} = F\delta(t) + L(pH_2 - pH) \tag{9.3}$$

where F represents the direct dilution effect of the meal on stomach pH, L represents the rate of decrease in stomach pH, and pH_2 is the equilibrium stomach pH. The model fit (Eq. 9.3) to the change in stomach pH after a 400 ml meal (pH 6) (Malagelada *et al.*, 1976) is shown in Figure 9.3. The model characterizes the change in pH after the 400 ml meal. The type of

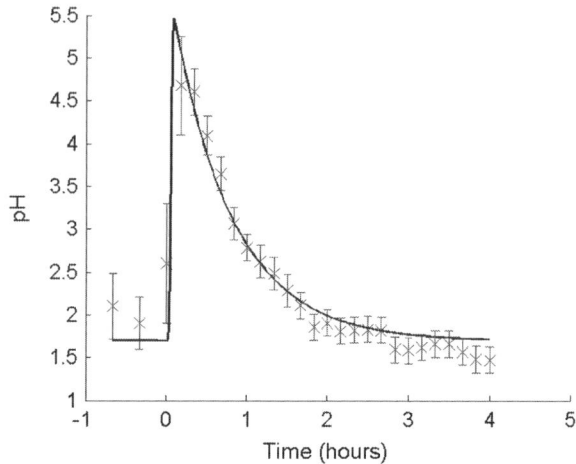

■ **FIGURE 9.3** The change in stomach pH after a 400 ml meal (pH 6) (Malagelada *et al.*, 1976) along with a model fit to the data (Eq. 9.3). Error bars denote standard errors.

meal will have an effect on the dynamics of the stomach pH and this model may not adequately describe all possible meal combinations. Additionally, the parameters of this model are likely to be meal dependent. This model does not describe the spatial changes in pH within the stomach.

MODELS OF THE TRANSPORT OF GASTRIC FLUID INTO FOOD PARTICLES

In the work of Van Wey *et al.* (2014), equations were derived from previous work in cheese brining, dehydration of carbohydrates and gelatin, absorption of water in canning processes, and, interestingly, soil science. In food processing, there have been a variety of models proposed to describe the diffusion of solutes or the leaching of solutes into a surrounding medium; however, Fickian diffusion is the most common (Schwartzberg and Chao, 1982; Turhan and Kaletunç, 1992; Diaz *et al.*, 1993). Most of these models used a constant effective diffusion coefficient (EDC), which is surprising given that it is well established that the EDC is dependent on moisture content (Crank, 1975). In both experimental work and modeling work of food processing, the EDC has been shown to vary in that the EDC increases concomitantly with an increase in moisture content. Furthermore, the theory of liquid transport in porous material, such as soil, has been shown to be relevant to food structures (Syarief *et al.*, 1987; Saguy *et al.*, 2005).

Following the modeling work of Davey *et al.* (2002), describing rice gelatinization during brewing processes, Van Wey *et al.* (2014) used a standard exponential function to describe the dependence of the EDC on gastric fluid (GF) content in conjunction with Darcy's law for unsaturated porous media (with no gravity effects) and conservation of mass:

$$D(\theta) = A \exp(b\theta) \tag{9.4}$$
$$\partial\theta/\partial t = \nabla(D(\theta)\nabla\theta) \tag{9.5}$$

where $D(\theta)$ is interpreted as the ratio of hydraulic conductivity to specific water capacity and A, $b \geq 0$ (see Figure 9.4). In addition to the exponential function (Crank, 1975), two other models have been proposed to describe

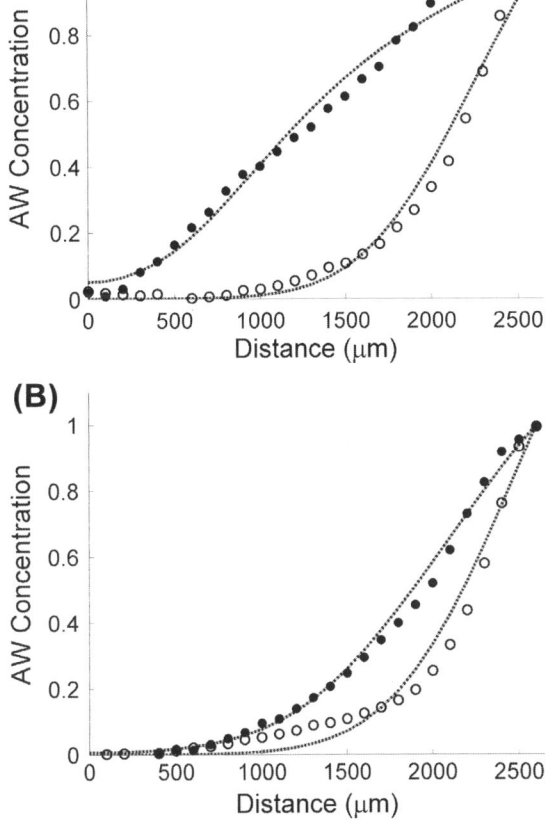

■ **FIGURE 9.4** Acidic water (AW) penetration front profiles for carrot (solid circles) and Edam cheese (open circles) soaked for 10 minutes at pH 1.51 (A) and at pH 7 (B). The model fit (Eqs. 9.4 and 9.5) for individual profiles is denoted by the dotted curves (Van Wey et al., 2014. The AW concentration is expressed as the ratio of the concentration of AW to the concentration of AW at the edge of the food sample). *Reprinted with permission from Elsevier (Van Wey et al., submitted).*

the dependence of the EDC on moisture content: sum of exponentials (Gomi *et al.*, 1996) and the power law (Yamamoto, 1999). Van Wey *et al.* (2014) found that the standard exponential function described the simulated gastric fluid front, in raw carrot core and Edam cheese, better than either the sum of exponentials or the power law. Equations 9.4 and 9.5 can also be used to describe the transport of pepsin into food particles and the transport of nutrients from food particles into the stomach.

THE EFFECT OF pH ON THE TRANSPORT OF GASTRIC FLUID INTO FOOD PARTICLES

In vitro digestion simulations of raw carrot have shown that the pH of the GF had a marked effect on the degradation and solid loss from the food matrix (Kong and Singh, 2011). This work, in conjunction with the fact that the pH of gastric contents is not static throughout the digestive period (Malagelada *et al.*, 1976), led Van Wey *et al.* (2014) to hypothesize that the rate of diffusion of GF into the food matrix was pH dependent. This pH dependence would, in turn, contribute to the lower dry solid loss of carrots at a higher pH.

Van Wey *et al.* (2014) found that the diffusion of acidic water (AW) into raw carrot core and Edam cheese was substantially different and that there were notable differences in the diffusion of AW at various pH values for each food matrix. As seen in Figure 9.5, the difference in the rate at which the AW diffused into the carrot core was far more dramatic than that found for Edam cheese. After 20 minutes of soaking in AW at a pH of 1.51, the AW had fully permeated the center of the carrot core, whereas at pH 4.3 the AW had penetrated less than 1 mm of the carrot (Figure 9.5A, B). While the difference in pH had a marked effect on the penetration rate of the AW for Edam cheese, it was of a lesser effect as compared to raw carrot core. This demonstrates that the food matrix itself (water, protein, fat, etc. content) affects the diffusion of AW into the food matrix (see Figures 9.6 and 9.7).

MODEL OF SOLID LOSS DUE TO FOOD PARTICLE TENDERIZATION

As noted earlier, tenderization of the food matrix results from the absorption of GF into the food matrix. During the tenderization process, solid loss occurs as various components of the food matrix are hydrolyzed and then leach into the stomach. To account for the role of GF in the solid loss of food matrices, the model depicting solid loss proposed by Van Wey and colleagues (2014) assumes that the rate at which the density of the dry food

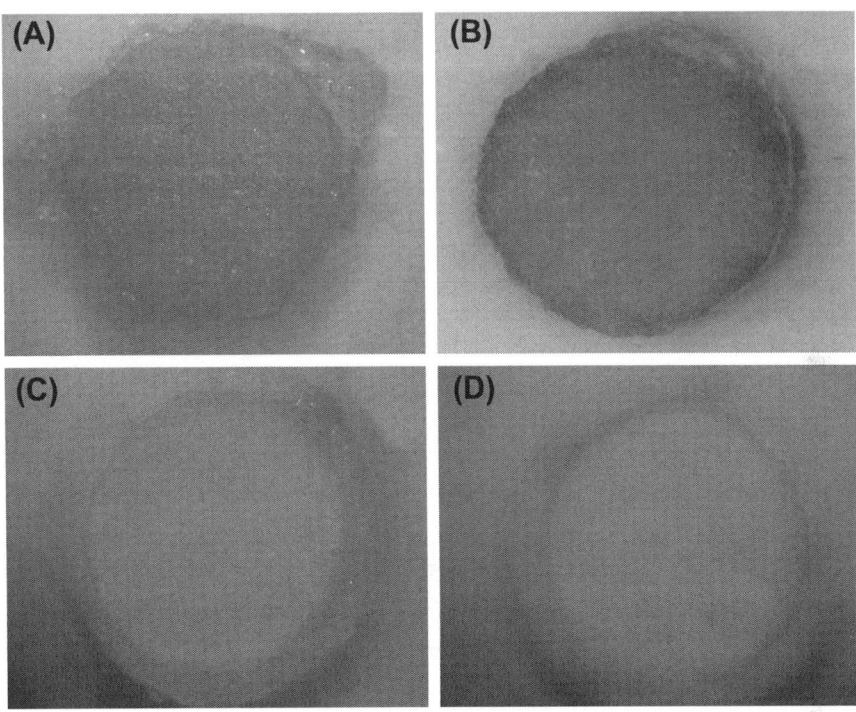

■ **FIGURE 9.5** Cross-section of raw carrot core (5.3 mm diameter) after 20 minutes in acidic water (AW) at pH 1.51 (A) and at pH 4.3 (B). Cross-section of Edam cheese (5.3 mm diameter) after 5 minutes in AW at pH 1.51 (C) and at pH 7 (D). The area in which AW has penetrated is shown by the presence of methylene blue (blue/purple) (Van Wey *et al.*, 2014). *Reprinted with permission from Elsevier (Van Wey* et al., *submitted).* [A color version of this figure is available online at www.booksite.elsevier.com/9780124046108].

particle changes is proportional to the concentration of GF, θ, and the density, ρ, of the food particle at time, t, with radius, r:

$$\frac{\partial \rho}{\partial t} = -k\theta\rho \tag{9.6a}$$

$$\rho(r, t = 0) = n\rho_0 \tag{9.6b}$$

$$\frac{\partial \rho(r = 0, t)}{\partial t} = \frac{\partial \rho(r = R(t), t)}{\partial t} = 0 \tag{9.6c}$$

$$\frac{S(t)}{S_0} = \frac{2\pi L_0}{S_0} \int_0^{R(t)} r\rho dr + (1 - n) \tag{9.6d}$$

where k is the rate of food degradation, the ratio, n, represents the expected percentage solid loss after a long period of soaking in gastric fluid (i.e., the fraction of digestible food matrix), and L_0 is the initial length of the cylindrical-shaped food sample. The rate of solid loss in this model is greater than that by surface erosion but less than that for bulk erosion.

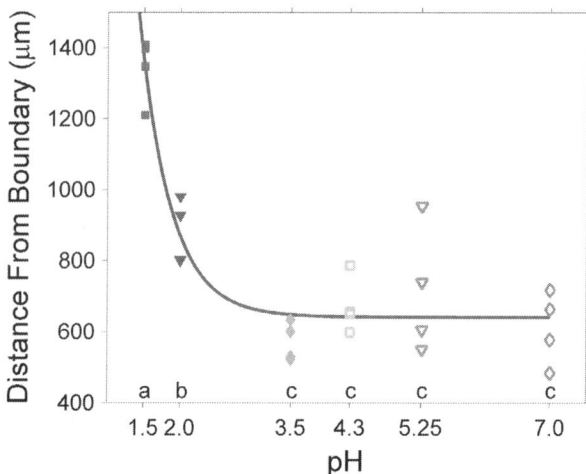

■ **FIGURE 9.6** Distance (μm) at which the acidic water (AW) concentration reaches 50% for carrot samples soaked in AW for 10 minutes, at pH 1.51 (solid square), 2.00 (solid triangles), 3.50 (solid diamonds), 4.30 (open square), 5.26 (open triangles), and 7.00 (open diamonds). The nonlinear regression model (Eq. 9.4) is represented by the curve. The different letters a, b, and c denote significant differences in mean values as determined by one-way ANOVA ($P = 0.05$). There is no significant difference in distance at which the AW concentration reaches 50% for pH >3.5, yet there is a dramatic difference for pH ≤ 3.5 (Van Wey et al., 2014). *Reprinted with permission from Elsevier (Van Wey et al., submitted).* [A color version of this figure is available online at www.booksite.elsevier.com/9780124046108].

This model was validated by comparing the model, which incorporated the diffusion coefficient as a function of pH, to the reported carrot solid loss (Kong and Singh, 2011). As seen in Figure 9.8A, the model adequately characterizes the expected solid loss at three different pH values. The dependence of the rate of degradation (k) on pH is shown in Figure 9.8B. The model demonstrates that solid loss is dependent upon the pH and the concentration of the GF in the food matrix.

The model was also fit to data on the effect of temperature on carrot solid loss (Kong and Singh, 2011). The model fit to the carrot solid loss is shown in Figure 9.8C (pH $= 1.8$) demonstrating that a change in the rate of food degradation (k) and percentage of food degraded (n) describes the difference in data as a result of temperature change. At a temperature of 37°C, $k = 0.10$ and $n = 0.47$, whereas at a temperature of 23°C, $k = 0.08$ and $n = 0.37$.

MODELS OF THE CHANGE IN TEMPERATURE WITHIN FOOD PARTICLES IN THE STOMACH

The transfer of heat into food matrices potentially plays a role on the rate of degradation of food particles (Kong and Singh, 2009b). The process of heat

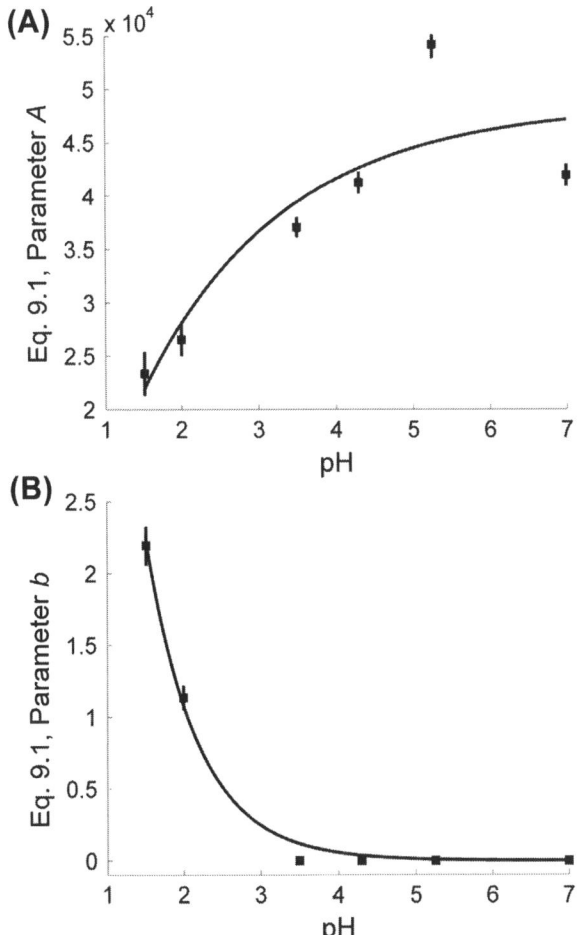

■ **FIGURE 9.7** Estimated parameters *A* and *b* (squares), with SEM, for the effective diffusion coefficient (Eq. 9.4) based on least squares best fit of raw carrot core multi-response data: four data sets for each pH. The relationships between the pH and the parameters (*A* and *b*) were determined by fitting a nonlinear regression model using maximum likelihood (curves). The relationship between pH and the respective parameters is consistent with Figure 9.6, in that for both parameters there is no significant difference in the values of the parameters for pH >3.5, with the exception of the outlier value at pH 5.3 for *A* (Van Wey *et al.*, 2014). *Reprinted with permission from Elsevier (Van Wey et al., submitted).*

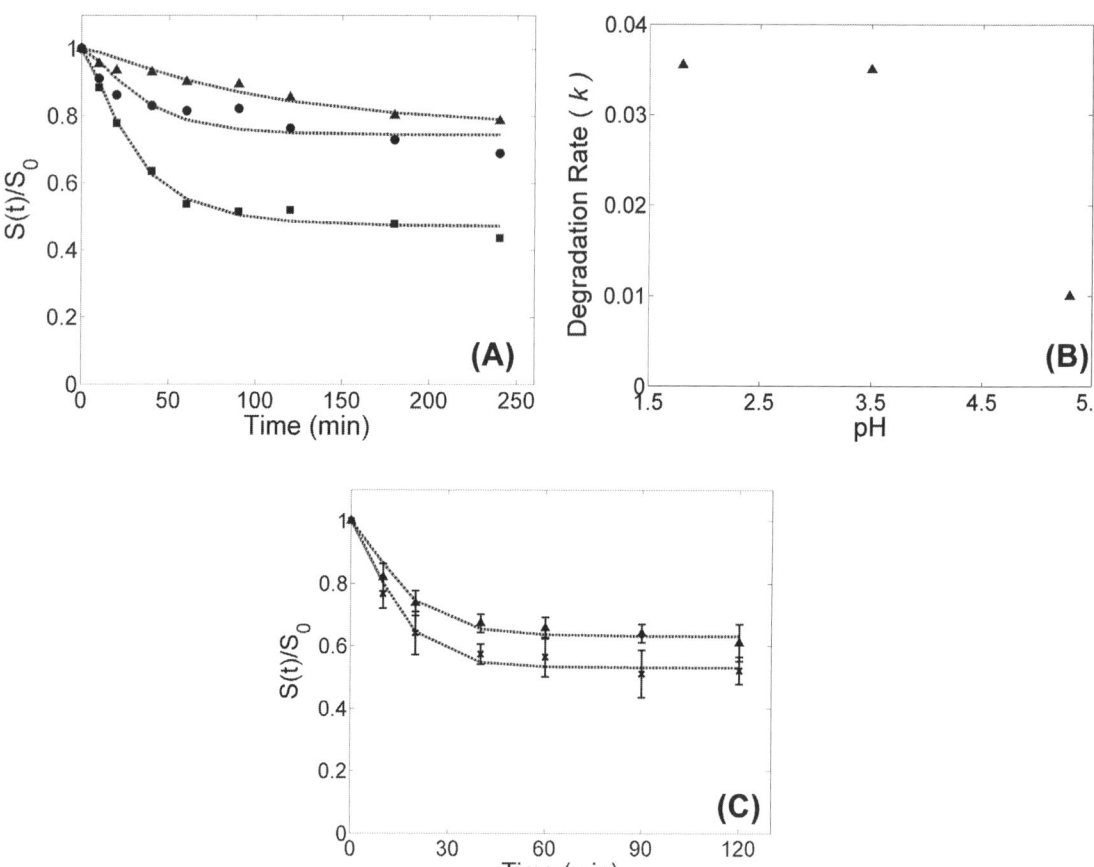

■ **FIGURE 9.8** (A) Solid loss of raw carrots (Fanbin Kong, personal communication) at fixed pH of 5.3 (triangles), 3.5 (circles), and 1.8 (squares) (Kong and Singh, 2011) and model fit (Eqs. 9.4—9.6) for the same fixed pH (curves) (Van Wey et al., 2014). (B) The dependence of the rate of solid degradation (*k*) of raw carrot on pH. (C) Solid loss of raw carrots at fixed pH of 1.8 at 23°C (triangles) and 37°C (circles) (Kong and Singh, 2011) and model fit (Eqs. 9.4—9.6) (curves). (A) *Reprinted with permission from Elsevier (Van Wey* et al., *submitted).*

transfer in the stomach modifies the food temperature to 37°C, which is optimal for the pepsin activity. The thermal diffusion of heat into a spherical food particle can be described by the diffusion equation for temperature, T (°C) (Farlow, 1993):

$$\frac{\partial T(r,t)}{\partial t} = \frac{1}{r^2 c \rho} \frac{\partial}{\partial r} \left(k r^2 \frac{\partial T(r,t)}{\partial r} \right)$$

$$\frac{\partial T(r = 0, t)}{\partial r} = 0$$

$$\frac{\partial T(r = R, t)}{\partial r} = -\frac{h}{k}(T(R,t) - 37), \quad \text{or} \quad T(r = R, t) = 37 \tag{9.7}$$

$$T(r, t = 0) = T_0$$

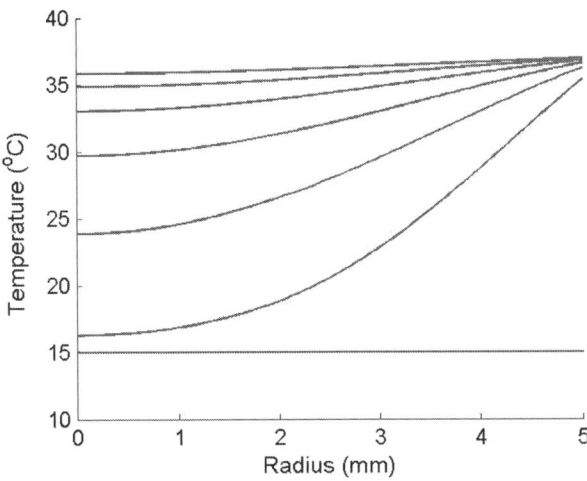

■ **FIGURE 9.9** The temperature distribution in a 5 mm radius spherical pepper particle at 0, 10, 20, 30, 40, 50, and 60 seconds after entering the stomach. The respective curves correspond to the consecutive times (i.e., the initial time corresponds to the lowest curve and 60 seconds corresponds to the uppermost curve). The temperature of the stomach is 37°C and the initial temperature of the pepper particle is 15°C.

where ρ is the density of the food, c is the thermal capacity of the food, h is the heat exchange coefficient, R is the radius of the particle, and k is the thermal conductivity. The boundary condition at the surface of the particle depends on the heat exchange coefficient and the level of mixing in the stomach (if the mixing is high then the second boundary condition is suitable). A model simulation for a 15°C pepper particle ($\rho = 200\ \text{kg m}^{-3}$, $c = 3.984\ \text{kJ kg}^{-1}\,^\circ\text{C}^{-1}$, $k = 0.136\ \text{W m}^{-1}\,^\circ\text{C}^{-1}$, $h = 500\ \text{W m}^{-2}\,^\circ\text{C}^{-1}$, $R = 5$ mm, $T_0 = 15°\text{C}$) is shown in Figure 9.9 for a spherical pepper particle at 0, 10, 20, 30, 40, 50, and 60 seconds after entering the stomach. The 5 mm radius pepper particle requires approximately 1 minute to equilibrate to 37°C, which is shorter than the time for gastric juice to diffuse into the food matrix. This result suggests that the transfer of heat into food matrices is not a limiting factor in the degradation of food particles.

MODELS OF FOOD PARTICLE EROSION

The degradation of food particles in the stomach can be described by the process of surface erosion. This can be described by the Noyes—Whitney equation which states that the rate of dissolution is proportional to the surface area of the solid, the diffusion coefficient, and the difference in concentration of the solid between the diffusion layer surrounding the solid and the bulk medium (the rate of dissolution is also inversely proportional to the diffusion layer thickness). This equation has successfully been used to describe the rate of drug dissolution (Willmann et al., 2010). For spherical food particles the rate of change in particle mass (M) is proportional to the surface of the particle:

$$\frac{dM}{dt} = -k\rho\left(4\pi r^2\right) = -k\left(\rho 4\pi \left(\frac{3M}{4\pi\rho}\right)^{2/3}\right)$$

$$= -k\left(\rho^{1/3} 4^{1/3} \pi^{1/3} 3^{2/3}\right) M^{2/3}, \quad M(0) = M_0 \tag{9.8a}$$

where M_0 is the initial mass of the food particle, ρ is the density of the food particle, r is the particle radius, and k is the rate of food particle erosion. The ability of this model to describe the degradation of barley food particles under static soaking conditions (Bornhorst and Singh, 2013) is shown in Figure 9.10A.

For cube-shaped particles the rate of solid loss also follows a two-thirds law:

$$\frac{dM}{dt} = -6k\rho^{1/3} M^{2/3}, \quad M(0) = M_0 \tag{9.8b}$$

where the rate of solid loss is greater than that for spherical particles due to the larger surface area to volume ratio for cube-shaped particles. For thin wafer particles or bulk erosion of food particles (bulk erosion occurs when gastric fluid quickly enters the food matrix and the particle degrades roughly at the same rate everywhere within the particle) the rate of solid loss is proportional to the amount of remaining solid:

$$\frac{dM}{dt} = -kM, \quad M(0) = M_0 \tag{9.8c}$$

This erosion model can be generalized to describe Eq. 9.8a–c:

$$\frac{dM}{dt} = -kM^\beta, \quad M(0) = M_0 \tag{9.9a}$$

$$M(t) = \left[-kt(1 - \beta) + M_0^{1-\beta}\right]^{(1/(1-\beta))} \tag{9.9b}$$

where $2/3 \leq \beta < 1$ characterizes the surface area as a function of particle mass, which in general will not be constant as the particle degrades. Note that the value of k in Eqs. 9.9a,b includes the constants that arise from converting from radius to mass as well as the particle density. Given that the density of the food particle may change as a result of soluble nutrients leaching from the food particle, the value of k may vary with time. When $\beta = 1$ this model simplifies to the exponential model. The ability of this model to describe the degradation of wheat food particles under static soaking conditions (Bornhorst and Singh, 2013) is shown in Figure 9.10B.

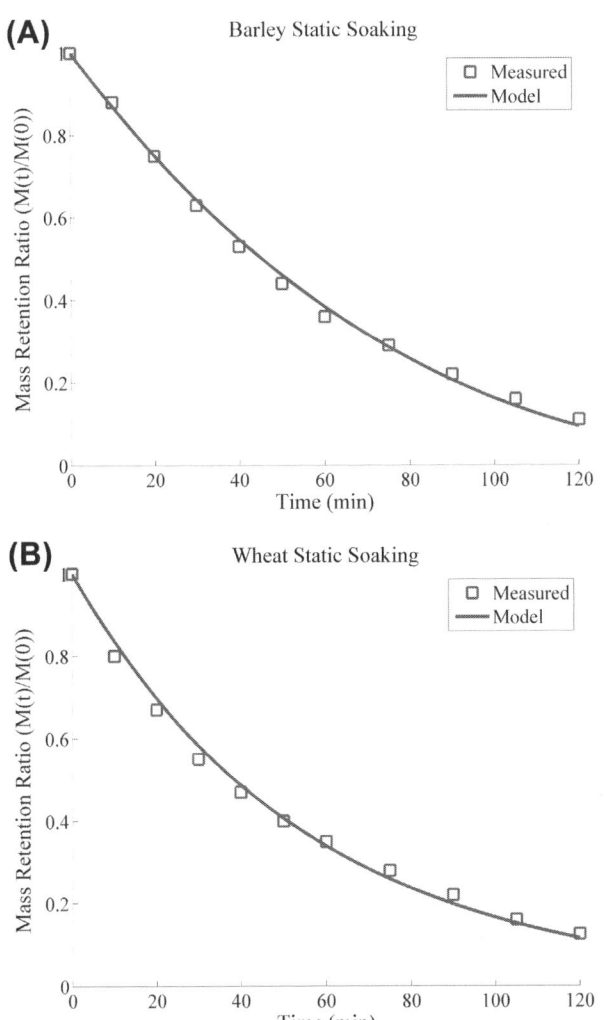

■ **FIGURE 9.10** (A) The degradation of barley particles under static soaking and erosion model fit to data ($\beta = 2/3$). (B) Degradation of wheat particles under static soaking and erosion model fit to data ($\beta = 0.97$).

The corresponding change in the particle radius for a spherical particle is given by:

$$\frac{dr}{dt} = -k, \quad r(0) = r_0 \tag{9.10a}$$

where the particle is completely degraded at time r_0/k. For Eqs. 9.8b, c−9.9a,b, the corresponding equation for the change in particle radius is:

$$\frac{dr}{dt} = -kr^{\alpha}, \quad r(0) = r_0 \tag{9.10b}$$

$$\alpha = 3\beta - 2 \tag{9.10c}$$

$$r(t) = \left[-kt(1 - \alpha) + r_0^{1-\alpha} \right]^{(1/(1-\alpha))} \tag{9.10d}$$

where $0 \leq \alpha \leq 1$, and a similar equation describes the change in particle diameter (with a modified parameter k).

MODELS OF STOCHASTIC ASPECTS OF FOOD PARTICLE EROSION

Food particle degradation is inherently a stochastic process. Identical food particles are not subject to the same physical and chemical forces within the stomach and will therefore differ in their rates of degradation. The probability of erosion of a surface element of a food particle in a given time interval can be described by a Poisson process (Göpferich and Langer, 1993). Surface site erosion is dependent on multiple Poisson processes that occur in parallel. If k occurrences of the Poisson event are required to erode a surface element then the waiting time, t, between k occurrences of the Poisson event are distributed according to an Erlang distribution:

$$p(t, k, \lambda) = \frac{\lambda^k t^{k-1} \exp(-\lambda t)}{(k - 1)!} \tag{9.11a}$$

where λ determines the erosion rate. When $k = 1$, the Erlang distribution simplifies to an exponential distribution:

$$p(t, \lambda) = \lambda \exp(-\lambda t) \tag{9.11b}$$

which has mean λ^{-1} and variance λ^{-2}. A model simulation of this stochastic process for a 3 mm diameter food particle is shown in Figure 9.11A, B. Evident is the irregular shape of the particle boundary, which increases the surface area to volume ratio and the rate of erosion, and the break-off of smaller particle fragments due to the stochastic erosion process.

(A)

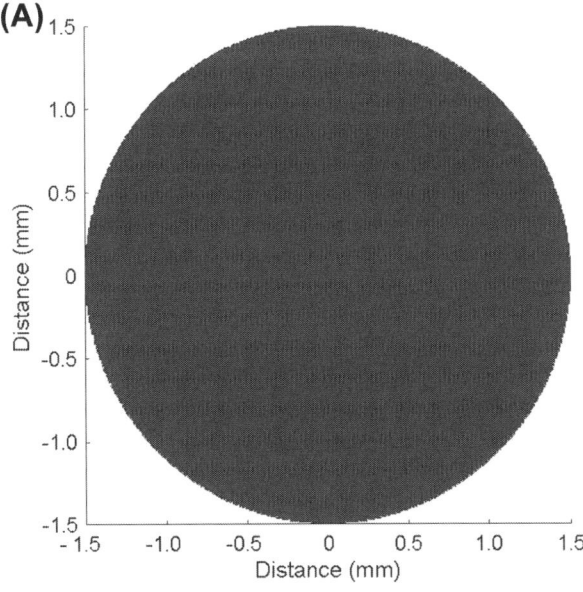

(B)

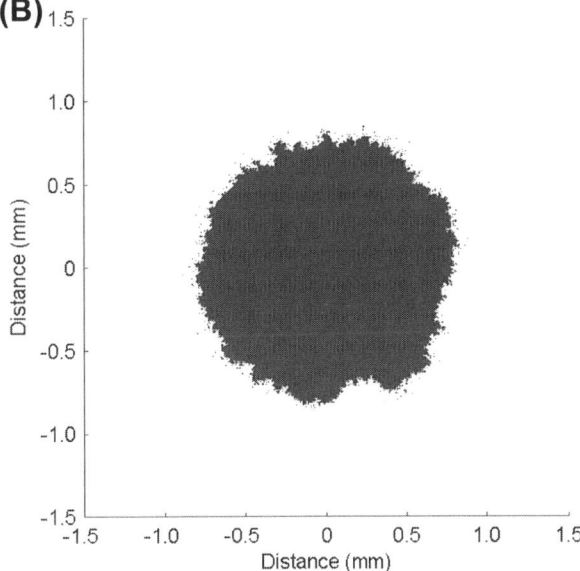

■ **FIGURE 9.11** (A) A 3 mm diameter food particle entering the stomach. (B) The breakdown in the 3 mm diameter in the stomach by stochastic erosion. [A color version of this figure is available online at www.booksite.elsevier.com/9780124046108].

MODELS OF THE ROLE OF FOOD PARTICLE GEOMETRY ON DEGRADATION

Most food particles are inherently non-spherical. The evolution in the surface erosion of a particle with time can be described using the level-set equation (Sethian, 1996):

$$\frac{d\Psi}{dt} + k|\nabla\Psi| = 0,$$

$$\Psi(\mathbf{x}, t) = 0, \ \mathbf{x} \in \Omega'$$

$$\Psi(\mathbf{x}, t) > 0, \ \mathbf{x} \in \Omega \tag{9.12}$$

$$\Omega'_t = \{\mathbf{x} : \Psi(\mathbf{x}, t) = 0\}$$

$$\Omega_t = \{\mathbf{x} : \Psi(\mathbf{x}, t) > 0\}$$

where t is time, $\mathbf{x}$ denotes the Cartesian coordinate, $\Psi(\mathbf{x}, t)$ represents the signed distance to the boundary of the food particle, k is the rate of erosion, Ω' is the initial boundary of the particle, Ω defines the initial particle geometry, Ω'_t is the boundary of the particle at time t, and Ω_t defines the particle geometry at time t. These can be solved numerically using fast-marching methods (Sethian, 1996). The breakdown of a non-spherical particle by erosion is shown in Figure 9.12. The particle eventually fragments into two smaller particles as a result of the non-spherical shape of the food particle. This approach can also be used to describe the erosion of multiple particles.

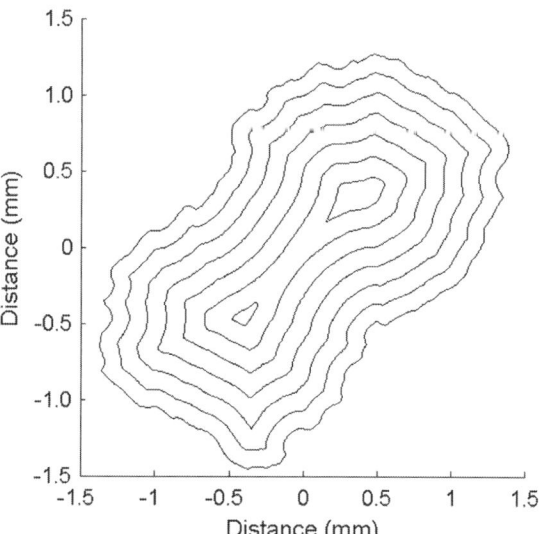

■ **FIGURE 9.12** The breakdown of a non-spherical particle by erosion. The outermost contour represents the initial particle boundary and each concentric contour represents the change at consecutive times. The time interval is food dependent.

MODELS OF FOOD PARTICLE FRAGMENTATION

The probability of a fragmentation event can be described by a Poisson process. The fracture probability is dependent on the particle composition, the level of tenderization, and geometry, as demonstrated in the previous section. Furthermore, fracture events that produce small-sized particles are more probable as they require less energy. Similarly, in general, fracture events that produce three or more particles are less probable than those that produce two particles as they require more energy. The particle size after a fragmentation event can therefore be described by a probability distribution.

The effect of particle fragmentation on the dynamics of the probability distribution for food particle mass, $P(M, t)$, can be described by the partial differential equation:

$$\frac{\partial P(M,t)}{\partial t} = -B(M,t)P(M,t) + \alpha^2 B(\alpha M,t)P(\alpha M,t),$$
$$P(M,t=0) = P_0(M),$$
$$P(0,t) = 0,$$

(9.13a)

where $\partial P/\partial t$ denotes a partial derivative with respect to time, $P_0(M)$ is the distribution in food particle mass entering the stomach, and $B(M, t)$ is the rate at which particles fragment into α equally sized daughter particles (generally $\alpha = 2$). These types of nonlocal equations have been used to describe the distribution in cell size as a result of cell division (Basse *et al.*, 2004). Asymmetric particle fragmentation can be described by the integro-partial differential equation:

$$\frac{\partial P(M,t)}{\partial t} = -B(M,t)P(M,t) + \int_M^\infty B(N,t)P(N,t)\kappa(M,N,t)dN$$
$$P(M,t=0) = P_0(M)$$
$$P(0,t) = 0$$

(9.13b)

where $\kappa(M, N, t)$ determines the likelihood of fragmentation of a particle of mass N into a daughter particle of mass M.

MODELS OF THE CHANGE IN FOOD PARTICLE SIZE DISTRIBUTION WITHIN THE STOMACH

The distribution in food particle size entering the stomach is generally described by lognormal and Weibull probability distributions (Jalabert-Malbos *et al.*, 2007; Flynn, 2012). However, during both mastication and digestion, some foods are known to generate bimodal distributions in particle size, which can be described by mixtures of such probability distributions (Kim *et al.*, 2011; Kong *et al.*, 2011). The initial food particle size influences the rate of particle degradation and is therefore related to the glycemic index of starchy foods (Bornhorst *et al.*, 2011). The change in the particle size distribution over time in the stomach can be described by an ordinary differential equation:

$$\frac{dr}{dt} = -f(r) \quad r(0) = r_0 \tag{9.14a}$$

where $f(r)$ is the rate of breakdown of particles of radius, r, over time, t.

Although most models of the breakdown of food particles are deterministic (i.e., Eq. 9.14a which will always produce the same answer for a given starting point) food breakdown is a stochastic process involving a population of particles. For example, the food particle may fragment into two smaller particles at an unknown time due to the random nature of mixing within the stomach. The food particle is also inhomogeneous and each layer of the particle will erode at different rates. The food particle may also be subject to random fluctuations in environmental conditions, such as pH within the stomach, that occur on a time scale faster than that of the model abstraction or that can be measured. Furthermore, due to imperfect knowledge of the process, the food particle degradation model components may be inaccurate. All of these processes can be interpreted as random perturbations and can be described by stochastic models of food particle breakdown that account for such random perturbations.

We assume that these perturbation processes occur on a time scale much faster than the characteristic time scale for food particle breakdown and that the perturbations are independent. The perturbations can be associated with a particular parameter or component of the differential equations describing food particle breakdown. For example, the effect of food particle inhomogeneity on food particle degradation can be described by:

$$f(r) = f(r) + \sigma f(r)\xi(t) \tag{9.14b}$$

where σ represents the amplitude of the perturbations and $\xi(t)$ denotes normal white noise, which is a Gaussian distributed process with zero mean and independent increments. The change in the particle size distribution over time in the stomach can be described by a stochastic ordinary differential equation:

$$\frac{dr}{dt} = -f(r) + \sigma g(r)dW, \quad r(0) = r_0$$
$$g(r) = f(r) \tag{9.14c}$$

where r is the particle radius, t is time, $f(r)$ is the rate of breakdown of particles of size r, $\sigma g(r)$ represents the amplitude of the perturbations, and dW is a Wiener stochastic process, i.e., a random process with independent normally distributed increments with zero mean and variance proportional to time (Gardiner, 2004). The Wiener process is the most natural continuous-time stochastic process to use without knowledge of the specific nature of the perturbations. This is because, according to the central limit theorem, the sum of a large number of these perturbations will tend to have a normal distribution regardless of the distribution of the individual small-scale perturbations (Arnold and Levfever, 1981). The requirement that $g(r) = f(r)$ holds only when the major stochastic process is the inhomogeneous erosion of each layer of the particle. In general, other stochastic processes will contribute and $g(r) \neq f(r)$.

The Fokker—Planck equation describing the dynamics of the probability distribution for food particle size, $P(r, t)$, for erosion takes the form (Gardiner, 2004):

$$\frac{\partial P(r,t)}{\partial t} = -\frac{\partial}{\partial r}(f(r)P(r,t)) + \frac{1}{2}\frac{\partial^2}{\partial^2 r}\left((\sigma g(r))^2 P(r,t)\right)$$

$$P(r, t = 0) = P_0(r)$$

$$P(0, t) = 0$$

(9.14d)

where $\partial P/\partial t$ denotes a partial derivative with respect to time and $P_0(r)$ is the distribution in food particle size entering the stomach, which is dependent on the mastication process. If the process of fragmentation is also incorporated (via Eq. 9.13a) then the dynamics of the probability distribution for food particle mass, $P(M, t)$, for erosion and fragmentation takes the form:

$$\frac{\partial P(M,t)}{\partial t} = -\frac{\partial}{\partial M}(f(M)P(M,t)) + \frac{1}{2}\frac{\partial^2}{\partial^2 M}\left((\sigma g(M))^2 P(M,t)\right)$$
$$- B(M,t)P(M,t) + \alpha^2 B(\alpha M,t)P(\alpha M,t)$$

$$P(M, t = 0) = P_0(M)$$

$$P(0, t) = 0.$$

(9.14e)

This model can be used to characterize the change in the distribution in rice grain size during 3 hours of simulated digestion (Kong *et al.*, 2011). The effect of digestion on the white rice grain size distribution is shown in Figure 9.13A—C with the model ($\alpha = 2$):

$$\frac{\partial P(M,t)}{\partial t} = -\frac{\partial}{\partial M}(kM^\beta P(M,t)) + \frac{1}{2}\frac{\partial^2}{\partial^2 M}\left((\sigma k M^\beta)^2 P(M,t)\right)$$
$$- BP(M,t) + 4BP(2M,t)$$

$$P(M, t = 0) = P_0(M)$$

$$P(0, t) = 0$$

(9.14f)

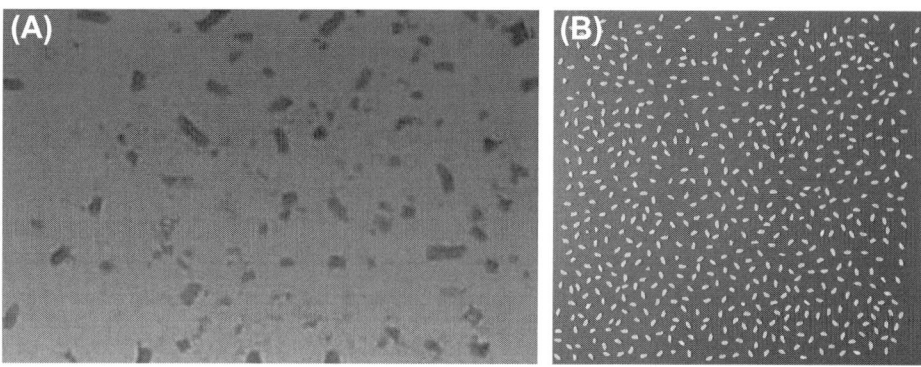

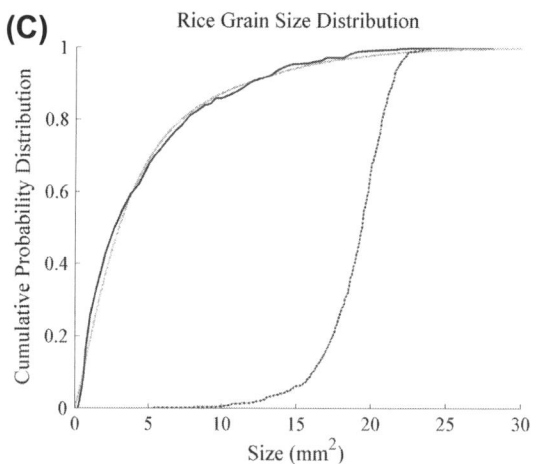

■ **FIGURE 9.13** (A) The effect of 3 hours of simulated digestion on the white rice grain size distribution (Kong *et al.*, 2011). (B) The distribution in rice grain size distribution prior to digestion. (C) The effect of 3 hours of simulated digestion on the white rice grain size distribution. Solid line is measured grain size distribution (Kong *et al.*, 2011), dotted line is model predicted (Eq. 9.14f) grain size distribution and dashed line is initial grain size distribution prior to digestion. *(A) Reprinted with permission from the Journal of Food Science.* [A color version of this figure is available online at www.booksite.elsevier.com/9780124046108].

which has four parameters k, β, σ, B. This demonstrates that the model is able to characterize the change in particle size distribution during digestion ($k = 0.0042$ mm$^{2(1-\beta)}$ min^{-1}, $\beta = 2/3$, $\sigma = 18$, $B = 0.0067$ min^{-1}). Furthermore, the model can be used to estimate the relative effects of erosion and fragmentation on the change in particle size distribution for different food matrices.

MODELS OF GASTRIC EMPTYING

Gastric emptying determines the time that food particles are resident within the stomach and therefore the duration of food particle degradation by chemical and physical processes within the stomach. There are three commonly used models of gastric emptying in the literature; the power exponential, the modified power exponential, and the linear exponential models.

The power exponential equation for gastric volume, $V(t)$, after a meal (Elashoff *et al.*, 1982) is:

$$V(t) = V_0 \exp\left(-\ln(2)(t/t_{half})^\alpha\right) \tag{9.15}$$

where V_0 is the initial postprandial volume, t_{half} is the time at which half the gastric content has emptied (also referred to as half emptying time in the literature), and α determines the duration of the lag phase in gastric emptying. This model was introduced as a substitute for the exponential function which does not characterize the biphasic nature of gastric emptying. The modified power exponential function also describes the biphasic nature of gastric emptying (Siegel *et al.*, 1988):

$$V(t) = V_0\left(1 - (1 - \exp(-kt))^\beta\right) \tag{9.16}$$

where V_0 is the initial postprandial volume, k describes the rate of gastric emptying, and β determines the length of the lag phase in gastric emptying. The linear exponential model (Eq. 9.1) has also been used to describe the biphasic change in gastric volume, V, after a meal (Goetze *et al.*, 2007):

$$V(t) = V_0(1 + kt/t_{empt})\exp\left(-t/t_{empt}\right) \tag{9.17}$$

where V_0 is the initial postprandial volume, k describes the rise in volume from V_0, and t_{empt} is the emptying time constant.

Gastric emptying can also be described using ordinary differential equations (Stubbs, 1977):

$$\frac{dV}{dt} = -kV^\alpha, \quad V(0) = V_0 \tag{9.18}$$

where k describes the rate of gastric emptying, α determines the kinetics of gastric emptying, and the meal is consumed at time zero. This equation yields gastric emptying curves given by Eq. 9.9. This model can be extended to account for the meal volume and the rate of secretion of gastric fluid into the stomach:

$$\frac{dV}{dt} = -kV + A\delta(t) + [at^n \exp(-Kt)H(t) + b] \tag{9.19}$$

where A denotes the meal volume, b is the basal rate of secretion of gastric fluid into the stomach, a is the meal-induced secretion of gastric fluid into the stomach, K determines the rate of decrease in the secretion of gastric fluid into the stomach after the meal, n determines the rate of increase in the secretion of gastric fluid into the stomach after the meal, $\delta(t)$ is the Dirac delta function, and $H(t)$ is the Heaviside switch function (which is zero for $t < 0$ and one for $t \geq 0$). The pre-meal/post-meal gastric volume is $V_0 = b/k$, which accounts for a deficiency in the models described by Eqs. 9.15–9.18 that have zero post-meal gastric volumes. The third and final term in square brackets denotes the rate of secretion of fluid into the stomach. This model also allows the rate of increase in gastric volume to be different from the rate of decrease in gastric volume ($k \neq K$) that is evident for some subjects (Goetze *et al.*, 2007). The fit of this five parameter model (Eq. 9.19, $n = 0$) to the gastric emptying dynamics after a 400 ml meal is shown in Figure 9.14A (Malagelada *et al.*, 1976). The five model parameters are readily identifiable from such experimental data where gastric volume was measured every 10 minutes for 4 hours after the meal. Malagelada *et al.* (1976) also measured the rate of secretion of fluid into the stomach and the fit of the six parameter model (Eq. 9.19, $n \neq 0$) to changes in both the gastric volume and rate of secretion of fluid into the stomach after a meal are shown in Figure 9.14A, B. The model is able to characterize the dynamics of both sets of data.

SUMMARY

There is still much work to be done in this field in order to accurately model food digestion in the stomach. For instance, enzymatic reactions are not accounted for as yet. The aforementioned models need to be extended to incorporate the effects of pepsin and gastric and salivary lipases on food degradation. Previous studies with kiwifruit proteins have shown that the effects of pepsin are also pH dependent (Lucas *et al.*, 2008). Even after an hour of digestion, there was remarkably little breakdown of proteins when the pH was greater than 2.5. This appears to be consistent with the effects of pH on diffusion of acidic water into carrot, in that there was no significant difference in the rate of penetration of acidic water for pH greater than 3.5. Other studies have shown that the digestibility of proteins in various foods (legumes, cereals, milk products, chicken, and kiwifruit) can be dependent upon the protein structure, in that the secondary structures (β-sheets) are resistant to proteolysis by pepsin (Guo *et al.*, 1995; Bublin *et al.*, 2008; Carbonaro *et al.*, 2012). The effect of pepsin will also depend on the amount of protein in the food matrix and the level of pepsin. As seen with studies with carrots (Kong and Singh, 2009b, 2011), the addition of pepsin to the simulated gastric fluid did not have a significant effect on the degradation of carrot due to the low protein content.

The food matrix composition has a marked effect on the degradation of the food particle and may also restrict or enhance enzymatic reactions. While it is expected that the amount of protein in a food matrix would have a significant effect

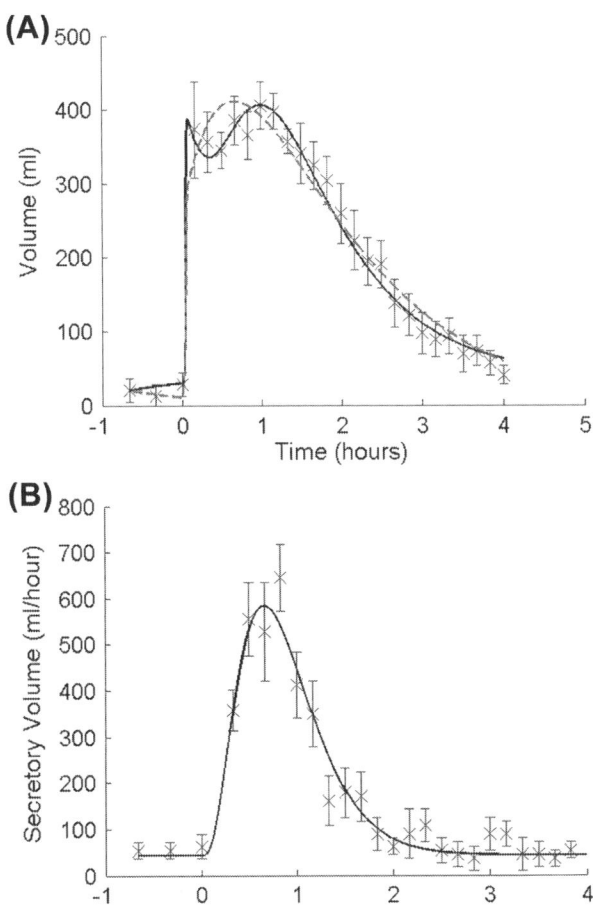

■ **FIGURE 9.14** (A) The gastric emptying dynamics after a 400 ml meal (Malagelada *et al.*, 1976) along with model fits to the data (Eq. 9.19). Error bars denote standard errors. The solid line is the fit of the six parameter model (Eq. 9.19) to changes in the gastric volume after a meal and the dotted line is the fit of the five parameter model (Eq. 9.19, $n = 0$). (B) The rate of secretion of fluid into the stomach after a 400 ml meal (Malagelada *et al.*, 1976) along with model fit to the data (Eq. 9.19, solid line). Error bars denote standard errors. [A color version of this figure is available online at www.booksite.elsevier.com/9780124046108].

on the ability of pepsin to hydrolyze peptides, the amount of fat in the food matrix can also have a marked effect on protein digestibility. In a double-blind, placebo-controlled, food challenge that tested for the allergenicity of peanuts in a recipe with low and high fat content, it was found that the higher fat content resulted in a lower oral allergenic effect (i.e., an itchy mouth) (Grimshaw *et al.*, 2003). It was surmised that the lack of early oral warning signs was due to a concealing effect of the fat in the high-fat food matrix, which reduced the bio-accessibility of the proteins. Thus, only after the fat was digested were the proteins available to interact with the epithelial cells of the stomach and small intestine; however, it is doubtful that there was much digestibility of the lipids in the stomach as only 10−30% of triglycerides have been found to be hydrolyzed in the stomach. The ability of fat to restrict the bioaccessibility of foods during gastric digestion has also been shown in studies with vitamin B_{12} encapsulated in a water-in-oil-in-water double emulsion (Giroux et al., 2013). In the study only 4.4% of vitamin B_{12} was released after 120 minutes of gastric digestion. Thus, the fat content of foods will have a significant effect on the digestibility of the food matrix. These findings may have an impact on the allergenicity of pre-packaged foods. Even though the allergenicity of foods are poorly understood, most allergens are proteins which have a molecular weight between 10 and 70 kDa that typically resist digestion (resistant to both pH effects and pepsin) and are often stable regardless of changes in temperature (Bublin et al., 2008); thus, foods high in fat may increase the risk of allergenicity as the fat may restrict breakdown in the stomach of protein that would otherwise be hydrolyzed (Moreno et al., 2005).

The human stomach is a very dynamic and complex system. Here we have outlined the few models in the literature that pertain to gastric digestion, as well as some new approaches to model erosion, fragmentation, changes in food particle distribution, and the effects of temperature on degradation. Only a few mathematical models have been developed to date; however, the building blocks to fully understand how the stomach works to degrade food particles are slowly being put together. Future models are likely to be very helpful in improving our understanding of these processes. Of course, future food degradation experiments are also critical in the advancement of mathematical models.

REFERENCES

Arnold, L., Levfever, R., 1981. Stochastic Nonlinear Systems in Physics, Chemistry and Biology. Springer, Berlin.

Basse, B., Baguley, B.C., Marshall, E.S., Wake, G.C., Wall, D.J.N., 2004. Modelling cell population growth with applications to cancer therapy in human tumour cell lines. Prog. Biophys. Mol. Biol. 85, 353−368.

Bornhorst, G.M., Singh, R.P., 2013. Kinetics of in vitro bread bolus digestion with varying oral and gastric digestion parameters. Food Biophys. 8, 50−59.

Bornhorst, G.M., Singh, R.P., Heldman, D.R., 2011. Rate kinetics of bread bolus disintegration during in vitro digestion. In: Taoukis, P.S., Stoforos, N.G., Karathanos, V.T., Saravacos, G.D. (Eds.), 11th International Congress on Engineering and Food, vol. II. Cosmosware, Athens, Greece.

Bublin, M., Radauer, C., Knulst, A., Wagner, S., Scheiner, O., Mackie, A.R., et al., 2008. Effects of gastrointestinal digestion and heating on the allergenicity of the kiwi allergens Act d 1, actinidin, and Act d 2, a thaumatin-like protein. Mol. Nutr. Food Res. 52, 1130−1139.

Carbonaro, M., Maselli, P., Nucara, A., 2012. Relationship between digestibility and secondary structure of raw and thermally treated legume proteins: a Fourier transform infrared (FT-IR) spectroscopic study. Amino Acids 43, 911−921.

Crank, J., 1975. The Mathematics of Diffusion, second ed. Oxford University Press, New York.

Davey, M.J., Landman, K.A., McGuinness, M.J., Jin, H.N., 2002. Mathematical modeling of rice cooking and dissolution in beer production. AIChE J. 48, 1811−1826.

Diaz, G., Wolf, W., Kostaropoulos, A.E., Spiess, W.E.L., 1993. Diffuion of low-molecular compounds in food model system. J. Food Process. Preserv. 17, 437−454.

Elashoff, J.D., Reedy, T.J., Meyer, J.H., 1982. Analysis of gastric emptying data. Gastroenterology 83, 1306−1312.

Farlow, S.J., 1993. Partial Differential Equations for Scientists and Engineers. Wiley, New York.

Ferrua, M.J., Kong, F., Singh, R.P., 2011. Computational modeling of gastric digestion and the role of food material properties. Trends Food Sci. Technol. 22, 480−491.

Ferrua, M.J., Singh, R.P., 2010. Modeling the fluid dynamics in a human stomach to gain insight of food digestion. J. Food Sci. 75, R151−R162.

Flynn, D.S., 2012. The particle size distribution of solid food after human mastication. Massey University, Palmerston North.

Gardiner, C.W., 2004. Handbook of Stochastic Methods for Physics, Chemistry and the Natural Sciences, third ed. Springer-Verlag, Berlin.

Giroux, H.J., Constantineau, S., Fustier, P., Champagne, C.P., St-Gelais, D., Lacroix, M., Britten, M., 2013. Cheese fortification using water-in-oil-in-water double emulsions as carrier for water soluble nutrients. Int. Dairy J. 29, 107−114.

Goetze, O., Steingoetter, A., Menne, D., Van Der Voort, I.R., Kwiatek, M.A., Boesiger, P., et al., 2007. The effect of macronutrients on gastric volume responses and gastric emptying in humans: A magnetic resonance imaging study. Am. J. Physiol.−Gastrointest. Liver Physiol. 292, G11−G17.

Gomi, Y.-I., Fukuoka, M., Takeuchi, S., Mihori, T., Watanabe, H., 1996. Effect of temperature and moisture content on water diffusion coefficients in rice starch/water mixtures. Food Sci. Technol. Int. 2, 171−173.

Göpferich, A., Langer, R., 1993. Modeling of polymer erosion. Macromolecules 26, 4105−4112.

Grimshaw, K.E.C., King, R.M., Nordlee, J.A., Hefle, S.L., Warner, J.O., Hourihane, J.O.B., 2003. Presentation of allergen in different food preparations affects the nature of the allergic reaction—a case series. Clin. Exp. Allergy 33, 1581−1585.

Guo, M.R., Fox, P.F., Flynn, A., Kindstedt, P.S., 1995. Susceptibility of β-lactoglobulin and sodium caseinate to proteolysis by pepsin and trypsin. J. Dairy Sci. 78, 2336−2344.

Jalabert-Malbos, M.L., Mishellany-Dutour, A., Woda, A., Peyron, M.A., 2007. Particle size distribution in the food bolus after mastication of natural foods. Food Qual. Pref. 18, 803−812.

Kim, E.H.J., Morgenstern, M.P., Bronlund, J.E., Fosterd, K.D., Le Gote, A., 2011. Food breakdown during human mastication—quantitative characterization. In: Taoukis, P.S.,

Stoforos, N.G., Karathanos, V.T., Saravacos, G.D. (Eds.), 11th International Congress on Engineering and Food, vol. II. Cosmosware, Athens, Greece.

Kong, F., Oztop, M.H., Singh, R.P., McCarthy, M.J., 2011. Physical changes in white and brown rice during simulated gastric digestion. J. Food Sci. 76, E450–E457.

Kong, F., Singh, R.P., 2008. A model stomach system to investigate disintegration kinetics of solid foods during gastric digestion. J. Food Sci. 73, E202–E210.

Kong, F., Singh, R.P., 2009a. Digestion of raw and roasted almonds in simulated gastric environment. Food Biophys. 4, 365–377.

Kong, F., Singh, R.P., 2009b. Modes of disintegration of solid foods in simulated gastric environment. Food Biophys. 4, 180–190.

Kong, F., Singh, R.P., 2011. Solid loss of carrots during simulated gastric digestion. Food Biophys. 6, 84–93.

Li, G., Wang, Y., He, R., Cao, X., Lin, C., Meng, T., 2009. Numerical simulation of predicting and reducing solid particle erosion of solid-liquid two-phase flow in a choke. Petroleum Sci. 6, 91–97.

Lucas, J.S.A., Cochrane, S.A., Warner, J.O., Hourihane, J.O.B., 2008. The effect of digestion and pH on the allergenicity of kiwifruit proteins. Pediatr. Allergy Immunol. 19, 392–398.

Malagelada, J.R., Longstreth, G.F., Summerskill, W.H.J., Go, V.L.W., 1976. Measurement of gastric functions during digestion of ordinary solid meals in man. Gastroenterology 70, 203–210.

Moreno, F.J., Mackie, A.R., Mills, E.N.C., 2005. Phospholipid interactions protect the milk allergen α-lactalbumin from proteolysis during in vitro digestion. J. Agric. Food Chem. 53, 9810–9816.

Saguy, I.S., Marabi, A., Wallach, R., 2005. New approach to model rehydration of dry food particulates utilizing principles of liquid transport in porous media. Trends Food Sci. Technol. 16, 495–506.

Schwartzberg, H.G., Chao, R.Y., 1982. Solute diffusivities in leaching processes. Food Technol. 36, 73–86.

Sethian, J.A., 1996. A fast marching level set method for monotonically advancing fronts. Proc. Natl. Acad. Sci. USA 93, 1591–1595.

Siegel, J.A., Urbain, J.L., Adler, L.P., Charkes, N.D., Maurer, A.H., Krevsky, B., et al., 1988. Biphasic nature of gastric emptying. Gut 29, 85–89.

Stubbs, D.F., 1977. Models of gastric emptying. Gut 18, 202–207.

Syarief, A.M., Gustafson, R.J., Morey, R.V., 1987. Moisture diffusion coefficients for yellow-dent corn components. Trans. Am. Soc. Agric. Eng. 30, 522–528.

Turhan, M., Kaletunç, G., 1992. Modeling of salt diffusion in white cheese during long-term brining. J. Food Sci. 57, 1082–1085.

Van Wey, A.S., Cookson, A.L., Roy, N.C., McNabb, W.C., Soboleva, T.K., Wieliczko, R.J., Shorten, P.R., 2014. A mathematical model of the effect of pH and food matrix composition on fluid transport into foods: an application in gastric digestion and cheese brining. Food Res. Int. http://dx.doi.org/10.1016/j.foodres.2014.01.002.

Willmann, S., Thelen, K., Becker, C., Dressman, J.B., Lippert, J., 2010. Mechanism-based prediction of particle size-dependent dissolution and absorption: Cilostazol pharmacokinetics in dogs. Eur. J. Pharm. Biopharm. 76, 83–94.

Yamamoto, S., 1999. Effects of glycerol on the drying of gelatin and sugar solutions. Drying Technol. 17, 1681–1695.

An Improved Understanding of Gut Function through High-Resolution Mapping and Multiscale Computational Modeling of the Gastrointestinal Tract

Timothy R. Angeli, Niranchan Paskaranandavadivel, Leo K. Cheng and Peng Du

Auckland Bioengineering Institute, The University of Auckland, Auckland, New Zealand

CONTENTS

Food Structures, Digestion and Health. http://dx.doi.org/10.1016/B978-0-12-404610-8.00010-4

INTRODUCTION

A vital component of the underlying control of gastrointestinal (GI) motility is the rhythmic bioelectrical "slow wave" activity, which is generated and propagated by networks of interstitial cells of Cajal (ICC), located within and between the smooth muscle layers of the GI wall (Lees-Green *et al.*, 2011). When combined with other co-regulating factors like neural and hormonal input, these slow waves serve to regulate GI motility (Huizinga and Lammers, 2009).

Slow waves were first discovered by Walter C. Alvarez in the early 1900s, when he attached silver-plated steel wires to the serosa of GI tissues, connected to a galvanometer, and captured pointer movements by the shadow cast onto moving bromide paper (Alvarez, 1922). Alvarez noted that these initial slow wave recordings "show[ed] that the electrograms obtained…from various parts of the stomach and bowel…resemble closely the kymographic records obtained by attaching light levers to those parts," thereby demonstrating a relationship between the electrical slow wave activity and mechanical contractile activity (Alvarez, 1922). These early studies by Alvarez effectively pioneered the foundation of organ-level GI electrophysiology, with Alvarez himself proclaiming, "we have, then, at our disposal a new method with which to study the activities of the digestive tract" (Alvarez, 1922).

Although the biological and electrophysiological characteristics of slow waves have been studied for nearly a century, many questions remain, and these topics persist as a focus of vigorous current research interest. Recent advances in experimental techniques, particularly the development of high-resolution GI electrical mapping and mathematical modeling, have provided a renewed understanding of slow wave activity. This chapter reviews the latest advancement in these techniques as well as the resultant current state of knowledge of GI electrical activity.

THE CELLULAR AND BIOPHYSICAL BASIS OF GASTROINTESTINAL ELECTRICAL ACTIVITY

Slow waves are a cyclical depolarization of the GI wall that is generated and propagated by ICC, typically characterized by an increase in the resting membrane potential of the smooth muscle cell (SMC) by about 30–35 mV, and primarily resulting from calcium influx. Slow waves serve to modulate the propensity of the GI tissue for contraction, and when combined with other co-regulatory inputs like nervous and hormonal signals, help regulate and coordinate the resultant contractile response (Huizinga and Lammers, 2009). Regenerative action potentials involving brief bursts of calcium influx also occur in some SMC of the GI tract, and are termed "spikes"

(Lammers and Slack, 2001). These spikes are typically superimposed on the plateau of the slow wave and lead to stronger contractile response.

The pacesetting system of the GI tract does not have specific point sources, as the heart does, but rather consists of a continuous network of ICC with inherent ability for pacemaking. Several specific types of ICC have been observed, classified on their location within the GI wall. For example, the myenteric plexus ICC are located in a layer between the smooth muscle layers, while intramuscular ICC are located within the layers of smooth muscle, effectively forming a three-dimensional network of ICC within the GI wall (Lees-Green *et al.*, 2011). In isolation, each ICC is capable of generating slow waves at an intrinsic frequency, but *in vivo*, ICC are connected via a complex network of gap junctions and synchronize to the single greatest frequency. This synchronization of ICC, called entrainment, allows ICC to generate and propagate organized slow wave patterns through the GI organs. The specific mechanisms responsible for ICC entrainment are unclear and under current investigation, but it is thought that entrainment results through voltage-dependent IP_3-mediated calcium release (van Helden *et al.*, 2010).

Although SMC are unable to actively generate slow waves, they have an underlying ability to regenerate slow waves on depolarization and are coupled to ICC through gap junctions (Huizinga and Lammers, 2009). Thereby, when initiated by the depolarization of ICC, SMC support the propagation of slow waves into the muscle layers of the GI wall. Intracellular recordings have verified that slow waves originate in ICC and conduct to SMC, effectively forming a common electrical syncytium (Cousins *et al.*, 2003).

MOTIVATION FOR HIGH-RESOLUTION MAPPING

Following on from the work of Alvarez in the early 1900s, experimental and methodological advances were made on slow wave recording techniques throughout the twentieth century, with improvements in electrode design, digitization of signals, amplification, and filtering. However, through the greater part of that century, studies were still limited to relatively few electrodes sparsely placed over the length of the stomach or intestine. Decades of these low-resolution gastric and intestinal studies progressively improved our understanding of slow wave characteristics, such as the inherent frequency gradient of the stomach (Kelly and Code, 1971; Hinder and Kelly, 1977) and small intestine (Diamant and Bortoff, 1969a; Code and Szurszewski, 1970; Szurszewski *et al.*, 1970), the location of the gastric pacemaker at a site along the upper greater curvature of the corpus (Kelly and Code, 1971; Hinder and Kelly, 1977), and the location of the initial intestinal pacemaker in the proximal duodenum (Hermon-Taylor and Code, 1971). However, the lack of spatial resolution in these studies forced assumptions to be made about the activity occurring between electrodes (Figure 10.1A) and precluded analysis of spatial slow wave characteristics like velocity, pacemaker profile, wavefront organization, propagation profiles, and pacemaker interaction. This desire and necessity for spatial resolution of slow wave propagation provided the motivation for the development of high-resolution GI mapping technology and subsequent high-resolution mapping studies, whereby spatially dense arrays of electrodes were placed on the serosa of the stomach or small intestine to simultaneously record the electrical activity and map its propagation in spatiotemporal detail (Lammers *et al.*, 1993; Du *et al.*, 2009).

The spatiotemporal understanding obtained through high-resolution electrical mapping has greatly increased our overall knowledge of GI electrical activity; this chapter thereby presents the techniques of high-resolution GI mapping, the subsequent renewed understanding of GI electrical activity, and the complementary computational modeling of this activity.

SCOPE

The scope of this chapter is limited to the electrical activity of the stomach and small intestine. Other organs, particularly the esophagus and colon, play imperative roles in digestion and are also influenced by underlying bioelectrical activity. However, readers are directed elsewhere for information on such organs (e.g., Furness, 2006). Although neural, hormonal, and mechanical pathways complement the bioelectrical regulation of motility (Huizinga and Lammers, 2009; Grundy and Brookes, 2012), it is the electrical component of motility that is the focus of this chapter, and particularly the organ-level electrical activity. Extracellular recordings are the preeminent technique for organ-level GI recordings (Angeli *et al.*, 2013a); this chapter is thereby focused on experimental data recorded using extracellular techniques, with preference on high-resolution recordings. Furthermore, this chapter is focused closely on human relevance, thus giving preference to *in vivo* studies in large animals, and particularly humans, when possible.

METHODS AND TECHNIQUES OF HIGH-RESOLUTION ELECTRICAL MAPPING

As introduced previously, recent high-resolution electrical mapping of the GI tract has begun to offer a comprehensive understanding of the underlying electrical activity associated with coordinating motility (Huizinga and Lammers, 2009). The major advantage of high-resolution mapping was

(A) Low-Resolution Recordings

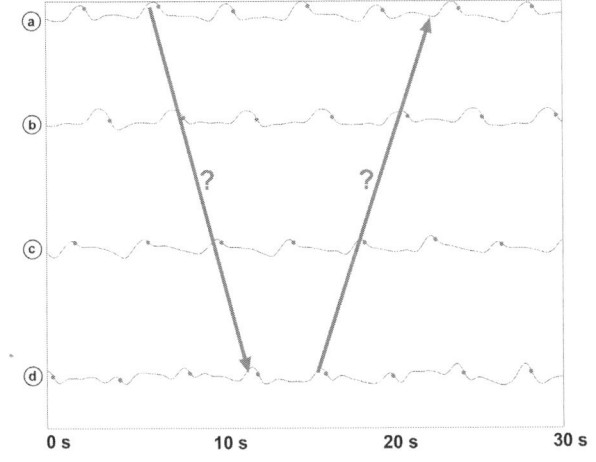
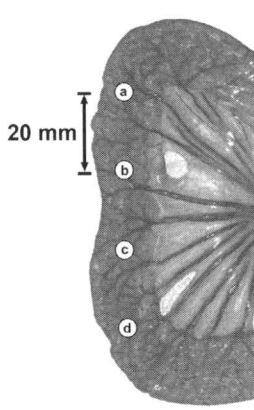

(B) High-Resolution Mapping

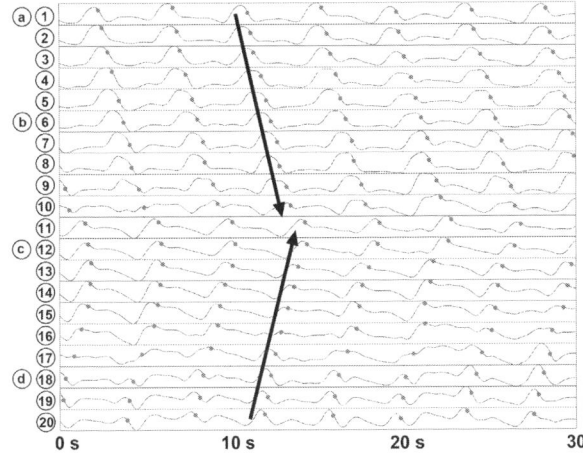
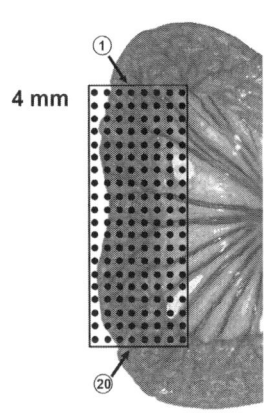

■ **FIGURE 10.1** A comparison of high-resolution mapping versus low-resolution electrical recordings, for the same segment of experimental data. (A) Low-resolution recordings prevent spatial analysis of slow wave propagation dynamics. (B) High-resolution recordings provide spatiotemporal definition of slow wave propagation. Both A and B show the same 30 s of experimental slow wave recordings, with electrograms corresponding to the labeled electrodes on the right. In this example, the propagation pattern is indeterminate from the low-resolution recordings (A), but is determined by high-resolution mapping to be colliding wavefronts (B). [A color version of this figure is available online at www.booksite.elsevier.com/9780124046108].

that it allowed previously assumed or estimated spatial slow wave characteristics to be definitively mapped and calculated. Figure 10.1 illustrates the spatial ambiguity resulting from sparse electrode techniques, whereas high-resolution mapping can accurately define the spatiotemporal propagation of slow wave fronts (Figure 10.1B). The development of high-resolution mapping allowed spatial patterns of slow wave propagation to be investigated in an experimental setting, in both normal and abnormal cases. However, high-resolution mapping presented a number of technological and experimental challenges, particularly relating to the vast amount of data generated. Therefore, robust methods for data collection, signal processing, and data analysis have been the focus of vigorous recent development. This section presents the current state of GI high-resolution mapping technology, equipment, acquisition, signal processing, and analysis.

Overview of High-Resolution Mapping

High-resolution electrical mapping typically employs more than 100 electrodes for the recording of extracellular slow wave activity from the GI serosa, as opposed to sparse electrode mapping where fewer than eight electrodes are typically utilized (Kelly *et al.*, 1969; Lammers *et al.*, 1996; Du *et al.*, 2009). For *in vivo* studies, a laparotomy is performed on the subject to give access to the stomach and/or small intestine (Du *et al.*, 2009; Egbuji *et al.*, 2010; Angeli *et al.*, 2013c). Slow wave activity is known to be influenced by temperature change (Ohba *et al.*, 1975; Gizzi *et al.*, 2010), so monitoring of vital signs and maintenance within the physiological ranges is essential. Temperature control can be accomplished by application of a heating pad, heat lamp, and/or thermal blanket in animal studies (Angeli *et al.*, 2013c), and is usually achieved by a forced-air warming blanket in humans. Anesthesia has been shown to affect slow wave activity, and opiates in particular have been associated with slow wave dysrhythmias, so should be avoided (Lammers *et al.*, 2008). However, many successful studies of slow wave activity have been performed under anesthetized conditions (e.g., Lammers *et al.*, 2009; Egbuji *et al.*, 2010; Angeli *et al.*, 2013c), and results closely match motility patterns observed by magnetic resonance imaging (MRI) in the awake fed state (O'Grady *et al.*, 2010).

High-resolution mapping involves an integrated series of recording and analysis procedures, as detailed in Figure 10.2, and further explained in the following sections.

High-Resolution Mapping Hardware

Lammers *et al.* were the first to perform high-resolution mapping in the GI tract, which they accomplished in 1993 in the rabbit duodenum

(A) High-resolution electrodes placed on the stomach

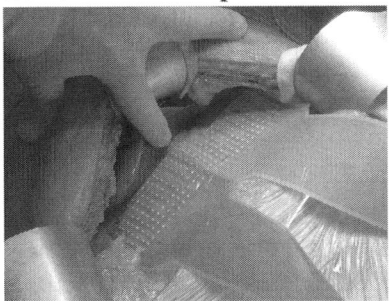

(B) Slow wave signals from electrodes in (a)

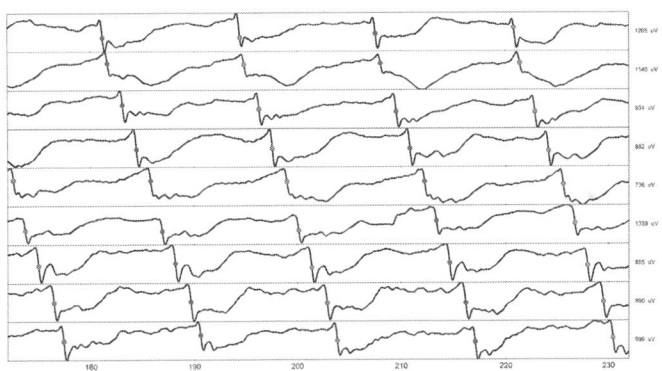

(C) Visualisation of slow wave activity

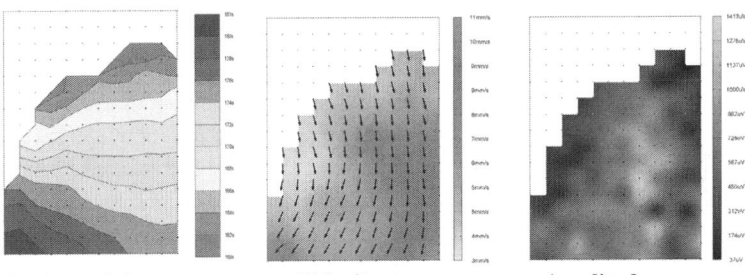

Isochronal time map   Velocity map   Amplitude map

■ **FIGURE 10.2** High-resolution mapping recording and analysis. (A) High-resolution PCB electrodes applied to the serosal surface of the stomach after a laparotomy to record gastric slow wave activity. (B) Slow wave signals after filtering of the raw data, with activation times marked with dots. (C) Visualization of slow wave activity. An isochronal time map shows the propagation of a single wavefront through time (red = early; blue = late), with each color band ("isochrones") representing the area of propagation per 2 s; a velocity map shows the direction and speed of the wavefront at each electrode; an amplitude map displays the slow wave amplitude at each electrode. [A color version of this figure is available online at www.booksite.elsevier.com/9780124046108].

(Lammers *et al.*, 1993). He and his colleagues then went on to apply high-resolution GI mapping to the *in vivo* canine stomach, outlining the gastric slow wave conduction system in spatiotemporal detail (Lammers *et al.*, 2009). Lammers *et al.* manufactured custom electrode arrays encompassing 240 electrodes housed in a rigid epoxy resin (Lammers *et al.*, 2009). More recently, Du *et al.* developed a different design of high-resolution mapping electrodes created with flexible printed circuit boards (PCBs) (Du *et al.*, 2009). With this PCB setup, base arrays of 32 electrodes could be tessellated in a variety of configurations based on the experimental objectives. These flexible PCB electrodes were validated in experimental porcine studies, but more importantly they were sterilizable, making them applicable for use in humans (Du *et al.*, 2009). PCB electrode arrays allowed for human translation of high-resolution GI mapping, a critical step forward toward improving our understanding of motility control through investigation of human slow wave conduction, in both the normal and diseased states (O'Grady *et al.*, 2010, 2012a).

A data acquisition system capable of simultaneously recording multiple channels was additionally required for high-resolution mapping. Lammers *et al.* used a custom mapping system adapted from the cardiac field, while GI high-resolution electrical mapping studies performed by the group in Auckland have utilized a modified ActiveTwo Biosemi system (Amsterdam, the Netherlands; http://www.biosemi.com/). The crucial technical aspects of any GI high-resolution recording system are: (1) ability to record a large number of channels simultaneous with a high bandwidth (e.g., 400 Hz); and (2) ability to record at an adequate sampling frequency (e.g., 30 Hz or greater).

Data Analysis Algorithms and Software Implementation

The large number of signals acquired through high-resolution mapping created a substantial data throughput problem, necessitating the development and implementation of computational algorithms for signal processing, data analysis, and visualization. Software packages have now been developed that offer a semi-automated pipeline for the analysis of high-resolution GI mapping data (Yassi *et al.*, 2012). At present, slow wave analysis algorithms are implemented as a post-processing step; however, recent advances have been made to adapt these methods to online analysis (Bull *et al.*, 2011).

Typically, the raw data are first filtered, whereby a combination of a moving median filter for baseline removal and low-pass filter for high-frequency noise removal has been found to be optimal (Paskaranandavadivel *et al.*, 2013). Slow wave events can then be detected and grouped into propagating wavefronts via automated algorithms (Erickson *et al.*, 2010, 2011), and activation maps can be created to visualize slow wave propagation (Erickson *et al.*, 2011). Quantitative slow wave metrics, such as velocity, amplitude, and frequency, can then be calculated and visualized (Paskaranandavadivel *et al.*, 2011, 2012; Yassi *et al.*, 2012).

A RENEWED UNDERSTANDING OF GASTROINTESTINAL ACTIVITY THROUGH HIGH-RESOLUTION MAPPING

Sparse electrode recordings have provided the historical foundation of our knowledge of GI slow wave initiation and conduction (e.g., Szurszewski *et al.*, 1970; Kelly and Code, 1971; Hinder and Kelly, 1977). However, it was not until the advent of high-resolution GI electrical mapping that the spatial dynamics of GI electrical activity could be comprehensively detailed (Lammers *et al.*, 1993; Du *et al.*, 2009; Angeli *et al.*, 2013c). The experimental application of these high-resolution techniques now spans two decades, over which time our understanding of GI electrical activity has vastly improved. This section presents and discusses the current knowledge of slow wave activity, focusing on the spatial dynamics elucidated by high-resolution mapping.

Gastric Slow Wave Activity

Low-resolution electrical recordings established the foundation of the gastric slow wave conduction system, including the fundamental understanding that the entire stomach normally has a single dominant pacemaker, with all tissue being entrained to that single point of slow wave origin. This single-pacemaker system was originally determined by arranging four to eight electrodes in lines along the stomach, in canines and humans (Kelly and Code, 1971; Hinder and Kelly, 1977), and has subsequently been verified and expanded on by high-resolution studies, as further explained below (Lammers *et al.*, 2009; O'Grady *et al.*, 2010). The dominant human pacemaker was found to be located on the upper greater curvature, 5–7 cm aboral to the cardia, with a lack of slow wave propagation proximally into the fundus (Hinder and Kelly, 1977).

Human gastric slow waves have been widely shown to occur at a frequency of around 3 cpm (cycles per minute), with a normal range often defined as 2–4 cpm (Parkman *et al.*, 2003). Gastric slow wave frequency reduces in the fed state, with the reduction of frequency partially dependent on the temperature of the ingested meal (Verhagen *et al.*, 1998). *In vivo* gastric slow wave entrainment is underpinned by a frequency gradient, intrinsic to the

ICC. The frequency gradient has been demonstrated through gastric transection studies, where a decreasing gradient from 3.0 to 0.8 cpm was observed when transections were progressively performed distally through the corpus and antrum (Kelly and Code, 1971; Hinder and Kelly, 1977). Gastric bisection down the midline of the organ also demonstrated a reduced frequency near the lesser curvature compared to that of the greater curvature. However, transections across the fundus were found to have no effect on the distal slow wave frequency.

High-resolution mapping studies have recently added substantial further detail to our knowledge of the gastric slow wave conduction system (Lammers et al., 2009; Egbuji et al., 2010; O'Grady et al., 2010). Canine, porcine, and human high-resolution studies have confirmed that the region of normal gastric pacemaking is located high on the greater curvature, and is characterized by high-amplitude and high-velocity activity (Lammers et al., 2009; Egbuji et al., 2010; O'Grady et al., 2010). This increased amplitude and velocity profile has been shown to be a result of the circumferential conduction that occurs at, and is isolated to, the pacemaker region, with the amplitude and velocity in this region returning to the normal lower corpus range when that tissue is entrained by retrograde-propagating wavefronts from a paced distal source (O'Grady et al., 2012b). The rapid circumferential conduction may likely be a result of different conduction rates through ICC layers, while the high amplitude is likely a consequence of the relationship between slow wave conduction velocity and the total transmembrane current entering the extracellular space (O'Grady et al., 2012b).

The rapid circumferential conductance at the pacemaker region quickly excites the circumferential tissue and establishes the slow wave as a ring wavefront, which travels organo-axially down the stomach with an amplitude and velocity that is approximately 66% lower than that at the pacemaker site. Slow wave propagation has been shown to be largely symmetrical over the anterior and posterior serosal surfaces (Egbuji et al., 2010).

Gastric slow waves propagate with a consistent velocity throughout the corpus (3 mm s^{-1} in humans), with a relatively distinct transition of $>2\times$ increase in amplitude and velocity at a region within the antrum (Lammers et al., 2009; O'Grady et al., 2010). The slow corpus conductance leads to the accrual of several wavefronts present in the corpus at one time (three to four in humans, at a spacing of 60 mm), which then spread out on entry into the distal rapid-conductance zone (Lammers et al., 2009; O'Grady et al., 2010; Du et al., 2010a). Gastric slow wave conductance terminates at the electrically quiescent pylorus (Wang et al., 2005; Lammers et al.,

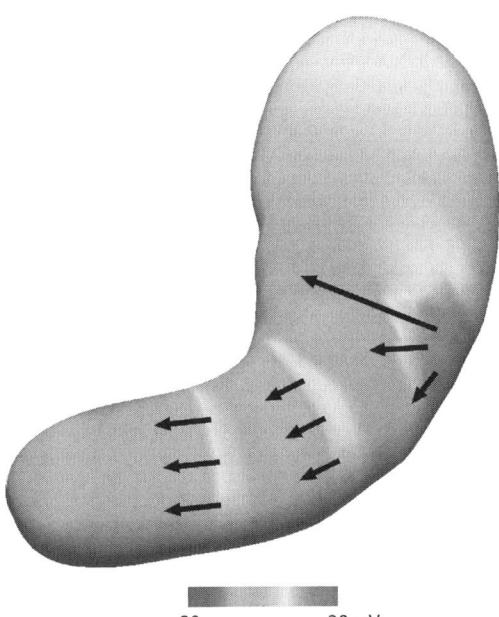

-80   -30 mV

■ **FIGURE 10.3** Summary of gastric slow wave propagation represented as a computational simulation informed through results of high-resolution gastric mapping. Gastric slow waves originate at a pace-maker site located on the upper greater curvature, rapidly propagating circumferentially across the stomach with high amplitude, forming a ring of activation that then propagates organo-axially down the stomach. Slow wave velocity accelerates at a location somewhere in the antrum of the stomach, as noted by a widening of the wavefront spacing in that region of this simulation, before terminating at the pylorus. Three to four wavefronts are simultaneously present on the stomach, and the fundus is typically in a permanently quiescent depolarized state. *Adapted from Du* et al. *(2010a).* [A color version of this figure is available online at www.booksite.elsevier.com/9780124046108].

2009); the pylorus thereby serves as an electrical barrier in addition to its function as a mechanical barrier during digestion, and gastric slow waves are effectively isolated from intestinal slow waves, allowing differentiated motility patterns and functions in those organs.

The overall pattern of gastric slow wave conductance is summarized in Figure 10.3, which displays an anatomically correct simulation of human gastric slow wave excitation and conductance based on high-resolution mapping data integrative through a computational modeling approach (O'Grady *et al.*, 2010; Du *et al.*, 2010a).

Small Intestine Slow Wave Activity

As in the stomach, historical sparse electrode recordings formed the foundation of our knowledge of small intestine slow waves (e.g., Alvarez,

1922; Diamant and Bortoff, 1969a; Szurszewski *et al.*, 1970), with high-resolution studies more recently adding spatial knowledge of pacemaking mechanisms, wave interactions, and propagation dynamics (e.g., Lammers *et al.*, 2005; Lammers and Stephen, 2008; Angeli *et al.*, 2013b). A fundamental declining slow wave frequency gradient has been demonstrated in multiple species using both low- and high-resolution techniques, with frequency decreasing from proximal duodenum to terminal ileum (Szurszewski *et al.*, 1970; Lammers *et al.*, 2005; Angeli *et al.*, 2013c). Human slow wave frequency has been shown to decrease by about 20–30% along the length of the small intestine (11–12 cpm in the duodenum to 8–9 cpm in the ileum) (Christensen *et al.*, 1966; Fleckenstein and Oigaard, 1978). The mechanisms of this frequency gradient, and thereby the nature of small intestine pacesetting, have been the subject of myriad studies, and remain uncertain.

Unlike the single-pacemaker conduction system of the stomach, the small intestine is governed by a hierarchy of pacemakers along its length. An initial pacemaker, located in the proximal duodenum, is followed by a series of transient peripheral pacemakers located along the subsequent length of the intestine (Diamant and Bortoff, 1969a; Suzuki *et al.*, 1986; Lammers *et al.*, 2005). It has been suggested through sparse electrode recordings that these peripheral pacemakers entrain local tissue and interact with adjacent pacemakers to determine propagation patterns and frequency across these segmental regions of the intestine (Diamant and Bortoff, 1969a; Szurszewski *et al.*, 1970). This traditional model of small intestine pacesetting suggests that proximal regions of high intrinsic ICC frequency entrain distal tissue until the disparity in driving frequency and intrinsic frequency is too great for the distal tissue to maintain (Furness, 2006). A new plateau of lower frequency is then established distal to the previous segment, with a region of irregular activity and "waxing and waning" (or "spindling") of the slow waves purportedly occurring at the interface of these frequency plateaus. This waxing and waning phenomenon is characterized by a cyclical increase and decrease of the slow wave amplitude, thought to be a result of the pacemakers in these segments, and thereby the frequencies, operating slightly out of phase (Suzuki *et al.*, 1986).

Entrainment of distal tissue by proximal segments of higher intrinsic frequency has been demonstrated by studies utilizing transections to uncouple the proximal and distal tissue, thereby uncovering the intrinsic frequency of that tissue (Diamant and Bortoff, 1969b; Code and Szurszewski, 1970). These studies have shown that the slow wave frequency drops immediately distal to an intestinal transection, and further declines at transections made distal to those previously performed, effectively elucidating a hierarchical entrainment of slow waves above the inherent local frequency along the length of the intestine (Diamant and Bortoff, 1969b; Code and Szurszewski, 1970).

The overall propagation of slow waves along the intestine progresses from oral to anal; however, variability in propagation direction is known to occur, as shown in Figure 10.4. In the canine, slow wave propagation was mainly aboral, with other propagation patterns making up approximately 15% of the activity, including oral propagation, propagation blocks, and circumferential activity (Lammers *et al.*, 2005). A recent porcine study further supported this variability of propagation direction, where a longer mapped segment showed that only about 33% of recordings displayed solely aboral propagation, with the remaining activity comprised, at least in part, of oral propagation and propagation blocks (Angeli *et al.*, 2013c).

Small intestine slow wave velocity, like frequency, decreases along the length of the intestine, and interspecies variability of slow wave velocity is pronounced. For example, canine slow wave velocity decreases from an

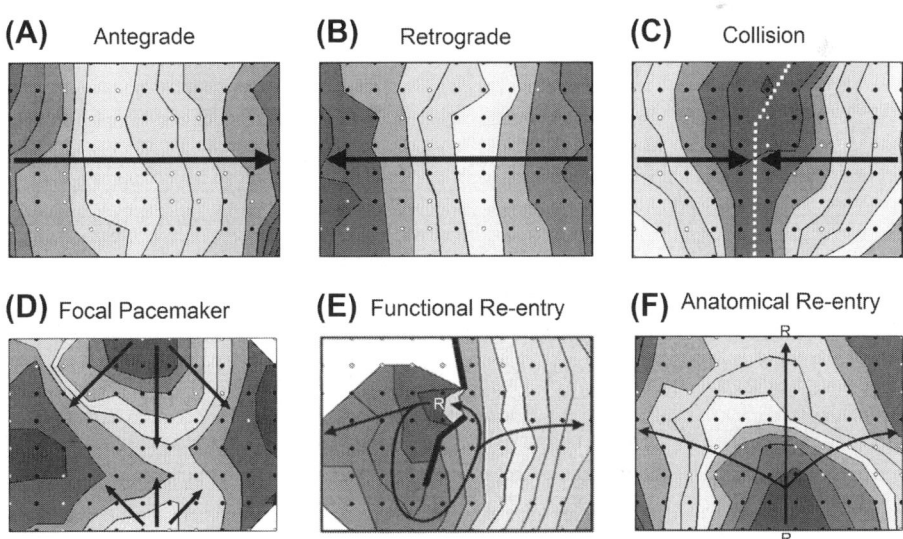

■ **FIGURE 10.4** Slow wave propagation patterns of the small intestine recorded by high-resolution mapping, including: (A) antegrade propagation; (B) retrograde propagation; (C) wavefront collision; (D) focal pacemaker; (E) functional re-entry; (F) circumferential (anatomical) re-entry. The maps are as described in Figure 10.2, with isochrones of: 0.25 s in C and D; 0.5 s in A, E, and F; and 1 s in B. The electrode array was wrapped around the intestine such that the two-dimensional maps illustrate the unfolded circumference of the intestine, with the top and bottom of each map located on adjacent sides of the mesentery. In all maps, orad is left and aborad is right. Each dot represents an electrode, with empty dots representing an electrode where activity was interpolated, likely due to poor contact at that electrode. In the cases of re-entry (E and F), the site of re-entry is arbitrary since the activation occurs in a circuit, but for illustrative purposes it is marked with an "R" in these maps. *Adapted from Angeli* et al. *(2013b, c).* [A color version of this figure is available online at www.booksite.elsevier.com/9780124046108].

average of $105\ mm\ s^{-1}$ in the duodenum to $8\ mm\ s^{-1}$ in the terminal ileum, with a steep decrease of velocity in the region of the duodenojejunal junction (Lammers *et al.*, 2005). Comparatively, porcine slow wave velocity decreases from an average of $15\ mm\ s^{-1}$ in the proximal duodenum to $9\ mm\ s^{-1}$ in the terminal ileum, with a relatively gradual decrease from proximal duodenum to mid-jejunum, followed by a relative plateau in velocity for the remainder of the intestinal length (Angeli *et al.*, 2013c). Small intestine slow waves have been shown to propagate in an anisotropic conduction pattern, propagating more rapidly in the circumferential direction than the longitudinal (ratio 1.3:1 in feline; 1.4:1 in porcine) (Lammers *et al.*, 2002; Angeli *et al.*, 2013b). As in the stomach, this anisotropy is likely to be a significant factor in causing wavefronts to rapidly orient into organo-axial rings of activation, promoting small intestine content transit.

Gastrointestinal Slow Wave Re-Entry

Re-entrant slow wave activity has recently been discovered to occur in both the stomach (Lammers *et al.*, 2008; O'Grady *et al.*, 2011) and small intestine (Lammers *et al.*, 2012; Angeli *et al.*, 2013b), and has become a topic of intense research interest (Lammers, 2013). Re-entry has been extensively investigated in cardiac electrophysiology, where it occurs exclusively as a dysrhythmic pattern associated with disorders like ventricular tachycardia (El-Sherif *et al.*, 1983; De Bakker *et al.*, 1988; Spach and Josephson, 1994). Re-entry transpires when electrical activation occurs over a loop of tissue, reactivating (i.e., "re-entering") that loop over successive cycles. This loop of tissue can establish either around a functional conduction block ("functional re-entry") or a tissue defect/anatomical structure like the GI lumen ("anatomical re-entry").

Functional re-entry has been mapped in the *in vivo* canine and porcine stomach (Figure 10.4E), and was usually associated with tachygastric frequency, which in turn was proposed to be a dysrhythmic pattern caused by intraoperative handling of the organ or the use of opiates as an anesthetic (Lammers *et al.*, 2008; O'Grady *et al.*, 2011). Intestinal re-entry has been observed *in vitro* in excised rat tissue, both normal and diabetic (Lammers *et al.*, 2012), and in the *in vivo* porcine intestine, where anatomical re-entry was observed to occur around the intestinal lumen, as shown in Figure 10.4F (Angeli *et al.*, 2013b). Re-entrant intestinal activity has also been simulated to occur as a result of temperature abnormalities in the intestine (Gizzi *et al.*, 2010). It is thought that re-entries can likely exist in the human, although this is yet to be experimentally observed in either the stomach or intestine (Angeli *et al.*, 2013b; Lammers, 2013).

MODELING GASTROINTESTINAL SLOW WAVE ACTIVITY

As previously presented, high-resolution mapping of the GI tract has yielded much information regarding the electrophysiology of the gut in the *in vivo* state, both in the normal state (Lammers *et al.*, 2009; Egbuji *et al.*, 2010; O'Grady *et al.*, 2010) and dysrhythmias (Lammers *et al.*, 2008; O'Grady *et al.*, 2011, 2012a). Some of these mapping results, and other types of measurements (e.g., anatomical), are beginning to have an integral role in the development of a new generation of mathematical models that are capable of performing predictive simulations of the mechanisms underpinning slow wave activity (Du *et al.*, 2013b). The transition from *in vivo* to *in silico* research is a rapidly expanding area of research and has benefited tremendously from precision measurement instruments and the ever-increasing high-performance computing power. Even though the development of GI mathematical models is still in its infancy compared to other well-established fields like cardiac and neural electrophysiology, some GI models have already usefully explored key mechanistic questions regarding propagation of slow waves and have generated significant hypotheses forming some of the present fundamental understanding of GI electrophysiology (Cheng *et al.*, 2010; Du *et al.*, 2010a, 2013b). This section briefly reviews the earlier models of GI electrophysiology and introduces the current biophysically-based multi-scale modeling approach. Examples of mathematical models informed by high-resolution mapping studies are also discussed in detail.

Modeling Slow Waves as Self-Excitatory Oscillators

The ability of the ICC to generate slow waves at a specific intrinsic frequency was quickly noticed and captured by investigators as early as the 1960s (Nelsen and Becker, 1968), even before the underlying intracellular pacemaking mechanisms were understood. Arguably, the pacemaking mechanisms are still yet to be completely described. Without knowing the intracellular processes that contribute to slow wave generation, investigators resorted to reproducing the phenomenological behaviors of slow waves only, by adopting mathematical oscillators such as the van der Pol equations (Nelsen and Becker, 1968; Sarna *et al.*, 1971, 1972). The mathematical oscillators are generally a system of equations, typically containing two ordinary differential equations that capture the rate of change of some dependent variables. By adjusting the parameters in the equations and causing a perturbation, such as an electrical stimulus, one could send the system into a stable oscillation that can be used to match the frequency of slow waves in different parts of the GI tract. The absolute values of these equations typically need to be scaled to match the membrane potential of slow waves. For example, below is an oscillator model proposed by Aliev *et al.* (Aliev *et al.*, 2000), which was derived from another oscillator equation used in early neural electrophysiology research, the Fitzhugh−Nagumo model (Fitzhugh, 1955):

$$\frac{du}{dt} = ku(u - a)(1 - u) - v \tag{10.1}$$

$$\frac{dv}{dt} = \varepsilon(\gamma(u - \beta) - v) \tag{10.2}$$

where u represents V_m (has to be scaled to match the experimentally recorded V_m values from the resting potential of $-70\,mV$ to a peak potential of $-30\,mV$ in the ICC), v is termed the recovery variable, k is the rate constant, a is the normalized threshold potential, ε controls the excitability of the system, γ is the recovery rate constant, and β is used to shift the cellular equilibrium from an excitatory to oscillatory state. The Aliev model was the first to recognize the electrophysiological coupling between ICC and SMC, notably, with ICC acting as a "source" of slow wave activity and SMC acting as a passive "sink" of the slow waves generated by the ICC (Aliev *et al.*, 2000). The different cell types were distinguished by employing different parameter values that send the system of equations into either a self-excitatory oscillation (Figure 10.5A) or passive tissue (SMC) (Aliev *et al.*, 2000).

Biophysically-Based Slow Wave Models

A decade of research into the kinetics of ion channels in ICC and SMC has promoted investigators to update the model description of slow waves. By this time, the Hodgkin-and-Huxley and Hill-type equations have been employed readily in the neural and cardiac fields to accurately reproduce the kinetics of a large number of ion channels, and promoted investigations into the effects of channelopathy at the molecular level on the whole-cell to whole-organ electrical activity (Coburn, 1989; Noble *et al.*, 2012). Investigators of GI science have also pointed out that one clear advantage of these biophysically-based models over the phenomenologically-based models is their ability to predict the outcome of perturbation of realistic physiological conditions (Sarna, 1990; Daniel *et al.*, 1994). For example, one of the most effective simulations in the cardiac modeling field has been to study the effects of sodium channelopathy resulting from mutation in the SCN5A gene on the development of the long-QT type-3 syndrome (Rudy and Silva, 2006). In contrast, biophysically-based slow wave cell models have been latent until Corrias and Buist published the first comprehensive descriptions of ICC and SMC slow wave models (Corrias and Buist, 2007, 2008), both of which were based on Hodgkin-and-Huxley and Hill-type equations that were

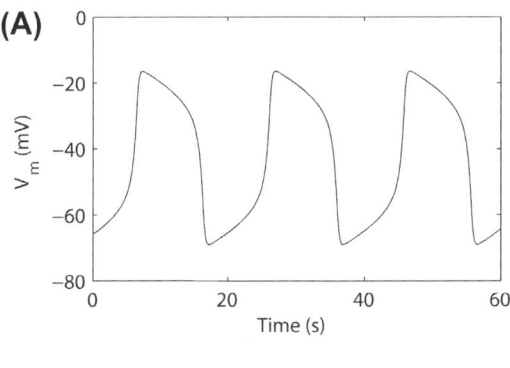

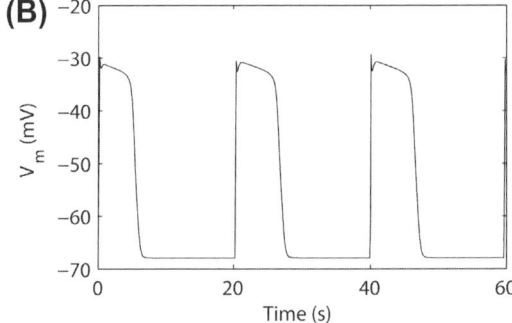

■ **FIGURE 10.5** Simulated gastric slow waves using cell models proposed by: (A) Aliev *et al.* (2000); (B) Corrias and Buist (2008).

fitted to the most recent understanding of the major ion channels in these cells at the time (Figure 10.5B).

The biophysically-based cell models were quickly adapted to simulate the coupling of ICC and SMC through a gap junction, as well as calibrating for the variations of slow wave resting membrane potentials (up to 10 mV across the gastric wall) reported by experimental recordings in the stomach (Szurszewski and Farrugia, 2004; Du *et al.*, 2010a). One possible explanation for these variations was the hyperpolarizing effect of agents like carbon monoxide (Szurszewski and Farrugia, 2004), the purpose of which is to exhibit a graded level of activation of smooth muscles in the stomach, for example, the muscles further away from the myenteric ICC layer are more hyperpolarized and therefore "easier" to activate; however, this hypothesis has yet to be proven experimentally.

Slow Wave Entrainment Model

The entrainment of electrical potential to a single pacemaker has been established in the cardiac pacemaker nodes (both the atrioventricular and sinoantrial nodes) (Podziemski and Zebrowski, 2013). In the GI field, a similar concept of entrainment governs the propagation mechanism of slow waves (Ward *et al.*, 2004); however, the ICC are distributed throughout the wall of the GI tract, rather than clustered to specific regions as in the heart. Functionally, the entrainment of slow wave serves as a loosely-coupled propagation system as each ICC is capable of generating slow waves without receiving inputs from the surrounding cells (Ordög *et al.*, 2002). In comparison, the tightly-coupled cardiac cells actively spread cardiac electrical potential at a significantly higher velocity (around $100 \, \text{mm s}^{-1}$) that is physiologically insensible in the gut (around $10 \, \text{mm s}^{-1}$).

From a modeling point of view, the entrainment mechanism of slow waves can be modeled as a voltage-dependent and loosely-coupled system, which involves a transition from single cell to multi-cell environment. Accordingly, new governing equations such as the cable equation, monodomain equation (Eq. 10.3), and bidomain equations are required to reconcile the slow waves at the cellular level to propagation of slow waves in the GI tissue (Cheng *et al.*, 2010; Du *et al.*, 2013b):

$$A_m \left(C_m \frac{\partial V_m}{\partial t} + I_{ion} \right) = \nabla \cdot (\sigma \nabla V_m) \qquad (10.3)$$

where A_m represents the cell surface to volume ratio, and σ presents the conductivity tensor of the tissue. The left-hand side of Eq. 10.3 is essentially the cell membrane equation that simulates the slow wave activity at the cellular level, whereas the right-hand side terms represent the rate of passive conduction as a diffusive process.

A number of investigators have attempted to reproduce the entrainment mechanisms at the cellular level by modifying the aforementioned biophysically-based ICC models (Buist *et al.*, 2010; Du *et al.*, 2010b). Interestingly, these two independent studies perturbed the same component of the ICC model, the IP_3 synthesis equation, inducing a strongly voltage-dependent IP_3 synthesis that led to an increase of intracellular calcium and therefore an onset of slow wave activity, largely independent of the intrinsic frequency of the modified ICC model (Figure 10.6).

It must be noted that the actual mechanism of entrainment is still inconclusive. In fact, direct voltage-dependent IP_3 synthesis is unlikely, as reported in other fields (Morgan *et al.*, 2013). A more likely mechanism has been proposed as a "pacemaker unit theory," in which a cluster of ion channels and intracellular organelles form a micro-domain of rapid calcium cycling in response to an extracellular input, causing a cascade of reactions in all

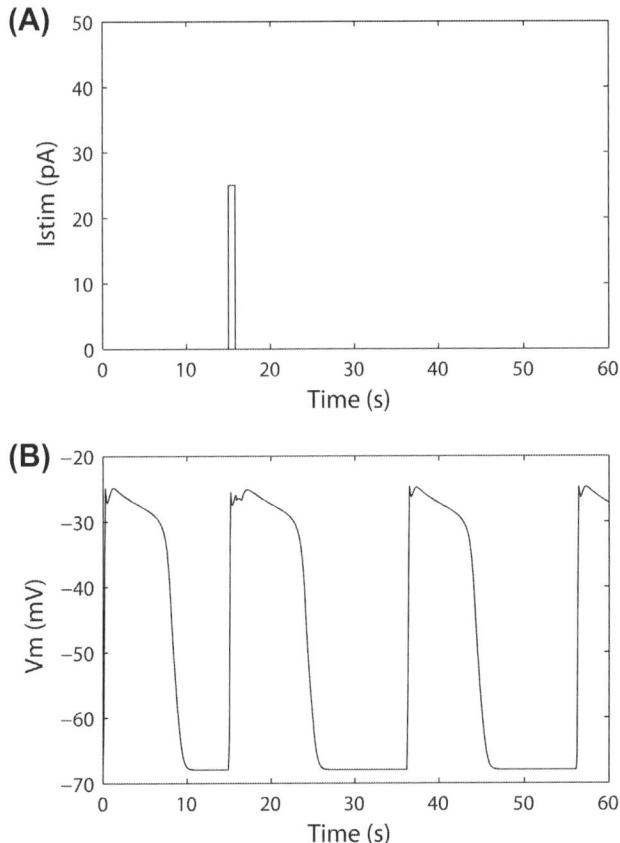

■ **FIGURE 10.6** Simulation of the response of an ICC model to a stimulus. (A) A single-pulse stimulus of 25 pA and 500 ms was applied to the cell model. (B) The slow wave generated by the cell model was phase-shifted forward due to the stimulus, without changing the frequency of the subsequent slow waves.

such micro-domains in the ICC and leading to the onset of slow wave activity (Faville *et al.*, 2009; Means and Sneyd, 2010). The pacemaker unit theory is supported by a cohort of experimental studies and has been modeled as both a "unitary potential" gastric ICC model and an intestinal ICC model (Lees-Green *et al.*, 2011). However, the unitary potential theory requires a highly compartmental approach to modeling the ICC and essentially models two levels of entrainment, one at the micro-domain level inside the cell and another one at the ICC—ICC level, both of which have proven to be challenging to achieve due to computational demand and a current lack of critical mechanistic details.

Whole-Organ Slow Wave Activation Model

The information produced by high-resolution mapping of the stomach motivated investigators to integrate the data in a dynamic manner at the whole-organ scale (Pullan *et al.*, 2004). To this end, a number of sources of information were required. First, activation timings of slow waves were obtained from high-resolution mapping data (O'Grady *et al.*, 2010). Because the high-resolution mapping arrays could only cover a limited region of the stomach at any one time, interpolation and offset of timings between recordings from the different areas of the stomach had to be performed (Du *et al.*, 2010a). Second, additional electrophysiological information was required, such as the gradient of gastric resting membrane potentials. In addition to the 10 mV voltage gradient across the gastric wall, experimental studies have shown that there is a resting membrane gradient from the fundus to the antrum of the stomach (Szurszewski and Farrugia, 2004). Previous intracellular studies have also shown different morphological features of gastric slow waves, in terms of upstroke rate and amplitude (Hirst and Edwards, 2006). Lastly, anatomical information of the stomach was required, which could be reconstructed from a three-dimensional image source, such as pre-operative CT, MRI, or high-resolution cross-sectional scans (e.g., the Visible Human Project; http://www.nlm.nih.gov/research/visible/). In one study, the outline of the stomach was manually segmented and a data cloud representing the outside of the stomach was generated; a finite element fitting method was then used to fit a geometric mesh to the data cloud until the mesh produced a sufficient agreement with the data cloud (Pullan *et al.*, 2004). In a separate study, the surface of the fitted mesh was then extruded 2.66 mm to represent the thickness of the gastric muscular layer (Huh *et al.*, 2003; Du *et al.*, 2010a).

With the aforementioned sources of information available, a coupled SMC-ICC model was used to solve for the slow waves in the stomach by prescribing initial onset activation times obtained from high-resolution mapping studies (Du *et al.*, 2010a). The solution procedure involved allocating multiple solution points inside each finite element of the stomach geometric mesh, with each solution point representing an independently working cell model. The result of this approach was that the activation times in the model could be matched to experimental recordings, and the propagation of slow waves could be visualized in a dynamic manner over time (Figure 10.3). However, a key drawback of this approach was that the activation pattern was essentially known *a priori*, thus eliminating the ability of the whole-organ model to perform predictive simulations of gastric electrical pacing.

Despite some of the drawbacks of this approach, the whole-organ model still offered a number of significant insights that could not be easily obtained from experimental recordings. First, the number of simultaneous wavefronts in the stomach could be easily determined using this approach, and related to gastric slow wave frequency. Second, by integrating electrophysiological and anatomical information from multiple recording modalities, the theoretical equivalent dipole of the stomach could be obtained, which was an important step toward understanding the relationship between slow wave activation and cutaneous recording (i.e., electrogastrogram, "EGG") (Du *et al.*, 2010a). A similar approach could be used to simulate propagation of intestinal slow waves, which had been previously modeled in an anatomically realistic intestine geometry using the Aliev model (Lin *et al.*, 2006). Finally, because of the reduced computational load, this approach is also appropriate for electromechanical coupling simulations.

Electromechanical Coupling of Gastrointestinal Tissue

Ultimately, the electromechanical coupling between slow waves and motility must be addressed for GI models to truly achieve functional and clinical translation. Initial attempts have been made to link slow wave propagation to tension generation in a short segment of the GI tract (Du *et al.*, 2011, 2013a). Theoretically, the tensions generated can be classified into either: (1) passive tension or "tone," which mainly depends on the material properties of the tissues opposing dilation in the lumen during peristalsis; or (2) active tension, which is generated by the smooth muscle itself, typically through calcium transient inside the SMC as slow waves develop. A recent modeling study has established the relationship between force generation and calcium transient in GI SMC (Gajendiran and Buist, 2011). This model was based on the view that depolarization of the cell membrane leads to activation of voltage-dependent calcium channels, such as iLVA and IL type. The slight increase in intracellular calcium through these ion channels triggers further release of calcium from intracellular stores like the sarcoplasmic reticulum, launching the cascade that results in myosin—actin force generation (Du *et al.*, 2013a). Another more direct way of modeling the calcium—tension relationship is to employ a Hill-type equation to fit known experimental data of calcium transient to tension development in the smooth muscle cells (Figure 10.7A) (Du *et al.*, 2011, 2013a).

The intestinal lumen can contract up to 70% of its "resting" diameter because the gut wall is under pre-stretch, and therefore the deformation involved is large and nonlinear (Du *et al.*, 2013a). Hence, in addition to a calcium—tension

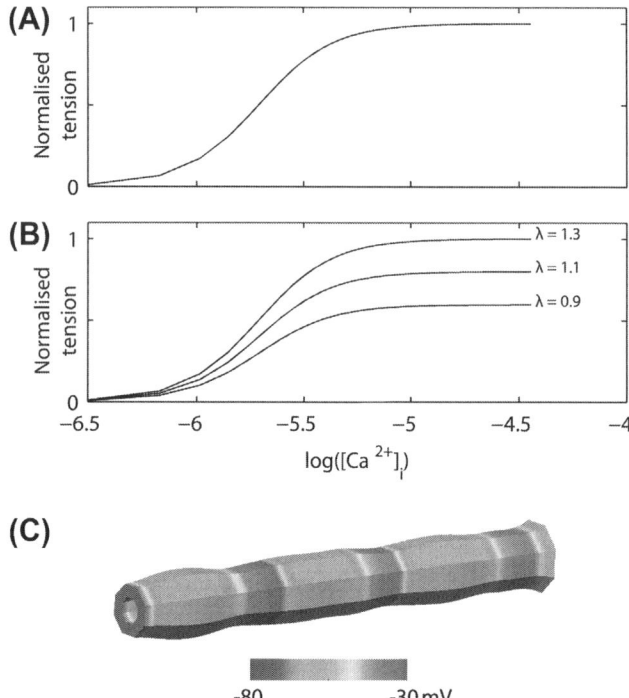

■ **FIGURE 10.7** Electromechanical coupling simulation. (A) A simulated trace of tension development in response to intracellular calcium. (B) Simulated traces of tension development as a function of pre-stretch. (C) Simulation of a peristaltic wave due to intestinal slow waves in the cylindrical model. *Adapted from Du* et al. *(2011, 2013a).* [A color version of this figure is available online at www.booksite.elsevier.com/9780124046108].

relationship at the cell level, material properties and additional governing equations, such as the finite deformation theory, are needed to account for the deformation of the gut wall as tension develops. In essence, the finite deformation "tracks" the movement of material during deformation in two co-ordinate systems: (1) a material system, in which the coordinate deforms with the material deformation so that the particle during each deformation stage is at a constant position according to the coordinate system; and (2) a spatial co-ordinate system, which remains constant throughout the deformation, serving as a spatial reference. Through mathematical tensor operations, a stress tensor (the second Piola–Kirchhoff stress tensor) can be derived to relate to the strain energy tensor of the system. The strain energy tensor is a theoretical representation that is used to develop the constitutive equations of the material (i.e., relating stress to strain). By combining a calcium–tension relationship and the required constitutive relationships, the deformation of a short segment of small intestine can be calculated (Figure 10.7B) (Du *et al.*, 2013a).

Simulations like that presented in Figure 10.7 provide a useful tool to investigate the development of intra-luminal pressure as the intestinal wall develops tension governed by the depolarization of slow waves. The developed pressure and deformation of the intestinal wall are essential information in the application of computational fluid dynamics studies used to predict the flow profile of particles in the gut lumen (Pal *et al.*, 2004; Miftahof and Akhmadeev, 2007; Ferrua and Singh, 2010).

CONCLUSIONS AND FUTURE DIRECTIONS

As seen throughout this chapter, great strides have been made in improving our understanding of the electrical component of GI control. High-resolution mapping has specifically added a spatiotemporal understanding about the propagation dynamics of slow waves in the stomach and small intestine; however, substantial research questions remain. It has been established, even from the very earliest studies of slow waves (Alvarez, 1922), that they play a key role in the regulation of GI contractile activity, and thereby help to regulate motility and digestion (Huizinga and Lammers, 2009). Yet, the exact mechanisms and relationships between slow waves and motility have not been fully elucidated. Future studies combining high-resolution mapping (e.g., Lammers *et al.*, 2005; O'Grady *et al.*, 2010; Angeli *et al.*, 2013c), spatiotemporal video analysis (e.g., Lentle *et al.*, 2012), and high-resolution manometry (e.g., Dinning *et al.*, 2010) would help to further understand the relationships and combined influence of slow waves on GI motility. GI dysrhythmias are an area of current vigorous research interest (Lammers, 2013), and high-resolution clinical mapping may likely elucidate underlying dysrhythmias in other GI disorders or serve to explain dysrhythmic electrical activity observed through low-resolution recordings, as with functional dyspepsia (Leahy *et al.*, 1999; Sha *et al.*, 2009). Future research on the effects of various fed states and different foods on GI electrical activity holds great potential not only to better understand the relationship between food and motility, but also to potentially utilize foods in a therapeutic capacity for GI disorders. For example, ginger has been shown to reduce gastric slow wave dysrhythmias while decreasing nausea (Lien *et al.*, 2003), effectively serving as a functional food with gastric therapeutic benefit.

The availability of experimental data has promoted development of a new generation of biophysically-based mathematical models that have been applied to answer, as well as form, important hypotheses regarding GI electrophysiology. While the advantages of such an expansive and integrative framework are obvious, many questions remain. Genetic and ion channel data have significantly contributed to the development of biophysically-based cell models of ICC and SMC. However, a "knowledge gap" has been found to exist relating to how the individual cells would behave *in vivo* in the intact organ in order to form a coherent propagating wavefront. This question was partially addressed by modeling the entrainment process of slow waves, but due to uncertainties in the actual entrainment mechanism, this area still remains an open question. At the whole-organ level, high-resolution mapping studies have promoted development of gastric slow wave propagation in an anatomically realistic stomach model, and to a certain degree in the small intestine as well. A next step will be to incorporate entrainment into these whole-organ models so the models can be used to predict the effect of an extrinsic factor, such as electrical pacing or pharmacological intervention. Furthermore, the development of a GI-specific electromechanical model is a truly exciting area, which is further highlighted by the recent discovery of the role of a stretch-sensitive sodium channel ($Na_V1.5$) in the gut (Beyder *et al.*, 2010), and the subsequent modeling study on the effect of stretch on an R76C mutation in the same channel (Poh *et al.*, 2012). However, more specific constitutive relationships of the GI tissue are required in order to improve the accuracy of these electromechanical models, which hold promise for future hypothesis testing of GI response in the fed state, potentially creating means to test the response of the GI tract to a variety of food parameters. Overall, the development of mathematical models concurrent with experimental research has facilitated rapid progression of GI science by allowing a more integrative and interdisciplinary approach.

REFERENCES

Aliev, R.R., Richards, W., Wikswo, J.P., 2000. A simple nonlinear model of electrical activity in the intestine. J. Theor. Biol. 204, 21−28.

Alvarez, W.C., 1922. The electrogastrogram and what it shows. J. Am. Med. Assoc. 78, 1116−1119.

Angeli, T.R., Du, P., Paskaranandavadivel, N., Janssen, P.W.M., Beyder, A., Lentle, R.G., et al., 2013a. The bioelectrical basis and validity of gastrointestinal extracellular slow wave recordings. J. Physiol. 591, 4567−4579.

Angeli, T.R., O'Grady, G., Du, P., Paskaranandavadivel, N., Pullan, A.J., Bissett, I.P., Cheng, L.K., 2013b. Circumferential and functional re-entry of in vivo slow-wave activity in the porcine small intestine. Neurogastroenterol. Motil. 25, e304−e314.

Angeli, T.R., O'Grady, G., Paskaranandavadivel, N., Erickson, J.C., Du, P., Pullan, A.J., et al., 2013c. Experimental and automated analysis techniques for high-resolution electrical mapping of small intestine slow wave activity. J. Neurogastroenterol. Motil. 19, 179−191.

Beyder, A., Rae, J.L., Bernard, C., Strege, P.R., Sachs, F., Farrugia, G., 2010. Mechanosensitivity of Nav1.5, a voltage-sensitive sodium channel. J. Physiol. 588, 4969−4985.

Buist, M.L., Corrias, A., Poh, Y.C., 2010. A model of slow wave propagation and entrainment along the stomach. Ann. Biomed. Eng. 38, 3022−3030.

Bull, S.H., O'Grady, G., Cheng, L.K., Pullan, A.J., 2011. A framework for the online analysis of multi-electrode gastric slow wave recordings. In: Conf. Proc. IEEE. Eng. Med. Biol. Soc., 1741–1744.

Cheng, L.K., O'Grady, G., Du, P., Egbuji, J.U., Windsor, J.A., Pullan, A.J., 2010. Gastrointestinal system. WIREs Syst. Biol. Med. 2, 65–79.

Christensen, J., Schedl, H.P., Clifton, J.A., 1966. The small intestinal basic electrical rhythm (slow wave) frequency gradient in normal men and in patients with variety of diseases. Gastroenterology 50, 309–315.

Coburn, B., 1989. Neural modeling in electrical stimulation. Crit. Rev. Biomed. Eng. 17, 133–178.

Code, C.F., Szurszewski, J.H., 1970. The effect of duodenal and mid small bowel transection on the frequency gradient of the pacesetter potential in the canine small intestine. J. Physiol. 207, 281–289.

Corrias, A., Buist, M.L., 2007. A quantitative model of gastric smooth muscle cellular activation. Ann. Biomed. Eng. 35, 1595–1607.

Corrias, A., Buist, M.L., 2008. Quantitative cellular description of gastric slow wave activity. Am. J. Physiol. Gastrointest. Liver Physiol. 294, G989–G995.

Cousins, H.M., Edwards, F.R., Hickey, H., Hill, C.E., Hirst, G.D.S., 2003. Electrical coupling between the myenteric interstitial cells of Cajal and adjacent muscle layers in the guinea-pig gastric antrum. J. Physiol. 550, 829–844.

Daniel, E.E., Bardakjian, B.L., Huizinga, J.D., Diamant, N.E., 1994. Relaxation oscillator and core conductor models are needed for understanding of GI electrical activities. Am. J. Physiol. 266, G339–G349.

De Bakker, J.M., Van Capelle, F.J., Janse, M.J., Wilde, A.A., Coronel, R., Becker, A.E., et al., 1988. Reentry as a cause of ventricular tachycardia in patients with chronic ischemic heart disease: electrophysiologic and anatomic correlation. Circulation 77, 589–606.

Diamant, N.E., Bortoff, A., 1969a. Nature of the intestinal slow-wave frequency gradient. Am. J. Physiol. 216, 301–307.

Diamant, N.E., Bortoff, A., 1969b. Effects of transection on the intestinal slow-wave frequency gradient. Am. J. Physiol. 216, 734–743.

Dinning, P.G., Arkwright, J.W., Gregersen, H., O'Grady, G., Scott, S.M., 2010. Technical advances in monitoring human motility patterns. Neurogastroenterol. Motil. 22, 366–380.

Du, P., Lim, J., Cheng, L.K., 2013a. A model of electromechanical coupling in the small intestine. Stud. Mechanobiol. Tissue Eng. Biomater. 14, 179–207.

Du, P., O'Grady, G., Cheng, L.K., Pullan, A.J., 2010a. A multiscale model of the electrophysiological basis of the human electrogastrogram. Biophys. J. 99, 2784–2792.

Du, P., O'Grady, G., Egbuji, J.U., Lammers, W.J.E.P., Budgett, D., Nielsen, P., et al., 2009. High-resolution mapping of in vivo gastrointestinal slow wave activity using flexible printed circuit board electrodes: methodology and validation. Ann. Biomed. Eng. 37, 839–846.

Du, P., O'Grady, G., Gao, J., Sathar, S., Cheng, L.K., 2013b. Toward the virtual stomach: progress in multiscale modeling of gastric electrophysiology and motility. WIREs Syst. Biol. Med. 5, 481–493.

Du, P., O'Grady, G., Gibbons, S.J., Yassi, R., Lees-Green, R., Farrugia, G., et al., 2010b. Tissue-specific mathematical models of slow wave entrainment in wild-type and

5-HT(2B) knockout mice with altered interstitial cells of Cajal networks. Biophys. J. 98, 1772–1781.

Du, P., Poh, Y.C., Lim, J.L., Gajendiran, V., O'Grady, G., Buist, M.L., et al., 2011. A preliminary model of gastrointestinal electromechanical coupling. IEEE Trans. Biomed. Eng. 58, 3491–3495.

Egbuji, J.U., O'Grady, G., Du, P., Cheng, L.K., Lammers, W.J.E.P., Windsor, J.A., Pullan, A.J., 2010. Origin, propagation and regional characteristics of porcine gastric slow wave activity determined by high-resolution mapping. Neurogastroenterol. Motil. 22, e292–e300.

El-Sherif, N., Mehra, R., Gough, W.B., Zeiler, R.H., 1983. Reentrant ventricular arrhythmias in the late myocardial infarction period. Circulation 68, 644–656.

Erickson, J.C., O'Grady, G., Du, P., Egbuji, J.U., Pullan, A.J., Cheng, L.K., 2011. Automated gastric slow wave cycle partitioning and visualization for high-resolution activation time maps. Ann. Biomed. Eng. 39, 469–483.

Erickson, J.C., O'Grady, G., Du, P., Obioha, C., Qiao, W., Richards, W.O., et al., 2010. Falling-edge, variable threshold (FEVT) method for the automated detection of gastric slow wave events in high–resolution serosal electrode recordings. Ann. Biomed. Eng. 38, 1511–1529.

Faville, R.A., Pullan, A.J., Sanders, K.M., Koh, S.D., Lloyd, C.M., Smith, N.P., 2009. Biophysically based mathematical modeling of interstitial cells of Cajal slow wave activity generated from a discrete unitary potential basis. Biophys. J. 96, 4834–4852.

Ferrua, M.J., Singh, R.P., 2010. Modeling the fluid dynamics in a human stomach to gain insight of food digestion. J. Food Sci. 75, R151–R162.

Fitzhugh, R., 1955. Mathematical models of threshold phenomena in the nerve membrane. Bull. Math. Biophys. 17, 257–278.

Fleckenstein, P., Oigaard, A., 1978. Electrical spike activity in the human small intestine. A multiple electrode study of fasting diurnal variations. Dig. Dis. Sci. 23, 776–780.

Furness, J.B., 2006. The Enteric Nervous System. Blackwell Publishing, Oxford.

Gajendiran, V., Buist, M.L., 2011. A quantitative description of active force generation in gastrointestinal smooth muscle. Int. J. Numer. Meth. Biomed. Eng. 27, 450–460.

Gizzi, A., Cherubini, C., Migliori, S., Alloni, R., Portuesi, R., Filippi, S., 2010. On the electrical intestine turbulence induced by temperature changes. Phys. Biol. 7, 16011.

Grundy, D., Brookes, S.J., 2012. Neural control of gastrointestinal function. In: Colloquium Series on Integrated Systems Physiology: From Molecule to Function. Morgan Claypool Life Sci. Publishers.

Hermon-Taylor, J., Code, C.F., 1971. Localization of the duodenal pacemaker and its role in the organization of duodenal myoelectric activity. Gut 12, 40–47.

Hinder, R.A., Kelly, K.A., 1977. Human gastric pacesetter potential. Site of origin, spread, and response to gastric transection and proximal gastric vagotomy. Am. J. Surg. 133, 29–33.

Hirst, G.D.S., Edwards, F.R., 2006. Electrical events underlying organized myogenic contractions of the guinea pig stomach. J. Physiol. 576, 659–665.

Huh, C.H., Bhutani, M.S., Farfán, E.B., Bolch, W.E., 2003. Individual variations in mucosa and total wall thickness in the stomach and rectum assessed via endoscopic ultrasound. Physiol. Meas. 24, N15–N22.

Huizinga, J.D., Lammers, W.J.E.P., 2009. Gut peristalsis is governed by a multitude of cooperating mechanisms. Am. J. Physiol. Gastrointest. Liver Physiol. 296, G1–G8.

Kelly, K.A., Code, C.F., 1971. Canine gastric pacemaker. Am. J. Physiol. 220, 112–118.

Kelly, K.A., Code, C.F., Elveback, L.R., 1969. Patterns of canine gastric electrical activity. Am. J. Physiol. 217, 461–470.

Lammers, W.J.E.P., 2013. Arrhythmias in the gut. Neurogastroenterol. Motil. 25, 353–357.

Lammers, W.J.E.P., Al-Kais, A., Singh, S., Arafat, K., El-Sharkawy, T.Y., 1993. Multielectrode mapping of slow-wave activity in the isolated rabbit duodenum. J. Appl. Phys. 74, 1454–1461.

Lammers, W.J.E.P., Slack, J.R., 2001. Of slow waves and spike patches. News. Physiol. Sci. 16, 138–144.

Lammers, W.J.E.P., Stephen, B., 2008. Origin and propagation of individual slow waves along the intact feline small intestine. Exp. Physiol. 93, 334–346.

Lammers, W.J.E.P., Stephen, B., Arafat, K., Manefield, G.W., 1996. High resolution electrical mapping in the gastrointestinal system: initial results. Neurogastroenterol. Motil. 8, 207–216.

Lammers, W.J.E.P., Stephen, B., Karam, S.M., 2012. Functional reentry and circus movement arrhythmias in the small intestine of normal and diabetic rats. Am. J. Physiol. Gastrointest. Liver Physiol. 302, G684–G689.

Lammers, W.J.E.P., Stephen, B., Slack, J.R., Dhanasekaran, S., 2002. Anisotropic propagation in the small intestine. Neurogastroenterol. Motil. 14, 357–364.

Lammers, W.J.E.P., Ver Donck, L., Schuurkes, J.A.J., Stephen, B., 2005. Peripheral pacemakers and patterns of slow wave propagation in the canine small intestine in vivo. Can. J. Physiol. Pharmacol. 83, 1031–1043.

Lammers, W.J.E.P., Ver Donck, L., Stephen, B., Smets, D., Schuurkes, J.A.J., 2008. Focal activities and re-entrant propagations as mechanisms of gastric tachyarrhythmias. Gastroenterology 135, 1601–1611.

Lammers, W.J.E.P., Ver Donck, L., Stephen, B., Smets, D., Schuurkes, J.A.J., 2009. Origin and propagation of the slow wave in the canine stomach: the outlines of a gastric conduction system. Am. J. Physiol. Gastrointest. Liver Physiol. 296, G1200–G1210.

Leahy, A., Besherdas, K., Clayman, C., Mason, I., Epstein, O., 1999. Abnormalities of the electrogastrogram in functional gastrointestinal disorders. Am. J. Gastroenterol. 94, 1023–1028.

Lees-Green, R., Du, P., O'Grady, G., Beyder, A., Farrugia, G., Pullan, A.J., 2011. Biophysically based modeling of the interstitial cells of Cajal: current status and future perspectives. Front. Physiol. 2, 1–19.

Lentle, R.G., De Loubens, C., Hulls, C., Janssen, P.W.M., Golding, M.D., Chambers, J.P., 2012. A comparison of the organization of longitudinal and circular contractions during pendular and segmental activity in the duodenum of the rat and guinea pig. Neurogastroenterol. Motil. 24, 686–695 e298.

Lien, H.-C., Sun, W.M., Chen, Y.-H., Kim, H., Hasler, W., Owyang, C., 2003. Effects of ginger on motion sickness and gastric slow-wave dysrhythmias induced by circular vection. Am. J. Physiol. Gastrointest. Liver Physiol. 284, G481–G489.

Lin, A.S.-H., Buist, M.L., Smith, N.P., Pullan, A.J., 2006. Modelling slow wave activity in the small intestine. J. Theor. Biol. 242, 356–362.

Means, S.A., Sneyd, J., 2010. Spatio-temporal calcium dynamics in pacemaking units of the interstitial cells of Cajal. J. Theor. Biol. 267, 137–152.

Miftahof, R., Akhmadeev, N., 2007. Dynamics of intestinal propulsion. J. Theor. Biol. 246, 377–393.

Morgan, A.J., Davis, L.C., Wagner, S.K.T.Y., Lewis, A.M., Parrington, J., Churchill, G.C., Galione, A., 2013. Bidirectional Ca^{2+} signaling occurs between the endoplasmic reticulum and acidic organelles. J. Cell. Biol. 200, 789–805.

Nelsen, T.S., Becker, J.C., 1968. Simulation of the electrical and mechanical gradient of the small intestine. Am. J. Physiol. 214, 749–757.

Noble, D., Garny, A., Noble, P.J., 2012. How the Hodgkin-Huxley equations inspired the Cardiac Physiome Project. J. Physiol. 590, 2613–2628.

O'Grady, G., Angeli, T.R., Du, P., Lahr, C., Lammers, W.J.E.P., Windsor, J.A., et al., 2012a. Abnormal initiation and conduction of slow-wave activity in gastroparesis, defined by high-resolution electrical mapping. Gastroenterology 143, 589–598 e1–3.

O'Grady, G., Du, P., Cheng, L.K., Egbuji, J.U., Lammers, W.J.E.P., Windsor, J.A., Pullan, A.J., 2010. Origin and propagation of human gastric slow-wave activity defined by high-resolution mapping. Am. J. Physiol. Gastrointest. Liver Physiol. 299, G585–G592.

O'Grady, G., Du, P., Paskaranandavadivel, N., Angeli, T.R., Lammers, W.J.E.P., Asirvatham, S.J., et al., 2012b. Rapid high-amplitude circumferential slow wave propagation during normal gastric pacemaking and dysrhythmias. Neurogastroenterol. Motil. 24, e299–e312.

O'Grady, G., Egbuji, J.U., Du, P., Lammers, W.J.E.P., Cheng, L.K., Windsor, J.A., Pullan, A.J., 2011. High-resolution spatial analysis of slow wave initiation and conduction in porcine gastric dysrhythmia. Neurogastroenterol. Motil. 23, e345–e355.

Ohba, M., Sakamoto, Y., Tomita, T., 1975. The slow wave in the circular muscle of the guinea-pig stomach. J. Physiol. 253, 505–516.

Ordög, T., Baldo, M., Danko, R., Sanders, K.M., 2002. Plasticity of electrical pacemaking by interstitial cells of Cajal and gastric dysrhythmias in W/W mutant mice. Gastroenterology 123, 2028–2040.

Pal, A., Indireshkumar, K., Schwizer, W., Abrahamsson, B., Fried, M., Brasseur, J.G., 2004. Gastric flow and mixing studied using computer simulation. Proc. R. Soc. Lond. B. 271, 2587–2594.

Parkman, H.P., Hasler, W.L., Barnett, J.L., Eaker, E.Y., 2003. Electrogastrography: a document prepared by the gastric section of the American Motility Society Clinical GI Motility Testing Task Force. Neurogastroenterol. Motil. 15, 89–102.

Paskaranandavadivel, N., Cheng, L.K., Du, P., O'Grady, G., Pullan, A.J., 2011. Improved signal processing techniques for the analysis of high resolution serosal slow wave activity in the stomach. Conf. Proc. IEEE Eng. Med. Biol. Soc. 1, 1737–1740.

Paskaranandavadivel, N., O'Grady, G., Du, P., Cheng, L.K., 2013. Comparison of filtering methods for extracellular gastric slow wave recordings. Neurogastroenterol. Motil. 25, 79–83.

Paskaranandavadivel, N., O'Grady, G., Du, P., Pullan, A.J., Cheng, L.K., 2012. An improved method for the estimation and visualization of velocity fields from gastric high-resolution electrical mapping. IEEE Trans. Biomed. Eng. 59, 882–889.

Podziemski, P., Zebrowski, J.J., 2013. A simple model of the right atrium of the human heart with the sinoatrial and atrioventricular nodes included. J. Clin. Monit. Comput. 27, 481–498.

Poh, Y.C., Beyder, A., Strege, P.R., Farrugia, G., Buist, M.L., 2012. Quantification of gastrointestinal sodium channelopathy. J. Theor. Biol. 293, 41–48.

Pullan, A.J., Cheng, L.K., Yassi, R., Buist, M.L., 2004. Modelling gastrointestinal bioelectric activity. Prog. Biophys. Mol. Biol. 85, 523−550.

Rudy, Y., Silva, J.R., 2006. Computational biology in the study of cardiac ion channels and cell electrophysiology. Q. Rev. Biophys. 39, 57−116.

Sarna, S.K., 1990. The challenge to the relaxation oscillator model. Am. J. Physiol. Gastrointest. Liver Physiol. 258, G994−G996.

Sarna, S.K., Daniel, E.E., Kingma, Y.J., 1971. Simulation of slow-wave electrical activity of small intestine. Am. J. Physiol. 221, 166−175.

Sarna, S.K., Daniel, E.E., Kingma, Y.J., 1972. Effects of partial cuts on gastric electrical control activity and its computer model. Am. J. Phys. 223, 332−340.

Sha, W., Pasricha, P.J., Chen, J.D.Z., 2009. Rhythmic and spatial abnormalities of gastric slow waves in patients with functional dyspepsia. J. Clin. Gastroenterol. 43, 123−129.

Spach, M.S., Josephson, M.E., 1994. Initiating reentry: the role of nonuniform anisotropy in small circuits. J. Cardiovasc. Electrophysiol. 5, 182−209.

Suzuki, N., Prosser, C.L., DeVos, W., 1986. Waxing and waning of slow waves in intestinal musculature. Am. J. Phys. 250, G28−G34.

Szurszewski, J.H., Elveback, L.R., Code, C.F., 1970. Configuration and frequency gradient of electric slow wave over canine small bowel. Am. J. Physiol. 218, 1468−1473.

Szurszewski, J.H., Farrugia, G., 2004. Carbon monoxide is an endogenous hyperpolarizing factor in the gastrointestinal tract. Neurogastroenterol. Motil. 16, 81−85.

van Helden, D.F., Laver, D.R., Holdsworth, J., Imtiaz, M.S., 2010. Generation and propagation of gastric slow waves. Clin. Exp. Pharmacol. Physiol. 37, 516−524.

Verhagen, M.A., Luijk, H.D., Samsom, M., Smout, A.J.P.M., 1998. Effect of meal temperature on the frequency of gastric myoelectrical activity. Neurogastroenterol. Motil. 10, 175−181.

Wang, X.-Y., Lammers, W.J.E.P., Bercik, P., Huizinga, J.D., 2005. Lack of pyloric interstitial cells of Cajal explains distinct peristaltic motor patterns in stomach and small intestine. Am. J. Physiol. Gastrointest. Liver Physiol. 289, G539−G549.

Ward, S.M., Dixon, R.E., De Faoite, A., Sanders, K.M., 2004. Voltage-dependent calcium entry underlies propagation of slow waves in canine gastric antrum. J. Physiol. 561, 793−810.

Yassi, R., O'Grady, G., Paskaranandavadivel, N., Du, P., Angeli, T.R., Pullan, A.J., et al., 2012. The gastrointestinal electrical mapping suite (GEMS): software for analyzing and visualizing high-resolution (multi-electrode) recordings in spatiotemporal detail. BMC. Gastroenterol. 12, 60.

Chapter

11

Novel Approaches to Tracking the Breakdown and Modification of Food Proteins through Digestion

Jolon M. Dyer[1,2,3,4] and Anita Grosvenor[1]

[1]*Food & Bio-Based Products, AgResearch Lincoln Research Centre, Christchurch, New Zealand,* [2]*Biomolecular Interaction Centre, University of Canterbury, Christchurch, New Zealand,* [3]*Wine, Food & Molecular Biosciences, Lincoln University, Canterbury, New Zealand,* [4]*Riddet Institute, based at Massey University, Palmerston North, New Zealand*

CONTENTS

INTRODUCTION

With the global population rapidly expanding, it is becoming increasingly important to derive optimum nutritional value from available food sources. Additionally, recent years have seen a significant increase in the amount of immunological food-related illnesses such as coeliac disease, leading to

Food Structures, Digestion and Health. http://dx.doi.org/10.1016/B978-0-12-404610-8.00011-6

303

strong interest in understanding how foods are digested. The extent to which positive or negative physiological effects are derived from food is directly related to its breakdown and digestion.

The human body has a complex system for breaking down food in order to derive nutrition and maintain health. First, food is crushed and broken down physically in the mouth while being mixed with saliva. The food is then gastrically digested at an acidic pH (between pH 1 and 5) for variable time periods before being neutralized in the small intestine. Enzymes are also mixed with the digesta during this process, including proteases, amylases, and lipases (Wickham *et al.*, 2009).

This process generates a complex mixture of food-derived components of varying molecular weight. The molecular composition of this digesta determines both physiological uptake and the biological effect of the food. Proteins are inherently more complex and variable than other macronutrients within food, and are linked to a wide range of nutritional and functional effects in the body (Anderson and Aziz, 2006). Characterizing and understanding the fate of proteins during digestion are therefore critical goals in food science and nutrition.

PROTEIN DIGESTION

Dietary proteins provide essential amino acids for the proper growth and repair of the human body, as well as energy. These proteins are complex food molecules that are structurally diverse (Syrbe *et al.*, 1998; Gerrard, 2002; Tornberg, 2005). Peptides are released from ingested foods throughout the digestive process as the proteins are broken down into increasingly smaller fragments hydrolytically, with the speed and nature of release dependent on structural and physical factors such as their solubility and degree of accessibility.

Hydrolytic enzymes from the stomach, pancreas, and small intestine play a key role in protein breakdown (Erickson and Kim, 1990). This breakdown results in a complex mixture of amino acids and peptides that can be absorbed both rapidly and efficiently in the small intestine. The way in which any given protein is broken down, along with how food processing, composition, and delivery influence the type, rate, and amount of peptides released, is of great importance to understanding food nutritional impact. Furthermore, this is of particular relevance to biological uptake and the effect of putative bioactive peptides and potential allergens.

Bioactive and Allergen Release

Dietary proteins have highly important functions in the body in addition to the provision of amino acids for nutrition. Biologically active peptide sequences inherently encrypted within proteins have been shown to have a range of diverse physiological functions (Anderson and Aziz, 2006). For proteinaceous foods such as mammalian milk, a wide range of additional functionality beyond simple nutrition is encoded deliberately within the proteins for specific roles, for instance in supporting infant growth and immune function (Kunz *et al.*, 1999). Some of these functions, such as the protective role of lactoferrin-derived peptides, are at least partially understood (Baker and Baker, 2005, 2009), while others are undoubtedly not yet characterized at all and remain to be discovered.

While a huge range of putative bioactive peptide sequences have been identified to date (Clare and Swaisgood, 2000), it is often not yet clear at all which of these play a real physiological role when ingested orally within a food, particularly when ingested in the form of the parent protein. This is due in large part to the high complexity and dynamism inherent in protein digestion, with any given single protein generating a massive number of potential peptides and this profile changing rapidly through the digestive process.

In order for a food-derived protein or peptide to deliver its potential bioactive function it must remain intact during digestion. For this reason, two key underpinning properties influencing bioavailability and physiological effect are dietary concentration and structural stability (Wickham *et al.*, 2009).

The same principles apply for the release and uptake of potential allergens. In order for a food allergen to sensitize someone it must also be released and preserved during digestion. Food allergens are therefore characterized by their structural stability (Wickham *et al.*, 2009).

Protein food functionality over and above nutrition requires the bioactive sequence of a peptide or protein to reach the target site of activity in the body; in other words, it must be bioavailable to achieve a beneficial effect. For most kinds of targeted food functionality this would usually mean that sufficient amounts of the active peptide need to survive digestion and be transported through the gut barrier into the bloodstream. The fact that intact food peptides and proteins can penetrate the gut barrier if preserved sufficiently through the digestive process is evidenced by the presence of antibodies to a wide range of food proteins in normal individuals (Dearman *et al.*, 2001; Dupont *et al.*, 2010). Further support comes from the observation of ovalbumin in the blood after oral consumption, even though the mechanism of this kind of transport is not yet well understood (Matsubara *et al.*, 2008).

It is clear, then, that understanding and validation of both nutritional and functional foods, as well as control of allergenicity in foods, require supporting analytical techniques that can profile and track food components at the molecular level through digestion. As food molecules are broken down and changed within the digestive tract, highly complex mixtures are generated. This has traditionally made monitoring the specific breakdown of food during digestion very difficult. Further in this chapter we present a powerful new redox proteomic-based approach to tracking protein truncation and peptide release during digestion. First, however, it is worthwhile to briefly review the *in vivo* and *in vitro* digestion models developed and utilized to date. A summary of model types and their principles is presented here.

Digestion Models

Mathematical models of digestion provide some information regarding the probable effect of, for example, different feeding strategies in animals (Krishnamoorthy *et al.*, 1983; Rivest *et al.*, 2000; Ferrua *et al.*, 2011), but physical simulations are required to monitor the specific behavior of food substances under the conditions encountered during digestion. Their complexity or simplicity, or whether they are performed *in vitro* or *in vivo*, is determined by the purpose of individual experiments, such as the section of the digestive tract of interest, or the type of analysis required.

In Vivo Models

The most accurate models of digestion are performed *in vivo* in the form of animal studies (e.g., with pigs) (Darragh and Moughan, 1995; Rivest *et al.*, 2000) through to human clinical trials (Tydeman *et al.*, 2010). These require controlled intakes and sampling of digestive fluids (Dellow *et al.*, 1988). The large number of participants required, the ethics approval required, and the expense and complexity involved in such models make these otherwise ideal digestive models more suitable to late-stage verification studies.

In Vitro Models

The most commonly used digestive models are *in vitro*. Models differ from each other in (1) the composition of the digestive fluids used, (2) the steps mimicked in the digestion sequence, (3) the mechanical manipulation of the substrate, and (4) the experimental parameters measured. Models vary in complexity from simple one-chamber static stomach simulations through to complex, multi-step dynamic digestive tract models. *In vivo* conditions and residence times in the different regions of the digestive tract (mouth, stomach, duodenum, jejunum, ileum, colon) vary widely depending on the fasting status of the individual, the type and particle size of the food, and the food composition (Oomen *et al.*, 2002). Which model is most appropriate is therefore dependent on the research question, the food type of interest (e.g., proteins, lipids, liquid, or solid), and the level of complexity required.

Some models focus one on component, such as the stomach, mouth and stomach, or small intestine (Kong and Singh, 2010, 2011; Tharakan *et al.*, 2010), while others encompass multiple sections of the digestive tract (Minekus *et al.*, 1995; Blanquet *et al.*, 2004; Anson *et al.*, 2010). *In vitro* digestion models may be validated by comparison to *in vivo* models (Chiang *et al.*, 2008). One notable challenge in this area is the number of differing *in vitro* models in the literature, and the lack of harmonization within these, which makes it difficult to determine correlations to human clinical trials.

EVALUATION OF PROTEIN MODIFICATION—REDOX PROTEOMICS APPROACHES

Proteolysis (protein cleavage) in model digestive simulations can be monitored at a holistic level using amino acid analysis (Adibi and Mercer, 1973) and/or polyacrylamide gel electrophoresis-based (Eiwegger *et al.*, 2006; Kaur *et al.*, 2010; Maldonado-Valderrama *et al.*, 2012) techniques. However, mass spectrometry-based approaches, which can sensitively and specifically identify proteins and their fragments within complex mixtures, are becoming the preferred analytical platform for evaluating protein breakdown (Moreno *et al.*, 2008; Picariello *et al.*, 2010). Recent advances in redox proteomics and quantitative peptide tracking (Dyer *et al.*, 2010, 2013; Grosvenor *et al.*, 2010, 2011, 2012) have provided powerful new tools for both qualitatively and quantitatively monitoring the modification of protein-based systems at the molecular level. Redox proteomics is an emerging subdiscipline of proteomics centered on the study of key reduction and oxidation events occurring within proteins (Dalle-Donne *et al.*, 2006). Application and customization of techniques adapted from classical proteomics have enabled protein modification to be evaluated in greater detail than previously possible. This field therefore provides an ideal underpinning set of tools for evaluating protein modification in food systems.

Characterizing Protein Primary Structural Modification

Evaluating process-induced protein modification at the primary structural level is made particularly challenging due to the fact that any given modification at a particular primary location in the protein is typically present in extremely low relative abundance within the sample as a whole. In addition, it is critical that any effective analytical technique not induce any artifactual residue modification. Proteomic-based approaches offer a powerful means to evaluate protein foods and their modification.

Protein mass spectrometry is the primary analytical tool in proteomics. The two most common modes utilized are electrospray ionization-mass spectrometry (ESI-MS) and matrix-assisted laser desorption ionization-mass spectrometry (MALDI-MS), which differ in the means by which the sample is introduced. Mass spectrometric-based approaches offer a significant advantage over other analytical technologies for analyzing complex proteinaceous mixtures, in that proteins and peptides can be both accurately and sensitively identified without the need for complete purification of any given component.

Protein samples are typically enzymatically digested and the generated peptides analyzed. Characterization of lower abundance peptides can be achieved through the use of liquid chromatography (LC)-ESI or LC-MALDI, with fractionation of the peptide mixture prior to mass spectrometric analysis. When different kinds of chromatographic media are utilized sequentially to optimize the peptide separation, this approach is known as MudPIT (multi-dimensional protein identification technology) (Thomas *et al.*, 2006).

For determination of peptide sequences, peptides are either manually or automatically selected and fragmented (MS/MS) via post-source decay or collision with an inert gas. This generates a characteristic profile of fragment ions that correspond to the amino acid residue sequence. Critically, this also allows characterization and location of sites of modification. In this way, a wide range of residue side-chain modifications that are induced during food protein production, processing, and handling can be characterized, including oxidative and Maillard-type modifications (Alomirah *et al.*, 2000; Léonil *et al.*, 2000; Hau and Bovetto, 2001; Careri *et al.*, 2002).

Mapping Modification Profiles

Detailed molecular-level modification profiling of protein foods is achieved through further refinement and advancement of the proteomic approaches described above.

Redox proteomic profiling of modifications through the protein system can be used to track the progressive degradation of native peptides, and the corresponding formation of modified peptides which may alter the food quality and function (Xiong, 2000). The relative levels of any given modification can be assessed through changes in relative abundance, as well as the overall profile of modifications. Tracking the variation in this profile allows the formation of specific peptide modifications to be monitored over time, or over variation in any other parameter. The usefulness of monitoring relative abundance can be limited by potential variation in ion abundance ratios between sample runs. One way to overcome this limitation is to utilize a novel application of stable isotope labeling. Proof-of-principle has been demonstrated for the use of stable isotope labeling techniques to effectively quantitate relative protein modification abundance with a high degree of reproducibility and sensitivity (Grosvenor *et al.*, 2012). This approach may be adapted to track protein truncation during digestion, as discussed in "Tracking Protein Truncation," below.

One notable development from the field of redox proteomics is the application of modification scoring. Scoring systems have been developed to give an overview of the degree of protein modification within a given system and how this changes (Dyer *et al.*, 2010). First, a modification cascade for a particular kind of processing insult is determined. One example of a modification cascade is the progressive oxidative modification of tryptophan to initial primary oxidation products, such as hydroxytryptophan, and then to further secondary oxidation products, such as *N*-formylkynurenine and tryptophandione derivatives (Dyer *et al.*, 2006). Each level of the modification cascade is assigned a score, with higher scores assigned to higher level modifications. Detailed redox proteomic mapping of the whole system can then be applied and an overall score generated for any particular kind of modification. This score can then be contrasted between samples to see how protein foods change under differing processing or storage conditions (Dyer *et al.*, 2010). This approach is particularly effective for validating damage mitigation strategies.

Protein modification can also be evaluated through the identification and tracking of marker peptides (Grosvenor *et al.*, 2012). Marker peptides are generally selected based on criteria such as reproducible extraction, high relative abundance, and the presence of a modifiable residue which is representative of overall modification within the food. For example, marker peptides for tracking oxidative damage would contain at least one oxidatively sensitive residue, such as tryptophan, while suitable marker peptides for tracking alkali modification would contain at least one alkali-sensitive residue, such as lysine (Struthers, 1981; Igarashi *et al.*, 2007). Once a range of suitable marker peptides is selected, they can be used to evaluate and follow the overall degree of modification, and to contrast between samples. In concert with the techniques for tracking protein truncation described in the following section, the utilization of marker peptides is envisaged to be a particularly effective approach for testing and validating the delivery of bioactives.

TRACKING PROTEIN TRUNCATION

The new techniques available for redox proteomic evaluation have been used in a novel approach for characterizing and tracking protein digestion, with potential application for both *in vivo-* and *in vitro*-based digestion studies.

In classical proteomic analysis, a proteolytic enzyme, typically trypsin, is used to break whole proteins down into peptides for subsequent analysis. However, to effectively track protein truncation and release of peptides during food digestion, it is very important that any protein hydrolytic cleavage occurring is only that attributable to physiological digestion process, so tryptic protein cleavage is inappropriate for this application. Monitoring the release of peptides from larger parent proteins requires the quantitative comparison of individual peptides, rather than of intact proteins. This may be achieved using a mass spectrometric quantitation technique called isobaric labeling.

Isobaric literally means "of the same mass." Accordingly, sets of isobaric tags, as used in mass spectrometry, all have the same mass and chemical formula when intact. An isobaric tag is constructed of three parts: (1) a reactive part to bind to peptides, (2) a reporter group, which breaks off for detection during peptide fragmentation, and (3) a balance group. The careful positioning of non-radioactive isomers of carbon, nitrogen, and oxygen in the reporter and balance groups results in each reporter within an isobaric set having a unique mass. A commercially available variant, iTRAQ, comes in eight forms, with reporter ions ranging in mass from 113 to 121 Da. The balance groups of these tags also vary in mass so that the total mass of all the tags is identical.

Up to eight peptide samples can be derivatized with iTRAQ reagents, allowing all the samples to be pooled together for analysis; the identical masses and chemistry of the labels means that all samples perform identically during separation steps and the first stage of mass spectrometry. During the fragmentation step of mass spectrometry, the reporter groups of the labels come free. A spectrum of the fragmented peptide will contain ion peaks that correspond to the fragments of the peptide (allowing peptide identification) and ion peaks that correspond to the reporter ions (allowing quantitative comparison between samples)—see Figure 11.1.

Isobaric quantitation information from peptides is normally used to generate quantitative information for the parent peptides. iTRAQ quantitative information may, however, also be used to track the formation or modification of individual peptides (Grosvenor *et al.*, 2010, 2012).

peptide sample mixes

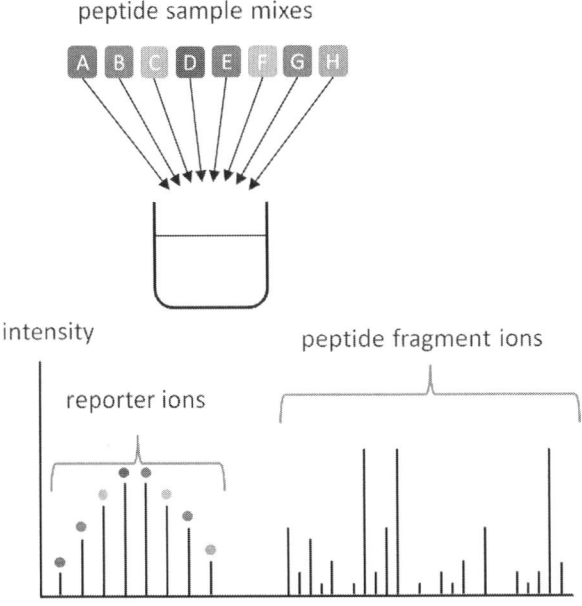

■ **FIGURE 11.1** A schematic of sample labeling and pooling, and a tandem (fragmentation) mass spectrum showing the ion peaks generated after fragmenting one peptide in a mixture containing eight isobaric-labeled samples. The mass/charge position of each reporter ion defines the sample identity, and the intensity defines the relative abundance of the particular peptide in each sample. In this example, the peptide would have been five times more abundant in samples D and E than in A. [A color version of this figure is available online at www.booksite.elsevier.com/9780124046108].

Case Study: Lactoferrin

This technique was applied to the study of the digestion of lactoferrin in a simulated stomach environment, with particular regard to the effects of milk pasteurization on lactoferrin digestion. Lactoferrin was chosen as a model because of its status as a high value dairy product. This iron-binding milk protein has attracted wide interest because of its antimicrobial (Van Der Kraan *et al.*, 2005; Murdock *et al.*, 2010) and other putative bioactivities (Lizzi *et al.*, 2009; Tomita *et al.*, 2009). Lactoferrin is affected at the subunit scale by pasteurization. However, until now, little has been known as to how lactoferrin breaks down into component peptides during digestion, and what influence prior processing has on this digestion.

Lactoferrin was purified (using cation exchange, filtration, and phenyl sepharose chromatography) from pasteurized (72°C for 15 sec) and non-pasteurized milk. A simple *in vitro* stomach digestion simulation was used to track the release of peptides from the parent proteins. Lactoferrin solutions (5 mg/ml) were added to 13.2 μg/ml pepsin, 150 mM NaCl/HCl (pH 2.0) at

37°C with stirring. Aliquots were removed at time points (0, 0.5, 2, 5, 10, 20, and 60 min) during the digestion and proteolysis was stopped by the addition of ice cold $NaHCO_3$ to 50 mM. All experiments were performed in triplicate.

Sample contaminants were removed by passage through C18 StageTips (Thermo Scientific), and the samples were reduced to dryness in a centrifugal concentrator. The digests were then reconstituted in 0.5 M triethylammonium bicarbonate. After combination with 100 μl isopropanol, iTRAQ labels were randomly assigned to time points and were added to the appropriate samples. Derivatization occurred as per the manufacturers' instructions. The labeled samples were pooled, evaporated to dryness, reconstituted in water, evaporated again, and reconstituted in LC-MALDI (liquid chromatography coupled to matrix-assisted laser desorption/ionization) buffer. Separation with a Proxeon Easy-nLC (Bruker, Bremen, Germany) utilized a trap column, then a gradient of 2% acetonitrile, 0.1% trifluoroacetic acid to 55% acetonitrile, 0.1% trifluoroacetic acid over 60 minutes across a 25 cm in-house packed C18 (Microsorb 300-5 media, Varian, Palo Alto, CA), 75 μm ID column. Sample was spotted onto an AnchorChip plate (Bruker) using a Proteineer fc fraction collector (Bruker). Automated tandem mass spectrometry was performed using an Ultraflex III MALDI-TOF/TOF mass spectrometer (Bruker, Bremen, Germany) in positive ion mode, with settings adjusted for stable isotope labeling.

The use of iTRAQ labels for each stage of digestion permitted the release of each peptide from lactoferrin to be tracked. Figure 11.2 demonstrates this: the peptide IAEKKADAVTLDGGM, located at position 68-82 in lactoferrin, was observed at eight time points during the simulated stomach digestion. Isobaric labeling allowed the abundance of specific peptides to be compared across samples (in this case, time points). Of the total signal attributed to IAEK-KADAVTLDGGM, about 8% of this was detected 30 seconds into digestion, and about 20% was detected 20 minutes into digestion. The abundance of the peptide falls during the digestion time course as it is digested into smaller lengths (see also Figure 11.3). The release can be visualized easily using a non-linear chart (Figure 11.2A) or in real time when rendered linearly (Figure 11.2B).

The release profiles of peptides can be compared: Figure 11.2 demonstrates that the smaller fragment, IAEKKADAVTL, was released later on during the digestion than its larger parent peptide. This type of comparison can show the relationship between protein digestion products and their peak times of bioavailability.

The application of sensitive qualitative and quantitative mass spectrometry to lactoferrin digestion enabled its characterization at a level of detail previously unobtainable. The release and subsequent degradation of each lactoferrin digestion peptide falling within the mass range scanned (700–3500 Da)

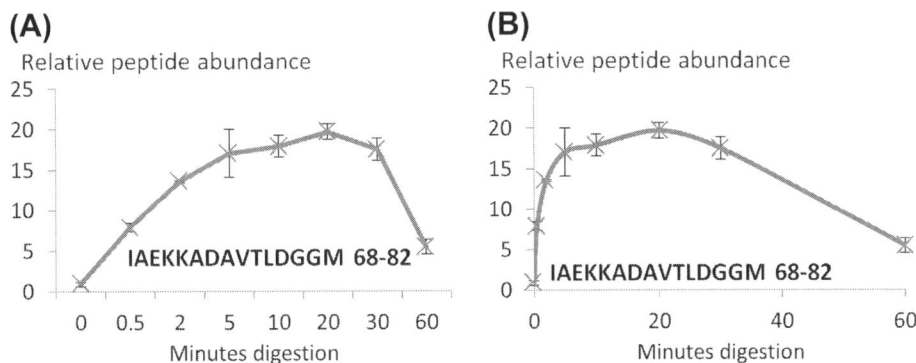

■ **FIGURE 11.2** Represented (A) non-linearly and (B) linearly, the abundance of peptide, IAEKKADAVTLDGGM, from position 68-82 in unpasteurized lactoferrin, as released during a simulated stomach digestion. Error bars represent SEM, where $n = 4$. [A color version of this figure is available online at www.booksite. elsevier.com/9780124046108].

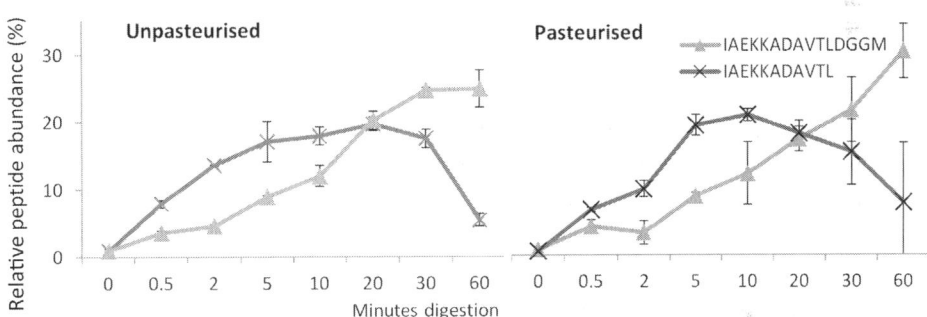

■ **FIGURE 11.3** Represented non-linearly, the abundance of IAEKKADAVTLDGGM (68-82) and its smaller fragment, IAEKKADAVTL (68-78), as released from un-pasteurized (left) or pasteurized (right) lactoferrin during a simulated stomach digestion. [A color version of this figure is available online at www.booksite. elsevier.com/9780124046108].

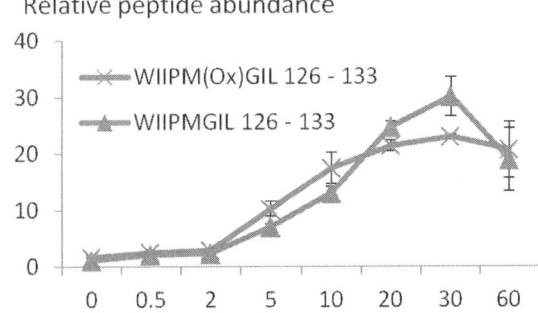

■ **FIGURE 11.4** The abundance of WIIPMGIL (126-133) and its oxidized equivalent, as released from unpasteurized lactoferrin during a simulated stomach digestion. [A color version of this figure is available online at www.booksite. elsevier.com/9780124046108].

was able to be captured in the intuitively understandable format represented in Figures 11.2 through 11.4.

FUTURE DIRECTIONS

These novel approaches and technologies offer the potential for accurate profiling and tracking of protein truncation modification and cascades within complex food systems, which traditionally has been very difficult due to the extreme complexity inherent in the breakdown process and products. They can be applied in conjunction with both *in vitro* (as exemplified here) and potentially *in vivo* digestion models. The qualitative and quantitative mass spectrometric approaches exemplified here have application to evaluating the digestive behavior of any food protein, determining and controlling the impact of food processing on nutrition and functionality, and developing new food delivery systems. This will also have important implications in understanding and controlling the release and uptake of potential allergens and bioactive components of protein foods.

REFERENCES

Adibi, S.A., Mercer, D.W., 1973. Protein digestion in human intestine as reflected in luminal, mucosal, and plasma amino acid concentrations after meals. J. Clin. Invest. 52, 1586–1594.

Alomirah, H.F., Alli, I., Konishi, Y., 2000. Applications of mass spectrometry to food proteins and peptides. J. Chromatogr. A. 893, 1–21.

Anderson, G.H., Aziz, A., 2006. Multifunctional roles of dietary proteins in the regulation of metabolism and food intake: application to feeding infants. J. Pediatr. 149, S74–S79.

Anson, N.M., Havenaar, R., Bast, A., Haenen, G.R.M.M., 2010. Antioxidant and anti-inflammatory capacity of bioaccessible compounds from wheat fractions after gastrointestinal digestion. J. Cereal Sci. 51, 110–114.

Baker, E.N., Baker, H.M., 2009. A structural framework for understanding the multifunctional character of lactoferrin. Biochimie 91, 3–10.

Baker, E.N., Baker, H.M., 2005. Molecular structure, binding properties and dynamics of lactoferrin. Cell. Mol. Life. Sci. 62, 2531–2539.

Blanquet, S., Zeijdner, E., Beyssac, E., et al., 2004. A dynamic artificial gastrointestinal system for studying the behavior of orally administered drug dosage forms under various physiological conditions. Pharm. Res. 21, 585–591.

Careri, M., Bianchi, F., Corradini, C., 2002. Recent advances in the application of mass spectrometry in food-related analysis. J. Chromatogr. A. 970, 3–64.

Chiang, C.-C., Croom, J., Chuang, S.-T., Chiou, P.W.S., Yu, B., 2008. Development of a dynamic system simulating pig gastric digestion. Asian–Australian J. Anim. Sci. 21, 1522–1528.

Clare, D.A., Swaisgood, H.E., 2000. Bioactive milk peptides: a prospectus. J. Dairy. Sci. 83, 1187–1195.

Dalle-Donne, I., Scaloni, A., Butterfield, D.A. (Eds.), 2006. Redox Proteomics: From Protein Modifications to Cellular Dysfunction and Diseases. John Wiley & Sons, Hoboken, New Jersey.

Darragh, A.J., Moughan, P.J., 1995. The three-week-old piglet as a model animal for studying protein digestion in human infants. J. Pediatr. Gastroenterol. Nutr. 21, 387—393.

Dearman, R.J., Caddick, H., Stone, S., Basketter, D.A., Kimber, I., 2001. Characterization of antibody responses induced in rodents by exposure to food proteins: influence of route of exposure. Toxicology 167, 217—231.

Dellow, D.W., Harris, P.M., Sinclair, B.R., 1988. Measurement of blood flow in the skin of sheep. Proc. Nutr. Soc. NZ 13, 130—133.

Dupont, D., Mandalari, G., Molle, D., et al., 2010. Comparative resistance of food proteins to adult and infant in vitro digestion models. Mol. Nutr. Food Res. 54, 767—780.

Dyer, J., Bell, F., Koehn, H., Vernon, J., Cornellison, C., Clerens, S., Harland, D., 2013. Redox Proteomic Evaluation of Bleaching and Alkali Damage in Human Hair. Int. J. Cosmetic Sci. 35 (6), 555—561.

Dyer, J.M., Bringans, S.D., Bryson, W.G., 2006. Determination of photo-oxidation products within photoyellowed bleached wool proteins. Photochem. Photobiol. 82, 551—557.

Dyer, J.M., Plowman, J., Krsinic, G., et al., 2010. Proteomic evaluation and location of UVB-induced photomodification in wool. Photochem. Photobiol. B. Biol. 98, 118—127.

Eiwegger, T., Rigby, N., Mondoulet, L., et al., 2006. Gastro-duodenal digestion products of the major peanut allergen Ara h 1 retain an allergenic potential. Clin. Exp. Allergy. 36, 1281—1288.

Erickson, R.H., Kim, Y.S., 1990. Digestion and absorption of dietary protein. Annu. Rev. Med. 41, 133—139.

Ferrua, M.J., Kong, F., Singh, R.P., 2011. Computational modeling of gastric digestion and the role of food material properties. Trends Food Sci. Technol. 22, 480—491.

Gerrard, J.A., 2002. Protein—protein crosslinking in food: methods, consequences, applications. Trends Food Sci. Technol. 13, 391—399.

Grosvenor, A.J., Morton, J.D., Dyer, J.M., 2012. Determination and validation of markers for heat-induced damage in wool proteins. Am. J. Anal. Chem. 3, 431—436.

Grosvenor, A.J., Morton, J.D., Dyer, J.M., 2010. Isobaric labelling approach to the tracking and relative quantitation of peptide damage at the primary structural level. J. Agric. Food. Chem. 58, 12672—12677.

Grosvenor, A.J., Morton, J.D., Dyer, J.M., 2011. Proteomic characterisation of hydrothermal redox damage. J. Sci. Food. Agric. 91, 2806—2813.

Hau, J., Bovetto, L., 2001. Characterisation of modified whey protein in milk ingredients by liquid chromatography coupled to electrospray ionisation mass spectrometry. J. Chromatogr. A. 926, 105—112.

Igarashi, N., Onoue, S., Tsuda, Y., 2007. Photoreactivity of amino acids: tryptophan-induced photochemical events *via* reactive oxygen species generation. Anal. Sci. 23, 943—948.

Kaur, L., Rutherfurd, S.M., Moughan, P.J., Drummond, L., Boland, M.J., 2010. Actinidin enhances protein digestion in the small intestine as assessed using an in vitro digestion model. J. Agric. Food Chem. 58, 5074—5080.

Kong, F., Singh, R.P., 2010. A human gastric simulator (HGS) to study food digestion in human stomach. J. Food. Sci. 75, E627—E635.

Kong, F., Singh, R.P., 2011. Solid loss of carrots during simulated gastric digestion. Food Biophys. 6, 84—93.

Krishnamoorthy, U., Sniffen, C.J., Stern, M.D., Van Soest, P.J., 1983. Evaluation of a mathematical model of rumen digestion and an *in vitro* simulation of rumen proteolysis to estimate the rumen-undegraded nitrogen content of feedstuffs. Br. J. Nutr. 50, 555—568.

Kunz, C., Rodriguez-Palmero, M., Koletzko, B., Jensen, R., 1999. Nutritional and biochemical properties of human milk, Part I: General aspects, proteins, and carbohydrates. Clin. Perinatol. 26, 307—333.

Léonil, J., Gagnaire, V., Mollé, D., Pezennec, S., Bouhallab, S.D., 2000. Application of chromatography and mass spectrometry to the characterization of food proteins and derived peptides. J. Chromatogr. A. 881, 1—21.

Lizzi, A.R., Carnicelli, V., Clarkson, M.M., Di Giulio, A., Oratore, A., 2009. Lactoferrin derived peptides: Mechanisms of action and their perspectives as antimicrobial and antitumoral agents. Mini-Rev. Med. Chem. 9, 687—695.

Maldonado-Valderrama, J., Wilde, P.J., Mulholland, F., Morris, V.J., 2012. Protein unfolding at fluid interfaces and its effect on proteolysis in the stomach. Soft Matter 8, 4402—4414.

Matsubara, T., Aoki, N., Honjoh, T., et al., 2008. Absorption, migration and kinetics in peripheral blood of orally administered ovalbumin in a mouse model. Biosci. Biotechnol. Biochem. 72, 2555—2565.

Minekus, M., Marteau, P., Havenaar, R., Huis in 't Veld, J.H.H., 1995. A multicompartmental dynamic computer-controlled model simulating the stomach and small intestine. Altern. Lab. Anim. 23, 197—209.

Moreno, F.J., Quintanilla-López, J.E., Lebrón-Aguilar, R., Olano, A., Sanz, M.L., 2008. Mass spectrometric characterization of glycated β-lactoglobulin peptides derived from galacto-oligosaccharides surviving the in vitro gastrointestinal digestion. J. Am. Soc. Mass Spec. 19, 927—937.

Murdock, C., Chikindas, M.L., Matthews, K.R., 2010. The pepsin hydrolysate of bovine lactoferrin causes a collapse of the membrane potential in Escherichia coli O157: H7. Probiot. Antimicr. Proteins 2, 112—119.

Oomen, A.G., Hack, A., Minekus, M., et al., 2002. Comparison of five *in vitro* digestion models to study the bioaccessibility of soil contaminants. Environ. Sci. Technol. 36, 3326—3334

Picariello, G., Ferranti, P., Fierro, O., et al., 2010. Peptides surviving the simulated gastrointestinal digestion of milk proteins: Biological and toxicological implications. J. Chromatogr. B. 878, 295—308.

Rivest, J., Bernier, J.F., Pomar, C., 2000. A dynamic model of protein digestion in the small intestine of pigs. J. Anim. Sci. 78, 328—340.

Struthers, B., 1981. Lysinoalanine: production, significance and control in preparation and use of soya and other food proteins. J. Am. Oil. Chem. Soc. 58, 501—503.

Syrbe, A., Bauer, W.J., Klostermeyer, H., 1998. Polymer science concepts in dairy systems—an overview of milk protein and food hydrocolloid interaction. Int. Dairy J. 8, 179—193.

Tharakan, A., Norton, I.T., Fryer, P.J., Bakalis, S., 2010. Mass transfer and nutrient absorption in a simulated model of small intestine. J. Food Sci. 75, E339—E346.

Thomas, S.N., Lu, B.-W., Nikolskaya, T., Nikolsky, Y., Yang, A.J., 2006. MudPIT (multidimensional protein identification technology) for identification of post-translational protein modifications in complex biological mixtures. In: Dalle-Donne, I., Scaloni, A., Butterfield, D.A. (Eds.), Redox Proteomics—From Protein Modifications to Cellular Dysfunctions and Diseases, pp. 233—252. Hoboken, John Wiley and Sons, Inc.

Tomita, M., Wakabayashi, H., Shin, K., Yamauchi, K., Yaeshima, T., Iwatsuki, K., 2009. Twenty-five years of research on bovine lactoferrin applications. Biochim. clin. 91, 52—57.

Tornberg, E., 2005. Effects of heat on meat proteins—implications on structure and quality of meat products. Meat Sci. 70, 493—508.

Tydeman, E.A., Parker, M.L., Faulks, R.M., et al., 2010. Effect of carrot (Daucus carota) microstructure on carotene bioaccessibility in the upper gastrointestinal tract. 2. In vivo digestions. J. Agric. Food Chem. 58, 9855—9860.

Van Der Kraan, M.I.A., Van Marle, J., Nazmi, K., et al., 2005. Ultrastructural effects of antimicrobial peptides from bovine lactoferrin on the membranes of Candida albicans and Escherichia coli. Peptides 26, 1537—1542.

Wickham, M., Faulks, R., Mills, C., 2009. In vitro digestion methods for assessing the effect of food structure on allergen breakdown. Mol. Nutr. Food Res. 53, 952—958.

Chapter

12

Dynamics of Gastric Contents During Digestion—Computational and Rheological Considerations

M.J. Ferrua,[1] Z. Xue[2] and R. Paul Singh[1,2]

[1]*Riddet Institute, Massey University, Palmerston North, New Zealand,* [2]*Department of Biological and Agricultural Engineering, University of California, Davis, CA, USA*

CONTENTS

Food Structures, Digestion and Health. http://dx.doi.org/10.1016/B978-0-12-404610-8.00012-8

319

INTRODUCTION

Functional foods are foods specifically designed to promote optimal human health and well-being by providing specific benefits beyond basic nutrition and energy needs (IFT, 2005). Health-conscious consumers are driving the growth of the global functional food market at a dramatic pace. Several sources have estimated that the global functional foods market has a value of US$33bn or more (Siró *et al.*, 2008; Granato *et al.*, 2010; Leatherhead Food Research, 2011). Since the 1990s, many leading food companies have invested and expanded their products in the functional foods sector. However, it is increasingly recognized that to further assist the development of these novel foods aim at improving the bioavailability of nutrients and bioactive compounds within the body, it is essential first to understand how food components and structures are transformed and absorbed during digestion.

The human digestive system begins with the mouth, where foods are chewed and mixed with salivary enzymes to form a cohesive mass commonly known as food bolus. During swallowing, food bolus is propelled into the esophagus and transported by the action of a progressive peristaltic wave into the stomach. As the wave sweeps along the esophagus, its lower sphincter relaxes to allow the boluses to pass into the stomach and closes immediately after that to prevent gastric contents from flowing back into the esophagus (Sherwood, 2010). Once in the stomach, foods are further digested by a series of chemical and mechanical effects, triggered and modulated by the secretionary and motor activities of the gastric wall. During gastric digestion, the ingested bolus is continuously transformed into a semisolid mixture of partially digested food and gastric fluids (called chyme), which is then slowly emptied into the small intestine. In the small intestine, the chyme is transported by a series of peristaltic, segmental, and longitudinal contractions of the intestinal wall, which also promote its continuous mixing with intestinal juices and its repeated contact with the absorptive surface of the wall (Seerden *et al.*, 2005). It is within this organ that the final digestion and subsequent absorption of nutrients occurs. As a result of this process, the digesta becomes increasingly less fluid until it finally reaches the large intestine. Once in the large intestine, most of the remaining water and minerals are absorbed, and a compact mixture of indigestible matter is obtained and excreted.

Despite the number of different processes to which foods are exposed upon ingestion, the conditioning and disintegration experienced during gastric digestion is becoming increasingly recognized to play a major role in the delivery of optimal nutrition (Wickham *et al.*, 2012). As a result, understanding the effect of gastric digestion on the final bioavailability of nutrients and pharmaceutical drugs within the body is an area of active research. In particular, the use of increasingly sophisticated *in vitro* models (aimed at reproducing the chemical and mechanical conditions that develop during the process) has found application as tools to predict the *in vivo* performance of different products during development.

These *in vitro* models, along with human and animal trials, are necessary for a good understanding of the link between the material properties of the food and *in vivo* performance during digestion. This knowledge is essential to assist the design of novel products of improved functionalities, and largely requires obtaining a good understanding of the local physicochemical conditions to which foods are exposed during digestion.

The goal of this chapter is to provide a brief overview of the stomach function during digestion, and attempts to understand the fluid mechanical conditions to which foods are exposed during the process by using computational and numerical tools.

GASTRIC FUNCTIONS

The stomach can be described as a J-shaped hollow organ, capable of expanding to receive and store a meal for a limited duration. Based on anatomical, histological, and functional differences, it can be divided into three sections: fundus, body, and antrum (Figure 12.1). Known to play a crucial role in the human digestive process, the stomach not only stores the ingested meal, inhibits the growth of microbial, promotes its chemical and mechanical disintegration, but also releases the digesta into the duodenum in a highly coordinated and controlled manner (Mayer *et al.*, 1994). These gastric functions are ultimately controlled by a complex and self-regulated interplay that occurs between the physicochemical properties of the ingested meal and the secretionary and motor response of the gastric wall.

As an immediate response to food ingestion, a reduction in gastric tone and an increase in the compliance of the proximal wall allow the stomach to receive and store the food intake without any significant increase in the

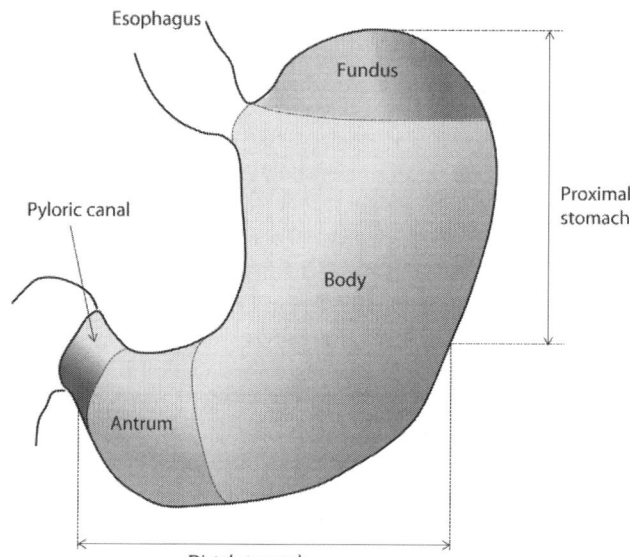

■ **FIGURE 12.1** Schematic diagram of stomach sections *(from Ferrua and Singh, 2010)*. [A color version of this figure is available online at www.booksite.elsevier.com/9780124046108].

gastric pressure, a process commonly known as "receptive relaxation" (Cannon and Lieb, 1911). This first response is then followed, and maintained, by a second response known as "adaptive relaxation" that continuously modulates the tone of the wall in response to the specific properties of the gastric contents (Jahnberg *et al.*, 1975). Together, these two responses are known as "gastric accommodation" and are suspected not only to facilitate the receptive and storage function of the stomach during digestion, but also to impact the distribution and emptying of gastric contents during the process (Schwizer *et al.*, 2002).

During its residence in the stomach, the swallowed meal is further disintegrated by a combined effect of chemical and mechanical forces. Gastric juices (consisting of HCl, digestive enzymes such as pepsinogen, HCO_3^-, mucin, salts, and intrinsic factor) are secreted by the gastric wall to promote the enzymatic splitting and chemical disintegration of food components and structures. Simultaneously, a series of periodic peristaltic antral contraction waves (ACWs) that continuously propagate along the distal region of the stomach are initiated. These ACWs begin as shallow indentations at the site of the gastric pacemaker, and their amplitude and velocity continuously increase as they propagate down towards the pylorus (O'Grady *et al.*, 2010). It is generally accepted that the mechanical forces and fluid motions generated by the propagation of these waves are the primary factors in promoting the mixing and breakdown of gastric contents during digestion (Schulze, 2006; Schwizer *et al.*, 2006). During gastric digestion, the pyloric sphincter relaxes and contracts periodically to facilitate the discharge of chyme into the small intestine (Dillard *et al.*, 2007). While it was determined earlier that there is a positive correlation between the emptying of solids and antral motility, the effect of ACWs on the gastric emptying of liquids is still not clear (Camilleri *et al.*, 1985).

In Vivo/Ex Vivo Assessment of Gastric Wall Functions

In general, while the composition and secretion of gastric juices are relatively well understood (Barret and Raybould, 2010; McClements and Li, 2010), the motor activity of the gastric wall has only been recently assessed by using new and advanced imaging technologies.

The use of magnetic resonance imaging (MRI) and single photon emission computed tomography (SPECT) has facilitated the analysis of gastric accommodation and emptying patterns in terms of gastric volume changes during digestion (Schwizer *et al.*, 2002; van den Elzen *et al.*, 2003; Kwiatek and Steingoetter, 2006; Goetze *et al.*, 2007). Spatiotemporal mapping results have also illustrated the significant role of the proximal region of the stomach in the "receptive relaxation" experienced by an *ex vivo* rat stomach when different fluids were perfused inside (Lentle *et al.*, 2010). In the same study, the authors also confirmed that the receptive relaxation of the fundus is the result of a tonal response of the wall (as all parts of the wall expanded simultaneously).

Similarly, MRI techniques have also been used to facilitate the *in vivo* analysis of the dynamics of the ACWs (Pal *et al.*, 2004, 2007; Steingoetter *et al.*, 2005; Kwiatek *et al.*, 2006; Treier *et al.*, 2006). In general, the waves have been found to have a frequency of 2.6 to 3.2 min^{-1}, travel at a horizontal velocity of 2.2 to 3.3 mm.s^{-1}, and die at about 1.2 cm from the pyloric sphincter. The maximum level of occlusion imposed by the waves varies between 60 and 90%, while their width ranges from 1.2 to 1.8 cm.

GASTRIC FLOW DYNAMICS—BRIDGING THE GAP BETWEEN DESIGN AND FUNCTIONAL BENEFITS

Despite the complex interaction of chemical and mechanical effects that drives the gastric digestion, a growing body of scientific evidence suggests that the fluid dynamics conditions that develop in the stomach play a major role on the final digestibility and metabolic activity of the meal (Marciani *et al.*, 2000; Dikeman *et al.*, 2006; Lentle and Janssen, 2010).

In particular, the local dynamics of gastric contents are expected to determine not only the mechanical forces that grind and rub food particles during digestion, but also the efficiency with which the ingested meal is mixed with gastric juices during the process. Unfortunately, the complex nature of the stomach motor activity and physics of gastric contents have so far prevented a good experimental assessment of their dynamics. To address this problem, the use of computational fluid dynamics (CFD) techniques has been recently identified as an alternative tool to investigate the dynamics of gastric flows under different physical and physiological conditions (Schulze, 2006).

Numerical Analysis of Gastric Flows

The first study that followed a numerical approach was carried out by Pal *et al.* (2004). Based on *in vivo* MRI data, the authors constructed a simplified two-dimensional (2D) model of the stomach geometry and motility during digestion. The model was then used to analyze how the amplitude and width of the ACWs, the rate of gastric emptying, and the relaxing period of the pyloric sphincter can affect the gastric behavior of a Newtonian fluid. Two major flow patterns were identified: a strong retropulsive jet-like motion that develops during the terminal propagation of the ACWs, and slow recirculation eddies in between successive waves. These flow patterns were found to be affected only by the geometrical characteristics of the ACWs (being enhanced by more occluding and narrower waves), and generate a well-defined mixing zone within the distal region of the stomach. Retropulsive velocities of up to $7 \, \mathrm{mm.s}^{-1}$ were predicted. In 2007, the authors extended this work and investigated the emptying pattern of a 10% glucose solution during the first 10 minutes of digestion (Pal *et al.*, 2007). Their results challenged the classical description of a plug emptying from the distal stomach, and suggested that the stomach can also empty along a narrow and long path close to its inner curvature, drawing gastric contents from the proximal stomach directly into the duodenum without too much mixing. However, up to now no validation of these results has been reported.

By using a simplified 2D model of the geometry and motility of the distal stomach, Kozu *et al.* (2010) numerically investigated the effect of fluid viscosity on the behavior of gastric flow and the extent of gastric mixing of a digestive enzyme secreted from the stomach wall (pepsinogen). Their results again showed the formation of retropulsive motions and eddy structures within the stomach, reporting slightly larger retropulsive velocities (of up to $12 \, \mathrm{mm.s}^{-1}$). In addition, they showed that the viscosity of the gastric fluid has little effect on the magnitude of highest retropulsive velocities that develop within the stomach, but it greatly affects the extent of these rapid retropulsive motions from the entire distal region to the location of the ter-minal ACW peak, as well as the extent of mixing experienced by the enzyme within the stomach.

It was Singh (2007) who created the first three-dimensional (3D) model of the stomach geometry and motility during digestion. The model was used to numerically investigate the effect of the occlusion and propagation speed of the ACWs on the dynamics of different Newtonian fluids. Similar to pre-vious studies, this work also indicated the formation of retropulsive veloc-ities and eddy structures, whose strength was significantly affected by the speed and occlusion of the ACWs, as well as the viscosity of the fluid. How-ever, the velocities predicted in this study were significantly higher (reaching values of up to 20 mm.s^{-1}) than those reported previously.

In another study, Imai *et al.* (2013) investigated the dynamics of antral recir-culation in gastric mixing using a 3D model of a human stomach. Using a free surface model, the authors also found that regardless of the posture and volume of gastric content, only the contents located in antral recircula-tion were well mixed. Their results showed that antral recirculation and the retropulsive flow motions generated by the peristaltic contractions of the distal wall are two major mechanisms promoting gastric mixing.

Following a numerical approach, our work at the Riddet Institute and Uni-versity of California, Davis, is aimed at developing a better understanding of the dynamics of gastric digestion by using computational fluid dynamic techniques (ANSYS Fluent$^{™}$). Based on this objective, a 3D computational model of the gastric geometry and motility during digestion was developed and used to analyze the behavior of gastric contents of different physical and rheological properties. In particular, we investigated the dynamics of gastric fluids with different rheological properties, as well as their effect on the distribution and fate of discrete food particles during digestion. In addition, preliminary work has been done to investigate the dynamics of more complex and realistic food digesta systems, characterized by a more densely packed mixture of solid food particles and digestive juices. In the following, a brief overview of the methodologies and findings achieved by these research efforts are presented.

COMPUTATIONAL MODEL OF A HUMAN STOMACH
Gastric Geometry

There is no unique description of the size and geometry of the human stomach. Its anatomy not only varies among different individuals, but it is also significantly affected by a number of physiological factors that generally include body posture, conditions of surrounding viscera, composition and amount of the ingested meal, and stage of digestive processes (Schulze, 2006).

In the absence of a unique description of the gastric anatomy, a computational model was developed based on the range of sizes reported in literature for an adult human stomach after a typical meal (Keet, 1993; Liao *et al.*, 2004; Schulze, 2006; Einhorn, 2009). As shown in Figure 12.2, the developed model depicts the stomach as a J-shaped hollow container, with a greater curvature of 34 cm, a lesser curvature of 15 cm, and a capacity of up to 0.9 L. The pyloric canal was 1.2 cm in diameter, while the widest transversal section was 10 cm in diameter and was located within the proximal region.

The geometric model was constructed by sweeping a series of circular rings of varying diameters with their centers located on a predefined centerline. The computational model was created using the preprocessing package Gambit. Since the geometric shape is symmetric with respect to the plane

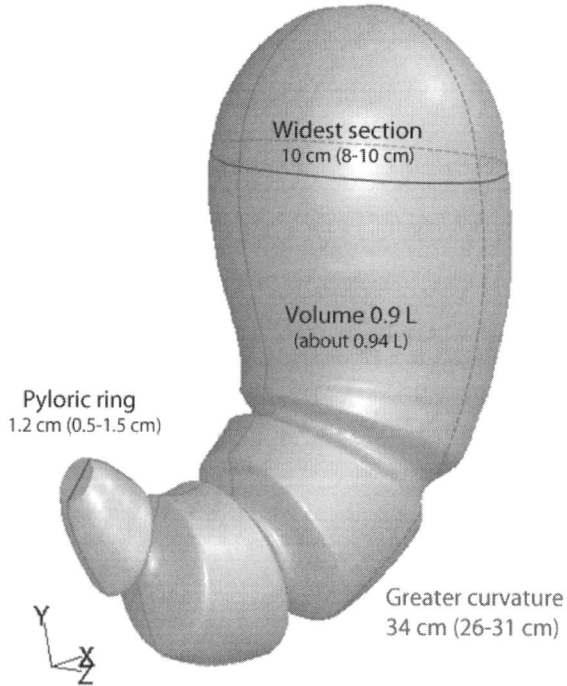

■ **FIGURE 12.2** Three-dimensional computational model of an average human stomach *(from Ferrua and Singh, 2011)*. [A color version of this figure is available online at www.booksite.elsevier.com/ 9780124046108].

that bisects the stomach along its lesser and greater curvatures, the computational domain was simplified to only one-half of the 3D geometry. Details regarding the parameters of the geometric model, its construction, and mesh scheme can be found in Ferrua and Singh (2010).

Gastric Motility

As previously discussed in "Gastric Functions" above, the motility of the gastric wall can be described in terms of two main muscular activities: the periodic propagation of antral contraction waves (ACWs) along the distal stomach, and the tonic contractions/expansions of the proximal region.

The dynamic activity of the ACWs was modeled based on *in vivo* MRI data previously reported and summarized in "Gastric Functions". In particular, the waves were initiated every 20 s at 15 cm from the pylorus and propagated down towards the pylorus with a constant horizontal velocity of 2.3 mm.s^{-1} (Figure 12.3). Their width was assumed to be 2.0 cm along

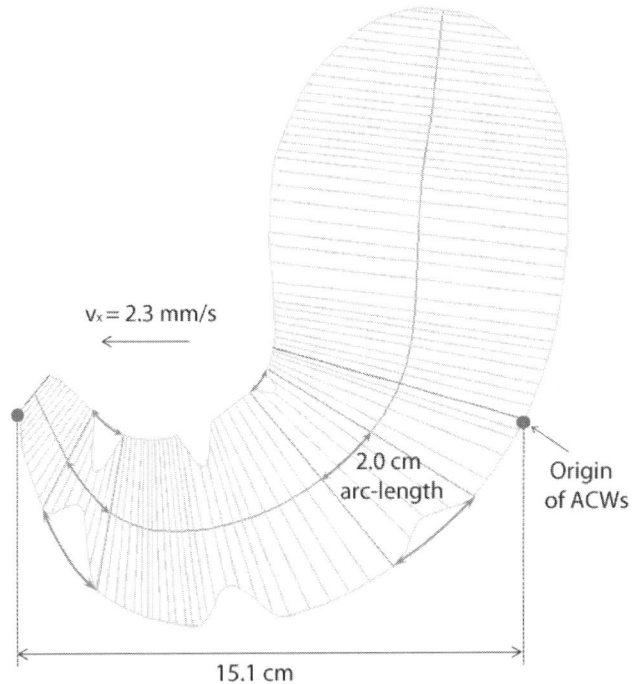

■ **FIGURE 12.3** Numerical characterization of the dynamics of the antral-contraction waves *(from Ferrua and Singh, 2011).* [A color version of this figure is available online at www.booksite.elsevier.com/9780124046108].

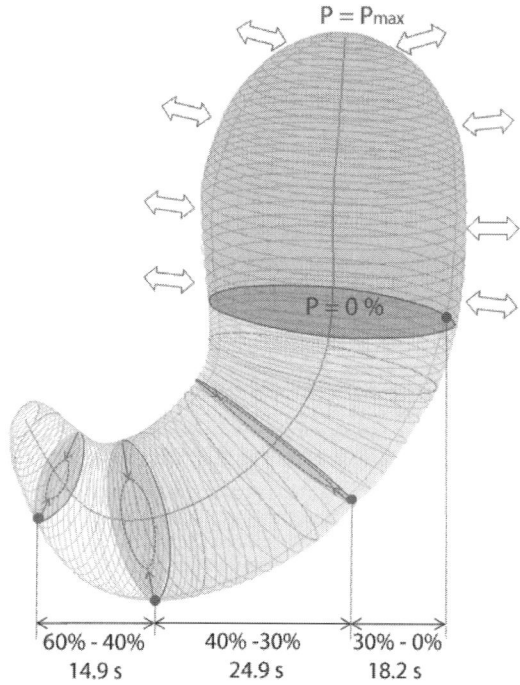

■ **FIGURE 12.3** cont'd.

the stomach centerline. Their amplitude was continuously increased as they traveled down, leading to a maximum occlusion of the luminal cavity of 60% immediately before disappearing at about 1.5 cm from the pylorus. The lifespan of each individual wave was 60 s, but their propagation pattern repeat itself every 20 s (i.e., period of overall ACW pattern $T_{ACW} = 20$ s). Gastric emptying was simulated by numerically opening the pyloric boundary during the first 14.7 s of each 20 s period (Dillard et al., 2007).

A series of tonic contractions/expansions in the fundus region was implemented to compensate for the temporal variations that occur in the stomach's volume as a result of the propagation of the ACWs and gastric emptying activities considered. These contractions/expansions were assumed to deform the fundus wall circumferentially, with percentages of contraction/expansion increasing or decreasing linearly from 0% (at the mid-corpus) to a time-dependent value at the top of the fundus.

NUMERICAL ANALYSIS OF GASTRIC FLOWS

To investigate the role of fluid rheology on the dynamics of gastric flows during digestion, the computational model was used to numerically analyze the flow behavior of two Newtonian fluids (with viscosities of 10^{-3} and 1 Pa.s) and a non-Newtonian shear-thinning fluid, as discussed by Ferrua *et al.* (2011). Examples of these fluids are water, honey, and a 5.8% TS (total solids) tomato juice, respectively.

Modeling the Dynamics of Gastric Flows (Flow Model)

The incompressible and laminar behavior of gastric fluids was modeled using the continuity and Cauchy momentum equations given by Eqs. 12.1 and 12.2, respectively:

$$\nabla \cdot \underline{u} = 0 \tag{12.1}$$

$$\rho \left(\frac{\partial \underline{u}}{\partial t} + \underline{u} \cdot \nabla \underline{u} \right) = -\nabla p + \nabla \underline{\underline{\tau}} + \rho \underline{g} \tag{12.2}$$

where, $\underline{u}$ is the velocity vector, ρ is the density, $\underline{\underline{\tau}}$ is the viscous stress tensor, p is the pressure and $\underline{g}$ is the body force vector. Depending on the rheological behavior of the gastric fluids, different expressions needed to be used to describe the viscous stresses developed within the stomach in terms of the dynamics of the flow.

Newtonian Fluids

In the case of an incompressible Newtonian fluid, viscous stresses are proportional to the rate of strain $(\underline{\underline{\Gamma}})$, with the constant of proportionality being the viscosity of the fluid (μ):

$$\underline{\underline{\tau}} = 2\mu \underline{\underline{\Gamma}} = \mu \left[\nabla \underline{u} + (\nabla \underline{u})^T \right] \tag{12.3}$$

By substituting this expression into Eq. 12.2, the momentum balance can then be written as the following:

$$\rho \left(\frac{\partial \underline{u}}{\partial t} + \underline{u} \cdot \nabla \underline{u} \right) = -\nabla p + \mu \Delta \underline{u} + \rho \underline{g} \tag{12.4}$$

In particular, the dynamics of the two Newtonian fluids were modeled by numerically solving the mass and momentum balances given by Eqs. 12.1 and 12.4.

Non-Newtonian Fluids

For some non-Newtonian fluids, viscous stresses can still be related to the rate of strain, but now in terms of an apparent viscosity (η), which itself also depends on the rate of strain of the fluid:

$$\underline{\underline{\tau}} = \eta(\underline{\underline{\Gamma}}) \, \underline{\underline{\Gamma}} \tag{12.5}$$

In the case of the non-Newtonian fluid investigated as part of this study (i.e., 5.8% TS tomato juice), its apparent viscosity can be directly related to just the magnitude of the rate of strain tensor (commonly known as the shear rate Γ given in Eq. 12.6):

$$\Gamma = \left[\frac{1}{2} (\underline{\underline{\Gamma}} : \underline{\underline{\Gamma}}) \right]^{1/2} \tag{12.6}$$

As a consequence, the apparent viscosity of tomato juice is usually characterized by means of the power law model given by Eq. 12.7:

$$\eta(\Gamma) = k\Gamma^{n-1} \tag{12.7}$$

where k is a measure of the average viscosity of the fluid (known as the consistency index) and n is the power law index whose value defines how the viscosity of the fluid changes with the shear rate. In particular:

$n = 1 \rightarrow$ Newtonian fluid
$n > 1 \rightarrow$ shear-thickening (dilatant fluids)
$n < 1 \rightarrow$ shear-thinning (pseudo-plastics)

The apparent viscosity of the tomato juice (5.8% TS) was modeled by a consistency index of $0.233 \, \mathrm{Pa.s}^n$ and a power law index of 0.59 (Steffe, 1996). Its dynamics were then simulated by numerically solving the system of equations given by Eqs. 12.2, 12.4, 12.5, and 12.7.

Boundary Conditions

Due to the lack of scientifically-based information regarding the flow conditions at the gastric walls, no-slip and no-penetration boundary conditions were imposed at all walls. A symmetry boundary condition was applied at the plane that dissects the stomach along its greater and lesser curvatures.

In the case of modeling gastric emptying, the boundary at the end of the pyloric canal was periodically open during the first 14.7 s of each 20 s period of ACW activities to allow 35% of the initial amount of gastric contents to be emptied from the stomach within 15 minutes of digestion (Pal *et al.*, 2004).

Numerical Solution

The complete numerical model was solved using the commercial CFD package ANSYS Fluent™. The total computational domain consisted of about 1 million elements, which were deformed at every time step according to the motility patterns previously described in the Gastric Motility section. Due to the periodic nature of the gastric motility, the flow field in the stomach becomes periodic after 48 to 80 s (depending on the fluid rheology) and repeats itself every 20 s (T_{ACW}). More details regarding the numerical scheme and convergence criteria used in the model can be found in Ferrua and Singh (2010).

Overall Flow Behavior and Gastric Functions (Numerical Results)

The numerical results illustrated not only the complex dynamics of gastric fluids, but also the formation of two distinct flow regions of widely different time scales (Figure 12.4). Regardless of the rheology of the fluid, the fastest and more irregular flow motions developed within the distal region, while the dynamics of the proximal stomach were generally characterized by much slower and regular motions that extended from the top of the fundus down to the lower body of the stomach.

In good agreement with the classical description of gastric functions, these overall flow features support the role of the proximal stomach as the region where the ingested meal is received, gently mixed with digestive juices, and slowly transported into the distal stomach. Once there, the stronger mechanical forces and fluid motions that develop in the distal region will further promote the mechanical disintegration of the digesta before being emptied into the small intestine.

While these overall dynamics of gastric flows were relatively independent of its rheological behavior, the local flow patterns and relative velocities developed within the stomach were not. Similar to previous CFD studies, the results showed that an increase in the viscosity of gastric contents reduces the strength of the fluid motions that develop during digestion. However, unlike previous studies, the results also showed that its effect on the local flow patterns and velocity magnitudes that develop within the proximal and distal regions was different.

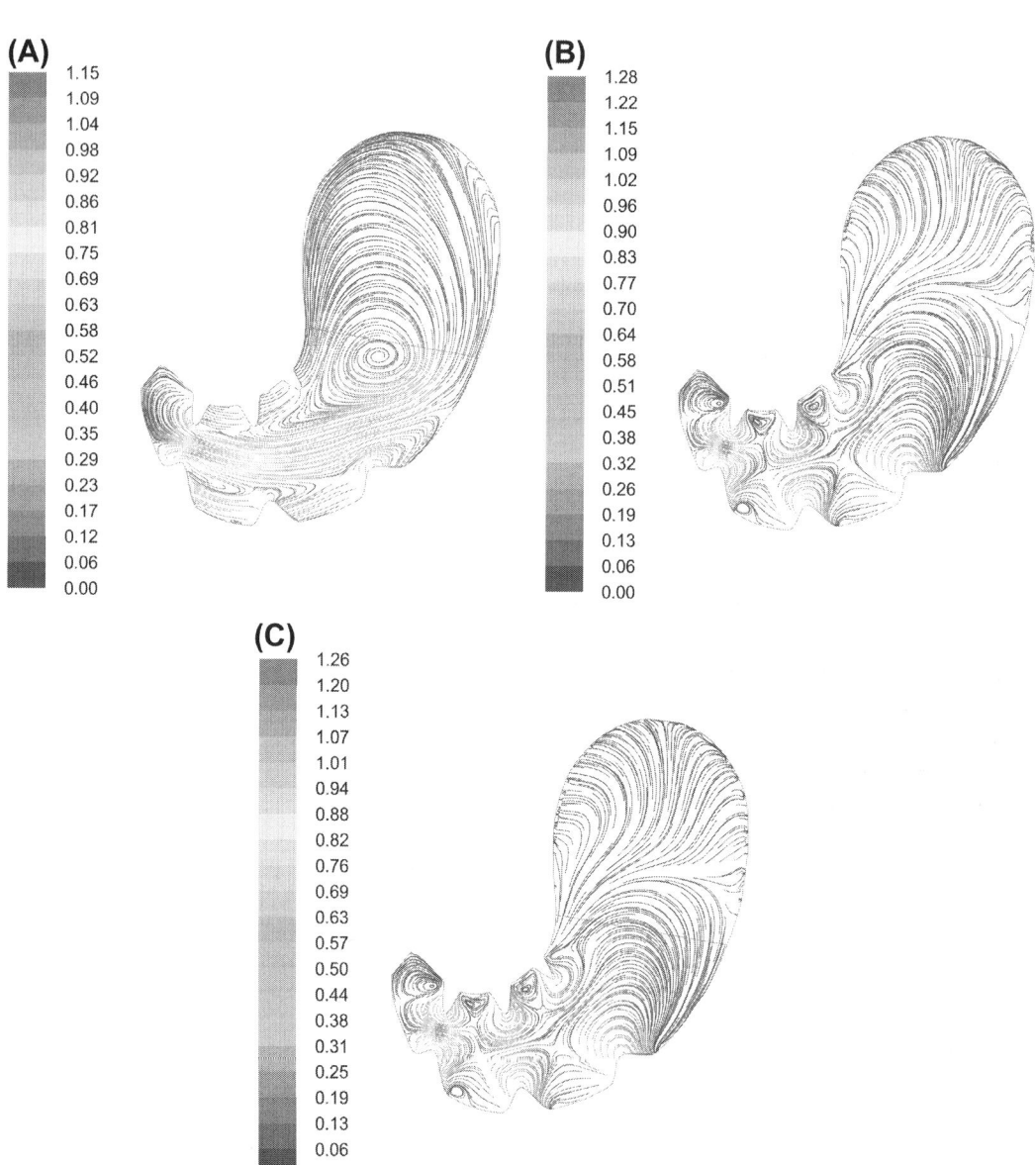

■ **FIGURE 12.4** Instantaneous streamlines at 57% of luminal occlusion (15 s out the 20 s period of the ACW dynamics), colored by velocity magnitude (cm/s). (A) Water (Newtonian 1×10^{-3} Pa.s). (B) Honey-like fluid (Newtonian 1 Pa.s). (C) Tomato juice (0.02 to 0.17 Pa.s). [A color version of this figure is available online at www.booksite.elsevier.com/9780124046108].

While the fluid viscosity did not significantly affect the development of regular flow motions in the proximal stomach, it did affect the local flow features generated within the distal region. As illustrated in Figure 12.4, in the case of a water-like fluid the propagation of ACWs led to the formation of rapid retropulsive motions that extended along the entire distal region back into the proximal stomach, and large circular motions (eddies) that extended across the region from the gastric wall deep into the luminal core. While these flow features have been largely associated with the mechanical disintegration of foods during digestion (Schulze, 2006), the results showed that their development is largely constrained to very thin, water-like, fluids. As illustrated in Figure 12.4, the simple addition of just a small amount of simple polysaccharides to water (as it is the case with the tomato juice) were enough to prevent the formation of large retropulsive motions, and to constrain eddies to regions closer to the wall. It is noteworthy that the same effects were observed regardless of the gastric emptying conditions being analyzed.

Conversely, the effect of the fluid viscosity on the velocity magnitude of the flow within the proximal region was 1.5 to 3 times larger than in the distal region (depending on whether gastric emptying was considered). As reported in Table 12.1, a 1000-fold increase in fluid viscosity led to a decrease of about 70% in the average velocity of proximal flows (regardless of gastric emptying activities), while the average velocity of the distal flows decreased by only 46% in the absence of gastric emptying, and by 26% in the presence of it.

By illustrating the detrimental role that gastric viscosity has on the flow features commonly regarded to drive the structural disintegration of foods within the distal stomach, as well as on the time scale of the proximal flow motions responsible for mixing and transport the ingested meal towards that region, these results provided new insights into the mechanisms through which meal viscosity can affect affect and delay gastric digestion (as reported *in vivo* by Marciani *et al.*, 2001).

Table 12.1 Average of the Mean Velocities Developed in the Distal and Proximal Region of the Stomach during the 20 s Periodic Period of the ACW Activity

Fluid	No Gastric Emptying		Gastric Emptying	
	V_{distal} (mm/s)	$V_{proximal}$ (mm/s)	V_{distal} (mm/s)	$V_{proximal}$ (mm/s)
Water (10^{-3} Pa.s)	1.48	0.23	1.07	0.18
Tomato juice (0.02–0.17 Pa.s)	0.81 (↓ 45.4 %)	0.07 (↓ 71.5 %)	0.80 (↓ 25.4 %)	4.7E-3 (↓ 71.9 %)
Honey (1 Pa.s)	0.80 (↓ 46.7 %)	0.06 (↓72.6 %)	0.80 (↓ 25.6 %)	5.0E-3 (↓ 73.6 %)

Experimental Insight into the Performance of the Model

Due to economical, ethical, and practical concerns, an actual validation of the developed model against *in vivo* data has been difficult to achieve. While advanced imaging technologies have facilitated a better understanding and characterization of a number of different gastric functions, their ability to assess the local dynamics of gastric flows is still limited (Schulze, 2006; Marciani, 2011). To the best of our knowledge, one of the few attempts made to measure intragastric velocities has been performed by Boulby *et al.* (1999). By using echo planar imaging techniques, the authors identified retrograde velocities of 3.6 to 5.2 cm.s^{-1}. While these results seem to be in good agreement with the orders of magnitude numerically predicted, the difficulty of characterizing the physiological conditions developed *in vivo* (gastric geometry, emptying activities, speed, and occlusion of ACWs) and the inherent variability of the data obtained prevent the use of these data to achieve a more quantitative assessment of the model performance.

Similarly, the development of an *in vitro* system capable of not only reproducing the geometry and motor activity of the human stomach, but also of facilitating a good experimental characterization of the flow conditions developed within it has proved to be difficult.

While not reproducing the actual geometry of the human stomach, Marra *et al.* (2011) developed an experimental unit aimed at replicating the main features driving the dynamics of gastric contents and facilitated, while facilitating the use of a non-intrusive flow measurement technique (particle image velocimetry) to trace the entire flow filled within it.

Aimed at mimicking the motion of terminal ACWs against a close pylorus, a stainless steel cylinder was periodically moved (3.1 ± 0.2 mm.s^{-1}) along a neoprene sheet wall to impose a 60% occlusion inside a transparent polycarbonate chamber ($20.5 \times 20.5 \times 2.0$ cm). The flow behavior of two Newtonian fluids (water: 10^{-3} Pa.s and 92% corn syrup solution: 0.98 Pa.s) and one shear thinning fluid (0.5% carboxymethyl cellulose solution: $k = 0.93$ Pa.s^n, $n = 0.68$) were experimentally investigated. Results were compared against the velocity field numerically predicted within this new system.

The good agreement found between experimental and numerical data confirmed the capability of the model to predict the fluid motions induced by peristalsis inside a confined domain (Figure 12.5). Similar to results predicted within the stomach model, the experimental results also indicated the

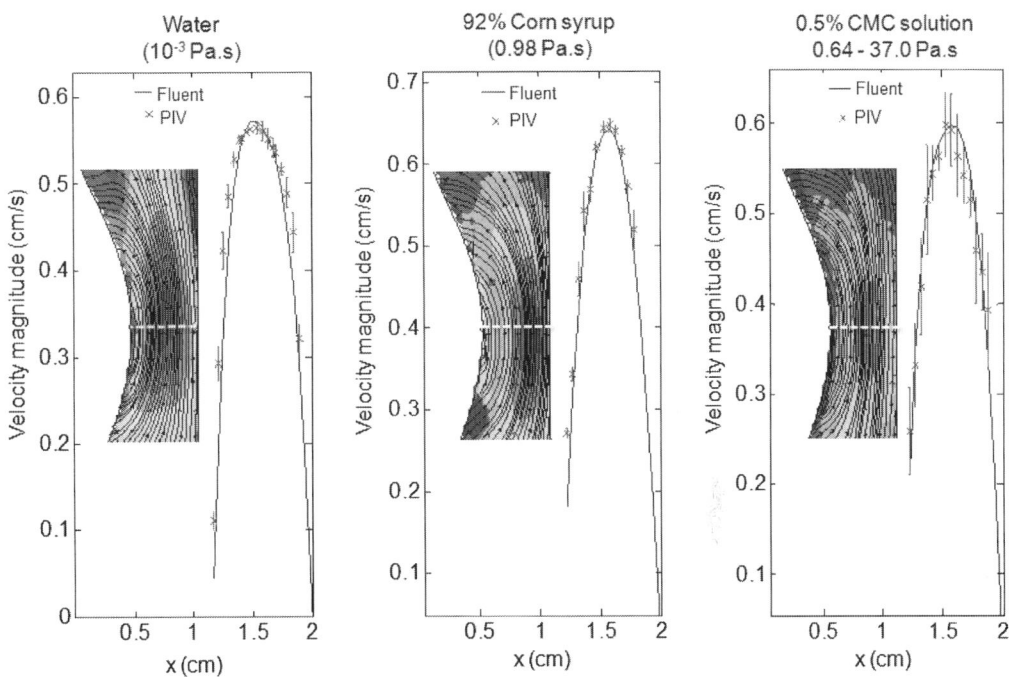

■ **FIGURE 12.5** Experimental analysis of model performance. Predicted versus experimental velocity profiles across the most occluded region of the experimental chamber *(from Ferrua* et al., *2011).* [A color version of this figure is available online at www.booksite.elsevier.com/9780124046108].

development of more confined retropulsive velocities within the chamber as the viscosity of the fluid increases. (Figure 12.5).

MIXING DYNAMICS OF DISTAL FLOWS DURING DIGESTION

While previous sections illustrate the role of fluid rheology on the different flow patterns and strength of gastric flows during digestion, little is known about how these differences can actually affect the dynamics of gastric mixing during the process.

To obtain a first insight into this dynamics, the elongation and dispersion experienced by different fluid segments as a result of the flow behavior of different gastric fluids were investigated. Characterizing the generation of interfacial area, and distribution of different fluid elements within the domain (primary mechanisms for mixing), these two material properties can be related to the dynamics and efficiency of bulk mixing.

Numerical Characterization of Gastric Mixing (Methodology)

Once the flow field reached a periodic solution in the absence of gastric emptying (i.e., after $4T_{ACW}$), three 5 mm long fluid segments (containing 10 fluid tracers each) were located distally to the most occluding ACW, and normal to the main flow direction (Figure 12.6). The trajectory of each individual fluid tracer was then tracked in time during the numerical simulation of the gastric flow.

The length of the segment was approximated by the linear distances between each consecutive pair of fluid tracers (i.e., using a piecewise linear approximation scheme). The stretching of each segment at any given time was then reported as the ratio between its current and initial lengths (L/L_0).

The dispersion of each fluid segment throughout the stomach domain was characterized by the root mean square radius proposed by Pal *et al.* (2004), which basically represents the standard deviation of the fluid tracers within each segment with respect to their center of mass (Eq. 12.8):

$$R_0^k\left(t,\, \overline{\mathbf{x}}_0^k\right) \;=\; \sqrt{\frac{1}{N} \sum_{i\in k=1}^{N_k} \left|\, \mathbf{x}_t^i - \overline{\mathbf{x}}_t^k \,\right|^2} \tag{12.8}$$

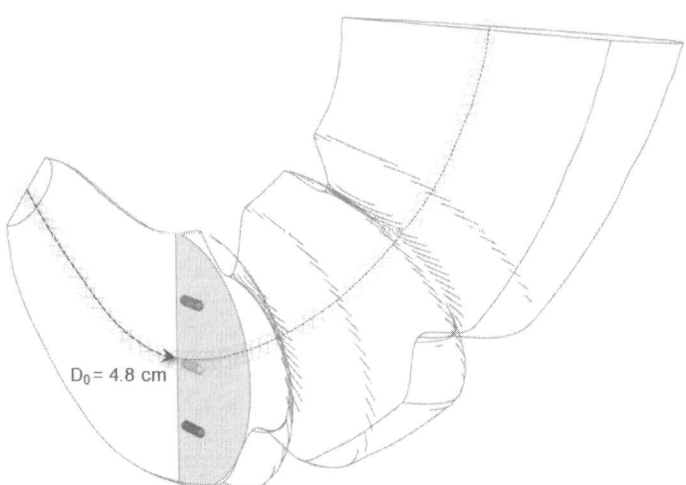

■ **FIGURE 12.6** Initial location of the fluid segments used to characterize the dynamics of mixing within the distal region of the stomach. [A color version of this figure is available online at www.booksite.elsevier.com/9780124046108].

where $\mathbf{x}_t^i$ is the location of the ith fluid tracer within the segment k at time t and $\bar{\mathbf{x}}_t^k$ is the center of mass of the N_k particles within the segment. Similar to their stretching, the dispersion of each fluid segment was reported as a ratio of their root mean square radius at time t to that at the moment of release(R_t^k/R_0^k).

Finally, to investigate the effect of fluid rheology on the trajectory and emptying pathways of gastric contents of different rheological properties, the displacement of each fluid segment was computed by tracking the location of its center of mass along the stomach centerline.

Dynamics of Gastric Mixing (Numerical Results)

The effect of fluid viscosity on the dynamics of gastric mixing and trajectory pathways generated during digestion were found to be significantly affected by gastric emptying (Figure 12.7).

In the absence of gastric emptying, mixing was generally enhanced along the inner and greater curvatures of the stomach, with the intensity and uniformity of the process being highly influenced by the rheology of the fluid (Figures 12.8 and 12.9). As viscosity increased, the slower and more ordered gastric motions that developed within the distal region led to a decrease in the stretching and dispersion of all the fluid segments analyzed. This detrimental effect was particularly important closer to the greater curvature and core regions of the stomach, with the dynamics of mixing becoming increasing heterogeneous and enhanced along the inner curvature of the stomach. With respect to the averaged displacement of fluid elements, in the absence of gastric emptying all the segments moved back into the stomach, with their relative motions being influenced by the viscosity of the fluid (Figure 12.10). In the case of a water-like fluid, despite the large retropulsive motions developed along the center of the stomach, the segments transported further back into the stomach were those located closer to the inner and greater curvatures of the stomach. Moreover, as the viscosity of the fluid increased, this pattern was inverted and the segment located along the stomach centerline became the one transported further back into the stomach (despite the fact that the retropulsive motions were now largely constrained to the location of the ACW peak).

Gastric emptying was found to have a profound effect not only on the overall dynamics of gastric mixing, but also on the role that the fluid rheology has on the intensity and uniformity of the process (Figure 12.7). In particular, emptying activities were found to have such a critical and detrimental effect on the intensity of gastric mixing that the role of fluid rheology on the dynamics of the process became largely negligible. As

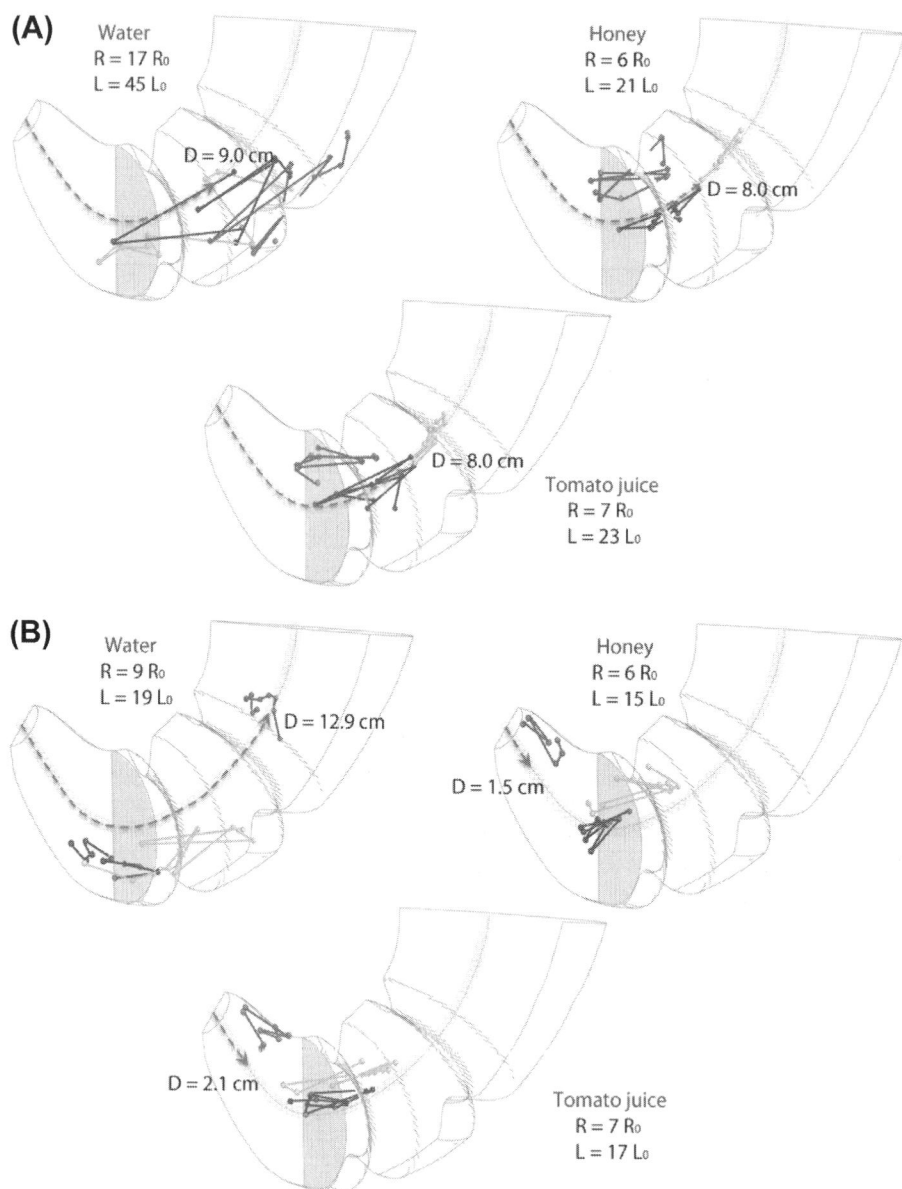

■ **FIGURE 12.7** The average dispersion (R), stretching (L), and distance from the pylorus (D) of the three fluid segments seeded in the flow after 40 s ($2T_{ACW}$) of release. (A) No gastric emptying. (B) Gastric emptying (in this case D represents the distance of only the segment that was initially released closer to the inner curvature of the stomach). [A color version of this figure is available online at www.booksite.elsevier.com/9780124046108].

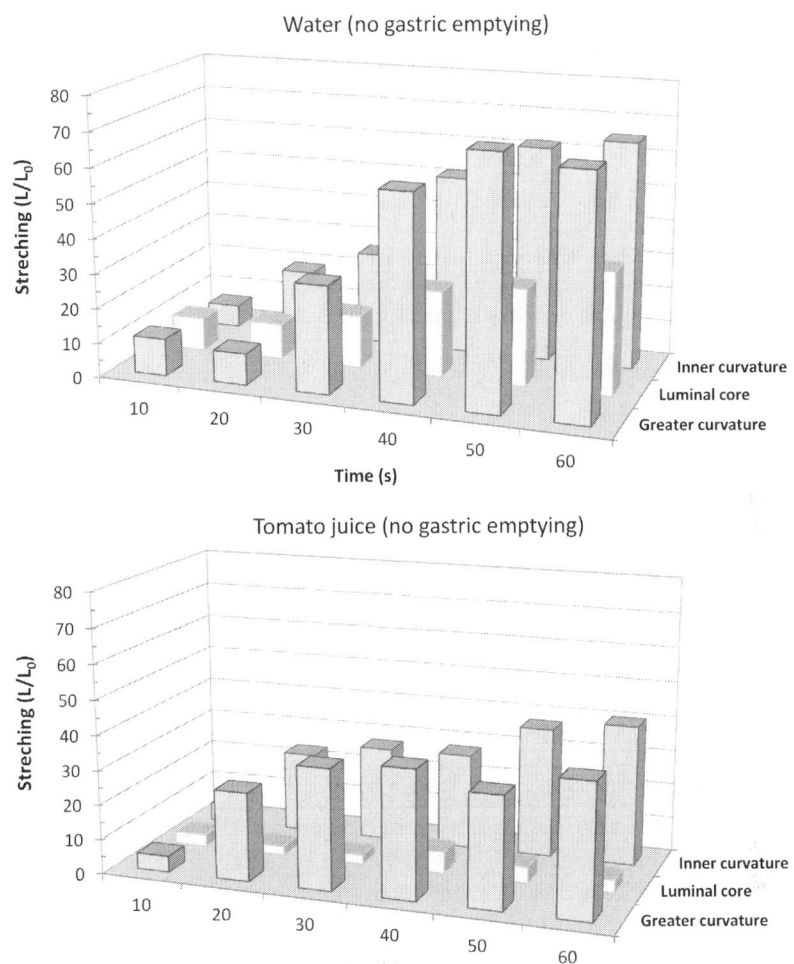

■ **FIGURE 12.8** Stretching of the different fluid segments initially located across a distal section of the stomach (no gastric emptying). [A color version of this figure is available online at www.booksite.elsevier.com/9780124046108].

illustrated in Figures 12.11 and 12.12, gastric emptying activities led to a more uniform but weaker mixing. It was also found that, in the presence of gastric emptying, fluid segments not only moved back into the stomach but also forward, with fluid viscosity having a particular role on the emptying pattern of different fluid tracers (Figure 12.13). In the case of a water-like fluid, emptying largely occurred along the greater curvature of the stomach, with fluid tracers located along the inner curvature of the stomach being transported further back into the proximal stomach. As the viscosity increased, this situation was reverted, with emptying

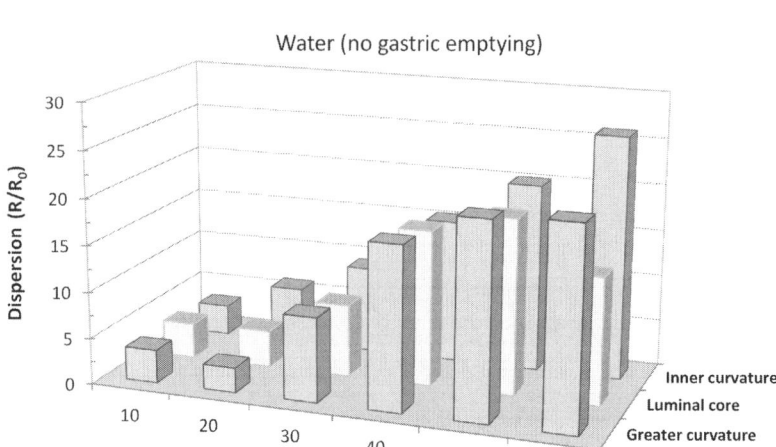

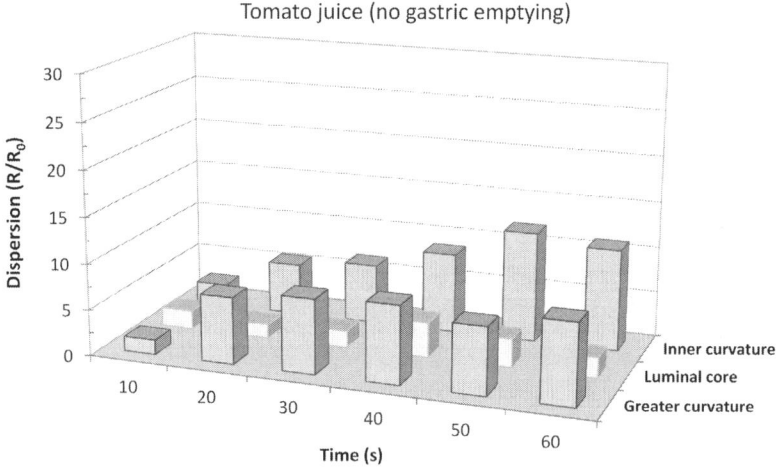

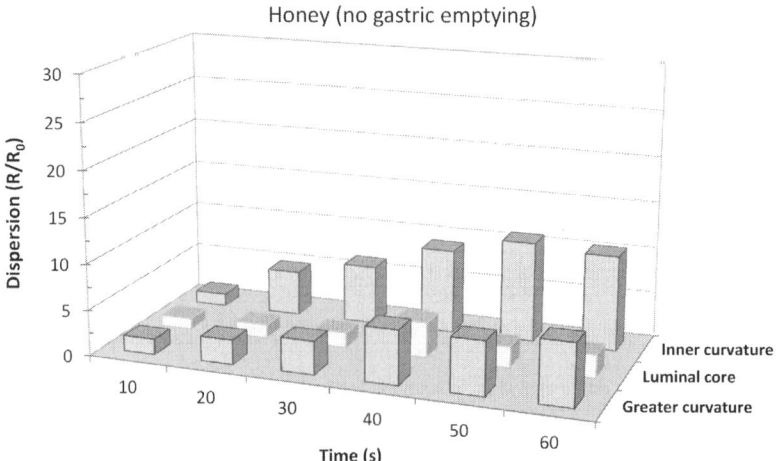

■ **FIGURE 12.9** Dispersion of the different fluid segments initially located across a distal section of the stomach (no gastric emptying). [A color version of this figure is available online at www.booksite.elsevier.com/9780124046108].

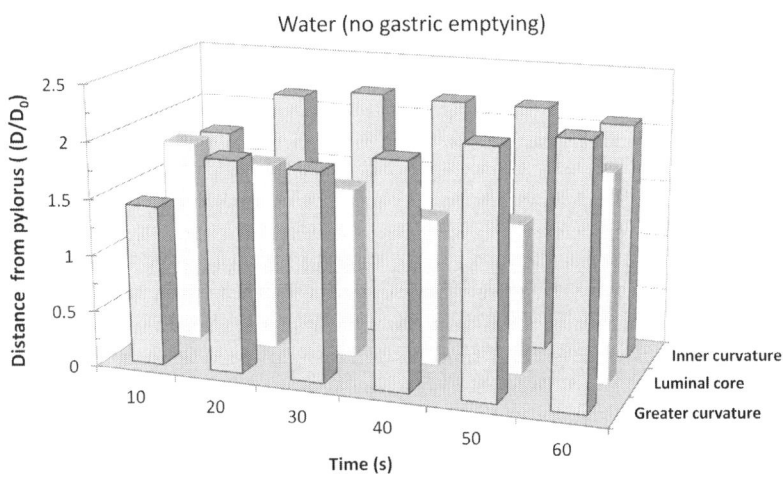

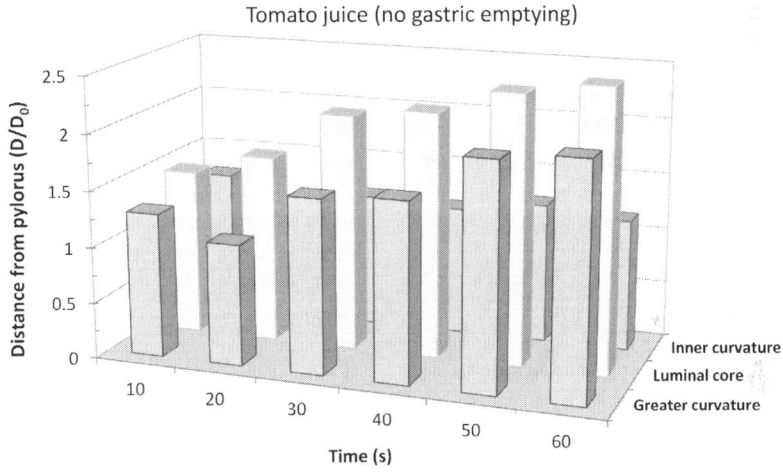

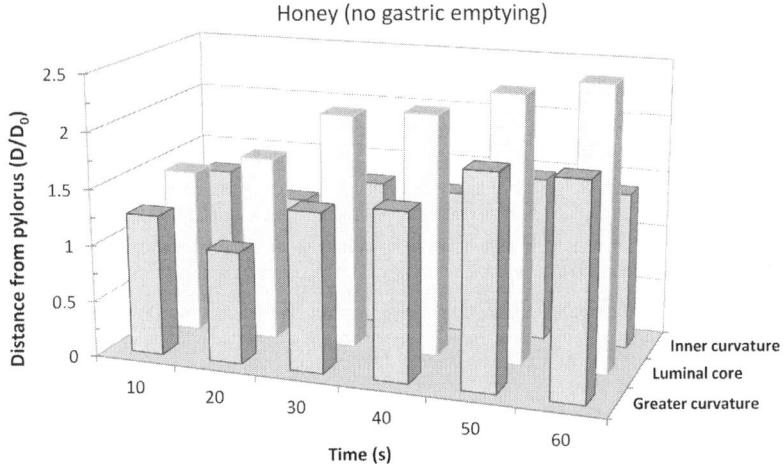

■ **FIGURE 12.10** Distance traveled along stomach centerline by the different fluid segments initially located across a distal section of the stomach (no gastric emptying). [A color version of this figure is available online at www.booksite.elsevier.com/9780124046108].

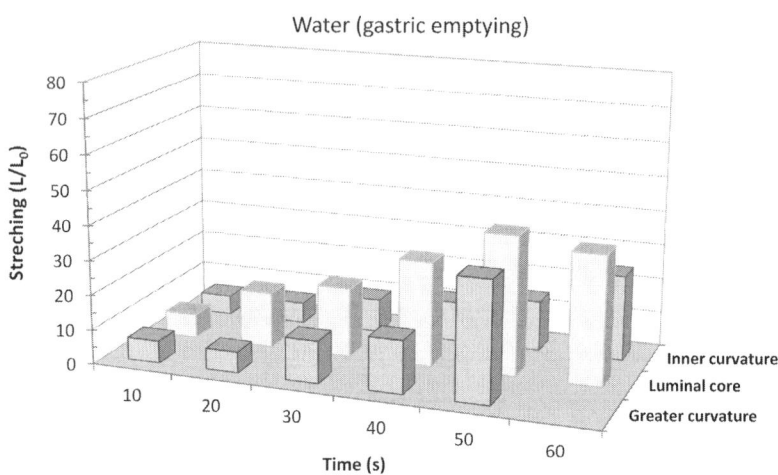

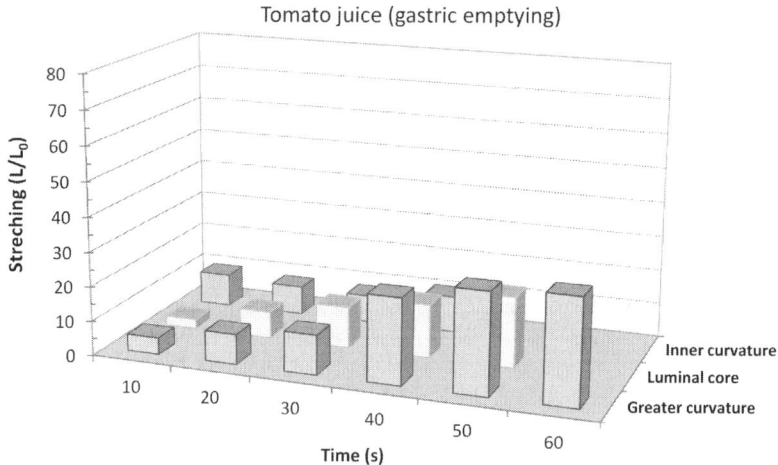

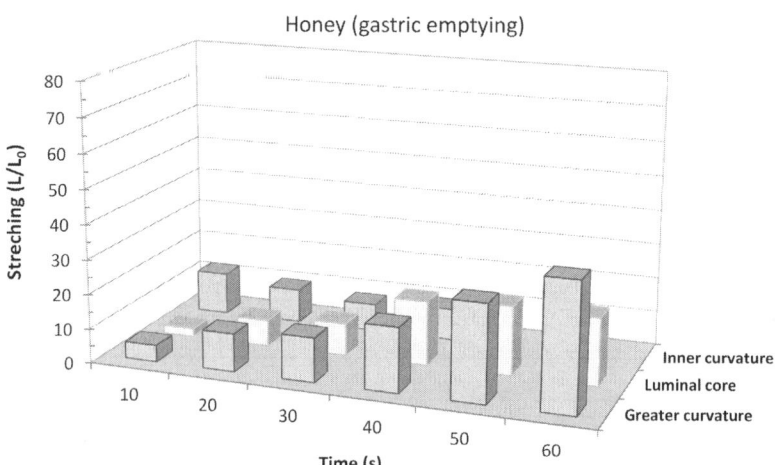

■ **FIGURE 12.11** Stretching of the different fluid segments initially located across a distal section of the stomach (gastric emptying). [A color version of this figure is available online at www.booksite.elsevier.com/9780124046108].

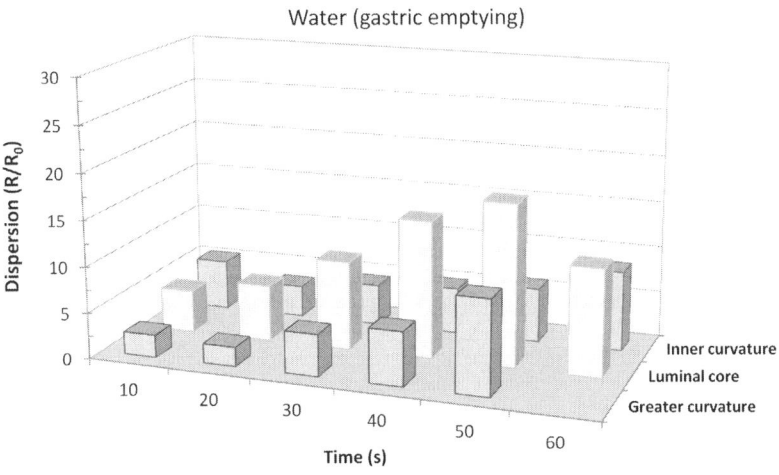

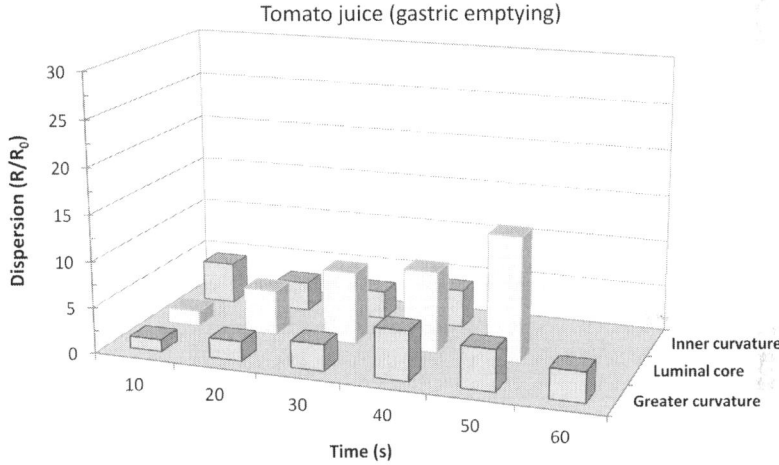

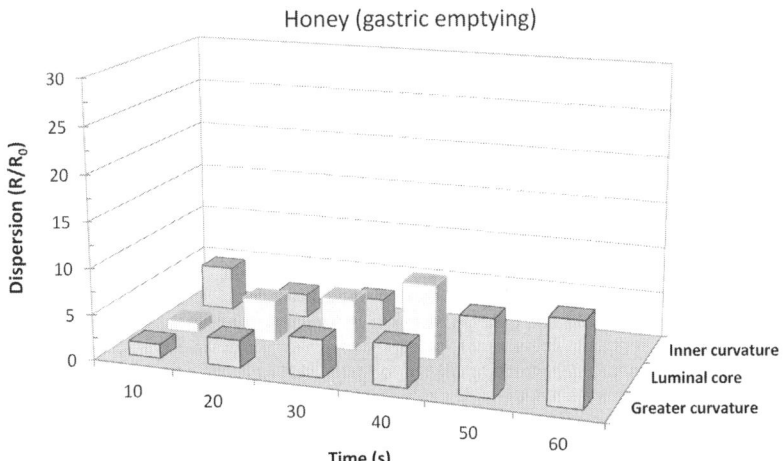

■ **FIGURE 12.12** Dispersion of the different fluid segments initially located across a distal section of the stomach (gastric emptying). [A color version of this figure is available online at www.booksite.elsevier.com/9780124046108].

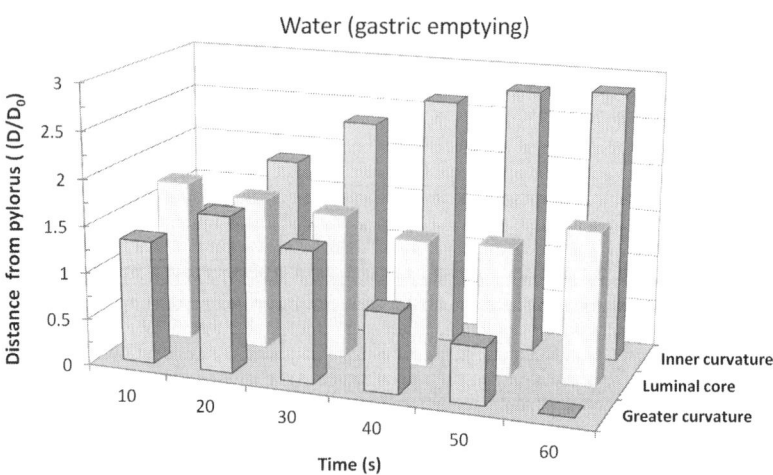

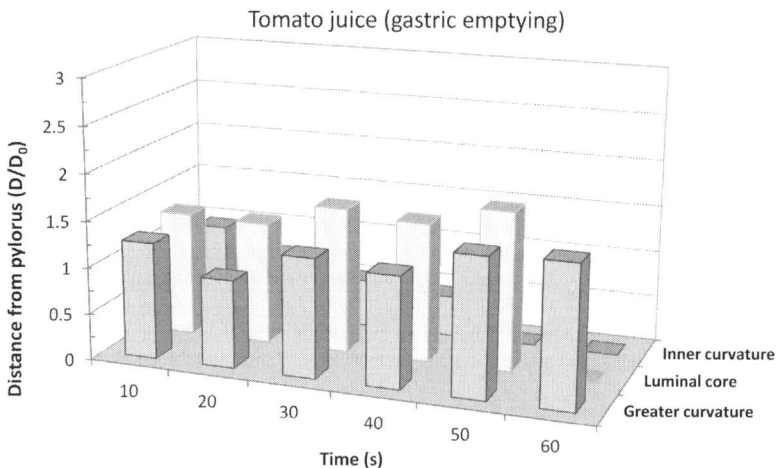

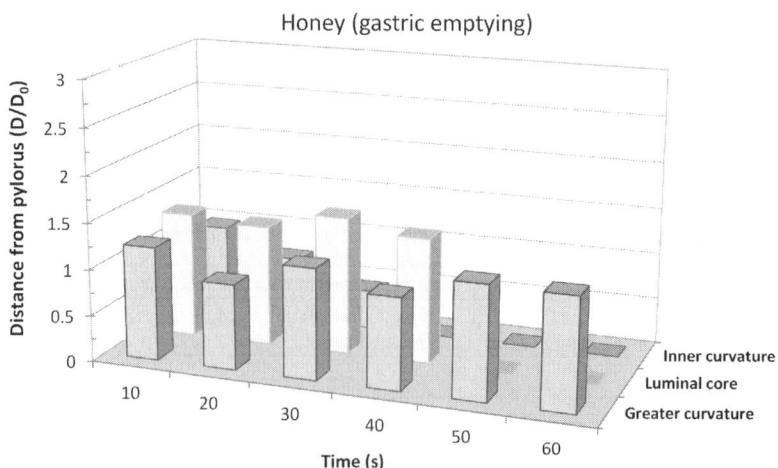

■ **FIGURE 12.13** Distance traveled along stomach centerline by the different fluid segments initially located across a distal section of the stomach (gastric emptying). [A color version of this figure is available online at www.booksite.elsevier.com/9780124046108].

largely occurring along the inner curvature and core regions of the stomach.

While this preliminary study illustrated the detailed insight that can be obtained into the dynamics of mixing and emptying pathways during digestion, current studies with more than 29,000 fluid tracers seeded inside the stomach model are being carried out to fully characterize the performance of gastric mixing under different physiological conditions.

THE DYNAMICS OF DISCRETE FOOD PARTICLES DURING DIGESTION

Having a better understanding of the role of gastric rheology on the flow behavior and mixing dynamics of gastric contents during digestion, the next step was to investigate how these properties can actually affect the motion and emptying of discrete food particles during digestion.

To achieve this goal, the dynamics of a series of discrete food particles associated with the ingestion of a small cylinder of raw carrot (dia. 2×1 cm) was numerically modeled (Ferrua and Singh, 2011, 2012).

Modeling the Dynamics of Discrete Food Particles

The number and size distribution of the particles released inside the stomach model were determined based on *in vivo* data of the mass and size distribution of carrot particles in human food boluses immediately before swallowing (Jalabert-Mabos *et al.*, 2007). In particular, 220 spherical particles with a density of 1054 kg/m^3 and sizes ranging from 0.4 to 4 mm (median size of 1.9 mm) were released at the level of the esophagogastric junction with a speed of 0.2 m/s (McMahon *et al.*, 2007).

The motion of the particles within three different gastric fluids under study was modeled using the discrete phase model implemented in ANSYS Fluent™. This method follows the Euler−Lagrange approach, where the continuum fluid phase is modeled by the continuity and momentum equations in a Eulerian frame of reference (Eqs. 12.1 and 12.2), while the trajectories of the dispersed particles $(d\mathbf{x}/dt)$ are individually computed at specified intervals of time by integrating the force balance on the particle in a Lagrangian frame of reference (Eqs. 12.9−12.10):

$$\frac{d\mathbf{u_p}}{dt} = F_D(\mathbf{u} - \mathbf{u_p}) + \frac{\mathbf{g}\,(\rho_p - \rho)}{\rho_p} \tag{12.9}$$

$$\frac{d\mathbf{x_p}}{dt} = \mathbf{u_p} \tag{12.10}$$

where $\mathbf{u_p}$ is the velocity of the particle p, $\mathbf{u}$ is the velocity of the fluid phase, $F_D(\mathbf{u}_p - \mathbf{u})$ is drag force per unit particle mass, ρ_p is the density of the particle p, ρ is the density of the fluid phase, $\mathbf{g}$ is the gravitational force, and $\mathbf{x_p}$ is the position of particle p.

The force balance given by Eq. 12.9 equates the particle inertia to the acceleration imposed on the particle by the drag (viscous) and buoyancy forces acting on it (first and second terms in the right-hand side of the equation, respectively). It is noteworthy that this modeling approach is constrained to a low volume fraction of particles within the flow domain (i.e., less than 10%), as it does not account for particle–particle interactions or the effect of the particle dynamics on the flow behavior of the continuous fluid phase.

The particles were released only after the flow field within the stomach reached a periodic solution in the absence of gastric emptying (i.e., after $4T_{ACW}$). Based on *in vivo* data, gastric emptying was not commenced until after 80 s of the particles being released inside the stomach (Hausken *et al.*, 1998). As in the analysis of gastric mixing, the dispersion experienced by the food particles during the process was computed based on standard deviation of their locations with respect to their center of mass (Eq. 12.8). The overall motion of the particles towards the pylorus was computed by tracking the location of their center of mass along the stomach centerline, and its average propagation speed from the moment of being released.

Dynamics and Fate of Discrete Food Particles (Numerical Results)

As illustrated by Figure 12.14, the rheology and behavior of gastric fluids has a marked effect on the dynamics, dispersion, and emptying of discrete food particles.

In the case of a water-like fluid, carrot particles moved down at an average speed of about 10 mm/s, and settled down over the greater curvature of the stomach within less than 30 s of being released (Figure 12.15A, B). As larger particles tended to settle down slightly faster than smaller ones, their dispersion during this initial period increased up to 15 times (Figure 12.15C). However, once they reached the greater curvature of the stomach, their motion rapidly became periodic and in synchrony with the propagation of the ACWs. Although the approaching ACWs kept moving the particles forward,

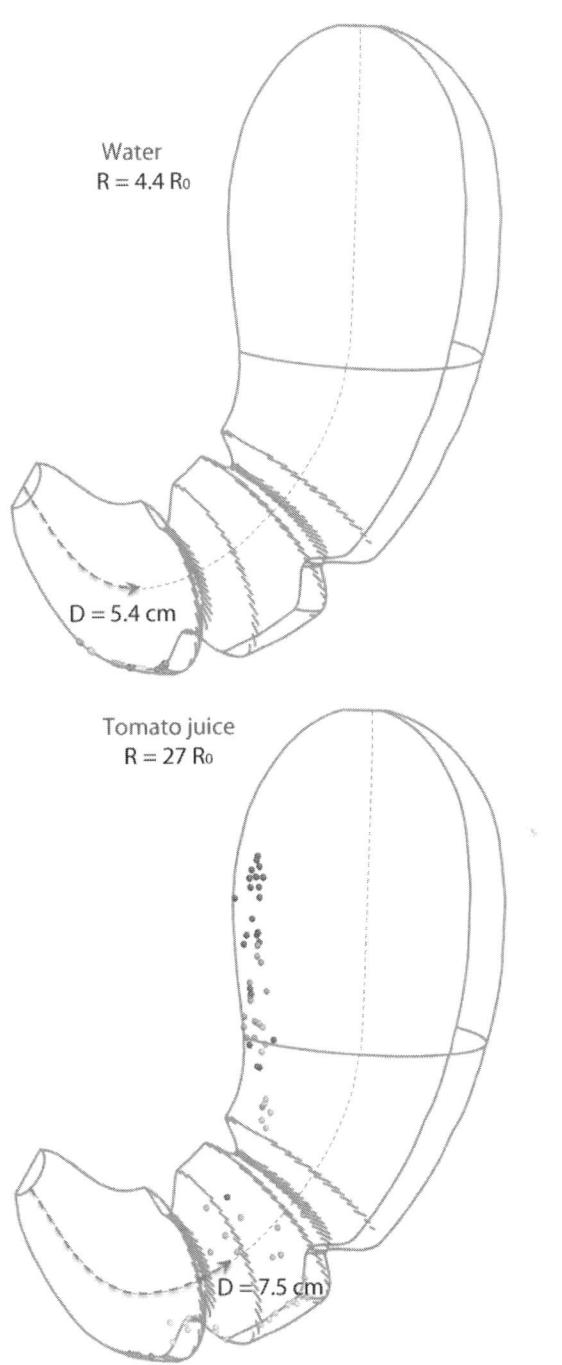

Water
R = 4.4 R$_0$

D = 5.4 cm

Tomato juice
R = 27 R$_0$

D = 7.5 cm

■ **FIGURE 12.14** Effect of the fluid rheology and gastric flow behavior on the dynamics of discrete carrot particles (snapshot after 2 min of their release). [A color version of this figure is available online at www.booksite.elsevier.com/9780124046108].

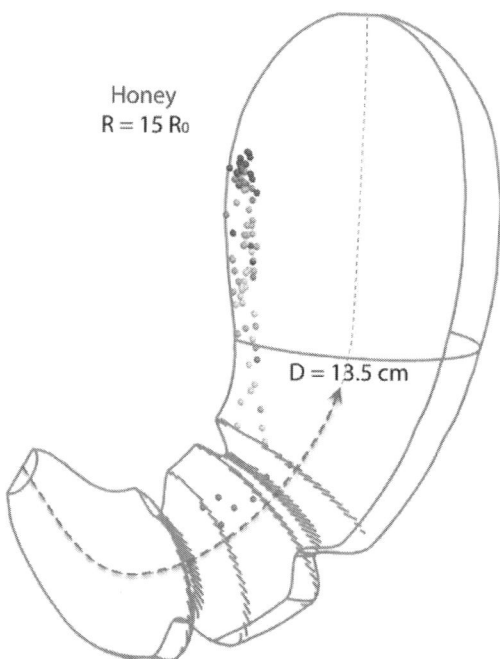

Honey
$R = 15\ R_0$

$D = 13.5\ \text{cm}$

■ **FIGURE 12.14** cont'd.

viscous forces were not strong enough to suspend the particles into the rapid fluid motions that develop in the core region of the stomach. As a consequence, the waves regularly outpaced the particles and gravity forces pulled them back into the dependent region. The average position of the particles became periodic within a distance of 4 to 5.4 cm from the pylorus, and their dispersion ranged from only 3 to 4 times their initial values. It is noteworthy that this dynamic was not affected by gastric emptying activities, with carrot particles never being emptied from the stomach.

These results illustrate the small effect that the strong and convoluted motions developed in the case of a water-like fluid have on the dynamics and fate of food particles during digestion. Further studies also showed that neither the development of more occluding ACWs (capable of imposing 80% of occlusion and retropulsive motions of up to 8 cm/s), nor a decrease in the size of the particles up to 60% (down to a size range of 1.5 to 0.2 mm) were able to facilitate the suspension of the particles within the rapid fluid motions that developed within the stomach (Ferrua and Singh, 2011). As illustrated in Figure 12.16, the motion of the particles was governed by buoyancy effects, with a 4.5% reduction of their density (i.e., to 1006 kg/m^3) being the only

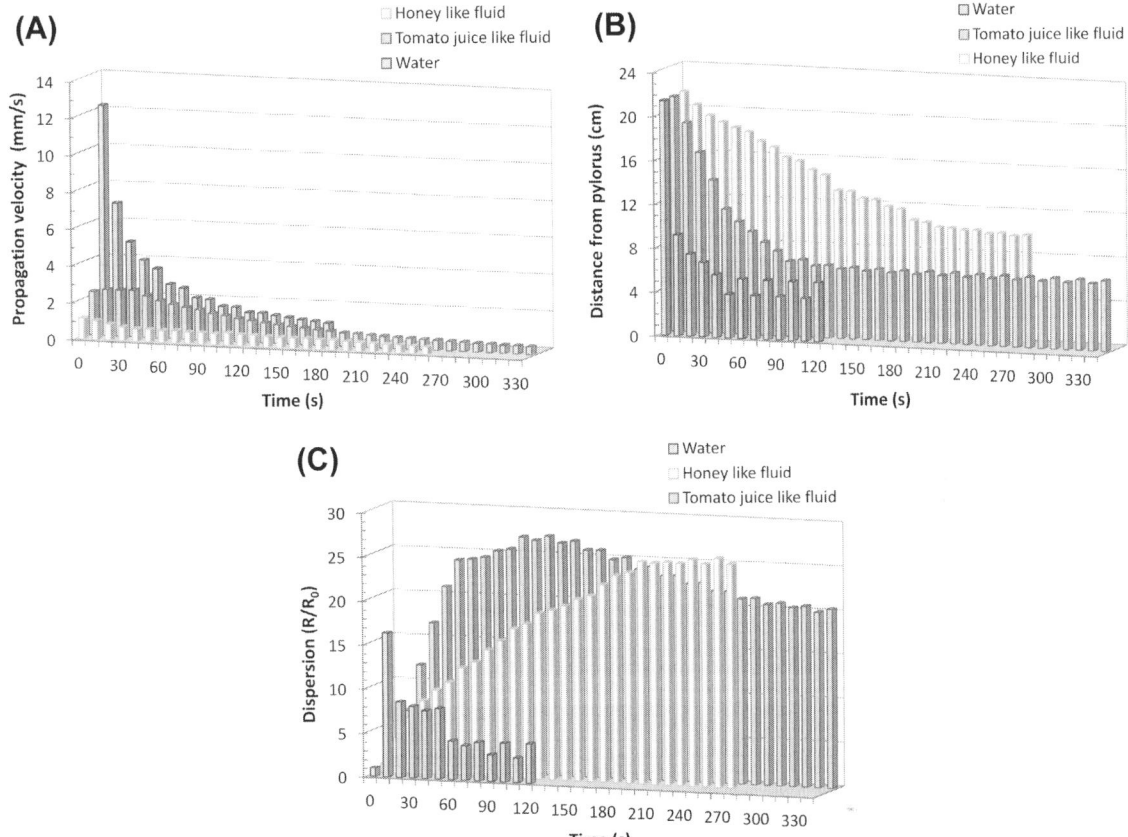

■ **FIGURE 12.15** Dynamics of discrete carrot particles within different gastric fluids until a periodic solution is obtained for a water-like fluid or the first particle is emptied from the stomach (honey at 4.5 min, tomato juice at 5.8 min). (A) Average propagation velocity of their center of mass along the stomach
centerline from moment of release. (B) Distance from pylorus to their center of mass along the stomach centerline. (C) Dispersion of carrot particles, R/R_0.
[A color version of this figure is available online at www.booksite.elsevier.com/9780124046108].

physical change that facilitated their suspension into the relatively faster fluid
motions of a water-like fluid. These results were in very good agreement with
in vivo data on the motion of a small number of garbanzo beans within the
stomach (Brown *et al.*, 1993), and supported the physiological role of the incisura angularis (i.e., angle of the lesser curvature that places the terminal
antrum relatively above the dependent region of the stomach) in the dynamics
and sieving of solid foods during digestion.

As the viscosity of the fluid increases, viscous forces were able to slow
down the motion of the particles immediately after their release (with
particles moving towards the pylorus at velocities lower than 2.5 mm/s,

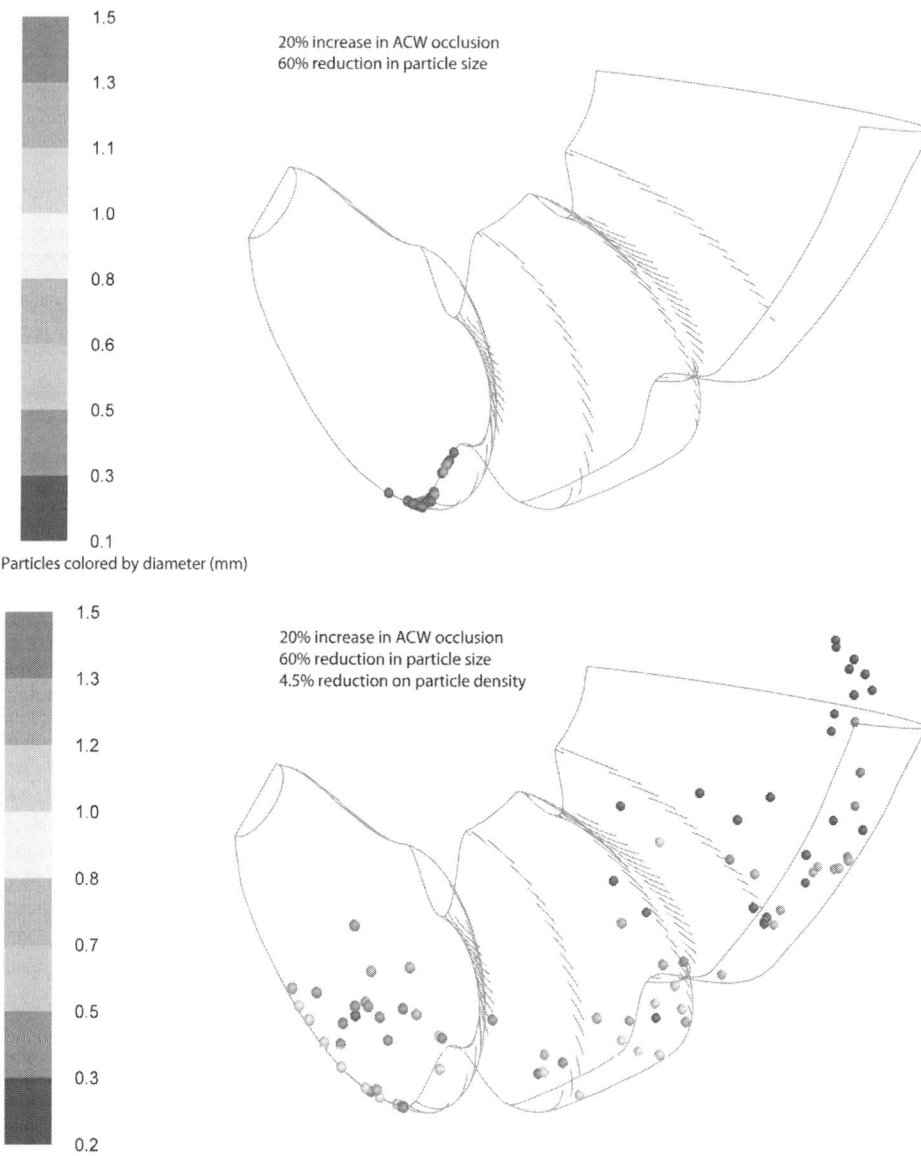

■ **FIGURE 12.16** Effect of particle size and density on the dynamics of carrot particles within a low viscous Newtonian fluid (water, 10^{-3} Pa.s) after 60 s of the physical changes being imposed. [A color version of this figure is available online at www.booksite.elsevier.com/9780124046108].

Figure 12.15A). Similarly, despite the slower and more ordered flow motions developed, the enhanced viscous forces facilitated the dispersion of the particles within the gastric lumen, particularly in the distal region of the stomach. Within 1 minute of release, the dispersion of the particles in the tomato juice fluid increased 24 times. The same level of dispersion was obtained in the very thick honey, but 2 min later (as it took longer for the particles to reach the more convoluted flow motions of the distal region, Figure 12.15B, C). The emptying of the food particles was facilitated by the fluid viscosity. The more viscous the fluid, the easier for the flow to keep the particles suspended within the core stream and to carry them through the pylorus sphincter. In the case of tomato juice (0.02 to 0.17 Pa.s), the first particle leaving the stomach was one of the smaller ones (0.4 mm in diameter) and its emptying occurred after 5.8 min. While in the case of honey (1 Pa.s) particles started emptying the stomach after 4.5 min, with the first one being 2.5 mm in diameter.

DYNAMICS OF MORE DENSELY PACKED FOOD DIGESTA SYSTEMS

While the previous study provided a first insight into the complex interaction of parameters affecting the dynamics and emptying of solid foods during digestion (beyond the simple behavior of the gastric fluids), it was limited to the analysis of just a small number of food particles (which for the most part cannot represent the physics of gastric contents associated with the ingestion of a regular meal).

To overcome this limitation, preliminary studies have been performed to investigate the dynamics of more complex and realistic food digesta systems, generally characterized by a relatively densely packed mixture of food particles and gastric fluids (Xue *et al.*, 2012). In particular, the dynamics of a continuous suspension of raw carrot particles (1 mm in size) in a continuous fluid phase (i.e., a Newtonian water-like fluid) was investigated under two different particle loading conditions (10 and 30%).

Modeling the Dynamics of Granular Food Digesta Systems (Methodology)

Modeling the dynamics of particulate flow systems (granular flows) is an area of current research. As the concentration of particles increases, the dynamics of their interactions become increasingly important, and their modeling becomes a key and limiting factor in the numerical analysis of this type of systems (Brennen, 2005).

Under the scope of a preliminary study, the dynamics of the carrot particles and gastric fluids were modeled by following an Euler—Euler approach that assumes both the liquid and solid phases are continuous and fully interpenetrating. The discrete nature of the dispersed phase is then modeled by a series of constitutive relations that are derived from either empirical information or the application of the kinetic theory of granular media.

In particular, as part of this study we implemented one of the most computationally cost-effective models available (Anonymous, 2012). Instead of solving the momentum and continuity equations for each phase, the model solves the continuity and momentum equations for the mixture and the volume fraction equation for the particulate phase (Eqs. 12.11−12.13), and the algebraic expressions for the velocity of the particulate phase relative to the velocity of the fluid phase:

$$\frac{\partial \rho_m}{\partial t} + \nabla \cdot (\rho_m \ \underline{\mathbf{u}}_m) = 0 \tag{12.11}$$

$$\frac{\partial}{\partial t}(\rho_m \ \underline{\mathbf{u}}_m) + \nabla \cdot (\rho_m \ \underline{\mathbf{u}}_m \ \underline{\mathbf{u}}_m) = -\nabla p_m + \nabla \cdot \mu_m \left[\nabla \underline{\mathbf{u}}_m + (\nabla \underline{\mathbf{u}}_m)^T \right]$$
$$+ \rho_m \ \underline{\mathbf{g}} + \underline{\mathbf{F}} + \nabla \cdot \left(\sum_{k=1}^{2} \alpha_k \rho_k \underline{\mathbf{u}}_{dr,k} \ \underline{\mathbf{u}}_{dr,k} \right) \tag{12.12}$$

$$\frac{\partial}{\partial t}(\alpha_p \ \rho_p) + \nabla \cdot (\alpha_p \ \rho_p \ \underline{\mathbf{u}}_m) = -\nabla \cdot (\alpha_p \ \rho_p \underline{\mathbf{u}}_{dr,k}) \tag{12.13}$$

where the sub index m refers to the properties of the mixture, p to the properties of the particulate phase, and k to the properties of each phase within the system. In particular:

$\rho_m = \sum_{k=1}^{2} \alpha_k \rho_k$ is the mixture density,
$\underline{\mathbf{u}}_m = (\sum_{k=1}^{2} \alpha_k \rho_k \underline{\mathbf{u}}_k)/\rho_m$ is the mass-averaged velocity,
p_m is the mixture pressure,
α_k is the volume fraction of phase k,
$\mu_m = \sum_{k=1}^{2} \alpha_k \mu_k$ is the mixture viscosity,
$\underline{\mathbf{u}}_{dr,k} = \underline{\mathbf{u}}_k - \underline{\mathbf{u}}_m$ is the drift velocity of phase k.

The drift velocity for the solid phase is defined in terms of the relative velocity (or slip velocity) between the solid and liquid phase ($\underline{\mathbf{u}}_{lp} = \underline{\mathbf{u}}_p - \underline{\mathbf{u}}_l$), generally modeled by a series of constitutive relations based on the kinetic theory of granular media:

$$\underline{\mathbf{u}}_{dr,p} = \underline{\mathbf{u}}_p - \underline{\mathbf{u}}_m = \left(1 - \frac{\alpha_p \rho_p}{\rho_m}\right) \underline{\mathbf{u}}_{lp} \qquad (12.14)$$

The particles were uniformly seeded within the stomach model only after the flow field reached a periodic solution in the absence of gastric emptying (i.e., after $4T_{ACW}$).

Dynamics of Granular Food Digesta Systems (Numerical Results)

The results suggested that while particle loading might not have a major role on the motion and fate of the food particles within the stomach, it does have a significant effect on the overall dynamics of the gastric flow.

As illustrated in Figure 12.17, carrot particles settle down over the greater curvature of the stomach within less than 10 s of their release, regardless of their volumetric loading. However, as the number of particles that settled increased, the section available for gastric fluids to move in response to the propagation of the ACWs became increasingly narrower and affected the overall dynamics of the mixture. While the particle loading was low enough to not fully block the distal region of the stomach, faster retropulsive motions that extended up to the fundus region along the inner curvature of the stomach developed. These retropulsive motions reached velocities of up to 7 cm/s and facilitated the continuous shearing of the surface sediment of the carrot particles within the distal stomach. On the other hand, as soon as the particle loading became large enough to fully block the distal region of the stomach, the velocities and region of influence of these retropulsive motions became increasingly smaller (Figure 12.18).

Similar to the case of discrete food particles, these results further support the significant role of buoyancy effects on the dynamics of food particles within low viscous fluids and highlight the significant effect that the particle loading can have on the overall dynamics of the system (in particular with respect to the development of strong retropulsive velocities and eddies within the distal region of the stomach). Further work is needed to fully understand how the concentration of food particles can actually affect the dynamic of mixing of gastric contents, as well as the level of shear and pressure forces to which foods are exposed during the process.

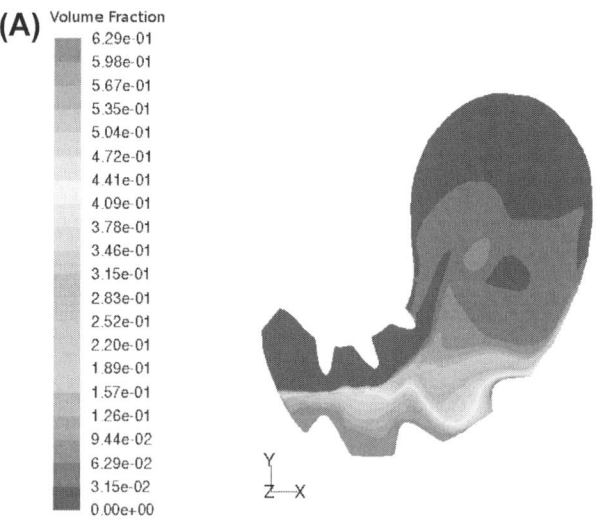

Contours of Volume fraction (particle)   (Time=1.0000e+01)   Jun 15, 2012
                     ANSYS FLUENT 12.0 (3d, dp, pbns, dynamesh, mixture, lam, transient)

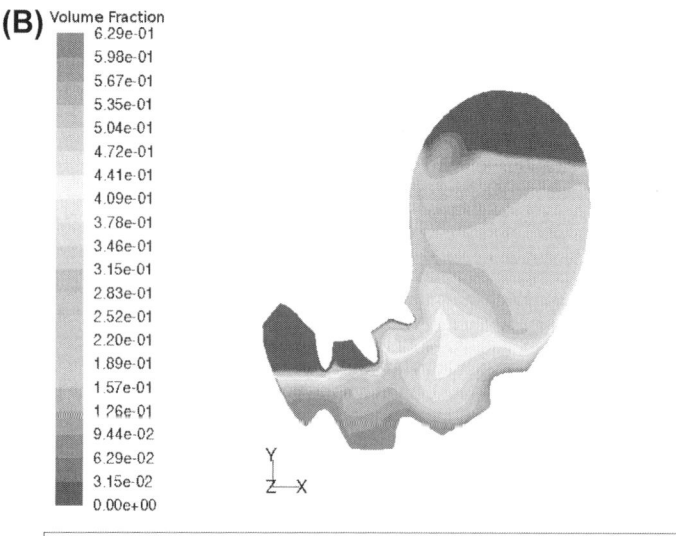

Contours of Volume fraction (particles)   (Time=1.0000e+01)   Jun 21, 2012
                     ANSYS FLUENT 12.0 (3d, dp, pbns, dynamesh, mixture, lam, transient)

■ **FIGURE 12.17** Contour plot of the volume fraction of raw carrots (1053 kg/m^3, 2 mm dia.) after 10 s of being uniformly released within a stomach containing a water-like fluid (1053 kg/m^3, 10^{-3} Pa.s). (A) 10% particle volume fraction, (B) 30% particle volume fraction. [A color version of this figure is available online at www.booksite.elsevier.com/9780124046108].

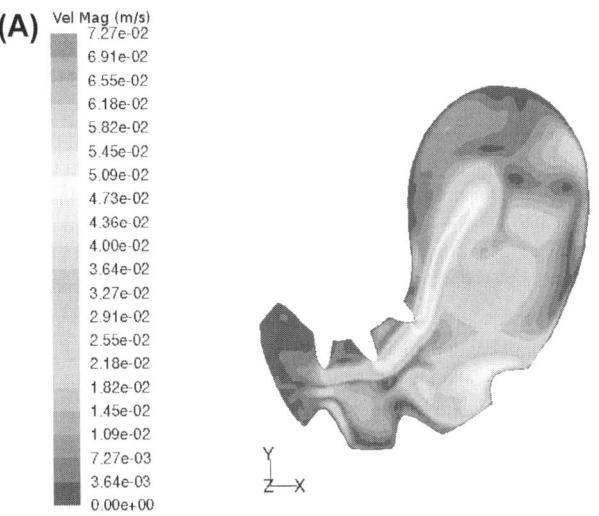

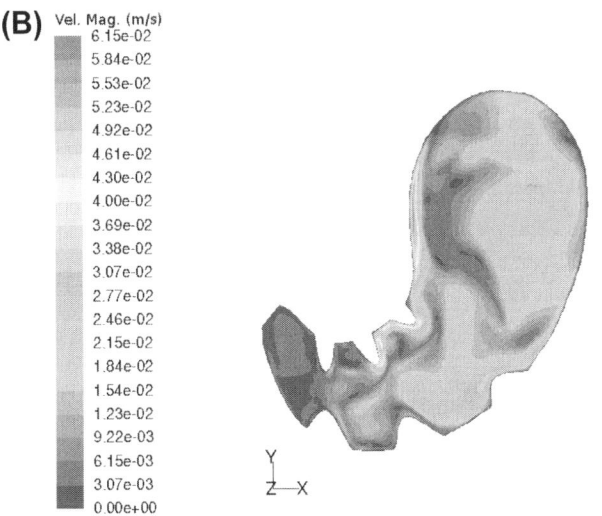

■ **FIGURE 12.18** Contour plot of the velocity field within a stomach full with a mixture of carrot particles (1053 kg/m³, 2 mm dia.) and gastric juices (1053 kg/m³, 10⁻³ Pa.s), after 10 s of the particles being uniformly released within the stomach. (A) 10% particle volume fraction, (B) 30% particle volume fraction. [A color version of this figure is available online at www.booksite.elsevier.com/9780124046108].

SUMMARY REMARKS AND FUTURE CHALLENGES

By providing a unique insight into the dynamic behavior of gastric contents during digestion, computational fluid dynamic techniques offer new opportunities to further understand the mechanisms driving food digestion.

Initial results have demonstrated that, contrary to expectations, the formation of strong retropulsive motions and eddy structures may not be the main causes driving the mechanical disintegration of food structures during digestion. These flow features were only predicted in the case of a very thin (water-like) fluid that cannot represent the typical physics of gastric contents, and also cannot transfer enough momentum into the motion of solid food particles to avoid their rapid sedimentation within the stomach. As the viscosity of the fluid increases, slower and more ordered flow motions develop within the stomach that not only compromise the formation of extended retropulsive motions and eddy structures within the distal region, but also hinder the transport and mixing of gastric contents within the stomach.

In addition, predicted results have facilitated a better understanding of the complex interaction of parameters driving the dynamics of particulated food systems during digestion. For example, it was found that even in the case of a discrete number of food particles, their dynamics were determined by a number of factors that extended far beyond the simple behavior of the gastric flow. Among these factors, the viscosity of the fluid, the amount and relative density of the particles, and the anatomical configuration of the terminal antrum (i.e., incisura angularis) seem to have a pivotal role in the distribution and emptying rates of solid foods during digestion.

Despite the ability of numerical tools to provide a better understanding of the fluid dynamic conditions driving the mixing and disintegration of gastric contents during digestion, much more work is needed to further advance the physiological relevance of these models.

Continuous advances in high performing computing and multiphysics simulation packages create unique opportunities to significantly enhance the physiological relevance and scope of this numerical approach. However, they also introduce unique challenges with one of the main limitations being our own understanding of how the fluid dynamic conditions predicted by the models relate to physical processes underlying gastric digestion. In particular, it will be essential to develop a good understanding not only of the complex interplay that occurs between the physicochemical properties of the digesta and the secretory and motor responses of the gastric wall, but also of the role of food material properties on their disintegration when exposed to the chemical and mechanical conditions expected to develop in vivo.

While capable of providing an invaluable tool to assist the design and development of novel foods and pharmaceuticals of enhanced functionalities, the future and successful implementation of this numerical approach hinges on a series of complex challenges that can only be achieved by the combined efforts across a number of different disciplines (including computational physics, fluid mechanics, material science, engineering, and physiology).

REFERENCES

Anonymous, 2012. ANSYS Fluent 12.1 documentation. N.H., ANSYS Inc, Lebanon.

Barret, K.E., Raybould, H.E., 2010. The gastric phase of the integrated response to a meal. In: Koeppen, B.M., Stanton, B.A. (Eds.), Berne & Levy Physiology, sixth edn. Elsevier Inc., Philadelphia.

Boulby, P., Moore, R., Gowland, P., Spiller, R.C., 1999. Fat delays emptying but increases forward and backward antral flow as assessed by flow-sensitive magnetic resonance imaging. Neurogastroenterol. Motil. 11, 27−36.

Brennen, C.E., 2005. Introduction to multiphase flow. In: Fundamentals of Multiphase Flow. Cambridge University Press, New York, pp. 1−28.

Brown, B.P., Schulze-Delrieu, K., Schrier, J.E., Abu-Yousef, M.M., 1993. The configuration of the human gastroduodenal junction in the separate of emptying of liquid and solids. Gastroenterology 105, 433−440.

Camilleri, M., Malagelada, J.R., Brown, M.L., Becker, G., Zinsmeister, A.R., 1985. Relation between antral motility and gastric emptying of solids and liquids in humans. Am. J. Phys. 249, G580−G585.

Cannon, W., Lieb, C., 1911. The receptive relaxation of the stomach. Am. J. Phys. 29, 267−273.

Dikeman, C.L., Murphy, M.R., Fahey Jr, G.C., 2006. Dietary fibers affect viscosity of solutions and simulated human gastric and small intestinal. Digesta J. Nutr. 136, 122−127.

Dillard, S., Krishnan, S., Udaykumar, H.S., 2007. Mechanics of flow and mixing at antroduodenal junction. World. J. Gastroenterol. 13, 1365−1371.

Einhorn, M., 2009. Diseases of the Stomach: A Text-book for Practitioners and Students, second edn. Cornell University Library. p. 110.

Ferrua, M.J., Kong, F., Singh, R.P., 2011. Computational modeling of gastric digestion and the role of food material properties. Trends Food Sci. Tech. 22, 480−491.

Ferrua, M.J., Singh, R.P., 2010. Modeling the fluid dynamics in a human stomach to gain insight of food digestion. J. Food. Sci. 75, R151−R162.

Ferrua, M.J., Singh, R.P., 2011. Understanding the fluid dynamics of gastric digestion using computational modeling. Procedia Food Science 1, 1465−1472.

Ferrua, M.J., Singh, R.P., 2012. Numerical analysis of fluid mixing and food particles distribution during gastric digestion. In: Institute of Food Technologists (IFT) Annual Meeting & Food Expo. Nevada, USA, Las Vegas, pp. 077−093.

Goetze, O., Steingoetter, A., Menne, D., van der Voort, I.R., Kwiatek, M.A., Bösiger, P., et al., 2007. The effect of macronutrients on gastric volume responses and gastric emptying in humans: a magnetic resonance imaging study. Am. J. Physiol. Gastrointest. Liver. Physiol. 292, G11−G17.

Granato, D., Branco, G.F., Nazzaro, F., Cruz, A.G., Faria, J.A.F., 2010. Functional foods and nondairy probiotic food development: trends, concepts, and products. Compr. Rev. Food Sci. Food Saf. 9, 292−302.

Hausken, T., Gilja, O.H., Odegaard, S., Berstad, A., 1998. Flow across the human pylorus soon after ingestion of food, studied with duplex sonography. Effect of glyceryl trinitrate. Scand. J. Gastroenterol. 33, 484−490.

IFT, 2005. Functional Foods: Opportunities and Challenges. IFT Expert Report. Institute of Food Technologists, Chicago, IL.

Imai, Y., Kobayashi, I., Ishida, S., Ishikawa, T., Buist, M., Yamaguchi, T., 2013. Antral recirculation in the stomach during gastric mixing. Am. J. Physiol. Gastrointest. Liver. Physiol. 304, G536−G542.

Jahnberg, T., Martinson, J., Hulten, L., Fasth, S., 1975. Dynamic gastric response to expansion before and after vagotomy. Scand. J. Gastroenterol. 10, 593−598.

Jalabert-Malbos, M.-L., Mishellany-Dutour, A., Woda, A., Peyron, M.-A., 2007. Particle size distribution in the food bolus after mastication of natural foods. Food Qual. Pref. 18, 803−812.

Keet, A.D., 1993. Infantile hypertrophic pyloric stenosis. In: The Pyloric Sphincteric Cylinder in Health and Disease. Springer-Verlag, Berlin Heidelberg, p. 107.

Kozu, H., Kobayashi, I., Nakajima, M., Uemura, K., Sato, S., Ichikawa, S., 2010. Analysis of flow phenomena in gastric contents induced by human gastric peristalsis using CFD. Food Biophys. 5, 330−336.

Kwiatek, M.A., Steingoetter, A., Pal, A., Menne, D., Brasseur, J.G., Hebbard, G.S., et al., 2006. Quantification of distal antral contractile motility in healthy human stomach with magnetic resonance imaging. J. Magn. Reson. Imaging. 24, 1101−1109.

Leatherhead Food Research, 2011. Future Directions for the Global Functional Foods Market. Leatherhead Food Research Market Report. Leatherhead Food Research, Surrey, UK.

Lentle, R.G., Janssen, P.W.M., 2010. Manipulating digestion with foods designed to change the physical characteristics of digesta. Crit. Rev. Food. Sci. Nutr. 50, 130−145.

Lentle, R.G., Janssen, P.W.M., Goh, K., Chambers, P., Hulls, C., 2010. Quantification of the effects of the volume and viscosity of gastric contents on antral and fundic activity in the rat stomach maintained ex vivo. Dig. Dis. Sci. 55, 3349−3360.

Liao, D., Gregersen, H., Hausken, T., Gilja, O.H., Mundt, M., Kassab, G., 2004. Analysis of surface geometry of the human stomach using real-time 3-D ultrasonography in vivo. Neurogastroenterol. Motil. 16, 315−324.

Marciani, L., 2011. Assessment of gastrointestinal motor functions by MRI: a comprehensive review. Neurogastroenterol. Motil. 23, 399−407.

Marciani, L., Gowland, P.A., Spiller, R.C., Manoj, P., Moore, R.J., Young, P., et al., 2000. Gastric response to increased meal viscosity assessed by echo-planar magnetic resonance imaging in humans. J. Nutr. 130, 122−127.

Marciani, L., Gowland, P.A., Spiller, R.C., Manoj, P., Moorel, R.J., Young, P., Fillery-Travis, A.J., 2001. Effect of meal viscosity and nutrients on satiety, intragastric dilution, and emptying assessed by MRI. Am. J. Phys. 280, G1227—G1233.

Marra, F., Ferrua, M.J., Singh, R.P., 2011. Experimental characterization of the fluid dynamics in an in-vitro system simulating the peristaltic movement of the stomach wall. Procedia Food Science 1, 1473—1478.

Mayer, E.A., 1994. The physiology of gastric storage and emptying. In: Johnson, L. (Ed.), Physiology of the Gastrointestinal Tract, third edn. Raven Press, New York, pp. 929—976.

McClements, D.J., Li, Y., 2010. Review of in vitro digestion models for rapid screening of emulsion-based systems. Food Funct 1, 32—59.

McMahon, B.P., Odie, K.D., Moloney, K.W., Gregersen, H., 2007. Computation of flow through the oesophagogastric junction. World. J. Gastroenterol. 13, 1360—1364.

O'Grady, G., Du, P., Cheng, L.K., Egbuji, J.U., Lammers, W.J.E.P., Windsor, J.A., et al., 2010. Origin and propagation of human gastric slow-wave activity defined by high-resolution mapping. Am. J. Physiol. Gastrointest. Liver. Physiol. 299, G585—G592.

Pal, A., Indireshkumar, K., Schwizer, W., Abrahamsson, B., Fried, M., Brasseur, J.G., 2004. Gastric flow and mixing studied using computer simulation. Proc. R. Soc. London, Ser. B 271, 2587—2594.

Pal, A., Brasseur, J.G., Abrahamsson, B., 2007. A stomach road or "Magenstrasse" for gastric emptying. J. Biomech. 40, 1202—1210.

Schulze, K., 2006. Imaging and modeling of digestion in the stomach and the duodenum. Neurogastroent. Motil. 18, 172—183.

Schwizer, W., Steingotter, A., Fox, M., Zur, T., Thumshirn, M., Bösiger, P., Fried, M., 2002. Non-invasive measurement of gastric accommodation in humans. Gut 51 (Suppl. 1), i59—i62.

Schwizer, W., Steingoetter, A., Fox, M., 2006. Magnetic resonance imaging for the assessment of gastrointestinal function. Scand. J. Gastroenterol. 41, 1245—1260.

Seerden, T.C., Lammers, W.J.E.P., De Winter, B.Y., De Man, J.G., Pelckmans, P.A., 2005. Spatiotemporal electrical and motility mapping of distension-induced propagating oscillations in the murine small intestine. Am. J. Physiol. Gastrointest. Liver. Physiol. 289, G1043—G1051.

Sherwood, L., 2010. The digestive system. In: Human Physiology: From Cells to Systems, seventh edn. Brooks/Cole, Belmont, CA, pp. 598—600.

Singh, S.K., 2007. Fluid flow and disintegration of food in human stomach. University of California, Davis. Ph.D. thesis.

Siró, I., Kápolna, E., Kápolna, B., Lugasi, A., 2008. Functional food. Product development, marketing and consumer acceptance—a review. Appetite 51, 456—467.

Steffe, J.F., 1996. Rheological Methods in Food Process Engineering. Freeman Press, East Lansing, MI.

Steingoetter, A., Kwiatek, M.A., Pal, A., Hebbard, G., Thumshirn, M., Fried, et al., 2005. MRI to assess the contribution of gastric peristaltic activity and tone to the rate of liquid gastric emptying in health. Proc. Int. Soc. Magn. Res. Med. 13, 426.

Steingoetter, A., Fox, M., Treier, R., Weishaupt, D., Marincek, B., Bösiger, P., et al., 2006. Effects of posture on the physiology of gastric emptying: a magnetic resonance imaging study. Scand. J. Gastroenterol. 41, 1155—1164.

Treier, R., Steinetter, A., Weishaupt, D., Goetze, O., Bösiger, P., Fried, M., et al., 2006. Gastric motor function and emptying in the right decubitus and seated body position as assessed by magnetic resonance imaging. J. Magn. Reson. Imaging. 23, 331−338.

van den Elzen, B.D.J., Bennink, R.J., Wieringa, R.E., Tytgat, G.N.J., Boeckxstaens, G.E.E., 2003. Fundic accommodation assessed by SPECT scanning: comparison with the gastric barostat. Gut 52, 1548−1554.

Wickham, M.J.S., Faulks, R.M., Mann, J., Mandalari, G., 2012. The design, operation, and application of a dynamic gastric model. Dissolut. Technol. August, 15−22.

Xue, Z., Ferrua, M.J., Singh, R.P., 2012. Computational fluid dynamics modeling of granular flow in human stomach. Alimentos Hoy 21, 3−14.

Food developments to meet the modern challenges of human health

Chapter 13

Applying Structuring Approaches for Satiety: Challenges Faced, Lessons Learned

David J. Mela[1] and Mike J. Boland[2]

[1]*Unilever R&D Vlaardingen, AC Vlaardingen, The Netherlands,* [2]*Riddet Institute, Palmerston North, New Zealand*

CONTENTS

Food Structures, Digestion and Health. http://dx.doi.org/10.1016/B978-0-12-404610-8.00013-X

INTRODUCTION
Satiety in Context

Obesity, together with its related health disorders, is recognized as a major problem in most developed regions and is a significant and increasing problem in developing regions as well. Foods and diets that can be advised and designed for improved weight management are an obvious potential solution. Structuring of foods gives the possibility of both managing the physical properties of a food and controlling release of nutrients during the digestive process, offering potential mechanisms for enhanced weight management benefits.

Potential targets for functional foods for weight management include:

- reduction of energy intake
- interference with energy uptake, and
- alteration of energy metabolism.

Interference with energy uptake from the total diet results in reduced absorption and increased excretory losses. Examples include inhibition of lipases and amylases to interfere with the digestive process, fat "trapping," formation of insoluble complexes and soaps in the gut, fat or carbohydrate modification, and possibly specific probiotics that interfere with energy availability. The observed effects of these approaches are, however, typically quite small (Boon *et al.*, 2005; Hsu *et al.*, 2006; Bendsen *et al.*, 2008). In practice this approach might perhaps reduce energy balance by up to 50−100 kcal/d, as larger effects might incur issues of safety and micronutrient availability. While these small effects may have a meaningful cumulative impact on body weight if sustained (and without compensation), such benefits would be slow and modest relative to what is achievable through dietary variations in energy intake.

Effects of changes in diet composition to alter energy metabolism could include increased thermogenesis, changes to substrate delivery and oxidation, and shifts in nutrient partitioning between fat and lean body mass. Except at extremes, diet composition has relatively small effects on energy expenditure (Buchholz and Schoeller, 2004). Furthermore, the increases in thermogenesis that could safely be achieved and sustained by active agents or compositions are probably at most around 2−3% of total energy expenditure (e.g., Hursel *et al.*, 2011; Ludy *et al.*, 2012; Whiting *et al.*, 2012; Gregersen *et al.*, 2013). As with altering energy availability, these effects if sustained may contribute meaningfully toward weight management, but benefits will be relatively slow and modest at an individual level.

Relative to other routes, reduction of energy intake can achieve the most substantial and immediate effects, and this approach is the primary target for most weight control foods and programmes (Mela, 2007, 2011). Although managing feelings of eating motivations (feelings of satiety, fullness, hunger, desire to eat) may offer many different benefits to consumers, one of these is clearly the potential to aid in the management of energy intake and body weight (Halford and Harrold, 2012; Hetherington *et al.*, 2013). A range of putative mediators of satiety occurs throughout the gastrointestinal tract, associated organs, and in the brain. Many of these could be suitable targets for satiating functional foods (Figure 13.1). Potential "functional" physiological targets for weight management and loss through reduction of energy intake could act by enhancing satiety or food intake control through either gut-based or post-absorptive actions. The latter would include possible effects of signals derived from post-meal nutrient metabolism (Langhans, 1996; Leonhardt and Langhans, 2005; Veldhorst *et al.*, 2008). In contrast, structural effects of foods will primarily act from within the gut, and these are considered further in this chapter.

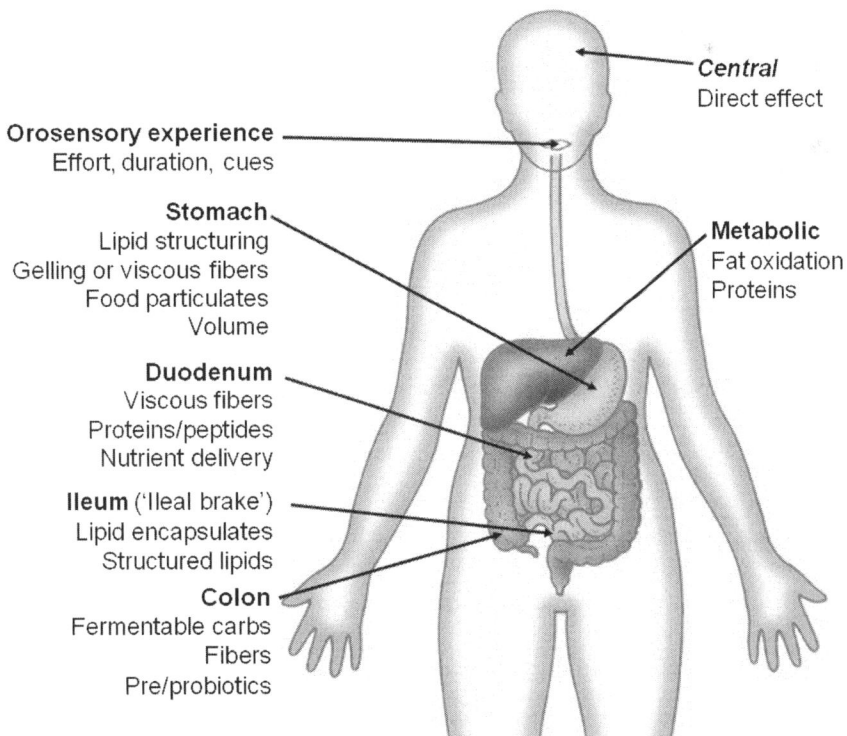

■ **FIGURE 13.1** Potential targets for foods and food ingredients for enhancing satiety, with some examples of possible mediators. *Adapted from Mela (2007). Copyrights Wiley-VCH Verlag GmbH & Co. KGaA and Dorling Kindersley. Reproduced with permission.*

SATIATION AND SATIETY EFFECTS OF FOODS

Within the field of appetite control, reference is made to processes of "satiation" and "satiety" (Blundell *et al.*, 2010; Figure 13.2), and it is important to be clear about the distinction between these. Satiation is generally regarded as the set of processes that lead to the termination of eating, which may be accompanied by a feeling of satisfaction that sufficient food has been consumed. Experimentally, satiation thus refers to how much is eaten within a meal, and is operationalized as food or energy intake. In contrast, satiety refers to the feelings of fullness (suppression of hunger or desire to eat) that persist after eating, and the processes that act to suppress further energy intake until hunger returns. This may also extend into an effect on subsequent energy intake (e.g., at the next meal). Satiation is generally considered to occur over the period of a meal, whereas satiety typically lasts for several hours thereafter. Satiation and satiety occur as a result of a range of interactions between the gut and brain, associated with various putative peptide signaling molecules, as well as cognition and perceptual cues. Approaches to reducing food intake could potentially act via either satiation or satiety, or both.

Approaches to the reduction of energy intake through satiation might include reducing portion sizes, volume dilution, or energy reduction so that we stop eating when we have consumed less energy. The feeling of post-meal satiety develops from an array of time-related physiological and cognitive factors. This ranges from the awareness of having eaten

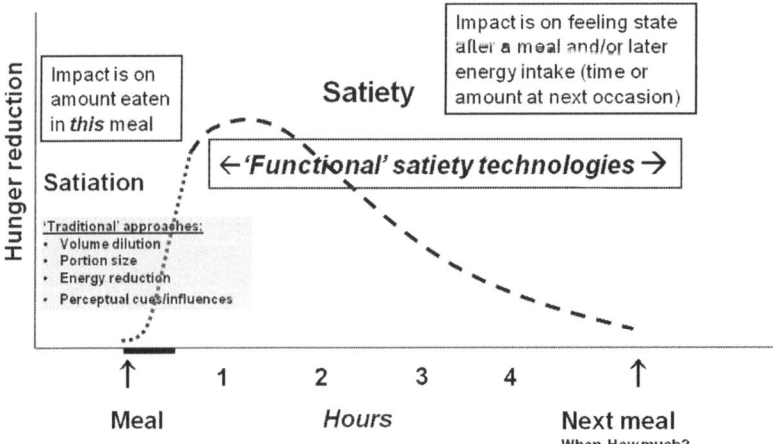

■ **FIGURE 13.2** Conceptualization and timeframes for processes of "satiation" and "satiety," indicating where traditional food-based approaches and potential "functional" technologies might be applied.

through gastric filling and intestinal transit, to a sequence of hormonal and metabolic responses to the presence of food breakdown products in the gut (Mela, 2007; Benelam, 2009; Smeets *et al.*, 2010).

There is increasing acceptance of a relatively standardized set of accepted procedures by which food effects on satiation or satiety (and related claims) can be assessed (Blundell *et al.*, 2010; EFSA Panel on Dietetic Products, Nutrition and Allergies, 2012). For feelings of hunger and satiety, line or "visual analogue" scales (VAS) are invariably used for collecting self-reported measures of eating motivations after consuming test foods. In most research sites this is now automated, with subjects prompted to record scores on hand-held devices at designated time points (Stubbs *et al.*, 2000), and thus relatively inexpensive and straightforward in execution. With the appropriate study design and controls in place, these methods provide for an effective, surprisingly sensitive, and objective measure of subjective sensations. Studies of satiety may also follow a period of self-reports with a later test meal, where *ad libitum* energy intake or time of the next meal request are recorded as additional measures. These behavioral measures may be accompanied by physiological measures, e.g., gastric emptying rates or satiety-related hormones. However, as discussed further in this chapter, such "biomarkers" offer mainly mechanistic insights, as they are often poorly related to the accepted behavioral measures of efficacy and claims support.

CURRENT STATUS OF "FUNCTIONAL" APPROACHES TO SATIETY

Most current product claims related to satiety or appetite control are based on applications of relatively standard, known ingredients, particularly macronutrient-type components such as fibers and proteins.

The difficulty with using these ingredients in a generic way to support claims is that they are variable in composition and structure, depending on source and processing, and are consumed in food matrices rather than in isolation. Thus, evidence derived from whole diets or very high levels of addition may not apply to use of lower levels in specific foods. A further difficulty with many fibers is that they may have very slow rates of hydration, and when delivered in low-moisture products may not hydrate rapidly enough to be effective during the time range of satiety effects (Kristensen and Jensen, 2011). Satiety effects are highly dependent on the level and *exact* physicochemical properties of the ingredients (Peters and Mela, 2008a; Ström *et al.*, 2009; Wanders *et al.*, 2011). These physicochemical properties are determined not only by the primary chemical structure, but can be affected profoundly by the food matrix and by *in-body* conditions, both physical and chemical. Physicochemical conditions in model systems are not the same as those in food, which in turn are not the same as body conditions. For example, in the stomach food is exposed to very low pH and to shearing effects, and is mixed with other components of the source food and of other foods eaten during the same meal. Thus, properties of macromolecular ingredients on their own or in whole or formulated foods can be poor predictors of how they will behave during digestion. This means that a great deal of caution is necessary when extrapolating generic claims such as that fiber or protein leads to satiety, and this needs to be tested for each particular food application. These issues are further considered in the subsequent section.

In addition to these generic ingredient approaches, a range of different "functional" technologies with putative appetite control efficacy have been developed by various food and ingredient manufacturers. The market potential for such (proprietary) solutions makes them highly attractive, but the costs and time needed for developing ingredients that are truly novel (in terms of source, chemistry, or biological target) and which can be brought to market with regulatory approval and substantiated claims, are barriers to commercial investment and feasibility. Worse, to date most of the agents proposed or currently sold as "functional" food ingredients for appetite control appear to lack consistent support for efficacy, or are unsuitable or ineffective in desired food formats (Wiseman *et al.*, 2008; Keogh *et al.*, 2010; Appleton *et al.*, 2011; Smit *et al.*, 2011; Blom *et al.*, 2011; Peters *et al.*, 2009, 2011a; Verhoef and Westerterp, 2011; Chan *et al.*, 2012; Márquez *et al.*, 2012).

STRUCTURING APPROACHES TO ENHANCING SATIETY FUNCTIONALITY IN FOODS

The challenges outlined above for "generic" and "functional" approaches suggest it may be more promising to focus on selectively amplifying known physiological effects of food components that contribute to satiety. There are a range of fairly well-characterized gastrointestinal targets that may be considered (Peters and Mela, 2008b; Delzenne *et al.*, 2010). The gastrointestinal tract has several advantages as a primary site for satiety effects of foods in terms of likely feasibility, safety, and fit with food-based concepts. Food structuring may be a particularly promising approach for the development of foods designed to significantly increase satiety. Two examples we focus on here are the selection of specific fibers to generate desired in-body structural effects, and the specific structuring of food lipids to delay their delivery and digestion.

Fibers for Satiety

The definition of fiber has always been problematic. Following recommendations of an expert panel in 2009, fiber has been defined by Codex:

Dietary fibre means carbohydrate polymers with ten or more monomeric units, which are not hydrolysed by the endogenous enzymes in the small intestine of humans and belong to the following categories:

- edible carbohydrate polymers naturally occurring in the food as consumed,
- carbohydrate polymers, which have been obtained from food raw material by physical, enzymatic or chemical means and which have been shown to have a physiological effect of benefit to health as demonstrated by generally accepted scientific evidence to competent authorities,
- synthetic carbohydrate polymers which have been shown to have a physiological effect of benefit to health as demonstrated by generally accepted scientific evidence to competent authorities.

In practice, fiber means hemicelluloses, pectins, and gums (soluble fiber), as well as cellulose, lignocellulose, and certain resistant starches (insoluble fiber). The fibers generally used in formulated foods are the soluble fibers, and include hemicelluloses, pectins, and gums from a broad selection of plant sources. These can have a wide range of physical functions in the food, such as stabilization of microstructure and increasing viscosity. The challenge in developing fibers for satiety is to find a defined fiber source and formulation that has:

- significant, consistent satiety effects in the feasible ranges,
- process and storage stability, and
- low sensory impact.

Viscous and Gelling Fibers

An example that highlights some of the challenges noted above is the application of fibers to enhance satiety in a dairy-based meal replacer shake (e.g., Peters *et al.*, 2011b). The proposed solution here is to find fiber systems that would be acceptable in a liquid product, yet develop more solid in-body structural characteristics (gelling or viscosity), and thus increase feelings of fullness.

Guar gum and alginate are types of fibers that have shown particular promise in producing satiety effects. Guar gum is a soluble carbohydrate from the Indian cluster bean (*Cyanopsis tetragonoloba*), a polymer of galactose and mannose in a ratio of about 2:1. The molecular weight ranges from 50,000 to 8,000,000, with the higher molecular weight polymers being more viscous in solution. At higher molecular weights and sufficient concentrations, guar gum gives viscous solutions when added in small amounts to water-based drinks, and retains its viscosity at the acid pH of the stomach. Published data with guar have generated mixed results with regard to satiety (Wanders *et al.*, 2011). However, as with many other fibers, the specific types of guar used in studies are variable and rarely sufficiently described and characterized, and conclusions cannot be generalized to all "guar." Not surprisingly, lower viscosity forms of guar gum—which are also much more food feasible—are unlikely to deliver the same level of effects on eating motivations as more viscous versions (Marciani *et al.*, 2001; Kristensen and Jensen, 2011).

Alginate (also called alginic acid) is an extract of various species of brown seaweed (*Phaeophyceae*) that is composed of linear copolymers of β-1-4 D-mannuronic acid and α-1-4-guluronic acid. Because alginate is prepared from many different seaweed sources, many specifications are available, for example one manufacturer offering more than 200 different types. Because of the highly charged nature of this polymer, it has low viscosity at neutral pH, but becomes viscous in the presence of divalent cations such as calcium, which can cross-link the polymer chains, or at low pH when the acidic side chains become protonated. The strength of the gels of alginate depends on the ratio of mannuronic to guluronic acid, with high guluronic acid alginates forming the strongest gels. Alginates are therefore a good candidate for use in a food that will have low viscosity as consumed, but can be expected to form a highly viscous system in the acid environment of the stomach. The theory and evidence for the use of alginates for appetite control and other health benefits has recently been comprehensively reviewed by Jensen *et al.* (2013).

The effects of both guar gum and strong-gelling and weak-gelling alginate in a milk-based meal replacer drink were explored *in vivo* using magnetic resonance imaging (Hoad *et al.*, 2004). Images of the stomach showed that guar remained homogeneous in the stomach, but the alginates formed lumps, with the strong-gelling alginate producing the largest volume. Compared with the control meal (which contained no added fiber) the meals all gave significantly greater scores for feelings of fullness at 115 minutes after ingestion, although stomach emptying times were not significantly different. This and other experiments have suggested that a formulation delivering a gel strength under gastric conditions of >2N is required for a reliable enhancement of satiety in a dairy-based shake format (Ström *et al.*, 2009).

The effects of alginate-containing beverages on satiety have been further explored by several groups. A comparison of high- and low-gulonate alginates in a preload resulted in increased feeling of fullness and reduced subsequent energy intake by 10% for the high-gulonate preload (Jensen *et al.*, 2012). Solah *et al.* (2010) compared the effects of whey protein and high- and low-gulonate alginate on eating motivations. This, too, showed that the high-gulonate alginate was more effective at reducing hunger, and that this reduced hunger more than the protein formulation (Solah *et al.*, 2010). Other studies showing beneficial satiety effects of alginates in liquid formats have been reported by Pelkman *et al.* (2007) and Paxman *et al.* (2008). In addition, a certain minimum product volume may be needed for an effect of alginates to be apparent. Ström *et al.* (2009) found that a mix that enhanced satiety in a 325 ml product was ineffective in 100 ml. Additional studies exploring the relationship between alginate concentrations and liquid volume suggest a volume of >200 ml may be required (unpublished data).

Applications Issues for the use of Gelling Fibers for Satiety

One of the key technical challenges with commercial use of alginates in ready-to-consume products has been to ensure good gel formation in the body, while avoiding thickening and gelling in foods or beverages during shelf-life. To prevent unwanted, spontaneous gelation in experimental products, alginate solutions have been combined with a separate source of soluble calcium or alginate powders hydrated only immediately prior to consumption (Pelkman *et al.*, 2007; Paxman *et al.*, 2008; Solah *et al.*, 2010). Peters *et al.* (2011b) used protein-based meal replacer drink with a

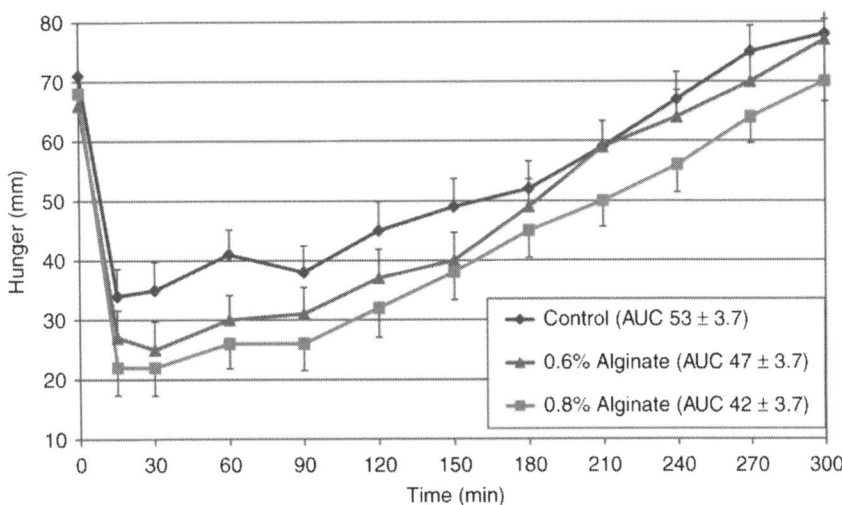

■ **FIGURE 13.3** Hunger ratings (least square (LS) means, $n = 23$) after consumption of control drink or drink with addition of 0.6 or 0.8% alginate. The area under the curve of 0.8% alginate significantly different from control ($P < 0.001$). AUC (LS mean ± s.e.) is shown as an insert in the legend. *Reproduced from Peters* et al. *(2011b), with permission.*

source of calcium that was insoluble under product conditions but soluble at gastric pH. An example of the effects of such a high gulonate alginate is shown in Figure 13.3 (from Peters *et al.*, 2011b). In this study, the alginate was added at either 0.6 or 0.8% in an acceptable, low-viscosity meal replacer drink. The drink was consumed by 23 volunteers in a crossover design trial and hunger and satiety were scored every 30 minutes following ingestion for 5 hours using the electronic visual analogue scoring system described above. The 0.8% alginate reduced "hunger" (area under the curve) by 20–30% ($p < 0.001$). Similar effects were seen for scoring for "fullness," "appetite for a meal," "appetite for a snack," and "how much do you want to eat?"

It is clear that a significant increase in satiety is achievable in drink formats through the use of gelling polymers such as alginate, but making a drink product with acceptable stability and sensory qualities is still a challenge.

Another strategy would be to use fibers in a solid format. The use of a combination of alginate and guar gum (sources and types unspecified) in a breakfast bar format was explored by Mattes (2007) in a double-blind randomized crossover study. Ratings of fullness, hunger, desire to eat, and prospective food consumption were recorded on visual analogue scales and dietary intake was recorded. No significant treatment effects were observed over each 5-hour post-loading period. There also was no evidence of a cumulative effect over the five treatment days. Daily energy intake was not different on the two treatments. A similar lack of efficacy results has been found in other studies adding high-guluronate alginate to bars, and rehydrated "instant" soup and pasta mixes (unpublished data). Poor efficacy in these formats may reflect relatively low product volumes or insufficient hydration of the alginates.

The use of dietary fibers and the relationship between satiety and physical measures has recently been reviewed in a meta-study of nine studies in which the dietary fibers had been adequately characterized in physicochemical terms, including the studies described above (Kristensen and Jensen, 2011). This review concluded that these studies indicate that a major increase in viscosity is necessary for a clinically relevant response. The review calls for more work to address the different effects of liquid and solid meals containing fibers on physical characteristics of digesta. It also highlights that most of the studies are short term, and the longer-term effect of meals containing fibers needs to be addressed to rule out the possibility of adaptation to gastric distension and viscosity.

It can be concluded that there is excellent proof of principle and demonstration of technical feasibility for the use of specific types of alginates in beverage formats to reduce hunger and increase satiety in the short term, but it must be noted that the effectiveness depends very much on the specific amount and type of material used (e.g., high-gulonate alginate) and the effect is volume and formulation dependent. Other fibers may also be effective if they meet relevant criteria for in-body physical effects. The longer-term effect of these materials is still to be determined.

Summary: Challenges for the Application of Fibers for Enhanced Satiety in Foods

- **In-body activation/functionality:** Need to ensure that the added ingredient mix has the desired functionality in the body. Because this is usually mediated by viscosity or gelling, steps need to be taken to minimize impact on the product and processing, e.g., by being "activated" only under in-body conditions.
- **Low dosing:** Allowable amounts of certain fibers may be limited by regulations, and lower levels will be more favorable in terms of product cost and quality.
- **Process and storage stability:** Additions of viscous or gelling fibers may affect processing (e.g., pumping, flow, and mixing) and the shelf-stability needs to be carefully checked.
- **Use in solid foods:** Under low moisture conditions, added fibers may have more limited feasibility and efficacy.
- **Sensory quality for desired formats:** In order to ensure desired sensory quality, fiber-based systems will likely have to be tailored for each specific food format.
- **Ownership of proprietary technologies:** Because of the very high level of prior art in this area, there may be relatively narrow opportunities for gaining a proprietary commercial advantage through exclusive technologies or formulations.

Lipid Structuring for Satiety

Lipids are an important component of most foods. They carry essential fatty acids and provide important sensory experiences, through mouthfeel and as carriers of many desirable flavor components. Long-chain fatty acids derived from digestion of common dietary fat sources are known to be potent stimulators for hormones involved in feedback into gastrointestinal processes and potentially also eating motivations and behavior (Little, *et al.*, 2007a, b; Maljaars *et al.*, 2008a). Although lipids are also a dense source of energy, changing the rates or sites of lipid digestion and uptake may offer routes to trigger satiety even at relatively low doses.

The "Ileal Brake"

Lipids are an attractive candidate for enhancing satiety through the "ileal brake" mechanism, which has been extensively investigated and proposed as a target for foods to help control appetite (Maljaars *et al.*, 2008a). The ileal brake, first described in 1984 (Spiller *et al.*, 1984), refers to a cascade of effects that occurs when the distal small intestine is exposed to nutrients, such as free fatty acids, leading to decreased gastrointestinal motility (the "brake") and an increase in feelings of satiety. The ileal brake is believed to be mediated by one or more circulating factors, although it is unclear how this works. Factors that have been proposed include a peptide that terminates in two tyrosines, known as peptide YY or PYY, cholecystokinin (CCK), and glucagon-like peptide 1 (GLP-1).

Proof of principle for the use of small amounts of lipids in stimulating the ileal brake effect on appetite measures has been well established through a recent set of studies in which lipid emulsions (as a source of triacylglycerides) have been directly introduced into the ileum via a naso-ileal tube (Maljaars *et al.*, 2008b, 2009, 2011). In the first of these trials (Maljaars *et al.*, 2008b), a fat-free meal replacer shake was given at 0 and 4 h (240 min), with an ileal infusion of 3 or 10 g oil emulsion from 105 to 150 min. The control condition was 3 g of oil in the shake itself, and a saline infusion. Key outcome measures were hunger and satiety ratings by visual analogue scales,

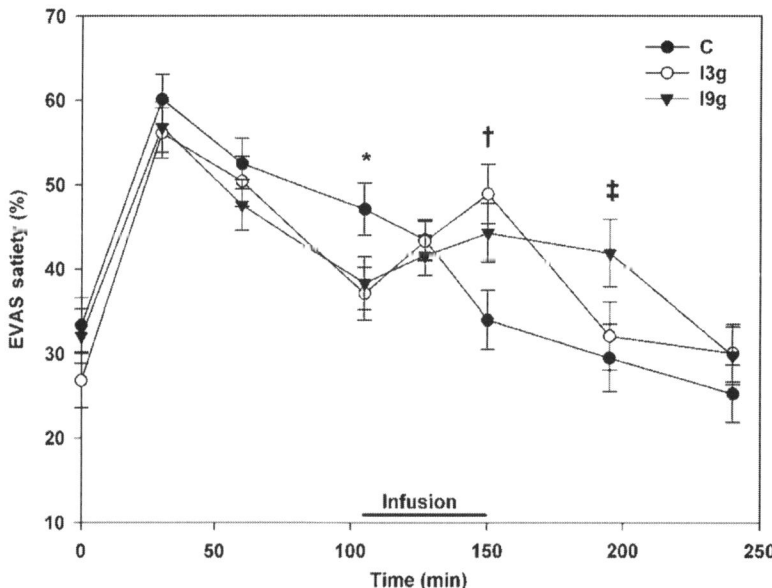

■ **FIGURE 13.4** Demonstration of lipid-induced ileal brake effects on eating motivations. Electronic visual analog scale (EVAS) scores for satiety, for subjects receiving either (●) a meal replacer containing 3 g lipid (control) followed by a saline infusion to the ileum, or a fat-free meal replacer followed by (○) 3 or (▼) 9 g lipid ileal infusion. Infusions took place over a 45 min period starting at 105 min after consuming the meal replacer. Statistically significant ($p < 0.05$) differences noted for control vs. both 3 and 9 g infusion*, 3 g infusion† only, and 9 g infusion‡ only. *Reproduced from Maljaars* et al. *(2008b) with permission.*

food intake (at 8 h), and blood levels of circulating CCK, PYY, and ghrelin. Three grams of lipid delivered to the ileum (vs. oral) had a significant effect on satiety (Figure 13.4) and food intake, and there was no additional effect with the 9 g infusion. During ileal fat perfusion, CCK increased in a dose-dependent manner, while PYY increased only at the high lipid level. Secretion of CCK but not PYY correlated with satiety scores.

In a further trial, 6 g of oil was delivered to different sites, the duodenum, the jejunum, and the ileum, sequentially or simultaneously, or the ileum alone (Maljaars *et al.*, 2011). Overall, ileal fat infusion had the most consistent and pronounced effects on food intake and satiety. Increasing the areas of exposure only affected hunger when fat was delivered simultaneously but not sequentially, and only ileal delivery significantly reduced food intake vs. control (oral fat). Interestingly, despite the behavioral effects, no statistically significant differences were observed for PYY, CCK, and GLP-1 compared with the control.

The Challenge of Delivering Ileal Brake Via Food Technologies

Triggering of the ileal brake by lipids requires that a meaningful dose of fatty acids is available for uptake in the more distal regions of the small intestine. These can be derived from local hydrolysis of triacylglycerides, as indicated by the infusion studies. Delivery of these in practice could be accomplished, for example, by use of encapsulates or structural designs to slow (but not inhibit) lipid digestion. A survey of the commercial and patent literature reveals a range of technology routes suggested to have potential application in achieving this (e.g., Mela, 2006; Chu *et al.*, 2009; Knutson *et al.*, 2010; Schellekens *et al.*, 2010; Li *et al.*, 2011; Müller *et al.*, 2012).

One of the difficulties in developing effective technologies is the lack of a validated *in vitro* model system for accurately predicting lipid delivery, and indeed the difficulty of validating actual delivery time and site clinically. While various lipid release models can be shown to be sensitive to manipulations, they do not necessarily give confidence as to what might happen *in vivo*. As a result, if a particular technology is not consistently effective, it is difficult and extremely costly to diagnose and correct the problem.

One structured oil-in-water emulsion claimed to act on appetite via the ileal brake has been commercialized and marketed (Fabuless™; DSM, 2014 [previously marketed as Olibra™ and Reducal™]). The structure of this emulsion, incorporating palm oil coated with oat galactolipids, is argued to protect the lipid from gastric digestion and allow release in the ileum (Knutson *et al.*,

2010; DSM, 2014). However, results for appetite-related measures in clinical trials with this ingredient have been highly variable (Appleton *et al.*, 2011). Initial studies reported that yoghurts containing this ingredient reduced self-reported appetite and markedly decreased subsequent food intake (Burns *et al.*, 2000, 2001, 2002). However, later studies—including those from the same research group—have been largely unable to replicate these findings in any consistent way, using a range of protocols and products, including yoghurts (Logan *et al.*, 2006; Diepvens *et al.*, 2007, 2008; Smit *et al.*, 2011, 2012; Chan *et al.*, 2012). Importantly, several of the studies have suggested that any possible efficacy of the emulsion may likely be lost when it is exposed to even mild food processing conditions and formats other than yoghurt (Smit *et al.*, 2011, 2012; Chan *et al.*, 2012).

Changing Rates of Gastric Lipid Delivery for Satiety?

Another approach to enhancing satiety using lipid structuring has aimed at changing the rates of lipid delivery from the stomach to the small intestine. Different experimental approaches have been used to manipulate this and evaluate potential impacts on appetite-related outcomes. Foltz *et al.* (2009) used a delivery by naso-gastric tube to compare delivery of oil in an emulsion with separate layering of oil on top of the aqueous phase. Others have looked at delivery of oil through emulsions that are unstable under gastric conditions and coalesce to give a lipid layer, compared with emulsions that are stable under gastric conditions (Marciani *et al.*, 2009; Keogh *et al.*, 2011). These studies have thus in different ways investigated the effect of intragastric layering: the formation of a layer of lipid material over the aqueous digesta in the fundus, in comparison with delivery of lipid as a (stable) emulsion. Parameters measured have included the rate of gastric emptying, blood lipid markers, blood peptides related to satiety, and self-reported hunger and satiety scores over a period of several hours post-ingestion. The results indicate inconsistent relationships between the physiological and behavioral measures. While significant changes in physiological function and biochemistry can often be brought about by layering the lipid phase or stabilizing it in an emulsion that can resist gastric conditions, the resulting effects on self-reported satiety and hunger, and subsequent food intake, showed at best small effects and in many cases no statistically significant effect at all. In effect, the various biophysical and biochemical mechanisms associated with the ingestion and digestion of fats and oils appear to be remarkably effective at dynamically normalizing the colloidal state of the fat during gastrointestinal transit. Accordingly, fine emulsions tend to display progressive coalescence, while free (or crudely dispersed) oil is able to undergo limited homogenization to produce coarse emulsion states. In support of this, it has been noted that phase-separated lipid structures were redispersed when they passed through the antrum and pylorus (Golding and Wooster, 2010).

Lipid Structuring and Satiety "Biomarkers"

As noted above, many clinical studies investigating lipid structuring for satiety and appetite control have included measures of possible hormonal mediators ("biomarkers") as adjuncts to behavioral outcomes. In developing functional foods and ingredients, the use of established biomarkers has been important where direct measure of outcomes (e.g., actual disease risk or incidence) is not readily feasible. Markers such as blood pressure and blood LDL-cholesterol have been invaluable as research tools and are accepted as evidence of efficacy for functional foods aimed at heart health claims, for example. In contrast, it is possible to directly measure acute variations in self-reported eating motivations and actual behaviors using methods that are accepted as sensitive, reliable, and robust (Blundell *et al.*, 2010). Thus, biomarkers are not *per se* needed to show efficacy or substantiate claims (EFSA, 2012). Nevertheless, they are potentially useful to identify putative mechanisms or confirm impacts on intended physiological targets. Application of markers for target confirmation may be especially valuable for research intended to manipulate lipid delivery, which is difficult to measure directly.

Peptides including CCK, peptide YY, GLP-1, and ghrelin have all been investigated as putative satiety biomarkers in many trials. However, while these and others have been implicated in the physiological signaling of hunger and satiety, their value as predictive markers of behavioral efficacy remains uncertain. There are many other examples of behavioral effects without evidence of corresponding changes in physiological markers and evidence of changes in markers but no significant behavioral effects (e.g., French *et al.*, 2000; Little *et al.*, 2005; Pilichiewicz *et al.*, 2006; Zijlstra *et al.*, 2009). The latter situation is more commonly observed, and particularly likely when biomarker evidence comes from *in vitro* testing of potential functional actives. Unfortunately, results obtained in a range of trials using intestinal perfusion of fat, i.e., using the ileal brake principle, also show these peptides to be inconsistently related to corresponding behavioral outcomes (Maljaars *et al.*, 2008b, 2009, 2011, 2012).

One of the problems is that the criteria for a "meaningful" change in proposed satiety biomarkers are not well established. Thus, "statistically" significant changes do not necessarily translate into biologically relevant effects. The degree of changes in the percentage or absolute response, the initial baseline value, threshold ceiling effects, and timing may all be relevant. A review covering 17 different studies of the effect of different foods on peptide markers (Mars *et al.*, 2012) concluded that GLP-1 and PYY are not very likely to contribute individually to a difference in the satiating capacity of foods, but that CCK is more likely to be relevant. The review further recommends that a combination of biomarkers should be used. From these and other studies, it is concluded that proposed biomarkers are not robust as proxy measures of satiety; they cannot be used as alternatives to behavioral measures and self-scoring, and they are not valid to support label claims. They are, however, a useful tool to provide insights and confirmation for possible mechanisms associated with satiety (Delzenne *et al.*, 2010).

Summary: Challenges for the Application of Lipid Structuring for Enhanced Satiety in Foods

- **Weak, inconsistent relationships *in vitro* data and biomarkers with *in vivo* functionality.** Although not needed as evidence of efficacy, poor ability to predict clinical efficacy of alternative technical approaches and confirm intended targets is a huge hurdle to the R&D process, making clinical trials an expensive gamble.
- **Stimulus control in gut.** An effective technical solution would need to offer a consistent rate or site of delivery and release of lipids, against a shifting environment, individual and diet variability, and product composition.
- **Efficacy needed at low levels of (preferably "healthy") lipids.** Because of the high energy density of lipids, any solution oriented toward appetite control and weight management needs to be effective at very low lipid loads, and preferably not requiring trans or saturated fatty acids.
- **Process/product stability.** For wider applications, the structure needed for efficacy needs to be robust to manufacturing and storage, and have a defined analytical specification so this can be monitored and verified.
- **Cost** (e.g., for encapsulation). Cost for many approaches to meet the challenges here may be prohibitive for wider food applications, though a highly effective technology could command a premium in the retail market.

FEASIBILITY ISSUES: FROM LABORATORY TO MARKET

The sections above have largely focused on gaining proof of principle for different physiological targets and technological approaches. While this is an important first step in commercial product development, it is nothing more than that. Together with clinical effectiveness, factory-scale production must be shown to be feasible and a viable market position must be identified (Mela, 2005).

Pilot and production-scale manufacture requires a reliable and consistent source of the active ingredients, and a way to describe and monitor required specifications to assure safety and efficacy. The decision to go into production requires appropriate cost models to show the product can be produced profitably, and an advantageous commercial position over possible competitors (e.g., through patents or control of supply or processing) will help to ensure an initial market lead can be sustained.

Products and ingredients also require regulatory and safety approval. While many ingredients are perceived to be "generally recognized as safe" because they have been consumed for many years, there is often still a need to assess levels of intake of a new product within the context of past history of consumption. Additional testing and assurances may be required for more novel targets, structures, or bioactivities to provide assurances of safety for sustained use in the general population.

For consumer acceptance, the product will of course need an adequate shelf-life and good sensory qualities. Crucially, the desired product claims and related communications for a food with added health functionality need to fulfill the applicable legal requirements for the regions where it is sold. In addition, many food manufacturers have their own internal standards for claims substantiation to ensure certain principles are adhered to globally, even if external requirements may be more lax. There may also be minimum nutritional requirements for products with health claims to ensure these do not appear on products that have a poor basic nutritional profile (Nijman *et al.*, 2007).

Satiety and appetite claims relating to specific technologies or formulations can be considered a type of "structure−function" claim. Most national or regional authorities have legislation addressing these types of claims in general, although detailed guidance or processes for scientific substantiation and evaluation are less common. However, there is increasing use of the "Process for the Assessment of Scientific Support for Claims on Foods" (PASSCLAIM; Aggett *et al.*, 2005; Gallagher *et al.*, 2011) system as a basis for evaluation the substantiation of claims. The general requirements this establishes for the scientific assessment of claims are summarized in Table 13.1. These principles are clearly embedded, for example, in the processes being used by the European Food Safety Authority (EFSA) as part of the process for health and nutrition claims authorization in the European Union (European Parliament, 2007).

Some aspects of the assessment merit particular attention as regards the potential use of food structuring or other approaches to deliver satiety, appetite, or weight control benefits. The first general requirement is for clear identification and complete characterization of the active component(s) and the food formulation in which they are to be used. This requirement, originally developed for

---

Table 13.1 Process for the Assessment of Scientific Support for Claims on Foods Criteria for the Scientific Substantiation of Claims

1. The food or food component to which the claimed effect is attributed should be characterized.

2. Substantiation of a claim should be based on human data, primarily from intervention studies, the design of which should include the following considerations:

  (a) Study groups that are representative of the target group.

  (b) Appropriate controls.

  (c) An adequate duration of exposure and follow-up to demonstrate the intended effect.

  (d) Characterization of the study group's background diet and other relevant aspects of lifestyle.

  (e) An amount of the food or food component consistent with its intended pattern of consumption.

  (f) The effect of the food matrix and dietary context on the functional effect of the component.

  (g) Monitoring of compliance with intake of food or food component under test.

  (h) The statistical power to test the hypothesis.

3. When the true endpoint of a claimed benefit cannot be measured directly, studies should use markers.

4. Markers should be:

  (a) Biologically valid in that they have a known relationship to the final outcome and their variability within the target population is known.

  (b) Methodologically valid with respect to their analytical characteristics.

5. Within a study the target variable should change in a statistically significant way and the change should be biologically meaningful for the target group consistent with the claim to be supported.

6. A claim should be scientifically substantiated by taking into account the totality of the available data and by weighing of the evidence.

From Aggett et al. (2005).

---

the pharmaceutical industry, is to ensure that the presence of the intact functional entity can be monitored. This is relatively straightforward with simple chemical compounds, but more difficult with complex molecules, such as the alginates. This becomes even more challenging when the activity is dependent on specific microstructures and the stability of those microstructures, as in the case of some of the lipid-based approaches. To gain regulatory approval, it will be necessary to have a precise definition of the active structure that can be the basis for analytical testing and product specification and that relates directly to its clinical effectiveness. Overcoming this hurdle presents a significant problem for foods that have their effect by means of structuring.

The second criterion is about effectiveness. This will always require one or more (usually several) human intervention studies, preferably double-blind randomized clinical trials. The qualifying notes (a to h) are all pertinent to such trials and establish the "gold standard" requirements for experimental design. Trials should also be carried out with representatives of the target population—this may mean carrying out several trials with different groups.

Regulatory authorities are prepared to accept evidence based on biomarkers, if they can be shown to accurately reflect the desired effect (criteria 3 and 4).

As discussed above, direct measurements of satiety and appetite control effects are possible, and putative physiological biomarkers have been found to be inconsistent indicators of potential benefits. Ironically, though, broad acceptance of "subjective" behavioral measures for claim substantiation was slow to develop, despite clear evidence of their greater sensitivity and reliability in comparison with most biological measures.

The final two criteria are about assuring consumer and commercial confidence. Demonstrated effects must be shown (with appropriate statistical designs) to be both statistically and biologically significant (EFSA Scientific Committee, 2011), the totality of studies and data in the literature must tell a consistent story, and the effect demonstrated in the research must be meaningful and achievable in the real world context. The practical assessment of these has been further considered in subsequent papers (e.g., Gallagher *et al.*, 2011).

At present, specific, official regulatory guidance for substantiation of satiety and appetite control claims is only available in the European Union (EFSA Panel, 2012). Draft guidance for satiety claims has gone out for public consultation in Canada, and a final version may be expected in the near future. Interestingly, both of these look for applicants to provide evidence that the effects are retained with sustained consumption of the product, "in order to exclude adaptation through compensatory mechanisms" (EFSA Panel, 2012). Evidence of changes in biochemical markers or mediators is *not* accepted as primary evidence for behavioral effects, though may support this. Contrary to popular beliefs, the United States Food and Drug Administration does not review or (pre-approve) structure–function claims for foods, although manufacturers may be required to provide substantiation to the Federal Trade Commission if challenged (United States Government Accountability Office, 2011). However, the US authorities also do not have any detailed criteria for the evaluation of claims specifically in this area.

CONCLUSIONS: LESSONS LEARNED

The title of this chapter is "Applying structuring approaches for satiety: challenges faced, lessons learned." The challenges are clear, and there are three key lessons that have been learned:

Structuring Approaches can be Effective in Enhancing Satiety

The approach with fibers has clearly shown that structures can promote satiety. The evidence is good for certain viscous fibers, and especially gelling fibers, although these effects are highly dependent on the specific fiber, product format, and formulation. The ileal brake approach for lipids has been shown to be efficacious in principle, but a consistently effective and feasible commercial application of this is not yet a practical reality. Other approaches based on lipid structuring clearly would need further development to be considered a viable route for satiety enhancement.

Proof of Principle is *not* Evidence of Feasibility

Feasibility in a real product is often the real barrier, rather than efficacy. Issues such as process stability and shelf-life are particularly important if a product is to make it to the supermarket shelves. The sensory attributes have to be right if the product is to succeed.

In addition, predictive models (rather than complete reliance on clinical testing) are needed in order to manage costs and timeframes. For fibers, certain instrumental measures may already give sufficient information to be used as a predictive step or criterion in the R&D process (Ström *et al.*, 2009). However, for lipid-based approaches, "screening" methods and *in vitro* tests have shown inconsistent relationships with clinical efficacy. Putative biomarkers have also been shown to be unreliable (and unnecessary) as indicators of efficacy, though they can be used to confirm mechanisms and targets.

One recommendation to overcome the lack of valid predictive models is to make more use of exploratory study designs at early stages, rather than the types of clinical testing required for claims confirmation. For example, partial factorial designs can be used as an efficient way to test large numbers of technical alternatives within a single protocol and response surface designs can be used to optimize treatment conditions. Exploratory research intended as a filtering or "go/no-go" decision-making step can also make use of different criteria (e.g., for effect size or statistical significance) than confirmatory trials.

The Name of the Game is Claims

For commercial purposes, a wide range of criteria need to be fulfilled in order to substantiate claims, to satisfy regulatory oversight, and to ensure consumer confidence. "If you can't define it, you can't claim it." This is a particular problem for "functional" technologies that rely on complex structures, and it remains to be seen how this hurdle will be overcome. Effective development and substantiation of products with satiety or appetite control claims also requires multidisciplinary teams, including physical, physiological, and behavioral expertise, as well as clinical and statistical support.

REFERENCES

Aggett, P.J., Antoine, J.-M., Asp, N.G., Bellisle, F., Contor, L., Cummings, J.H., et al., 2005. PASSCLAIM—process for the assessment of scientific support for claims on foods. Consensus on criteria. Eur. J. Nutr. 44, 1—30.

Appleton, K.M., Smit, H.J., Rogers, P.J., 2011. Review and meta-analysis of the short-term effects of a vegetable oil emulsion on food intake. Obesity Rev. 12, e560—e572.

Bendsen, N.T., Hother, A.L., Jensen, S.K., Lorenzen, J.K., Astrup, A., 2008. Effect of dairy calcium on fecal fat excretion: a randomized crossover trial. Int. J. Obesity 32, 1816—1824.

Benelam, B., 2009. Satiation, satiety and their effects on eating behaviour. Nutr. Bull. 34, 126—173.

Blom, W.A.M., Abrahamse, S.L., Bradford, R., Duchateau, G.S.M.J.E., Theis, W., Orsi, A., et al., 2011. Effects of 15-d repeated consumption of *Hoodia gordonii* purified extract on safety, *ad libitum* energy intake, and body weight in healthy, overweight women: a randomized controlled trial. Am. J. Clin. Nutr. 94, 1171—1181.

Blundell, J., de Graaf, C., Hulshof, T., Jebb, S., Livingstone, B., Lluch, A., et al., 2010. Appetite control: methodological aspects of the evaluation of foods. Obesity Rev. 11, 251—270.

Boon, N., Hul, G.B., Viguerie, N., Sicard, A., Langin, D., Saris, W.H., 2005. Effects of 3 diets with various calcium contents on 24-h energy expenditure, fat oxidation, and adipose tissue message RNA expression of lipid metabolism-related proteins. Am. J. Clin. Nutr. 82, 1244—1252.

Buchholz, A.C., Schoeller, D.A., 2004. Is a calorie a calorie? Am. J. Clin. Nutr. 79, 899S—906S.

Burns, A.A., Livingstone, M.B., Welch, R.W., Dunne, A., Robson, P.J., Lindmark, L., et al., 2000. Short-term effects of yoghurt containing a novel fat emulsion on energy and macronutrient intakes in non-obese subjects. Int. J. Obesity (London) 24, 1419—1425.

Burns, A.A., Livingstone, M.B., Welch, R.W., Dunne, A., Reid, C.A., Rowland, I.R., 2001. The effects of yoghurt containing a novel fat emulsion on energy and macronutrient intakes in non-overweight, overweight and obese subjects. Int. J. Obesity (London) 25, 1487—1496.

Burns, A.A., Livingstone, M.B., Welch, R.W., Dunne, A., Rowland, I.R., 2002. Dose—response effects of a novel fat emulsion (Olibra) on energy and macronutrient intakes up to 36 h post-consumption. Eur. J. Nutr. 56, 368—377.

Chan, Y.-K., Strik, C.M., Budgett, S.C., McGill, A.-T., Proctor, J., Poppitt, S.D., 2012. The emulsified lipid Fabuless (Olibra) does not decrease food intake but suppresses appetite when consumed with yoghurt but not alone or with solid foods: a food effect study. Physiol. Behav. 105, 742—748.

Chu, B.S., Rich, G.T., Ridout, M.J., Faulks, R.M., Wickham, M.S., Wilde, P.J., 2009. Modulating pancreatic lipase activity with galactolipids: effects of emulsion interfacial composition. Langmuir 25, 9352—9360.

Delzenne, N., Blundell, J., Brouns, F., Cunningham, K., de Graaf, K., Erkner, A., et al., 2010. Gastrointestinal targets of appetite regulation in humans. Obesity Rev. 11, 234—250.

Diepvens, K., Soenen, S., Steijns, J., Arnold, M., Westerterp-Plantenga, M., 2007. Long-term effects of consumption of a novel fat emulsion in relation to body-weight management. Int. J. Obesity 31, 942—949.

Diepvens, K., Steijns, J., Zuurendonk, P., Westerterp-Plantenga, M., 2008. Short-term effects of a novel fat emulsion on appetite and food intake. Physiol. Behav. 95, 114—117.

DSM, 2014. http://www.dsm.com/markets/foodandbeverages/en_US/products/nutraceuticals/fabuless.html. Accessed 15 January 2014.

EFSA Panel on Dietetic Products, Nutrition and Allergies, 2012. Guidance on the scientific requirements for health claims related to appetite ratings, weight management, and blood glucose concentrations. EFSA J. 10, 2604. http://dx.doi.org/10.2903/j.efsa.2012.2604 [11 pp.].

EFSA Scientific Committee, 2011. Statistical significance and biological relevance. EFSA J. 9, 2372. http://dx.doi.org/10.2903/j.efsa.2011.2372 [17 pp.].

European Parliament and the Council of the European Union, 2007. Corrigendum to Regulation (EC) No. 1924/2006 of the European Parliament and of the Council of 20 December 2006 on nutrition and health claims made on foods. OJEU 18 January, 1—16.

Foltz, M., Maljaars, J., Schuring, E.A.H., van der Wal, R.J.P., Boer, T., Duchateau, G.S.M., et al., 2009. Intragastric layering of lipids delays lipid absorption and increases plasma CCK but has minor effects on gastric emptying and appetite. Am. J. Physiol.—Gastric and Liver Physiology 296, G982—G991.

French, S.J., Conlon, C.A., Mutuma, S.T., Arnold, M., Read, N.W., Meijer, G., Francis, J., 2000. The effects of intestinal infusion of long-chain fatty acids on food intake in humans. Gastroenterology 119, 943—948.

Gallagher, A.M., Meijer, G.W., Richardson, D.P., Rondeau, V., Skarp, M., Stasse-Wolthuis, M., et al., 2011. A standardised approach towards proving the efficacy of foods and food constituents for health CLAIMs (PROCLAIM): providing guidance. Br. J. Nutr. 106 (Suppl. 2), S16—S28.

Golding, M., Wooster, T., 2010. The influence of emulsion structure and stability on lipid digestion. Curr. Opin. Coll. Inter. Sci. 15, 90—101.

Gregersen, N.T., Belza, A., Jensen, M.G., Ritz, C., Bitz, C., Hels, O., et al., 2013. Acute effects of mustard, horseradish, black pepper and ginger on energy expenditure, appetite, ad libitum energy intake and energy balance. Br. J. Nutr. 109, 556—563.

Halford, J.C.G., Harrold, J.A., 2012. Satiety-enhancing products for appetite control: science and regulation of functional foods for weight management. Proc. Nutr. Soc. 71, 350—362.

Hetherington, M.M., Cunningham, K., Dye, L., Gibson, E.L., Gregersen, N.T., Halford, J.C.G., et al., 2013. Potential benefits of satiety to the consumer: scientific considerations. Nutr. Res. Rev. 26, 22—38.

Hoad, C.L., Rayment, P., Spiller, R.C., Marciani, L., de Celis Alonso, B., Traynor, C., et al., 2004. In vivo imaging of intragastric gelation and its effect on satiety in humans. J. Nutr. 134, 2293—2300.

Hsu, T.F., Kusumoto, A., Abe, K., Hosoda, K., Kiso, Y., Wang, M.F., Yamamoto, S., 2006. Polyphenol-enriched oolong tea increases fecal lipid excretion. Eur. J. Clin. Nutr. 60, 1330—1336.

Hursel, R., Viechtbauer, W., Dulloo, A.G., Tremblay, A., Tappy, L., Rumpler, W., Westerterp-Plantenga, M.S., 2011. The effects of catechin rich teas and caffeine on energy expenditure and fat oxidation: a meta-analysis. Obesity Rev. 12, e573—e581.

Jensen, M.G., Knudsen, J.C., Viereck, N., Kristensen, M., Astrup, A., 2012. Functionality of alginate based supplements for application in human appetite regulation. Food Chem. 132, 823—829.

Jensen, M.G., Pedersen, C., Kristensen, M., Frost, G., Astrup, A., 2013. Review: efficacy of alginate supplementation in relation to appetite regulation and metabolic risk factors: evidence from animal and human studies. Obesity Rev. 14, 129—144.

Keogh, J.B., Woonton, B.W., Taylor, C.M., Janakievski, F., Desilva, K., Clifton, P.M., 2010. Effect of glycomacropeptide fractions on cholecystokinin and food intake. Br. J. Nutr. 104, 286—290.

Keogh, J.B., Wooster, T.J., Golding, M., Day, L., Otto, B., Clifton, P.M., 2011. Slowly and rapidly digested fat emulsions are equally satiating but their triglycerides are differentially absorbed and metabolized in humans. J. Nutr. 141, 809—815.

Knutson, L., Koenders, D.J.P.C., Fridblom, H., Viberg, A., Sein, A., Lennernäs, H., 2010. Gastrointestinal metabolism of a vegetable-oil emulsion in healthy subjects. Am. J. Clin. Nutr. 92, 515—524.

Kristensen, M., Jensen, M.G., 2011. Dietary fibres in the regulation of appetite and food intake. Importance of viscosity. Appetite 56, 65—70.

Langhans, W., 1996. Metabolic and glucostatic control of feeding. Proc. Nutr. Soc. 55, 497—515.

Leonhardt, M., Langhans, W., 2005. Fat oxidation, appetite and weight control. In: Mela, D.J. (Ed.), Food, Diet and Obesity. Woodhead Publishing Ltd., Cambridge, pp. 356—378.

Li, Y., Hu, M., Du, Y., Xiao, H., McClements, D.J., 2011. Control of lipase digestibility of emulsified lipids by encapsulation within calcium alginate beads. Food Hydrocoll. 25, 122—130.

Little, T.J., Feltrin, K.L., Horowitz, M., Smout, A.J.P.M., Rades, T., Meyer, J.H., et al., 2005. Dose-related effects of lauric acid on antropyloraoduodenal motility, gastrointestinal hormone release, appetite and energy intake in healthy men. Am. J. Physiol.: Reg. Comp. Physiol. 289, R1090—R1098.

Little, T.J., Horowitz, M., Feinle-Bisset, C., 2007a. Modulation by high-fat diets of gastrointestinal function and hormones associated with the regulation of energy intake: implications for the pathophysiology of obesity. Am. J. Clin. Nutr. 86, 531—541.

Little, T.J., Russo, A., Meyer, J.H., Horowitz, M., Smyth, D.R., Bellon, M., et al., 2007b. Free fatty acids have more potent effects on gastric emptying, gut hormones, and appetite than triacylglycerides. Gastroenterology 133, 1124—1131.

Logan, C.M., McCaffrey, T.A., Wallace, J.M.W., Robson, P.J., Welch, R.W., Dunne, A., Livingstone, M.B.E., 2006. Investigation of the medium-term effects of Olibra™ fat emulsion on food intake in non-obese subjects. Eur. J. Clin. Nutr. 60, 1081—1091.

Ludy, M.J., Moore, G.E., Mattes, R.D., 2012. The effects of capsaicin and capsiate on energy balance: critical review and meta-analyses of studies in humans. Chem. Senses 37, 103—121.

Maljaars, P.W., Peters, H.P., Mela, D.J., Masclee, A.A., 2008a. Ileal brake: a sensible food target for appetite control. A review. Physiol. Behav. 95, 271—281.

Maljaars, P.W., Symersky, T., Kee, B.C., Haddeman, E., Peters, H.P., Masclee, A.A., 2008b. Effect of ileal fat perfusion on satiety and hormone release in healthy volunteers. Int. J. Obesity (London) 32, 1633—1639.

Maljaars, J., Romeyn, E.A., Haddeman, E., Peters, H.P., Masclee, A.A., 2009. Effect of fat saturation on satiety, hormone release, and food intake. Am. J. Clin. Nutr. 89, 1019—1024.

Maljaars, P.W., Peters, H.P., Kodde, A., Geraedts, M., Troost, F.J., Haddeman, E., Masclee, A.A., 2011. Length and site of the small intestine exposed to fat influences hunger and food intake. Br. J. Nutr. 106, 1609—1615.

Maljaars, P.W., van der Wal, R.J., Wiersma, T., Peters, H.P., Haddeman, E., Masclee, A.A., 2012. The effect of lipid droplet size on satiety and peptide secretion is intestinal site-specific. Clin. Nutr. 31, 535—542.

Marciani, L., Gowland, P.A., Spiller, R.C., Manoj, P., Moore, R.J., Young, P., Fillery-Travis, A.J., 2001. Effect of meal viscosity and nutrients on satiety, intragastric dilution, and emptying assessed by MRI. Am. J. Physiol. 280, G1227—G1233.

Marciani, L., Faulks, R., Wickham, M.S.J., Bush, D., Pick, B., Wright, J., et al., 2009. Effect of intragastric acid stability of fat emulsions on gastric emptying, plasma lipid profile and postprandial satiety. Br. J. Nutr. 101, 919—928.

Márquez, F., Babio, N., Bulló, N., Salas-Salvadó, J., 2012. Evaluation of the safety and efficacy of hydroxycitric acid or garcinia cambogia extracts in humans. Crit. Rev. Food Sci. Nutr. 52, 585—594.

Mars, M., Stafleu, A., de Graaf, C., 2012. Use of satiety peptides in assessing the satiating capacity of foods. Physiol. Behav. 105, 483—488.

Mattes, R.D., 2007. Effects of a combination fiber system on appetite and energy intake in overweight humans. Physiol. Behav. 90, 705—711.

Mela, D.J., 2005. A commercial R&D perspective on weight control foods and ingredients. In: Mela, D.J. (Ed.), Food, Diet and Obesity. Woodhead Publishing Ltd., Cambridge, pp. 492—510.

Mela, D.J., 2006. Novel food technologies: enhancing appetite control in liquid meal replacers. Obesity 14, 179S—181S.

Mela, D.J., 2007. Foods design and ingredients for satiety: Promises and proof. Lipid Technol. 19, 180—183.

Mela, D.J., 2011. Weight management product innovation: Where are we? Food Sci. Technol. 25, 26—28.

Müller, M., Bell, D., Horlacher, P., 2012. Particles. European Patent EP 2407034 issued January 18, 2012.

Nijman, C.A., Zijp, I.M, Sierksma, A., Roodenburg, A.J., Leenen, R., van den Kerkhoff, C., et al., 2007. A method to improve the nutritional quality of foods and beverages based on dietary recommendations. Eur. J. Clin. Nutr. 61, 461—471.

Paxman, J.R., Richardson, J.C., Dettmar, P.W., Corfe, B.M., 2008. Daily ingestion of alginate reduces energy intake in free-living subjects. Appetite 51, 713—719.

Pelkman, C.L., Navia, J.L., Miller, A.E., Pohle, R.J., 2007. Novel calcium-gelled, alginate-pectin beverage reduced energy intake in nondieting overweight and obese women: interactions with dietary restraint status. Am. J. Clin. Nutr. 86, 1595—1602.

Peters, H.P.F., Mela, D.J., 2008a. Weight control and satiety—understanding the science and opportunity. Agro Food Industry Hi-tech 19, 20—22.

Peters, H.P.F., Mela, D.J., 2008b. The role of the gastrointestinal tract in satiation, satiety and food intake: evidence from research in humans. In: Harris, R.B.S., Mattes, R. (Eds.), Appetite and Food Intake: Behavioral and Physiological Considerations. CRC Press, Boca Raton, pp. 187—211.

Peters, H.P., Boers, H.M., Haddeman, E., Melnikov, S.M., Qvyjt, F., 2009. No effect of added β-glucan or of fructooligosaccharide on appetite or energy intake. Am. J. Clin. Nutr. 89, 58—63.

Peters, H.P.F., Foltz, M., Kovacs, E.M.R., Mela, D.J., Schuring, E.A.H., Wiseman, S.A., 2011a. The effect of protease inhibitors derived from potato formulated in a minidrink on appetite, food intake and plasma cholecystokinin levels in humans. Int. J. Obesity 35, 244—250.

Peters, H.P.F., Koppert, R.J., Boers, H.M., Ström, A., Melnikov, S.M., Haddeman, E., et al., 2011b. Dose-dependent suppression of hunger by a specific alginate in a low-viscosity drink formulation. Obesity 19, 1171—1176.

Pilichiewicz, A.N., Little, T.J., Brennan, I.M., Meyer, J.H., Wishart, J.M., Otto, B., et al., 2006. Effects of load, and duration, of duodenal lipid on antropyloroduodenal motility, plasma CCK and PYY, and energy intake in healthy men. Am. J. Physiol.—Reg. Integr. Comp. Physiol. 290, R668—R677.

Schellekens, R.C., Stellaard, F., Olsder, G.G., Woerdenbag, H.J., Frijlink, H.W., Kosterink, J.G., 2010. Oral ileocolonic drug delivery by the colopulse-system: a bioavailability study in healthy volunteers. J. Controlled Release 146, 334—340.

Smeets, P.A.M., Erkner, A., De Graaf, C., 2010. Cephalic phase responses and appetite. Nutr. Rev. 68, 643—655.

Smit, H.J., Keenan, E., Kovacs, E.M.R., Wiseman, S.A., Peters, H.P.F., Mela, D.J., Rogers, P.J., 2011. No efficacy of processed Fabuless (Olibra) in suppressing appetite or food intake. Eur. J. Clin. Nutr. 65, 81—86.

Smit, H.J., Keenan, E., Kovacs, E.M., Wiseman, S.A., Mela, D.J., Rogers, P.J., 2012. No appetite efficacy of a commercial structured lipid emulsion in minimally processed drinks. Int. J. Obesity 36, 1222—1228.

Solah, V.A., Kerr, D.A., Adikara, C.D., Meng, X., Binns, C.W., Zhu, K., et al., 2010. Differences in satiety effects of alginate- and whey protein-based foods. Appetite 54, 485—491.

Spiller, R.C., Trotman, I.F., Higgins, B.E., Ghatei, M.A., Grimble, G.K., Lee, Y.C., et al., 1984. The ileal brake. Inhibition of jejunal motility after ileal fat perfusion in man. Gut 25, 365—374.

Ström, A., Boers, H.M., Koppert, R., Melnikov, S.M., Wiseman, S., Peters, H.P.F., 2009. Physico-chemical properties of hydrocolloids determine their appetite effects. In: Williams, P.A., Phillips, G.O. (Eds.), Gums and Stabilisers for the Food Industry, No. 15. Royal Society of Chemistry, Cambridge, UK.

Stubbs, R.J., Hughes, D.A., Johnstone, A.M., Rowley, E., Reid, C., Elia, M., et al., 2000. The use of visual analogue scales to assess motivation to eat in human subjects: a review of their reliability and validity with an evaluation of new hand-held computerized systems for temporal tracking of appetite ratings. Br. J. Nutr. 84, 405—415.

United States Government Accountability Office, 2011. FDA Needs to reassess its approach to protecting consumers from false or misleading claims. GAO-11-102. www.gao.gov/assets/320/314473.pdf.

Veldhorst, M., Smeets, A., Soenen, S., Hochstenbach-Waelen, A., Hursel, R., Diepvens, K., et al., 2008. Protein-induced satiety: effects and mechanisms of different proteins. Physiol. Behav. 23, 300—307.

Verhoef, S.P., Westerterp, K.R., 2011. No effects of Korean pine nut triacylglycerol on satiety and energy intake. Nutr. Metab. 8, 79.

Wanders, A.J., van den Borne, J.J.G.C., de Graaf, C., Hulshof, T., Jonathan, M.C., Kristensen, M., et al., 2011. Effects of dietary fibre on subjective appetite, energy intake and body weight: a systematic review of randomized controlled trials. Obesity Rev. 12, 724–739.

Whiting, S., Derbyshire, E., Tiwari, B.K., 2012. Capsaicinoids and capsinoids. A potential role for weight management? A systematic review of the evidence. Appetite 59, 341–348.

Wiseman, S., Mennen, L., Haddeman, E., Peters, H., Mela, D.J., 2008. No measurable effect of pine nut oil between or before meals on satiety or energy intake. Obesity 16 (Suppl. 1), 357–P.

Zijlstra, N., Mars, M., de Wijk, R.A., Westerterp-Plantenga, M., Holst, J.J., de Graaf, C., 2009. Effect of viscosity on appetite and gastro-intestinal hormones. Physiol. Behav. 97, 68–75.

Chapter 14

Technological Means to Modulate Food Digestion and Physiological Response

L. Donato-Capel, C.L. Garcia-Rodenas, E. Pouteau, U. Lehmann, S. Srichuwong, A. Erkner, E. Kolodziejczyk, E. Hughes, T.J. Wooster, and L. Sagalowicz

Food Science and Technology Department, Nutrition and Health Department, Nestec Ltd, Nestlé Research Center, Lausanne, Switzerland

CONTENTS

Food Structures, Digestion and Health. http://dx.doi.org/10.1016/B978-0-12-404610-8.00014-1

389

INTRODUCTION

Increasing awareness of the impact of the diet not only on nutrition, but also on health and wellness, have led to an increase in the development of food products with health benefits, the so-called functional foods. Functional foods can be defined as dietary items that, besides providing nutrients and energy, beneficially modulate one or more targeted functions in the body, by enhancing a certain physiological response and/or by reducing the risk of disease (Nicoletti, 2012).

The journey of a food product through the human body follows different stages. At first, the product composition and structure will impact on its sensorial attributes—such as texture, taste, and flavor—and on its acceptance by the consumer. Upon ingestion, the food will be digested in the gastrointestinal tract (GI tract) by a series of mechanical, chemical, and enzymatic processes in which the food is converted into nutrients suitable for use by the body. The next step consists of the absorption of nutrients from the gut lumen to the enteric circulation, through the intestinal epithelium. Nutrient absorption rate is partially dependent on both nutrient concentration and its physical properties (Lentle and Janssen, 2010). Finally, nutrients will be dispatched to the targeted tissues through the systemic circulation. The physiological response to a food can be modulated by the different stages of this journey.

In recent years, a growing understanding of the destructuring of food through the process of digestion and of the impact of this process on the biological response has been observed (Norton *et al.*, 2007). Subsequently, efforts have been made in structuring food towards controlled destructuring, to promote targeted biological responses associated with health benefits (Turgeon and Rioux, 2011). Modulating the composition and physicochemical properties of the food product could improve its physiological effects. In order to address specific biological responses, there is a great interest in tailoring designed functional properties for the macronutrients: proteins, lipids, and carbohydrates. Indeed, in addition to their nutritional profile, the structural and physicochemical characteristics of macronutrients within a food matrix may lead to different digest compositions, with different delivery outcomes in the gastrointestinal tract, and therefore may trigger different physiological responses.

The present chapter aims to illustrate through different examples, mainly derived from studies done with the Nestlé Research Center of Lausanne, how the macronutrient composition and structure of a meal can modulate digestion and trigger specific biological responses in the intestine and the whole body. In addition, we will refer to the development of different technological means that, by modifying macronutrient structure, can alter their digestion and metabolism. Through these examples, the close link between food science and technology and nutrition and health science will be illustrated.

MODULATION OF PROTEIN DIGESTION AND BIOLOGICAL RESPONSES

Following their ingestion, dietary proteins are digested in the gut providing amino acids that are absorbed through the epithelium cells of the intestine. The amino acids reach the splanchnic, and thereafter the systemic blood circulation. They integrate into the body amino acid pool and are used for synthesis of body proteins, peptides, and amino acid derivatives.

During digestion, proteins in the diet can beneficially modulate various physiological processes. Proteins enhance diet-induced energy expenditure and satiety more than other macronutrients, which supports the idea of high-protein diets for body weight control (Anderson and Moore, 2004; Acheson *et al.*, 2011). Moreover, proteins are involved in the modulation of insulin production and thereby the regulation of glucose metabolism (Tremblay *et al.*, 2007) and some peptides affect blood pressure (Saito, 2008). However, protein digestion can be responsible for detrimental physiological responses, for example by generating peptides with allergenic potential (Mackie and Macierzanka, 2010).

The digestion of proteins follows three sequential processes: food mastication, proteolysis, and oligopeptide and amino acid absorption. Proteins are degraded into oligopeptides and, to a lesser extent, amino acids in the gut lumen, with a first hydrolysis step in the stomach, through the action of pepsin, and then in the duodenum, through the action of pancreatic proteases. Protein hydrolysis continues at the surface of the intestinal epithelium by the so-called brush border peptidases. Further digestion can be observed after absorption into the intestinal epithelial cells by the action of di- and tri-peptidases located intracellularly.

Impact of Protein Source, Composition, and Structure on Protein Digestion Rate

Protein composition varies by its amino acid sequence. The latter determines not only its nutritional value but also the specific spatial arrangement of the molecule leading to a unique structure, charge partition, and surface hydrophobicity. Upon ingestion of the food, these unique physicochemical features affect protein behavior in the environmental conditions of the gastrointestinal tract, and ultimately regulate the kinetics of its digestion.

A concept of slow and fast proteins was developed with the milk protein fractions that separate the casein micelles and the whey proteins. Whey proteins are digested quickly, leading to an elevated but short-in-time amino acid concentration in the peripheral plasma, while micellar casein is slowly digested and results in a moderate but prolonged increase in the plasma amino acid concentration (Boirie *et al.*, 1997). These differences in the kinetics of amino acid appearance in plasma have been attributed to the differences of *in vivo* digestion rate in the gastric phase. While micellar casein becomes insoluble and forms a curd in the stomach that slows down protein release in the intestine, whey proteins are soluble in gastric acidic conditions, which decreases their transit time in the stomach and favors a fast digestion in the intestine (Mahé *et al.*, 1996).

In past years, the kinetics of protein digestion has been related to specific physiological outcomes. In other terms, proteins that were digested slowly led to metabolic responses that significantly differed from proteins displaying a fast digestion rate. Thus, this concept of slow/fast proteins appears interesting in the context of functional foods as a means to improve the physiological condition of specific populations. Pioneering this research, Beaufrere's group showed that during the postprandial period, fast proteins induced higher stimulation of body protein synthesis and of amino acid oxidation compared to slow proteins. On the other hand, slow proteins induced a stronger and more prolonged inhibition of body protein degradation and resulted in higher postprandial protein retention in the body than fast proteins (Boirie *et al.*, 1997, Dangin *et al.*, 2001). Nevertheless, further results from the same group suggested that the metabolic response to slow and fast proteins is more complex and is dependent on factors such as the age of the consumer or the macronutrient composition of the meal. Thus, contrary to previous data with single protein meals, a complete meal containing a fast protein induced higher net protein retention than an otherwise identical meal containing a slow protein (Dangin *et al.*, 2003). These results were qualitatively similar in young and elderly volunteers, but the differences in protein balance induced by the slow and fast protein meals were stronger in the elderly population (Dangin *et al.*, 2003). These results suggest that complete meals containing fast proteins might be particularly adapted to limit body protein losses associated with ageing, although confirmatory studies with long-term feeding protocols are still missing.

More recently, Tessari *et al.* (2007) demonstrated differential incretin and insulin responses to fast and slow proteins in type 2 diabetic patients.

A casein-containing meal resulted in lower postprandial insulin and glucagon-like polypeptide 1 (GLP-1) levels than an otherwise identical meal containing fast proteins (i.e., whey protein or a mix of L-amino acids mimicking casein composition). Furthermore, compared with casein, the meal with free amino acids led to lower postprandial glycemia. These results suggest that modulating the *in vivo* protein digestion rate could help the glycemic control of these patients. A further example of how different protein sources can differently modulate metabolic processes was provided by Acheson *et al.* (2011). This group showed that replacing 10% of the carbohydrate of a complete meal with protein promoted thermogenesis with more pronounced effects when whey protein was used compared to soy protein or casein. Contrary to the previous examples, the design of this study does not distinguish the effect of protein digestion rate from that of its amino acid composition. Nevertheless, the results suggest that different dietary proteins could be used to modulate energy balance and that some protein sources could be particularly indicated in weight control diets.

The above-mentioned studies illustrate how the source and the digestion rate of a protein can lead to different physiological responses, which in turn may trigger different health benefits. Several factors are to be considered, however, regarding the behavior of a protein in the gastrointestinal tract. Not only may the age of the consumer (Dangin *et al.*, 2003) and the time of meal consumption (Arnal *et al.*, 1999, 2002) trigger different physiological responses, but also the composition, mineral contents, processing conditions, and matrix contents of the starting protein material may play an important role in the protein digestion profile. However, these factors are most of the time neglected in physiological investigations. As an example, we found in a rat model that the molecular structure of caseins had a great impact on their gastric clotting properties, gastric emptying, and digestion rates. When casein was structured as micellar casein or in the form of calcium caseinate, a hard clot was formed in the stomach, which then was slowly emptied. However, other individual casein salt forms, such as sodium and potassium caseinates, formed a flocculated and smooth clot that was emptied and digested at a rate close to that of whey proteins (unpublished results).

Impact of Protein Microstructure

Besides its composition, protein functionality can be used as a technological means to modulate its behavior during digestion in the GI tract. Indeed, proteins are used in many food products for their functional properties, such as emulsifying, gelling, foaming, and chelating properties. Their potential to interact with different components of the food matrix has often been used to tune the structure and texture of food products (Kinsella, 1984). Exploiting the effect of such functionalities on protein digestion could be an interesting and flexible approach to trigger a variety of targeted physiological responses. Next, we report some examples of the structural modification of micellar casein and whey proteins and the consequences of this change on protein digestion profiles and the subsequent postprandial biological responses.

The structure of micellar casein can be modified by treatment with transglutaminase, an enzyme favoring cross-linking between the molecules of casein, without affecting the size of casein micelles (Huppertz and de Kruif, 2008). The effect of such treatment on digestion kinetics was studied *in vivo* in a rat model. Interestingly, we found that when micellar casein was treated with transglutaminase, no coagulum was visible in the stomach and a faster transit time and overall digestion rate was measured from the kinetics of nitrogen disappearance from the stomach and intestine (Figure 14.1). Changes in the kinetics of digestion of micellar casein can therefore be expected to lead to different physiological responses.

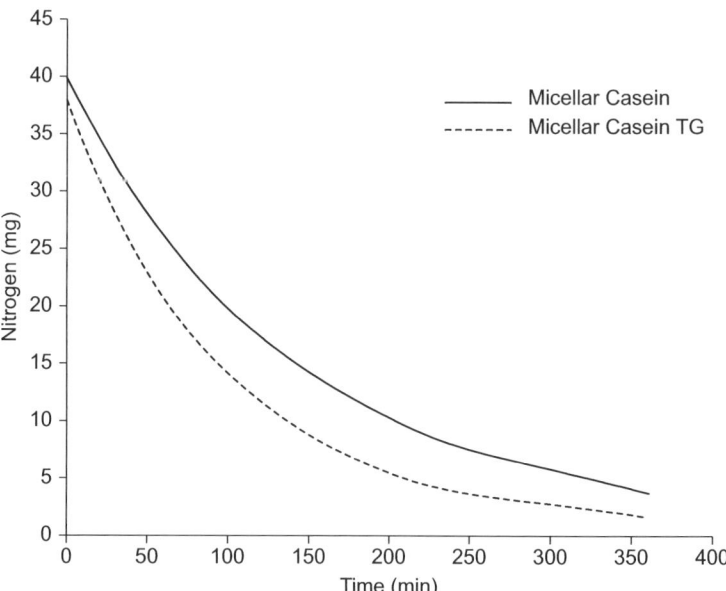

■ **FIGURE 14.1** Kinetics of nitrogen disappearance from the gut upon ingestion of native or transglutaminase-treated micellar casein. Residual nitrogen was measured by Kjeldhal analysis of the luminal contents in the rat stomach and intestine as function of digestion time.

Other authors, using *in vitro* and minipig models, have found no effect of transglutaminase on the digestion rate when casein was in the form of sodium caseinate instead of micellar casein (Roos, 2003). This highlights the fact that similar treatments can lead to different responses, depending on the characteristics of the starting material. In this case, we can speculate that the initial structure of the casein could influence its reactivity to the transglutaminase leading to different structures that can behave differently through the digestion process.

The functional properties of whey proteins can be also exploited to modulate their digestion kinetics. Under heat treatment, whey proteins generate aggregates with structural features that vary according to environmental conditions, contributing to various functionalities in food products (Nicolai *et al.*, 2011). It has been found that when whey proteins were heated at a pH close to the isoelectric point of the main whey proteins (around pH 5.9), stable suspensions of rather monodisperse spherical particles, called microgels, were formed (Schmitt *et al.*, 2009, 2010). The solubility curve of these aggregates showed a minimum value around pH 4−5 where the surface charge is decreased, leading to interaction of the microgels (Schmitt *et al.*, 2010) with the formation of a curd-like structure (Donato *et al.*, 2011).

Our recent clinical study has compared the plasma concentrations of amino acids after ingestion of high-protein meals containing either whey protein isolate (WPI) or whey protein microgels (WPM) (Pouteau *et al.*, 2012). Interestingly, it was found that the profile for changes in plasma amino acids following consumption of the WPM was delayed compared to that of WPI (Figure 14.2). This result indicates that mild heat treatment applied to WPI in controlled pH conditions, leading to WPM, can modulate the kinetics of whey protein digestion. Since minimum WPM solubility is observed in the acidic gastric pH conditions, we can speculate that the delay observed in plasma amino acid appearance with this material might be due to the formation of a WPM clot in the stomach, which could slow its transit time to the intestine. In association with the delayed appearance of amino acids in plasma, the hormonal response, as exemplified by the plasma insulin peak, was lower 30 min after ingestion of the WPM meal compared to that of the WPI meal (Figure 14.3). The plasma glucagon response was also lower after the WPM meal. The lower levels of plasma amino acids at 30 min probably induced lower hormonal peak responses which consequently modulated the whole body metabolism. Additionally, the delayed peak of plasma amino acids after the WPM meal compared to the WPI may have triggered the whole body metabolism differently, in particular protein metabolism. The outcomes of this clinical study illustrate the major

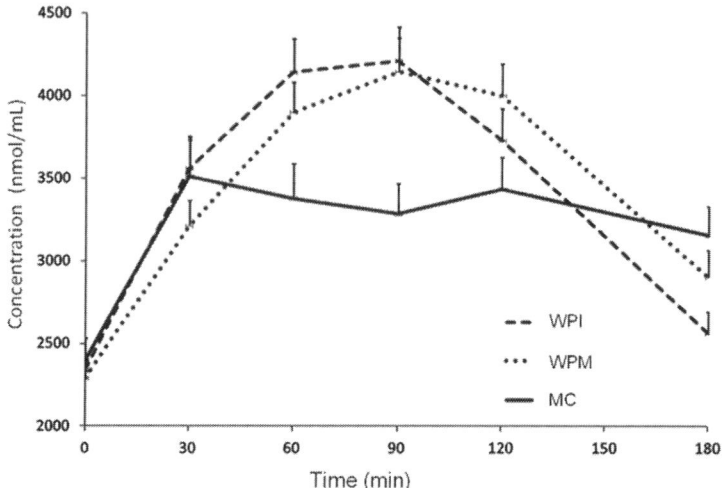

■ **FIGURE 14.2** Plasma concentrations (mean ± s.e., $n = 16$) of total amino acids in healthy adults after ingestion of isocaloric and isonitrogenous high-protein meals differing in the protein source: whey protein isolate (WPI), whey protein microgels (WPM), or micellar casein (MC) (Pouteau *et al.*, 2012).

impact of the physicochemistry of the ingested proteins, independent of the amino acid composition, on whole body metabolism and physiology.

Contrasting results can be found in the literature regarding the impact of heat treatment on either promoting, delaying, or maintaining whey protein kinetics of digestion either *in vivo* (Caugant *et al.*, 1992; Kitabake *et al.*, 2001) or *in vitro* (Carbonaro *et al.*, 1997; Stanciuc *et al.*, 2008; Inglinstad

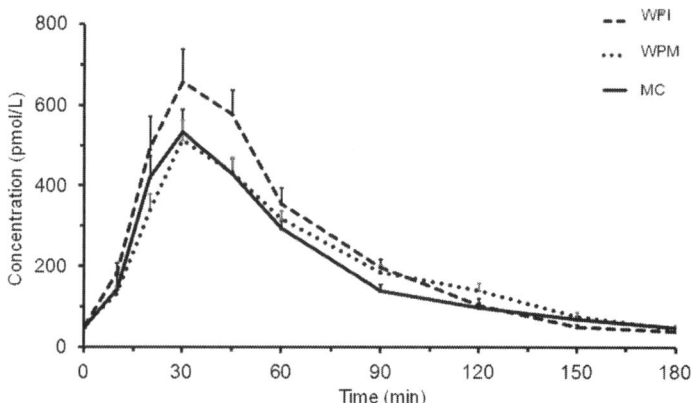

■ **FIGURE 14.3** Plasma concentrations (mean ± s.e., $n = 23$) of insulin in healthy adults after ingestion of isocaloric and isonitrogenous high-protein meals differing in the protein source: whey protein isolate (WPI), whey protein microgels (WPM), or micellar casein (MC) (Pouteau *et al.*, 2012).

et al., 2010). Again, the conditions of heat treatments, the composition of the matrix and the type of structure formed are key factors that modulate protein structure and enzyme accessibility during digestion. The conditions of measurements (*in vitro* or animal models) have also shown some limitations regarding the prediction of human behavior.

Some of the preceding examples illustrate that prediction of protein digestion rates and the associated physiological responses based only on protein amino acid composition can have important limitations and that other factors associated with the protein material need to be accounted for. Nevertheless, clinical trials are too time and resource consuming to properly assess the effect of these factors. The design of functional products based on the kinetics of protein digestion would require the availability of predictive but simplified screening systems.

In vitro models of protein digestion exist, but their predictive value is limited when it comes to mimicking *in vivo* digestion kinetics. As an illustration, many groups report micellar casein to be digested more rapidly than whey protein *in vitro* (Tomé, 1991; Carbonaro *et al.*, 1997; Lindberg *et al.*, 1998; Almaas *et al.*, 2006; Salami *et al.*, 2008; Inglinstad *et al.*, 2010), which is opposite to the above and to *in vivo* data described in literature (Mahé *et al.*, 1996). The most likely reason of this discrepancy is that none of the *in vitro* methods developed until now has been able to mimic the gastric emptying process, which is actually the rate-limiting step of milk proteins *in vivo* (Mahé *et al.*, 1996). The *in vitro* results may be of more value when the objective is to predict the global accessibility of the protein to the digestive enzymes and therefore its digestibility (Boye *et al.*, 2012), it may have some limitation when investigating the whole digestion of proteins for accurate extrapolation to the complex human physiology. Animal models appear to be an alternative tool to mimic the protein digestion kinetics in humans (Caugant, 1992). However, our internal data suggest that the predictive value of rat models for the kinetics of amino acid appearance in plasma could be relatively weak (unpublished results). Other species, such as the pig, with intestinal physiology that is closer to that of humans, may be more predictive of this particular outcome. Nevertheless, proper validation studies are required.

Cumulative evidence indicates that the rate of protein digestion is an important factor in the modulation of the postprandial physiological response to dietary proteins, which could be exploited in the design of functional foods. However, there are still many open questions regarding the effect of interactions between the protein macromolecules and their environment, in both the food matrix and the gastrointestinal tract, on the digestion rate. Better understanding of these phenomena is a prerequisite to the design of protein-based foods with consistent health effects. Systematic research in this field would require high throughput and predictive model systems, which are not yet available. The development of such models will lead to an important step forward in this area. Alternatively, joint efforts between food and technological science and human physiology, in particular gastroenterology, will continue to improve our understanding of the physiological effects of modulating protein digestion.

MODULATION OF CARBOHYDRATE DIGESTION AND BIOLOGICAL RESPONSE

Carbohydrates are a class of chemical compounds based on the elements carbon, hydrogen, and oxygen. In human nutrition, carbohydrates play a critical part in supplying metabolic energy as well as other physiological benefits that enable the body to perform its regular functions. Food carbohydrates can be classified into two categories according to their digestibility. The first comprises carbohydrates which are digested, absorbed, and metabolized in the upper gastrointestinal tract. These digestible carbohydrates cover a wide range of mono-, di-, oligo-, and polysaccharides, e.g., glucose, fructose, lactose, sucrose, maltodextrins, and starch. The second category consists of indigestible carbohydrates which can be partially or completely fermented by bacteria in the large intestine. This group comprises oligosaccharides and complex polysaccharides, such as raffinose, stachyose, β-glucan, arabinoxylans, inulin, cellulose, pectin, and resistant starch. Resistant starch can be further divided into four types depending on its origin and structure (Englyst et al., 1992).

Starch is the major carbohydrate in the human diet and provides technological properties such as thickening and gelling to food products, through its physical transformation (Biliaderis, 2009). Native and modified starches are used in the food industry due to their high availability, comparatively low cost, and unique properties. According to its digestibility, starch can be part of two categories, i.e., digestible and indigestible carbohydrates, depending on botanical source and physical state. Digestion of starch in the upper gastrointestinal tract and fermentation of starch in the large intestine define its nutritional quality and associated physiological benefits.

In the next sections, emphasis will be on the effect of starch composition and modification of starch structure on modulation of its digestion and biological response.

Physiological Effects of Rapidly Digestible, Slowly Digestible, and Resistant Starches

The nutritional quality and health effects of starch depend strongly on the processing and structure of the starch. The timeline of digestion and the release of energy for the body in the form of glucose are major physiological properties of starch. Starch digestion is initiated in the mouth by salivary α-amylase, but the major part of digestion occurs in the small intestine, where starch is broken down by pancreatic α-amylase into maltose, maltotriose, and α-limit dextrins. These products, together with lactose and sucrose, if any, are further hydrolyzed by brush border enzymes to yield monosaccharides before absorption in the small intestine. A recent report demonstrates that brush border enzymes also hydrolyze large starch molecules *in vitro* (Lin *et al.*, 2012). Englyst *et al.* (1992) classified starch into three fractions according to their digestion kinetics measured *in vitro* as a marker for the *in vivo* glycemic response, although there is still some controversy about the correlation between the *in vitro* data and the *in vivo* digestion rate in humans. This method allows an estimation of the digestibility of starch in the human gastrointestinal tract and gives an indication of the physiological benefits. According to Englyst's classification, rapidly digestible starch (RDS) is completely digested *in vitro* within 20 min and results *in vivo* in a fast postprandial glucose response. Slowly digestible starch (SDS) induces a more sustained blood glucose response as the *in vitro* digestion takes 20 to 120 min. The resistant starch (RS) fraction, in contrast, escapes the digestion processes in the upper gastrointestinal tract and reaches the large intestine where it is fermented.

In food products, all three forms commonly coexist, but the ratios can be changed via suitable pre-processing, processing, and post-processing methods. However, while both RS and RDS can be measured accurately with *in vitro* and *in vivo* methods (e.g., Englyst *et al.*, 1996a, b; McCleary *et al.*, 2002), no suitable measurement for SDS has been established. There are several methods to measure the *in vitro* digestion kinetics of carbohydrates with the aim of simulating glycemic responses. The biggest limitation is the complexity of the physiological processes involved in the digestion and absorption of carbohydrate, which cannot be fully mimicked by *in vitro* methods. A major gap is an official, approved, and validated method, e.g., by the Association of Analytical Communities (AOAC).

Main Physiological Effects of RDS

The intake of RDS delivers a rapid increase of postprandial glucose which, in turn, induces rapid insulin secretion. It can be used as a fast energy source, e.g., for energy-requiring tasks. However, a high postprandial glycemia and insulinemia might be risk factors for the onset of type 2 diabetes and other metabolic disorders in vulnerable populations (Aller *et al.*, 2011).

Main Physiological Effects of SDS

The intermediate fraction, defined as slowly digestible starch (SDS), is fully digested and absorbed in the small intestine but at a slower rate than RDS, leading to a slower and more sustained release of glucose, and thus a lower postprandial insulin peak (Figure 14.4) (Lehmann and Robin, 2007; Lehmann, 2009). A decreased glycemic response, in line with a lower insulin response, may be beneficial for people with impaired glucose tolerance, common in the general population (EFSA, 2011). A recent study in healthy men found no differences in glycemic response after SDS intake compared to a rapidly digestible starch product, due to a slower glucose clearance rate but a lower postprandial insulin and glucose-dependent insulinotropic polypeptide concentration, offering beneficial metabolic long-term health effects (Eelderink *et al.*, 2012). A lower postprandial insulinemia and better lipid metabolism were found by Harbis *et al.* (2004) in obese insulin-resistant subjects. Ells *et al.* (2005) proposed a reduction of potential risk factors for the metabolic syndrome by replacing RDS by SDS. Several authors have showed that RDS and SDS differ in their ability to stimulate secretion of gut incretin hormones and suggested therefore a beneficial effect of SDS

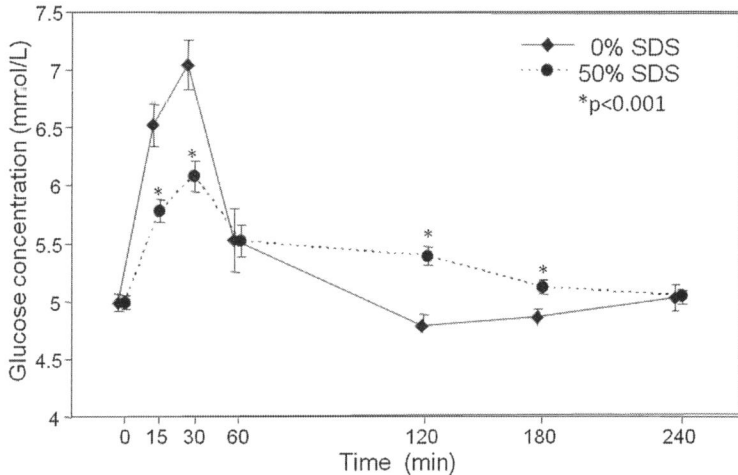

■ **FIGURE 14.4** Plasma glucose levels (mean ± s.e., $n = 26$) of healthy subjects after consumption of 20 g available carbohydrates including 0% (glucose) or 50% SDS (slowly digestible starch). Consumption of SDS induced a significantly lower glucose peak during the first hour and sustained blood glucose levels after 120 and 180 min compared to the 0% SDS (Lehmann, 2009).

in the late postprandial phase, for glucose homeostasis and energy storage (Wachters-Hagedoorn *et al.*, 2006; Eelderink *et al.*, 2012). Consumption of SDS has been shown to reduce the insulin requirement and prevent evening hypoglycemia in diabetic patients undergoing insulin treatment (Golay *et al.*, 1992; Axelsen *et al.*, 1999). In conclusion, SDS consumption may be beneficial in reducing the risk of type 2 diabetes and metabolic syndrome. Both are risk factors for cardiovascular diseases. While there are good reasons to assume that SDS could play a role in body weight management, and to prolong or increase mental and physical performance, clinical studies investigating physiological effects of SDS beyond these metabolic effects are limited (Lehmann and Robin, 2007; Zhang and Hamaker, 2009). This is mainly due to the fact that there are currently no commercial and palatable SDS products available.

Main Physiological Effects of RS

Resistant starch (RS) is the portion of starch that cannot be hydrolyzed by the human digestive enzymes and thus does not raise the blood glucose concentration. RS acts as a specific dietary fiber. However, in contrast to most other dietary fibers, RS is fermented to produce a high amount of butyrate, with suggested anti-carcinogenic and anti-inflammatory health effects (Brouns *et al.*, 2002). Other health benefits of RS, mainly resulting from the fermentation into short-chain-fatty acids and bulking activity in the gastrointestinal tract, have been reviewed recently by several authors (e.g., Topping *et al.*, 2003; Sajilata *et al.*, 2006).

In the past years, efforts have been made to develop food applications containing an increased amount of SDS. In the next paragraph, we summarize a few routes to promote SDS delivery.

Impact of Structure on the Digestibility of Native Starch

Starch-containing products are, in general, thermally processed before consumption. The industrial processes commonly used include baking, extrusion, pasteurization, and sterilization. It is known that the simplest sources of SDS are certain native starch granules (Lehmann and Robin, 2007). However, SDS structure is not well retained during thermal processing, where starch gelatinization occurs and enzyme susceptibility drastically increases (Lehmann and Robin, 2007; Zhang and Hamaker, 2009). The intermediate digestibility property of SDS in native starch is attributed to certain semi-crystalline features and granular morphology (Lehmann and Robin, 2007; Srichuwong and Jane, 2007). These factors vary considerably with botanical source. In general, densely-packed crystalline regions of starch granules are more resistant to enzyme hydrolysis than the loosely-packed or amorphous regions (Gallant *et al.*, 1992). It is known that the A-type polymorph starches (e.g., normal cereals, tapioca) are more readily digested by enzymes than the B-type polymorph starches (e.g., high amylose cereals, potato) (Srichuwong *et al.*, 2005; Srichuwong and Jane, 2007). The B-type polymorph, high amylose maize starch, is highly resistant to enzymes and gelatinized only at very high temperature. These features can be attributed to its high amylose content, long amylopectin branch-chain and homogeneous internal structure (Jane *et al.*, 1999; Jiang *et al.*, 2010). In contrast, the greater enzyme susceptibility of A-type starches is attributed to their weak double-helical structures comprising a larger proportion of short amylopectin chains, and presence of surface pores and channels (Fannon *et al.*, 1992). A larger proportion of the short amylopectin chains are also associated with the higher digestion rates observed among different A-type starches (Srichuwong *et al.*, 2005). The presence of amylose in non-waxy starch may also strengthen the outer layer of the granule against enzyme hydrolysis (Pan and Jane, 2000).

Tuning Starch Structure Toward Promoting SDS Formation through Different Technological Means

Although native commercial starches such as normal or waxy maize, wheat, barley, and rice yield substantial SDS content in their native forms, their crystalline features are lost during food processing, where heat treatment is commonly employed. Thus, their SDS, as well as the RS fraction, is converted to the RDS accordingly. There are many commercial RS ingredients available with good process properties and proven physiological benefits. However, SDS, with a more fragile structure, does not, in general, survive food processing and no commercially available ingredients of slowly digestible starch (in contrast to a few slowly digestible carbohydrates) exist. It has been shown that most extruded and bakery products yield mostly RDS and rarely retain the SDS fraction (Englyst *et al.*, 2003). Yet, challenges remain in the manufacture of SDS-enriched products at large scale. During past decades, several technological means have been evaluated to create both SDS ingredients and SDS-enriched products. They are briefly explained in the following section.

Branching of Starch or Carbohydrate Polymers

Since α-1,6 linkages are slowly digested, an increase in branching density of starch molecules might slow down overall digestibility. Several patents describe the formation of highly-branched glucose polymers (e.g., Fuertes *et al.*, 2004) and highly-branched starch (e.g., Kawabata *et al.*, 2001; Komae and Kato, 2002; Backer and Saniez, 2005), as well as branching enzymes. Pullulan, a highly branched extracellular polysaccharide excreted by the fungus *Aureobasidium pullulansis*, was found in a clinical trial to be slowly digestible (Wolf *et al.*, 2003). It is composed of three α-1,4-linked glucose molecules that are repeatedly polymerized by α-1,6 linkages on the terminal glucose.

Recrystallization

Retrogradation of starch molecules occurs during the aging of starch dispersions. Amorphous starch molecules reassociate to develop double-helical structures, which may favor crystallite formation depending on the conditions (Srichuwong and Jane, 2007). Selection of substrate properties (such as amylose content and chain lengths) as well as retrogradation conditions (such as time, storage temperature, and water content) can modify the rate of retrogradation, crystalline structure, and subsequent digestibility. Depending on these factors, retrogradation can primarily generate SDS (Guraya *et al.*, 2001a, b; Robin *et al.*, 2008) or RS type III (Leloup *et al.*, 1992; Eerlingen and Delcour, 1995; Lehmann *et al.*, 2002, 2003; Sajilata *et al.*, 2006). Accordingly, the process conditions and resulting crystalline structures are different between SDS and RS productions.

Hydrothermal treatment (heating beyond the gelatinization temperature under an excess of water) and enzymatic treatments were approaches used to promote SDS formation (Guraya *et al.*, 2001a, b; Shin *et al.*, 2004). Guraya *et al.* (2001b) showed that a partial debranching of starch was most favorable for SDS formation. The formation of SDS was also favored by low cooling temperatures, which favored the nucleation step of crystallization. In contrast, higher temperatures favored the propagation and maturation steps, resulting in less digestible material and higher RS formation (Eerlingen and Delcour, 1995). It was hypothesized that the resulting SDS fraction might result from the formation of imperfect B-type crystallites, which are more prone to digestion (Guraya *et al.*, 2001b; Shin *et al.*, 2004). This was confirmed by Robin *et al.* (2008), investigating the influence of the chain lengths on retrogradation and SDS formation. Short-chain α-(1,4)-D-glucan samples were generated by debranching of potato amylopectin and fractionation on gel permeation chromatography. The collected fractions were gathered to generate two samples with an average numerical degree of polymerization (DPn) of 32 and 42, respectively. The results revealed that the SDS content was affected by both chain length and recrystallization temperature. The highest SDS level (49%) was obtained at $-80°C$ from the sample with the lowest average DPn. Its structure showed B-type polymorphic crystallites with a substantial amorphous part and a melting temperature of 85°C, which is in agreement with other studies (Zhang *et al.*, 2008a, b). These retrograded B-type crystallites were thought to be responsible for SDS. In native waxy starch, a parabolic relationship between SDS content and the weight ratio of amylopectin short chains (degree of polymerization (DP) <13) to long chains (DP ≥13) can be observed (Zhang *et al.*, 2008a, b). Both short and long amylopectin chains can lead to high SDS contents via different mechanisms. Long amylopectin branch chains form imperfect crystallites during retrogradation, leading to increased SDS content; while the short branch chains are relatively high in the slowly digestible α-1,6 linkages.

For RS type III formation, linear α-(1,4)-D-glucan chains are used as substrate, either derived from amylose-rich starches (Haralampu, 2000) or from debranching of waxy starches (Lehmann *et al.*, 2002, 2003). In a first step, following hydrothermal treatment, a net-like polymer structure is formed, followed by double-helix formation of the linear chains with a DP 26 to 73. During further storage, the helices further aggregate by further hydrogen-bound formation. This network also includes shorter α-(1,4)-D-glucan with a DP of 6–30 (Leloup *et al.*, 1992). X-ray crystallographs of the resulting products show B-type polymorphism. A following heat-moisture treatment can result in RS contents up to 84% (Lehmann *et al.*, 2003). Peak temperatures of about 145°C found in DSC measurements pointed to a high thermal stability of the RS products, suitable for food processing (Eerlingen and Delcour, 1995; Lehmann *et al.*, 2002).

Annealing and Heat-Moisture Treatment

The two hydrothermal treatments, annealing and heat-moisture treatment (HMT), can be used to modify the physicochemical properties and digestibility of starch. In both processes, the starch granule remains intact. Annealing is performed in excess of water (>60% w/w) or at intermediate water content (40 to 55% w/w), while HMT is carried out at low levels of water (generally below 35% w/w water). The treatment temperature lies between the glass transition temperature and the gelatinization temperature of the starch.

Annealing and HMT have been widely used to generate RS (Lehmann *et al.*, 2003; Sajilata *et al.*, 2006). An increase in SDS compared to native starch following annealing or HMT in sweet potato starch (Shin *et al.*, 2005), in a range of flours (Niba, 2003), and rice starch (Anderson *et al.*, 2002) was observed, depending on the process conditions. Various structural changes occur and might be the reason for the modified digestion properties (Lehmann and Robin, 2007).

Creation of SDS in Polymer Matrices

Entrapment of starch in a dense polymer network with limited water access hinders the enzyme accessibility and consequently lowers the digestibility of starch. This can lead to the formation of SDS or RS type I, for example as an entrapped starch in a protein network of pasta (Colonna *et al.*, 1990; Fardet *et al.*, 1998). A casein network formed around gelatinized starch lowered levels of incretin hormone secretion and released less glucose than after starch digestion alone as indicated by significantly lower levels of glucose transporter mRNA transcripts, measured in cell culture experiments (Bruen *et al.*, 2012). In some biscuits with very low moisture levels during the treatment, the extent of gelatinization is reduced and partially intact granules and gelatinized starch coexist. This can result in a higher content of SDS (Englyst *et al.*, 2003). Starch can also be entrapped by biopolymers (e.g., alginate) to create defined networks with modified slow and sustained glucose release (Venkatachalam *et al.*, 2009).

Starch digestion kinetics can lead to different physiological responses, which could contribute to different health benefits. Special efforts have been made to design carbohydrates with tailored digestion in food matrices. The matrix composition and processing conditions greatly affect the starch structure and resulting digestion properties. While the benefits and applications of RS are already well established, the emerging concept of SDS enrichment in foods is gaining more and more attention. The benefits beyond modulating glycemic response and other metabolic effects should be further explored. In conclusion, an increased incorporation of SDS or RS into starchy foods at the expense of RDS could increase the nutritional quality of food products and thus allow food manufacturers to offer healthier options to their consumers.

MODULATION OF LIPID DIGESTION AND BIOLOGICAL RESPONSE

There has been a resurgence in the interest in fat digestion over the past decade, particularly from the food and nutrition research communities (Feinle-Bisset *et al.*, 2005; Mun *et al.*, 2007; Chu *et al.*, 2009; Maljaars *et al.*, 2009; Reis *et al.*, 2009; Singh *et al.*, 2009; Golding and Wooster, 2010). This interest has stemmed from the paradoxical role the lipids play in our diet. Lipids are a key source of essential fatty acids and facilitate the absorption of hydrophobic vitamins (Carey *et al.*, 1983; Borel *et al.*, 2001; Fave *et al.*, 2004; Carriere *et al.*, 2005). However, fat is the most energy dense macronutrient, contributes strongly to the hedonic aspects of food, and is commonly cited as a leading cause of obesity (Lissner and

Heitmann, 1995). A number of research groups have sought to understand the relationship between food structure and fat digestion in an effect to balance this dual role of fat (Borel *et al.*, 2001; Fave *et al.*, 2004; Carriere *et al.*, 2005; Armand, 2007; Mun *et al.*, 2007; Chu *et al.*, 2009; Maljaars *et al.*, 2009; Reis *et al.*, 2009; Singh *et al.*, 2009; Golding and Wooster, 2010; Mullertz *et al.*, 2012). In the case of triglycerides, it may be desirable to slow down absorption while for other nutrients, such as carotenoids and vitamins, it is often necessary to increase it. Typical populations, where it is desirable to ensure or increase lipid absorption, include small and premature babies, the elderly, and people suffering from a disease leading to poor digestion and absorption such as cystic fibrosis (Fave *et al.*, 2004; Aloulou and Carriere, 2008). In order to use food structure to control nutrient uptake, it is first necessary to understand digestion and absorption.

The digestion of triglycerides, from a molecular point of view, is relatively well known (Carey *et al.*, 1983; Lowe, 1997). Lipases, which are a class of interfacially-activated enzymes, are responsible for the hydrolysis of triglycerides (Lowe, 1997; Carriere *et al.*, 1998). The action of lipase(s) starts to take place in the stomach where acid-stable gastric lipase can convert between 5 and 30% of fat into 1,2-diglycerides, liberating a fatty acid from the sn-3 position (Armand *et al.*, 1996). The majority of fat digestion occurs in the small intestine where co-lipase-dependent pancreatic lipase, in conjunction with co-lipase and bile salts, finishes the hydrolysis and produces 2-monoglycerides and free fatty acids (Armand *et al.*, 1996). In healthy adults the combined lipolytic capacity is in slight excess relative to the amount of fat ingested, and hence fat digestion is highly efficient (>97%) (Carriere *et al.*, 2005; Armand, 2007). Fat digestion and absorption in infants is less efficient because of the immaturity of the digestive tract, which results in lower pancreatic lipase concentrations and much lower bile salt concentrations (Hamosh, 1996). The lower pancreatic lipase activity is in part compensated for by greater activity of gastric lipase and the activity of pancreatic lipase related protein type 2 (PLRP2) (Armand *et al.*, 1996; Hamosh, 1996; Lindquist and Hernell, 2010). In breast-fed infants, a lipase secreted in mothers' breast milk (bile salt stimulated lipase, BSSL) is thought also to contribute to fat digestion (Lindquist and Hernell, 2010).

A fundamental aspect of fat digestion is that it occurs at the interface between the insoluble fat substrate and the aqueous digestive milieu (Carey *et al.*, 1983; Fave *et al.*, 2004). Hence, food structure is critical to both the digestion and absorption of fat, because it controls enzyme access and digesta departure (Carey *et al.*, 1983; Reis *et al.*, 2009; Golding and Wooster, 2010). A key step in fat digestion is the formation of emulsion droplets which provides a surface for lipase adsorption. Much of the fat we eat is already emulsified, and the breast milk that infants consume is an emulsion that is created naturally in the endoplasmic reticulum and secreted through the cell membrane of alveolar cells (Neville, 1995). Modern eating patterns mean that a significant portion of fat in the adult diet is derived from processed foods where it is already emulsified. However, highly emulsified fat is a relatively new concept in the evolution of the human diet. A significant amount of fat was in the past obtained in a relatively unemulsified form from the visceral fat from meat and natural oil bodies in seed crops and fruits (Speth, 2012). The body has highly-evolved mechanisms to facilitate the emulsification of such "free" fat via the mechanical actions of the mouth and the stomach (Golding and Wooster, 2010).

A particular focus of recent research into fat digestion has been on how to regulate the kinetics or timing of fat digestion (Fave *et al.*, 2004; Maljaars *et al.*, 2008; Chu *et al.*, 2009; Reis *et al.*, 2009; Golding and Wooster, 2010). In general the lipid digestion capacity in healthy adults is in slight excess relative to its substrate, hence the main mechanisms that have been used to regulate fat digestion are to control the ability of lipases to bind to the interface of emulsified fat droplets, by stopping enzyme activity (Daher *et al.*, 1997; Carriere *et al.*, 1999); controlling the composition of the interface (Wickham *et al.*, 1998; Sandra *et al.*, 2008; Chu *et al.*, 2009; Maldonado-Valderrama *et al.*, 2011); controlling the area of the interface (Seimon *et al.*, 2009; Golding *et al.*, 2011); or encapsulation (Li *et al.*, 2010). A number of groups have demonstrated *in vitro* and/or *in vivo* that such approaches can have considerable effects on the rate (Seimon *et al.*, 2009; Golding *et al.*, 2011) and/or the extent (Sjostrom *et al.*, 1998; Carriere *et al.*, 1999) of fat digestion.

Impact of Emulsion Structure on Lipid Digestion

In our own research, we have been very interested in these concepts and have focused on how the composition of the interface and bulk of an emulsion affects its digestion and digesta structure.

An important tool that we have used to follow the digestion kinetics of different emulsions is the Lipid Droplet Digestion Experiment (LiDDEx). The LiDDEx is a versatile technique allowing *in vitro* investigation of various aspects of the digestion of lipid-based model foods. The LiDDEx monitors the structural biotransformations, the establishment of lipid digestion kinetics, as well as the elucidation of the interactions of lipase with model foods, under simulated small intestinal conditions. The digestions are carried out in the chamber shown in Figure 14.5A and B. A fixed amount of an emulsion is diluted in a fixed volume of reconstituted bile salts. Pancreatin is added at the last moment. An aliquot of this mixture is then transferred

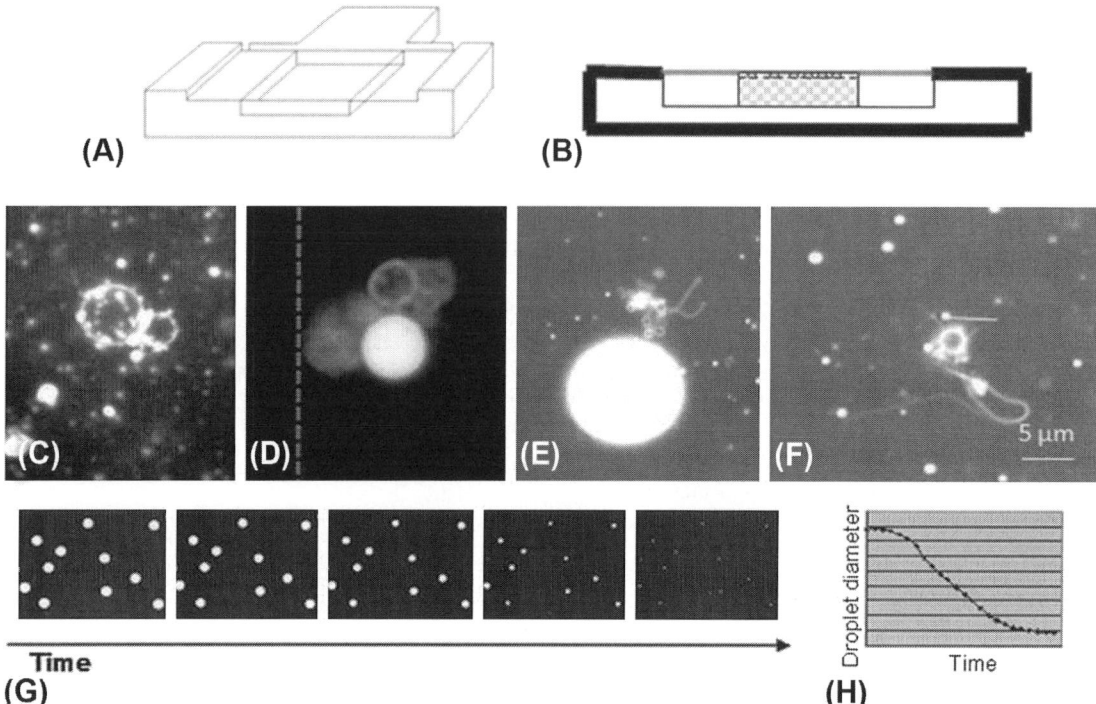

■ **FIGURE 14.5** LiDDEx digestion experiment. (A) Plexiglass digestion chamber: the clearance is adapted to different applications (0.1 to 5 mm). (B) Cross-sectional scheme of the digestion chamber with the digestion mixture in place. (C) to (F) Structures generated during the simulated small intestinal digestion of an MCT emulsion stabilized by sodium caseinate. (G) Time lapse series of microscopy images recorded from stained O/W emulsion droplets under *in vitro* intestinal medium. (H) Kinetics of emulsion digestion obtained as the decrease in the average droplet diameter over time. [A color version of this figure is available online at www.booksitc.clscvicr.com/9780124046108].

into the chamber, covered with a glass cover slide coated with a dry film of Nile red to image lipids by fluorescence microscopy. To establish digestion kinetics with the LiDDEx, time lapse images (Figure 14.5G) are captured using digital camera software (typically every 5 minutes for 2 hours or more). Once the images have been captured, the image analysis software allows computation of the size changes of droplet during digestion, as shown in Figure 14.5H.

Figures 14.5C to f illustrate the application of the LiDDEx to the investigation of the structural biotransformations of an MCT emulsion stabilized by sodium caseinate. During the early stages of simulated intestinal digestion, vesicular structures can be seen on aggregating oil droplets (Figure 14.5C). As digestion proceeds, multivesicular structures appear (Figure 14.5D), then tubular assemblies form from the build-up of the mono- and diglycerides generated by the digestion of the MCT oil (Figure 14.5E and F). The

LiDDEx design allows rapid screening of the structural biotransformations of both simple and sophisticated food matrices. Moreover, when used in combination with confocal microscopy, spectral analysis of the emission spectrum of Nile red during the digestion provides insight into the type of digestion products that are created and their role in the formation of these self-assembled structures. Examples of digestion kinetics provided by the LiDDEx are shown in Figure 14.6. Figure 14.6A demonstrates the influence

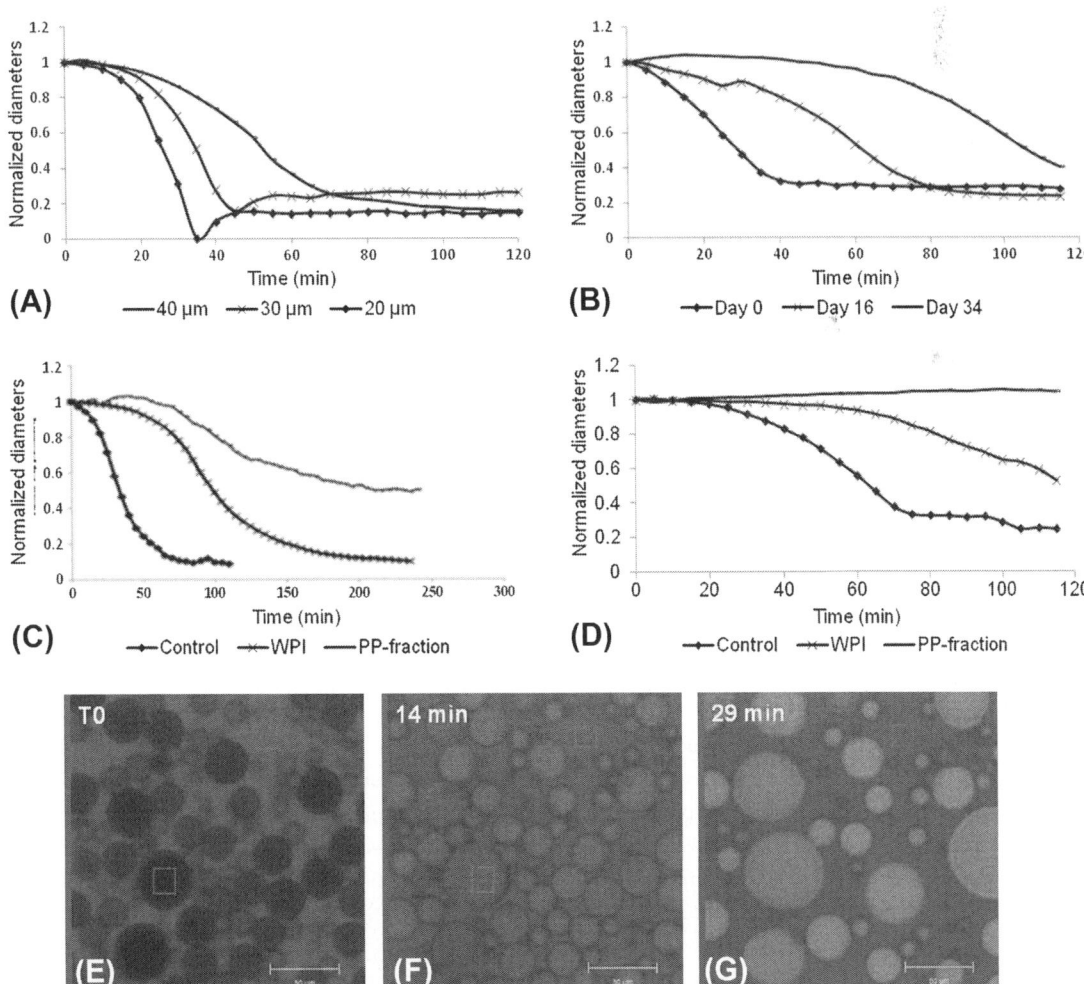

■ **FIGURE 14.6** Normalized diameters as function of time obtained with LiDDEx digestion experiment setup. (A) Influence of droplet size: digestion rates of monodispersed MCT-BLG emulsions of three different droplet sizes. (B) Influence of interface ageing: digestion kinetics of MCT-BLG emulsion at different times after their processing. (C) Influence of the type of interface: digestion of monodispersed MCT emulsion (control) and same emulsion stabilized by whey protein isolate (WPI) or proteose-peptone (PP). (D) Influence of the bulk composition: digestion of MCT-WPI emulsion (control) and same emulsion with WPI or PP present in the bulk. (E) to (G) Diffusion of RITC-labeled *Rhizopus* lipase into MCT oil droplets stabilized by BLG, during the early stage of their digestion. [A color version of this figure is available online at www.booksite.elsevier.com/9780124046108].

of emulsion droplet size on the kinetics of fat digestion. The gradients of the LiDDEx plots represent the kinetics of digestion of the different MCT emulsions, the rate of which increases as the emulsion size decreases because of the increase in surface area available for lipase adsorption. This result stresses the interest in using monodisperse emulsions for assessment and interpretation of fat digestion kinetics. All subsequent kinetics experiments were carried using the same emulsion size (30 μm). Figure 14.6B highlights another aspect of the digestion of emulsions stabilized by a protein (β-lactoglobulin, BLG). For these experiments, the digestion kinetics of emulsions stabilized by BLG were assessed at different times after they had been produced. It is apparent that the rate of digestion of these emulsions changes with the age of the emulsion. It can be seen that the time taken for digestion to start becomes longer with increasing age of the emulsion; however, the gradients of the LiDDEx plot once digestion has started are quite similar. The delay in the start of digestion suggests that reorganization of the protein layer at the interface slows down the action of the lipase. This teaches us that the kinetics of fat digestion of foods containing emulsions stabilized by BLG (or whey protein isolate) are likely to critically depend on their storage time. Such a time-dependent behavior does not occur with emulsions stabilized by a surfactant (i.e., Tween 80—data not shown). Figure 14.6C shows the effect of two different interfaces on the digestion rate of MCT oil in comparison to MCT stabilized by bile salts (control). For these experiments, the oil droplets had the same initial size and the emulsions were digested within the same timeframe after processing. Clearly, the presence of the emulsifier slows down the digestion rate of the MCT oil, and proteose peptone has a stronger inhibiting effect than WPI. Overall, different parts of the LiDDEx graphs can help to understand different aspects of digestion: a delay in the start of digestion might indicate that the interface presents a barrier to digestion, while the slope of the LiDDEx plot indicates the kinetics of fat digestion of the lipid droplets.

Finally, Figure 14.6D illustrates the use of the LiDDEx to investigate the digestion of emulsions in a more realistic context, i.e., depending on the composition of the bulk of the emulsions. For these experiments, emulsions stabilized by WPI (control) were digested in the presence of WPI or proteose peptone fraction in the bulk phase. These graphs demonstrate unambiguously that, when present in the bulk phase, both emulsifiers slow down the rate of digestion of emulsions stabilized by WPI and that the proteose peptone fraction is a far more potent inhibitor than WPI. As can be seen in Figure 14.6C, in the presence of the proteose peptone fraction, the decrease in size of the oil droplets is preceded by a size increase. Moreover, in Figure 14.6D, the digestion of an MCT-WPI emulsion in the presence of the proteose peptone fraction in the bulk phase shows that the size increase lasts until the end of the experiment (2 hours). Such behavior was observed in the presence of other ingredients in the bulk phase such as chitosan and minerals (calcium, magnesium, zinc). This strongly suggests that the observed swelling might result from the flux of water into the droplet. To investigate this phenomenon, lipase from *Rhizopus oryzae* was covalently labeled with Rhodamine isothiocyanate and desalted on a G-25M Sephadex column. The LiDDEx experiment was carried out by adding the labeled lipase to pancreatin in the digestion medium and the behavior of the system was investigated by confocal microscopy. The results of the experiment are shown in Figures 14.6E to G and clearly demonstrate the diffusion of labeled lipase into the core of the oil droplet. In the control experiment, which consisted of adding free Rhodamine label to the pancreatin instead of the labeled *Rhizopus* lipase, no significant increase in the fluorescence intensity inside the oil droplet was observed over the same period of time (and even longer).

Emulsion Structuring During Digestion

The experiments with the LiDDEx technique highlight not only that food structure impacts on fat digestion, but that the digestion of fat can create a number of colloidal structures. The colloidal transformations of lipid emulsions during *in vitro* intestinal digestion were studied in greater detail using advanced scattering techniques. Intestinal lipid digestion was analyzed using time-resolved small-angle X-ray scattering (SAXS), polarized and depolarized dynamic light scattering, time-resolved optical microscopy, and cryogenic transmission electron microscopy as complementary methods. The system that was chosen for study was a mixture of oleic acid (OA) and monoolein (MO). The full digestion reaction, in the absence of preferential adsorption of molecules into the bloodstream, will result in two free fatty acids and one monoglyceride. For this reason, the phase behavior of OA−MO mixtures has been studied by several groups (Borne *et al.*, 2001; Nakano *et al.*, 2002).

We examined emulsion droplets that contained mixtures of OA and MO at pHs up to 9, which therefore simulated the conditions of the gastrointestinal tract from the stomach to the end of the small intestine (Salentinig *et al.*, 2010). Figure 14.7 shows a phase diagram of the internal structure of the oil droplets as a function of temperature and the ratio of OA to MO. Digestion ultimately produces two molecules of OA for one molecule of monoglyceride, giving the following phase sequence. When going from pH 3 to 8, the interior of the oil droplet develops structure, absorbing water by

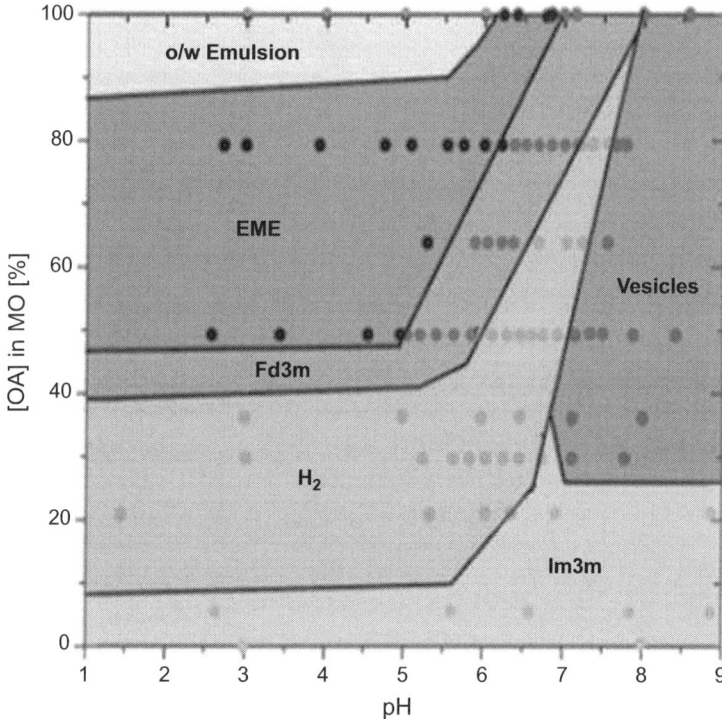

■ **FIGURE 14.7** Phase behavior of mixtures of oleic acid (OA) and monolein (MO) present in the oil droplet of an oil in PBS buffer emulsion. The phase observed is represented as a function of the OA: MO ratio and pH. The phases were determined by small angle X-ray scattering (SAXS) at a temperature of 25°C. O/W emulsion is an oil-in-water emulsion, EME is a reversed microemulsion, Fd3m is the micellar cubic phase, H_2 is the reversed hexagonal phase, and Im3m is a particular inverted bicontinuous cubic structure. Each point in the diagram represents a separate SAXS measurement. *Reprinted with permission from Salenting et al. (2010), Copyright 2010, American Chemical Society.*

forming self-assembled structures. Such structures form spontaneously when water, surfactants (which can be the products of digestion), and oils such as triglycerides are mixed. The first self-assembled structure to appear is a reversed microemulsion phase (EME in Figure 14.7). At a pH of about 6, the reversed micelles present within the oil droplet start to order and the structure becomes an inverted micellar cubic phase (also called Fd3m due to the symmetry of the space group). Finally, at pH higher than 6.5, a reversed hexagonal structure is observed (H_2) and at pH higher than 8 only vesicles are present (Salentinig *et al.*, 2010). This phase diagram may help to predict the potential structures that appear during the digestion of fat. However, certain assumptions need to be made. For example, the presence of other components, such as bile acids, diglycerides, and triglycerides, is neglected. Note that the phase behavior of the monoolein (monoglyceride)—diolein (diglyceride) (Borne *et al.*, 2000) and the phase behavior of unsaturated monoglyceride—triglyceride (Engstroem, 1990; Sagalowicz *et al.*, 2013) were studied and no additional phase was found compared to the phase diagram of Figure 14.7.

In separate experiments the *in vitro* digestion of triolein droplets was performed using simulated intestinal conditions, i.e., in the presence of pancreatic lipase, bile salts, and at pH 6.8. A setup was designed so that small angle X-ray scattering (SAXS) could be done every 2 minutes to follow digestion dynamically (Figure 14.8). During digestion, the triolein droplets underwent a sequence of phase transitions over 420 minutes:

oil droplets → reversed microemulsion(EME) → reversed micellar cubic(Fd3m)

When the pH of the simulated intestinal digestion media was increased the following transitions in structure were observed:

reversed micellar cubic(Fd3m) → reversed hexagonal(H$_2$) → finally vesicles

(Salentinig *et al.*, 2011). It must be added that micelles formed by bile acids were very likely present; however, in the SAXS spectra their low signal was hidden by the signal of other structures present during *in vitro* digestion. The presence of water inside the oil droplets likely favors penetration

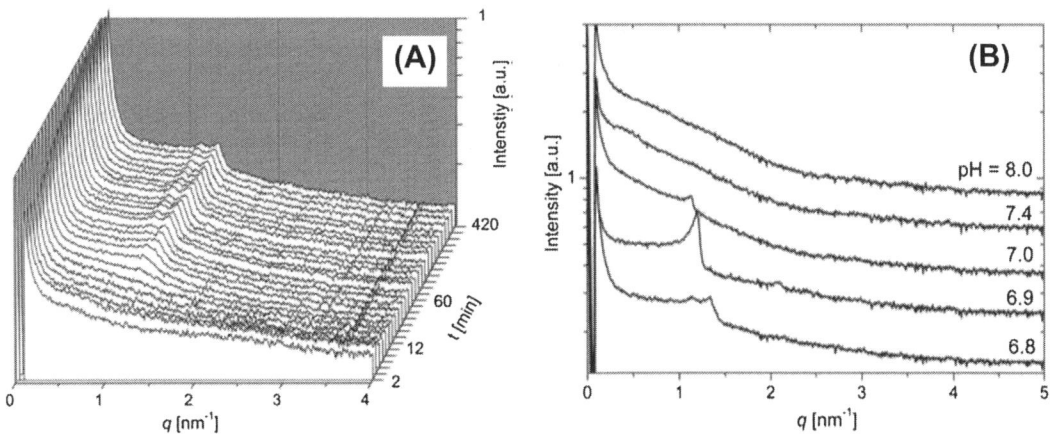

■ **FIGURE 14.8** (A) *In vitro* digestion of a triolein emulsion stabilized by proteins, in the presence of bile salts and pancreatin, as observed by time-resolved SAXS at 25°C showing intensity as function of q, which is the length of the scattering vector, defined by $q = (4\pi/\lambda) \sin \theta/2$, λ being the wavelength of the X-ray radiation and θ the scattering angle. A transition from structureless emulsion to a reversed microemulsion (EME) and then reversed micellar cubic (Fd3m) can be observed. A doublet peak, corresponding to Fd3m, is present at q about 1.3 and after about 60 minutes. (B) SAXS at 25°C showing intensity as function of q for pH readjustment after the digestion of 420 min. Increasing pH leads to a transition from Fd3m to reversed hexagonal (H$_2$) and vesicles above pH 7 (Salentinig *et al.*, 2011).

of lipase into the droplets and may accelerate fat digestion, a result which reflects our earlier findings with the LiDDEx, where covalently-labeled lipase was seen to penetrate the interior of the emulsion droplets. Similar phase transitions were observed by Warren *et al.* (2011). More recently, Müllertz *et al.* (2012) found that the intestinal fluids of humans contained vesicular structures and micelles after ingestion and digestion of a mixture rich in olive oil. The presence of lipid droplets was also observed; however, Müllertz *et al.* could not identify their internal structure. Understanding of the transitions undergone by emulsions during digestion is important because these structures participate in the solubilization and transport of lipophilic nutrients and pharmaceuticals (Porter *et al.*, 2007). This is most evident in the effect of phytosterols on cholesterol and vitamin absorption (Ling and Jones, 1995; Jones *et al.*, 2000). When present in the lumen, phytosterols help to reduce cholesterol absorption by competing for solubilization within bile salt micelles. Since the digestive milieu has a finite bile concentration, cholesterol, and to a lesser extent lipophilic vitamins and carotenoids, are not solubilized within the bile micelle and hence are not absorbed. Overall, it can be seen that an understanding of the structural transitions undergone by emulsions during digestion is important for the design of foods that regulate digestion, but is also critical for understanding absorption and limiting the side effects of anti-nutrients such as lipase inhibitors.

CONCLUSIONS

In order to assess physiological response, modulation of digestion of the main macronutrients can be done through their selection based on their composition and/or through modifying their structural properties. Examples above have illustrated that many different approaches can be followed to modulate the structure of macronutrients and most often one technology can lead to different biological responses. These differences can be explained by the fact that multiple factors impact on the physiological response from the consumed food product. Indeed, most often food products are composed of complex matrices with different macro- and micronutrients involved. The composition, structure, and combination of the different elements of the food product are key elements to understand and to consider when addressing a specific physiological response through modulation of its digestion profile. In addition, it should be noted that other factors also play a key role in this process, such as quantity ingested, time of ingestion, hormone regulation, but also factors related to the consumer, such as age and physiological conditions.

Targeting a functional behavior in the human gastrointestinal tract requires an integrated product and process design which includes, besides the research and development component, the legal, regulatory, processing, and communication counterparts that are necessary to fulfill consumer needs.

REFERENCES

Acheson, K.J., Blondel-Lubrano, A., Oguey-Araymon, S., Beaumont, M., Emady-Azar, S., Ammon-Zufferey, C., et al., 2011. Protein choices targeting thermogenesis and metabolism. Am. J. Clin. Nutr. 93, 525–534.

Aller, E.E.J.G., Abete, I., Astrup, A., Martinez, J.A., van Baak, M.A., 2011. Starches, sugars and obesity. Nutrients 3, 341–369.

Almaas, H., Cases, A.-L., Devold, T.G., Holm, H., Langsrud, T., Aabakken, L., et al., 2006. In vitro digestion of bovine and caprine milk by human gastric and duodenal enzymes. Int. Dairy J. 16, 961–968.

Aloulou, A., Carriere, F., 2008. Gastric lipase: an extremophilic interfacial enzyme with medical applications. Cell. Mol. Life Sci. 65, 851–854.

Anderson, A.K., Guraya, H.S., James, C., Salvaggio, L., 2002. Digestibility and pasting properties of rice starch heat-moisture treated at the melting temperature (Tm). Starch/Stärke 54, 401–409.

Anderson, G.H., Moore, S.E., 2004. Dietary proteins in the regulation of food intake and body weight in humans. J. Nutr. 134, 974–979.

Armand, M., 2007. Lipases and lipolysis in the human digestive tract: where do we stand? Curr. Opin. Clin. Met. Care 10, 156–164.

Armand, M., Hamosh, M., Mehta, N.R., Angelus, P.A., Philpott, J.R., Henderson, T.R., Dwyer, N.K., Lairon, D., Hamosh, P., 1996. Effect of human milk or formula on gastric function and fat digestion in the premature infant. Pediatr. Res. 40, 429–437.

Arnal, M.A., Mosoni, L., Boirie, Y., Houlier, M.-L., Morin, L., Verdier, E., et al., 1999. Protein pulse feeding improves protein retention in elderly women. Am. J. Clin. Nutr. 69, 1202–1208.

Arnal, M.A., Mosoni, L., Dardevet, D., Ribeyre, M.C., Bayle, G., Prugnaud, J., Patureau Mirand, P., 2002. Pulse protein feeding pattern restores stimulation of muscle protein synthesis during the feeding period in old rats. J. Nutr. 5, 1002–1008.

Axelsen, M., Arvidsson, L.R., Lonmroth, P., Smith, U., 1999. Breakfast glycaemic response in patients with type 2 diabetes: effects of bed time dietary carbohydrates. Eur. J. Clin. Nutr. 53, 706–710.

Backer, D., Saniez, M.H., 2005. Soluble highly branched glucose polymers and process for their preparation. Patent EP1369432.

Biliaderis, C.G., 2009. Structural transitions and related physical properties of starch. In: BeMiller, J., Whistler, R. (Eds.), Starch: Chemistry and Technology. Academic Press, Amsterdam, pp. 293–372.

Boirie, Y., Dangin, M., Gachon, P., Vasson, M.-P., Maubois, J.-L., Beaufrère, B., 1997. Slow and fast dietary proteins differently modulate postprandial protein accretion. Physiology 94, 14930–14935.

Borel, P., Pasquier, B., Armand, M., Tyssandier, V., Grolier, P., Alexandre-Gouabau, M.C., et al., 2001. Processing of vitamin A and E in the human gastrointestinal tract. Am. J. Physiol. Gastrointest. Liver Physiol. 280, 95–103.

Borne, J., Nylander, T., Khan, A., 2000. Microscopy, SAXD, and NMR studies of phase behavior of the monoolein-diolein-water system. Langmuir 16, 10044—10054.

Borne, J., Nylander, T., Khan, A., 2001. Phase behavior and aggregate formation for the aqueous monoolein system mixed with sodium oleate and oleic acid. Langmuir 17, 7742—7751.

Boye, J., Wijesinha-Bettoni, R., Burlingame, B., 2012. Protein quality evaluation twenty years after the introduction of the protein digestibility corrected amino acid score method. Br. J. Nutr. 108, 183—211.

Brouns, F., Kettlitz, B., Arrigoni, E., 2002. Resistant starch and "the butyrate revolution." Trends Food Sci. Technol. 13, 251—261.

Bruen, C.M., Kett, A.P., O'Halloran, F., Chaurin, V., Fenelon, M.A., Cashman, K.A., Giblin, L., 2012. Effect of gelatinisation of starch with casein proteins on incretin hormones and glucose transporters in-vitro. Br. J. Nutr. 107, 155—163.

Carbonaro, M., Cappelloni, M., Sabbadini, S., Carnovale, E., 1997. Disulfide reactivity and in vitro protein digestibility of different thermal-treated milk samples and whey proteins. J. Agric. Food Chem. 45, 95—100.

Carey, M.C., Small, D.M., Bliss, C.M., 1983. Lipid digestion and absorption. Annu. Rev. Physiol. 45, 651—677.

Carriere, F., Grandval, P., Gregory, P.C., Renou, C., Henniges, F., Sander-Struckmeier, S., Laugier, R., 2005. Does the pancreas really produce much more lipase than required for fat digestion? J. Pancreas 6, 206—215.

Carriere, F., Renou, C., Ransac, S., Lopez, V., De Caro, J., Ferrato, F., et al., 1999. Effect of Orlistat on lipolysis following standard test meals. Am. J. Physiol. Gastrointest. Liver Physiol. 281, 16—28.

Carriere, F., Withers-Martinez, C., van Tilbeurgh, H., Roussel, A., Cambillau, C., Verger, R., 1998. Structure-function relationships of pancreatic lipases. Fett-Lipid 100, 96—102.

Caugant, I., Petit, H.V., Charbonneau, R., Savoie, L., Toullec, R., Thirouin, S., Yvon, M., 1992. In vivo and In vitro gastric emptying of protein fractions of milk replacers containing whey proteins. J. Dairy Sci. 75, 847—856.

Chu, B.S., Rich, G.T., Ridout, M.J., Faulks, R.M., Wickham, M.S.J., Wilde, P.J., 2009. Modulating pancreatic lipase activity with galactolipids. effects of emulsion interfacial composition. Langmuir 25, 9352—9360.

Colonna, P., Barry, J.L., Cloarec, D., Bornet, F., Gouilloud, S., Galmiche, J.P., 1990. Enzymic susceptibility of starch from pasta. J. Cereal Sci. 11, 59—70.

Daher, G.C., Cooper, D.A., Zorich, N.L., King, D., Riccardi, K.A., Peters, J.C., 1997. Olestra ingestion and dietary fat absorption in humans. J. Nutr. 127, 1694S—1698S.

Dangin, M., Boirie, Y., Garcia-Rodenas, C.L., Gachon, P., Fauquant, J., Callier, P., et al., 2001. The digestion rate of protein is an independent regulating factor of postprandial protein retention. Am. J. Physiol. Endocrinol. Metab. 280, 340—348.

Dangin, M., Guillet, C., Garcia-Rodenas, C.L., Gachon, P., Bouteloup-Demange, C., Reiffers-Magnani, C., et al., 2003. The rate of protein digestion affects protein gain differently during aging in humans. J. Physiol. 549, 635—644.

Donato, L., Kolodziejczyk, E., Rouvet, M., 2011. Mixtures of whey protein microgels and soluble aggregates as building blocks to control rheology and structure of acid induced cold-set gels. Food Hydrocoll. 25, 734—742.

Eelderink, C., Schepers, M., Preston, T., Vonk, R.J., Oudhuis, L., Priebe, M., 2012. Slowly and rapidly digestible starchy foods can elicit a similar glycemic response because of differential tissue glucose uptake in healthy men. Am. J. Clin. Nutr. 96, 1017−1024.

Eerlingen, R.C., Delcour, J.A., 1995. Formation, analysis, structure and properties of type III enzyme resistant starch. J. Cereal Sci. 22, 129−138.

EFSA Journal 2011;9(7):2292. [15 pp.].

Ells, L.J., Seal, C.J., Kettlitz, B., Bal, W., Mathers, J.C., 2005. Postprandial glycaemic, lipaemic and haemostatic responses to ingestion of rapidly and slowly digested starches in healthy young women. Br. J. Nutr. 94, 948−955.

Englyst, H.N., Kingman, S.M., Cummings, J.H., 1992. Classification and measurement of nutritionally important starch fractions. Eur. J. Clin. Nutr. 46, 33−50.

Englyst, H.N., Kingman, S.M., Hudson, G., Cummings, J.H., 1996a. Measurement of resistant starch in vitro and in vivo. Br. J. Nutr. 75, 749−755.

Englyst, H.N., Veenstra, J., Hudson, G.J., 1996b. Measurement of rapidly available glucose (RAG) in plant foods: a potential in vitro predictor of the glycaemic response. Br. J. Nutr. 75, 327−337.

Englyst, K.N., Vinoy, S., Englyst, H.N., Lang, V., 2003. Glycaemic index of cereal products explained by their content of rapidly and slowly available glucose. Br. J. Nutr. 89, 329−339.

Engstroem, L., 1990. Aggregation and structural changes in the L2-phase in the system water/soybean oil/sunflower oil monoglycerides. J. Dispers. Sci. Technol. 11, 479−489.

Fannon, J.E., Hauber, R.J., BeMiller, J.N., 1992. Surface pores of starch granules. Cereal Chem. 69, 284−288.

Fardet, A., Hoebler, C., Baldwin, P.M., Bouchet, B., Gallant, D.J., Barry, J.L., 1998. Involvement of the protein network in the in vitro degradation of starch from spaghetti and lasagne: a microscopic and enzymic study. J. Cereal Sci. 27, 133−145.

Fave, G., Coste, T.C., Armand, M., 2004. Physicochemical properties of lipids: new strategies to manage fatty acid bioavailability. Cell. Mol. Biol. 50, 815−831.

Feinle-Bisset, C., Patterson, M., Ghatei, M.A., Bloom, S.R., Horowitz, M., 2005. Fat digestion is required for suppression of ghrelin and stimulation of peptide YY and pancreatic polypeptide secretion by intraduodenal lipid. Am. J. Endocrinol. Metab. 289, 948−953.

Fuertes, P., Roturier, J.M., Petitjean, C., 2004. Soluble highly branched glucose polymers. Patent US 2005/0159329 A1.

Gallant, D.J., Bouchet, B., Builéon, B., Pérez, S., 1992. Physical characteristics of starch granules and susceptibility to enzymatic degradation. Eur. J. Clin. Nutr. 46, 3−16.

Golay, A., Koellreutter, B., Bloise, D., Assal, A.-P., Wursch, P., 1992. The effect of muesli or cornflakes at breakfast on carbohydrate metabolism in type 2 diabetic patients. Diabetes Res. Clin. Prac. 15, 135−142.

Golding, M., Wooster, T.J., 2010. The influence of emulsion structure and stability on lipid digestion. Curr. Opin. Colloid Interface Sci. 15, 90−101.

Golding, M., Wooster, T.J., Day, L., Xu, M., Lundin, L., Keogh, J., Clifton, P., 2011. Impact of gastric structuring on the lipolysis of emulsified lipids. Soft Matter 7, 3513−3523.

Guraya, H.S., James, C., Champagne, E.T., 2001a. Effect of cooling and freezing on the digestibility of debranched rice starch and physical properties of the resulting material. Starch/Stärke 53, 64−74.

Guraya, H.S., James, C., Champagne, E.T., 2001b. Effect of enzyme concentration and storage temperature on the formation of slowly digestible starch from cooked debranched rice starch. Starch/Stärke 53, 131–139.

Hamosh, M., 1996. Digestion in the newborn. Clin. Perinatol. 23, 191–209.

Haralampu, S.G., 2000. Resistant starch—a review of the physical properties and biological impact of RS3. Carbohydr. Polymers 41, 285–292.

Harbis, A., Perdreau, S., Vincent-Baudry, S., Charbonnier, M., Bernard, M.C., Raccah, D., et al., 2004. Glycemic and insulinemic meal responses modulate postprandial hepatic and intestinal lipoprotein accumulation in obese, insulin resistant subjects. Am. J. Clin. Nutr. 80, 896–902.

Huppertz, T., de Kruif, G.G., 2008. Structure and stability of nanogel particles prepared by internal cross-linking of casein micelles. Int. Dairy J. 18, 556–565.

Inglinstad, R.A., Devold, T., Ericksen, E.K., Holm, H., Jacobsen, M., Liland, K.H., et al., 2010. Comparison of the digestion of caseins and whey proteins in equine, bovine, caprine and human milks by human gastrointestinal enzymes. Dairy Sci. Technol. 90, 1–15.

Jane, J.L., Chen, Y.Y., Lee, L.F., McPherson, A.E., Wong, K.S., Radosavljevic, M., Kasemsuwan, T., 1999. Effects of amylopectin branch chain length and amylose content on gelatinization and pasting properties of starch. Cereal Chem. 76, 629–637.

Jiang, H., Campbell, M., Blanco, M., Jane, J.L., 2010. Characterization of maize amylose-extender (ae) mutant starches: Part II. Structures and properties of starch residues remaining after enzymatic hydrolysis at boiling-water temperature. Carbohydr. Polymers 80, 1–12.

Jones, P.J., Raeini-Sarjaz, M., Ntanios, F.Y., Vanstone, C.A., Feng, J.Y., Parsons, W.E., 2000. Modulation of plasma lipid levels and cholesterol kinetics by phytosterol versus phytostanol esters. J. Lipid Res. 41, 697–705.

Kawabata, Y., Toeda, K., Takahashi, T., Shibamoto, N., 2001. Highly branched starch and method for producing the same. Patent JP2001294601.

Kinsella, J.E., 1984. Milk proteins: physicochemical and functional properties. Crit. Rev. Food Sci. Nutr. 21, 197–262.

Kitabake, N., Wada, R., Fujita, Y., 2001. Reversible conformational change in b lactoglobulin A modified with N-ethylmaleimide and resistance to molecular aggregation on heating. J. Agric. Food Chem. 49, 4011–4018.

Komae, K., Kato, T., 2002. Method for producing branched starch. Patent JP2002078497.

Lehmann, U., 2009. Industrial viewpoint of slowly digestible carbohydrates and potential applications. Whistler Center, Purdue University, USA.

Lehmann, U., Jacobasch, G., Schmiedl, D., 2002. Characterization of resistant starch Type III from banana (Musa acuminata). J. Agric. Food Chem. 50, 5236–5240.

Lehmann, U., Robin, F., 2007. Slowly digestible starch—its structure and health implications: a review. Trends Food Sci. Technol. 18, 346–355.

Lehmann, U., Rössler, C., Schmiedl, D., Jacobasch, G., 2003. Production and physicochemical characterization of resistant starch type III derived from pea starch. Nahrung/Food 47, 60–63.

Leloup, V.M., Colonna, P., Ring, S.G., 1992. Physico-chemical aspects of resistant starch. J. Cereal Sci. 16, 253–266.

Lentle, R.G., Janssen, P.W.M., 2010. Manipulating digestion with foods designed to change the physical characteristics of digesta. Crit. Rev. Food Sci. Nutr. 50, 130–145.

Li, Y., Hu, M., Du, Y.M., Xiao, H., McClements, D.J., 2010. Control of lipase digestibility of emulsified lipids by encapsulation within calcium alginate beads. Food Hydrocoll. 25, 122−130.

Lin, A.H.M., Lee, B.H., Nichols, B.L., Quezada-Calvillo, R., Rose, D.R., Naim, H.Y., Hamaker, B.R., 2012. Starch source influences dietary glucose generation at the mucosal α-glucosidase level. J. Biol. Chem. 287, 36917−36921.

Lindberg, T., Engberg, S., Sjoberg, L.B., Lonnerdal, B., 1998. In vitro digestion of proteins in human milk fortifiers and in preterm formula. J. Pediatr. Gastroenterol. Nutr. 27, 30−36.

Lindquist, S., Hernell, O., 2010. Lipid digestion and absorption in early life: an update. Curr. Opin. Clin. Nutr. Metab. Care 13, 314−320.

Ling, W.H., Jones, P.J.H., 1995. Dietary phytosterols: a review of metabolism, benefits and side effects. Life Sci. 57, 195−206.

Lissner, L., Heitmann, B.L., 1995. Dietary fat and obesity—evidence from epidemiology. Eur. J. Clin. Nutr. 49, 79−90.

Lowe, M.E., 1997. Structure and function of pancreatic lipase and colipase. Annu. Rev. Nutr. 17, 141−158.

Mackie, A., Macierzanka, A., 2010. Colloidal aspects of protein digestion. Curr. Opin. Colloid Interface Sci. 15, 102−108.

Mahé, S., Roos, N., Benamouzig, R., Davin, L., Luengo, C., Gagnon, L., et al., 1996. Gastrojejunal kinetics and the digestion of [15N] b-lactoglobulin and casein in humans: the influence of the nature and quantity of the protein. Am. J. Clin. Nutr. 63, 546−552.

Maldonado-Valderrama, J., Wilde, P., Macierzanka, A., Mackie, A., 2011. The role of bile salts in digestion. Adv. Colloid Interface Sci. 65, 36−46.

Maljaars, J., Romeyn, E.A., Haddeman, E., Peters, H.P.F., Masclee, A.A.M., 2009. Effect of fat saturation on satiety, hormone release, and food intake. Am. J. Clin. Nutr. 89, 1019−1024.

Maljaars, P.W.J., Peters, H.P.F., Mela, D.J., Masclee, A.A.M., 2008. Ileal brake: a sensible food target for appetite control. A review. Physiol. Behav. 95, 271−281.

McCleary, B.V., McNally, M., Rossiter, P., 2002. Measurement of resistant starch by enzymatic digestion in starch and selected plant materials: a collaborative study. J. AOAC Int. 85, 1103−1111.

Mullertz, A., Fatouros, D.G., Smith, J.R., Vertzoni, M., Reppas, C., 2012. Insights into intermediate phases of human intestinal fluids visualized by atomic force microscopy and cryo-transmission electron microscopy ex vivo. Mol. Pharm. 9, 237−247.

Mun, S., Decker, E.A., McClements, D.J., 2007. Influence of emulsifier type on in vitro digestibility of lipid droplets by pancreatic lipase. Food Res. Int. 40, 770−781.

Nakano, M., Teshigawara, T., Sugita, A., Leesajakul, W., Taniguchi, A., Kamo, T., et al., 2002. Dispersions of liquid crystalline phases of the monoolein/oleic acid/pluronic F127 system. Langmuir 18, 9283−9288.

Neville, M.C., 1995. Lactogenesis in women: a cascade of events revealed by milk composition. In: Jensen, R. (Ed.), The Composition of Milks. Academic Press, San Diego, pp. 87−98.

Niba, L.L., 2003. Processing effects on susceptibility of starch to digestion in some dietary starch sources. Int. J. Food Sci. Nutr. 54, 97−109.

Nicolai, T., Schmitt, C., Britten, M., 2011. β-Lactoglobulin and WPI aggregates: formation, structure and applications. Food Hydrocoll. 25, 1945−1962.

Nicoletti, M., 2012. Nutraceuticals and botanicals: overview and perspectives. Int. J. Food Sci. Nutr. 63, 2–6.

Norton, I., Moore, S., Fryer, P., 2007. Understanding food structuring and breakdown: engineering approaches to obesity. Obesity Rev. 8, 83–88.

Pan, D., Jane, J.L., 2000. Internal structure of normal maize starch granules revealed by chemical surface gelatinization. Biomacromolecules 1, 126–132.

Porter, C.J.H., Trevaskis, N.L., Charman, W.N., 2007. Lipids and lipid-based formulations: optimizing the oral delivery of lipophilic drugs. Nat. Rev. Drug Discov. 6, 231–248.

Pouteau, E., Bovetto, L., Schlup-Ollivier, G., Grathwohl, D., Macharia, H., Beaumont, M., Macé, K., 2012. Microgel formation of whey protein reduces its insulinogenic index without modifying glycemic response in healthy men. Poster presented at EPSEN, Barcelona.

Reis, P., Holmberg, K., Watzke, H., Leser, M.E., Miller, R., 2009. Lipases at interfaces: a review. Adv. Colloid Interface Sci. 147–148, 237–250.

Robin, F., Merinats, S., Simona, A., Lehmann, U., 2008. Influence of chain length an alpha-1,4-D-glucan recrystallization and slowly digestible starch. Starch/Stärke 60, 551–558.

Roos, N., 2003. Cross-linking by transglutaminase changes neither the in vitro proteolysis nor the in vivo digestibility of caseinate. Kieler Milchwirtschaftliche Forschungsberichte 55, 261–276.

Sagalowicz, L., Guillot, S., Acquistapace, S., Schmitt, B., Maurer, M., Yaghmur, A., De Campo, L., Rouvet, M., Leser, M., Glatter, O., 2013. Influence of vitamin e acetate and other lipids on the phase behavior of mesophases based on unsaturated monoglycerides. Langmuir 29 (26), 8222–8232.

Saito, T., 2008. Antihypertensive peptides derived from bovine casein and whey proteins. Adv. Exp. Med. Biol. 606, 295–317.

Sajilata, M.G., Singhal, R.S., Kulkarni, P.R., 2006. Resistant starch—a review. Comp. Rev. Food Sci. Food Safety 5, 1–17.

Salami, M., Yousefi, R., Ehsani, R.M., Dalgalarrondo, M., Chobert, J.-M., Haertlé, T., et al., 2008. Kinetics characterization of hydrolysis of camel and bovine milk proteins by pancreatic enzymes. Int. Dairy J. 18, 1097–1102.

Salentinig, S., Sagalowicz, L., Glatter, O., 2010. Self-assembled structures and pKa value of oleic acid in systems of biological relevance. Langmuir 26, 11670–11679.

Salentinig, S., Sagalowicz, L., Leser, M.E., Tedeschi, C., Glatter, O., 2011. Transitions in the internal structure of lipid droplets during fat digestion. Soft Matter 7, 650–661.

Sandra, S., Decker, E.A., McClements, D.J., 2008. Effect of interfacial protein cross-linking on the in vitro digestibility of emulsified corn oil by pancreatic lipase. J. Agric. Food Chem. 56, 7488–7494.

Schmitt, C., Bovay, C., Vuilliomenet, A.M., Rouvet, M., Bovetto, L., Barbar, R., Sanchez, C., 2009. Multiscale characterization of individualized β-lactoglobulin microgels formed upon heat treatment under narrow pH range conditions. Langmuir 25, 7899–7909.

Schmitt, C., Moitzi, C., Bovay, C., Rouvet, M., Bovetto, L., Donato, L., et al., 2010. Internal structure and colloidal behaviour of covalent whey protein microgels obtained by heat treatment. Soft Matter 6, 4876–4884.

Seimon, R.V., Wooster, T., Otto, B., Golding, M., Day, L., Little, T.J., et al., 2009. The droplet size of intraduodenal fat emulsions influences antropyloroduodenal motility, hormone release, and appetite in healthy males. Am. J. Clin. Nutr. 89, 1729−1736.

Shin, S.I., Choi, H.J., Chung, K.M., Hamaker, B.R., Park, K.H., Moon, T.W., 2004. Slowly digestible starch from debranched waxy sorghum starch: preparation and properties. Cereal Chem. 81, 404−408.

Shin, S.I., Kim, H.J., Ha, H.J., Lee, H.S., Moon, T.W., 2005. Effect of hydrothermal treatment on formation and structural characteristics of slowly digestible non-pasted granular sweet potato starch. Starch/Stärke 57, 421−430.

Singh, H., Ye, A., Horne, D., 2009. Structuring food emulsions in the gastrointestinal tract to modify lipid digestion. Proq. Lipid Res. 48, 92−100.

Sjostrom, L., Rissanen, A., Andersen, T., Boldrin, M., Golay, A., Koppeschaar, H.P.F., Krempf, M., 1998. Randomized placebo-controlled trial of Orlistat for weight loss and prevention of weight regain in obese patients. Lancet 352, 167−173.

Speth, J.D., 2012. Middle Palaeolithic subsistence in the Near East: zooarchaeological perspectives—past, present and future. Before Farming 2012, 1.

Srichuwong, S., Jane, J.L., 2007. Physicochemical properties of starch affected by molecular composition and structures: a review. Food Sci. Biotechnol. 16, 663−674.

Srichuwong, S., Sunarti, T.C., Mishima, T., Isono, N., Hisamatsu, M., 2005. Starches from different botanical sources. I. Contribution of amylopectin fine structure to thermal properties and enzyme digestibility. Carbohydr. Polymers 60, 529−538.

Stanciuc, N., van der Plancken, I., Rotaru, G., Hendrickx, M., 2008. Denaturation impact in susceptibility of beta-lactoglobulin to enzymatic hydrolysis: a kinetic study. Revue Roumaine de Chimie 53, 921−929.

Tessari, P., Kiwanuka, E., Cristini, M., Zaramella, M., Enslen, M., Zurlo, C., Gracia-Rodenas, C., 2007. Slow versus fast proteins in the stimulation of beta-cell response and the activation of the entero-insular axis in type 2 diabetes. Diabetes/Metab. Res. Rev. 23, 378−385.

Tomé, D., 1991. Digestibilité des protéines en fonction de la taille des fractions protéiques. Médecine & Nutrition 27, 129−132.

Topping, D.L., Fukushima, M., Bird, A., 2003. Resistant starch as prebiotic and symbiotic: state of the art. Proc. Nutr. Soc. 62, 171−176.

Tremblay, F., Lavigne, C., Jacques, H., Marette, A., 2007. Role of dietary proteins and amino acids in the pathogenesis of insulin resistance. Annu. Rev. Nutr. 27, 293−310.

Turgeon, S.L., Rioux, L.-E., 2011. Food matrix impact on macronutrients nutritional properties. Food Hydrocoll. 25, 1915−1924.

Venkatachalam, M., Kushnick, M.R., Zhang, G., Hamaker, B.R., 2009. Starch-entrapped biopolymer microspheres as a novel approach to vary blood glucose profiles. J. Am. Coll. Nutr. 28, 583−590.

Wachters-Hagedoorn, R.E., Priebe, M.G., Heimweg, J.A.J., Heiner, A.M., Englyst, K.N., Holst, J.J., et al., 2006. The rate of intestinal glucose absorption is correlated with plasma glucose-dependent insulinotropic polypeptide concentrations in healthy men. J. Nutr. 136, 1511−1516.

Warren, D.B., Anby, M.U., Hawley, A., Boyd, B.J., 2011. Real time evolution of liquid crystalline nanostructure during the digestion of formulation lipids using synchrotron small-angle X-ray scattering. Langmuir 27, 9528−9534.

Wickham, M., Garrood, M., Leney, J., Wilson, P.D.G., Fillery-Travis, A., 1998. Modification of a phospholipid stabilized emulsion interface by bile salt: effect on pancreatic lipase activity. J. Lipid Res. 39, 623−632.

Wolf, B.W., Garleb, K.A., Choe, Y.S., Humphrey, P.M., Maki, K.C., 2003. Pullulan is a slowly digestible carbohydrate in humans. J. Nutr. 133, 1051−1055.

Zhang, G., Ao, Z., Hamaker, B.R., 2008a. Nutritional property of endosperm starches from maize mutants: a parabolic relationship between slowly digestible starch and amylopectin fine structure. J. Agric. Food Chem. 56, 4686−4694.

Zhang, G., Hamaker, B.R., 2009. Slowly digestible starch: concept, mechanism, and proposed extended glycemic index. Crit. Rev. Food Sci. Nutr. 49, 852−867.

Zhang, G., Sofyan, M., Hamaker, B.R., 2008b. Slowly digestible state of starch: mechanism of slow digestion property of gelatinized maize starch. J. Agric. Food Chem. 56, 4695−4702.

Chapter

15

Describing Dietary Energy—Towards the Formulation of Specialist Weight-Loss Foods

Paul J. Moughan

Riddet Institute, Massey University, Palmerston North, New Zealand

CONTENTS

INTRODUCTION

The Almond Board of California recently petitioned food authorities in the USA to reduce the number of calories officially allocated to almonds on food labels (Addy, 2012), a move highlighting the inadequacies of current approaches to describing "available" energy in foods. This followed publication by Novotny *et al.* (2012) of the results of a human metabolic study that showed that the metabolizable energy (ME) content of almonds is considerably overestimated (around 30% overestimation) after application of either

Food Structures, Digestion and Health. http://dx.doi.org/10.1016/B978-0-12-404610-8.00015-3

Atwater general factors or Atwater specific factors. The "available" energy content of a food is related to the often complex inter- and intra-molecular structures of foods and the macro-architecture of these molecules, and represents the net outcome of a set of interactions between the food structure and the mammalian processes of food assimilation, absorption, and utilization.

The means by which energy values are assigned to foods in practice are variable and dependent upon the jurisdiction. In the USA, for example, and according to the US Code of Federal Regulations, the energy content of a food may be calculated by the Atwater general factors, by the Atwater general factors minus a correction for fiber, by the Atwater specific factors, by data for specific food factors for particular foods, or by bomb calorimetry after adjustment for loss of nitrogen through urea (Novotny et al., 2012). Most commonly in the USA, the food energy value is derived by application of the Atwater general factors (4 kcal/g for protein; 9 kcal/g for fat; 4 kcal/g for carbohydrate, i.e., 4/9/4). The Atwater approach is also used in the European Union for legislative purposes (where the term "carbohydrate" is defined as total carbohydrate minus total dietary fiber).

The Atwater general factors (Merrill and Watt, 1973) were not originally intended to be used in this manner, and Atwater understood full well the implications of dietary-induced thermogenesis for "available" energy, discussing the concept of "physiological fuel value." Nevertheless, Atwater general factors, although they ignore differential dietary-induced thermogenesis across the absorbed nutrients and involve a number of simplifications, are applied widely in practice.

These simplifications include that the ME of a food can be accurately predicted based on three chemical components (crude protein, fat, and carbohydrate) and that the ME of each component does not vary greatly across diverse foods. This reductionist approach largely ignores the effects of food structures and the interplay between these in the assimilation of a complex diet. Implicit in the systems is that the ME content of a food is affected by the fat, carbohydrate, and protein contents only; that the ratio of gross energy (heat of combustion) to either fat, protein, or carbohydrate is constant across foods; that differences among foods and food ingredients in the degree of digestibility of each of fat, carbohydrate, and protein are quantitatively unimportant; that differences in urinary excretion of energy per unit dietary protein are quantitatively unimportant; and that interactive effects among dietary components on nutrient assimilation can be ignored. There is considerable published evidence, however, that these simplifications are untenable, and as discussed above these factorial-based ME systems do not account for what are quite large differences in the biochemical efficiencies of utilization of absorbed nutrients. It is not surprising then that the Atwater method can lead to inaccuracies. Human metabolism studies conducted at our own center (Zou *et al.*, 2007) clearly demonstrate significant ($P < 0.05$) differences between ME calculated based on Atwater or similar systems and determined ME values. Adults ($n = 27$) were given either a high-fat low-fiber diet, a low-fat high-fiber fruit and vegetable-based diet, or a low-fat high-fiber cereal-based diet. Differences (Atwater predicted vs. actual) of up to 4% for the refined diet and up to 11% for the low-fat/high-fiber diets were found. Such differences highlight the inaccuracy of the factorial systems for predicting ME, especially in weight-loss foods and diets specifically formulated to have lower "available" energy contents, whereby even small differences in ME are likely to be of practical significance. The inclusion of such weight-loss foods in the diet is an important strategy for combating obesity in humans, and the "available" calorific values of such foods need to be known with a high degree of accuracy. In particular, the accuracy of the relative rankings of the available energy contents of such foods is of importance.

SMALL- AND LARGE-BOWEL DIGESTION, ABSORBED NUTRIENTS AND BIOCHEMICAL EFFICIENCY OF ADENOSINE TRIPHOSPHATE PRODUCTION

During digestion in humans, alcohol is absorbed from the stomach and small intestine; glucose, simple sugars, fatty acids, and amino acids from the small intestine; and volatile (short chain) fatty acids, after fermentation, which occurs mainly in the large intestine. In a study with humans (Coles *et al.*, 2010), estimates were obtained for the digestibility of energy and nutrients from the upper digestive tract (mouth to terminal ileum) and large intestine separately for low- and high-fiber diets. Total tract energy digestibility ranged from 92% (high-fiber diet) to 96% (pectin-based and low-fiber diets), while at the ileal level the digestibility of energy ranged from 79 to 86% for the high-fiber to the low-fiber diet. The upper tract (ileal) digestion of starch, sugars, fatty acids, and amino acids was high but differed with diet type. Non-starch polysaccharides (NSP) were poorly digested in the upper tract for all diets except for the pectin-based (PE) diet where more than half of the NSP appeared to be fermented in the upper digestive tract. The large differences between fecal and ileal nutrient loss (see Table 15.1) highlight that fecal (total tract) digestibility data alone provide incomplete information on nutrient uptake. There is a need to be able to determine the uptake of nutrients and energy in the upper and lower digestive tracts separately. Traditional apparent fecal (total tract) digestibility data are

Table 15.1 Actual Daily Intakes and Predicted Upper-tract Uptake and Colonic Loss for Several Dietary Components for Adult Humans ($n = 21$) Given Four Different Diets

	Diet			
	Wheat Bran	**Pectin**	**Low Fiber**	**High Fiber**
Energy (MJ/d)				
Daily intake	9.4	8.1	10.0	8.7
Upper-tract uptake	7.9	6.7	8.6	6.9
Colonic loss	0.9	1.1	1.0	1.1
Protein (g/d)				
Daily intake	134.8	113.7	87.0	79.8
Upper-tract uptake	114.2	88.2	72.0	62.6
Colonic loss	13.0	20.5	9.2	7.4
Fat (g/d)				
Daily intake	65.6	51.3	85.4	66.9
Upper-tract uptake	63.5	48.0	77.2	60.3
Colonic loss	0.6	2.3	6.6	4.6
Starch (g/d)				
Daily intake	39.7	16.2	135.8	121.2
Upper-tract uptake	39.4	16.1	134.2	119.9
Colonic loss	0.2	0.03	1.5	1.0
Sugars (g/d)				
Daily intake	162.1	176.4	127.2	95.8
Upper-tract uptake	138.5	163.3	108.6	75.0
Colonic loss	23.3	13.0	18.2	20.4
Non-starch polysaccharides (g/d)				
Daily intake	28.3	17.9	20.2	23.1
Upper-tract uptake	0.3	10.0	1.6	0
Colonic loss	12.8	5.9	13.0	10.2

Values from Coles et al. (2010).

unable to differentiate between upper-tract and colonic nutrient uptake for different diets despite the fact that it is important to be able to predict both the site (upper tract or colon) and degree of absorption of each nutrient, because nutrients vary in the efficiency with which they yield energy that is ultimately useful to the body (net ATP gains) and the energy made available to the body via short chain fatty acids from nutrients fermented mainly in the hindgut (colon) is less than that obtained from direct nutrient uptake in the upper tract. The capture of energy (conversion to ATP) from food is less than completely efficient in intermediary metabolism (Flatt and Trenblay, 1997).

Ideally, food energy values should reflect the amount of energy in the respective food components (protein, fat, carbohydrate, alcohol, polyols, organic acids, and novel compounds) and their ultimate utilization by the human body. To achieve this with accuracy, information on the uptake of each component from each compartment of the digestive tract is required. Such information cannot be obtained routinely with human subjects, so models of digestion are needed.

Further, an entirely different problem with the reductionist approach to determining tabulated values for nutrient digestibility in foods arises in that the digestibility of a particular nutrient in a particular food or food ingredient is influenced by the total food matrix (the meal) that is consumed. Different food structures from different ingredients interact with each other and with the organism to alter the digestive milieu and influence the outcome of the digestive processes. Digestibility values from individual dietary components may not be additive for a mixed diet, thus further highlighting the need for digestion models.

MODELS OF DIGESTION AND FERMENTATION
Digestion

There are many *in vitro* and *ex vivo* models of gastric and small-intestinal digestion in humans and simple-stomached mammals, which are useful for studying aspects of food breakdown under defined chemical and physical conditions, but none of these has been consistently demonstrated of value for predicting *in vivo* nutrient digestibility with quantitative accuracy (Butts *et al.*, 2012). However, both the laboratory rat and the growing pig have similar upper digestive tracts (mouth to end of small intestine) to the adult human, both anatomically and physiologically. The laboratory rat is a convenient experimental animal and digesta samples can be readily obtained upon euthanasia of the animal. However, only small quantities of digesta are sampled, and with only one sample of digesta obtained per animal. Also, the nocturnal rat is a selective eater with a nibbling habit. The growing pig, being a meal-eating omnivore, is the preferred animal model for digestion studies with humans (Miller and Ullrey, 1987; Moughan and Rowan, 1989; Deglaire *et al.*, 2009; Deglaire and Moughan, 2012), and the cannulated pig model enables multiple sampling over time from the free-living animal, with a considerably larger sample size than rat, and with less dependence on indigestible dietary markers.

Fermentation

The rat and pig both have different hindgut structures from those of humans, and on this basis alone are not good models for fermentation in humans, though the adult dog has been suggested as a better animal model (Hendriks *et al.*, 2012).

Over the last two decades a considerable amount of work has been undertaken in developing *in vitro* fermentation models based on the use of fresh human fecal innocula (Coles *et al.*, 2005) and these models show considerable promise. Such an *in vitro* fermentation assay for humans (an adaptation of the method developed by Edwards *et al.*, 1996) has been optimized for its key variables (enzyme/inoculum concentration, pH, mixing, and duration of incubation) by Coles *et al.* (2011, 2013a) and has been shown to give realistic values for hindgut fermentation. The latter method uses ileal digesta (see animal models above) as the substrate and provides an estimate of the proportion of dry matter fermented in the colon.

Dual Assay

The Riddet Institute has developed a dual assay for food digestion based on combining the upper-tract digestion animal model (see above) with the optimized *in vitro* fermentation assay (see above). The assay predicts absorption in the adult human of food macronutrients from the upper digestive tract, the amount of material entering the colon, and the fermentability (and thus short chain fatty acid production) of that material. The dual digestibility assay has been validated (Coles *et al.*, 2013b) by comparing predicted (model) total tract organic matter digestibility (OMD) with that determined in adult humans in a metabolic study. The results for a wide range of diets are shown in Table 15.2.

The dual digestion model can be applied to foods to predict the uptake of all energy-containing nutrients from the digestive tract of the adult human.

DIFFERENTIAL EFFICIENCIES OF UTILIZATION OF ABSORBED NUTRIENTS FOR ENERGY (ATP) SUPPLY

In a state of negative energy balance (weight loss), all absorbed nutrients will eventually be catabolized, with the production of ATP to meet the body's energy demands. The production of ATP energy per unit energy present in the dietary nutrient is not the same, however, for all nutrients. For example,

Table 15.2 Comparison of Predicted (Dual *in vivo/in vitro* Digestion Model) Total Tract Organic Matter Digestibility (OMD) with OMD Determined in Adult Humans

	OMD %	
Diet	Predicted	Observed
Wheatbran	92.3	93.5
Pectin	96.9	96.9
Low fiber	94.8	96.7
High fiber	91.5	92.8
Fruit/vegetable	91.4	92.3
Cereal	89.6	90.0

From Coles et al. (2013b).

when material is fermented in the hindgut, a considerable amount of energy (heat of fermentation) is expended in the degradative process. Amino acids are deaminated before oxidation with synthesis of urea and the urea being excreted in the urine. This involves energy costs. Further, different absorbed nutrients are oxidized using different biochemical (stoichiometric) pathways with different efficiencies. Approximate values for the efficiency of the generation of ATP in the cell following the oxidation of selected nutrients are shown in Table 15.3, highlighting the need to take these efficiencies into account when comparing the uptake and utilization of different nutrients.

AN OVERALL MODEL OF DIGESTION AND POST-ABSORPTIVE NUTRIENT UTILIZATION

Ultimately, "available" energy for work processes should be expressed as ATP equivalents (ATP being the universal currency of energy in the body). A model has been developed (Coles *et al.*, 2013c) that allows prediction of the net ATP yield (i.e., ATP generated during nutrient catabolism net of direct ATP costs of catabolism), arising from the metabolism of absorbed energy-yielding food substrates. The quantities of nutrients predicted to be absorbed from the upper digestive tract and colon of the adult human and subsequently to be available for cellular metabolism are determined by applying the dual *in vivo/in vitro* digestibility assay (described above). The potential ATP yield from the catabolism of the predicted absorbed nutrients is then estimated using a series of mathematical equations, derived using documented stoichiometric relationships. The objective of the model is to allow a relative ranking of foods designed specifically to achieve and

Table 15.3 Approximate Values for the Biochemical Efficiency of Conversion (%) of Gross Energy (kJ) to Available Energy (ATP, kJ) for Selected Nutrients

	%
Glucose	68
Lysine	50
Phenylalanine	39
Histidine	30
Glutamine	55
Glycerol	62
Lauric acid	64
Stearic acid	60
Butyric acid[1]	54
Acetic acid[1]	48
Ethanol	44

[1]Note these are the efficiencies of utilization of the absorbed volatile fatty acids. Considerable energy is also lost during bacterial fermentation (heat of fermentation), making the net energy yield from the fermented materials (mainly fiber) especially low.
Values from Coles et al. (2013c).

maintain body weight loss in adult humans, and is a tool for the formulation and testing of novel weight-loss foods.

The model relates to a healthy adult human consuming food at or below maintenance energy levels, and is based on a number of assumptions (Coles *et al.*, 2013c). The dual digestibility assay of the overall model has been validated and found to accurately predict nutrient uptakes from the gut. The stoichiometric relationships for nutrient catabolism, moreover, are well known; however, there is some uncertainty as to quantitative nutrient flows through alternative pathways (different catabolic pathways and pathways for synthesis, storage, and subsequent catabolism) and there is uncertainty also surrounding the mitochondrial efficiency of ATP production (the free energy of ATP hydrolysis *in vivo*).

The absolute accuracy of the overall model predictions remains to be demonstrated using human calorimetry, but relative differences obtained using the model for a "standardized" set of model parameters should provide accurate rankings of foods and diets, and it is in this capacity that the overall model is currently used.

A schematic diagram describing the overall model is given in Figure 15.1.

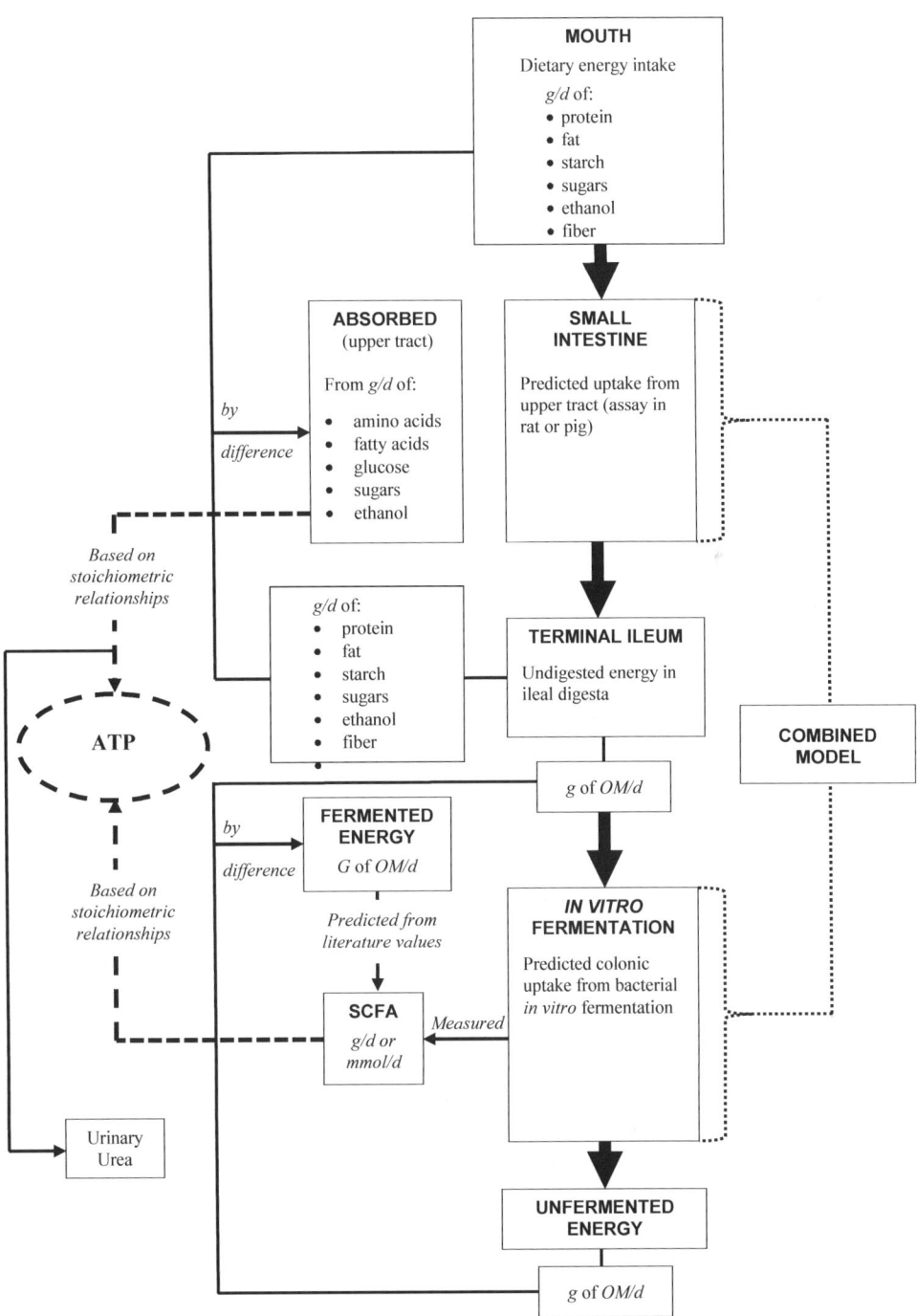

■ **FIGURE 15.1** Schematic representation of the overall model.

EXAMPLES OF MODEL APPLICATION

The model has been applied to determine the available (ATP) energy content of several foods and two examples are given below:

1. *Available energy content of two proprietary meal replacement formulas.* Two commercially available meal replacement formulas (Formula A, a weight-gain supplement, Formula B, a weight-loss formulation; chemical composition as shown in Table 15.4) were subjected to the dual *in vivo/in vitro* digestibility assay (nutrient digestibility and organic matter fermentability data shown in Table 15.5) and ATP production was simulated. The predicted ATP yields are given in Table 15.6, and highlight that the formulas had very different available (ATP) energy contents. As expected, Formula B would be the most effective for weight loss and weight maintenance. When the modeled available energy values were compared with predicted available energy values for dextrin (a baseline compound), the model ranked the available energy relative to dextrin of the two formulas very differently from conventional ME models such as the Atwater ME (Coles, 2010).

2. *Available energy content of kiwifruit.* Gold kiwifruit (chemical composition as shown in Table 15.7) was subjected to the dual *in vivo/in vitro* digestibility assay (nutrient digestibilities and organic matter fermentability shown in Table 15.8) and ATP production simulated. The predicted ATP yield was 109 mol ATP/kg dry matter, a considerably lower value than that found for the meal replacement formulas in the example discussed above (Table 15.6), indicating that kiwifruit is a very suitable weight-loss food. The metabolizable energy (ME) content of kiwifruit was also calculated along with the ME and available (ATP) energy contents of a model food, dextrin (used in the study as a baseline reference point). The values for dextrin were calculated, assuming that

Table 15.4 Determined Chemical Composition (g/kg Dry Matter) of Two Meal Replacement Formulas

	Formula	
	A	**B**
Crude protein	144.4	382.4
Fat	121.2	14.3
Starch	173.6	4.7
Sugars	399.6	380.6
Non-starch polysaccharides	22.1	132.3
Ash	65.0	91.4

Table 15.5 Predicted Upper-Tract (Ileal) Digestibility of Nutrients and Colonic Fermentation in Humans[1] for Two Meal Replacement Formulas

	Apparent Digestibility (%)	
	Formula A	Formula B
Crude protein	89.3	92.7
Total fat	98.1	94.4
Starch	99.9	93.0
Sugars	91.0	90.3
Organic matter	93.0	81.5
	Hindgut Fermentability (%)	
	Formula A	Formula B
Organic matter	88.7	84.2

[1]Predicted values using the dual in vivo/in vitro assay and with the growing rat as an animal model for upper-tract digestion in humans.

Table 15.6 Predicted ATP Yields[1] (mol/kg Dry Matter) for Two Meal Replacement Formulas

	Formula	
	A	B
Predicted ATP yield	160.1	121.4

[1]ATP predictions from the overall model.

Table 15.7 Determined Chemical Composition of Gold Kiwifruit (Hort 16A) Flesh[1]

	Content (g/kg Dry Matter)
Crude protein	68
Fat	27
Starch	3.7
Sugars	533
Non-starch polysaccharides	134
Ash	44

[1]From Henare et al. (2012).

Table 15.8 Upper-Tract (Ileal) Digestibility of Nutrients and Colonic Fermentation in Humans[1] for Gold Kiwifruit (Hort 16A) Flesh

True Digestibility (%)

Crude protein	65

Apparent Digestibility (%)

Total fat	34
Starch	83
Sugars	95.8
Organic matter	64.0

Fermentability (%)

Organic matter	48.6

[1]Predicted values using the dual in vivo/in vitro assay and with the growing pig as an animal model for upper-tract digestion in humans. From Henare et al. (2012). The digestibility values for protein and fat appear low, but have been corroborated in further testing (P. Moughan, unpublished).

Table 15.9 Metabolizable or Available (ATP) Energy Contents of Dextrin and Kiwifruit Expressed as a Proportion of the Respective Energy Content of Dextrin

	ME	Available Energy
Dextrin	1	1
Gold kiwifruit	0.73	0.61

From Henare et al. (2012).

the digestion of dextrin and the absorption of released glucose were complete. The respective ME and available energy values were expressed relative to those of dextrin (Table 15.9). The ATP available energy measure demonstrates that kiwifruit provide considerably less energy, compared to the dextrin baseline, than would be concluded based on the ME system of evaluation. Such differences are of practical significance. Kiwifruit is a more effective weight-loss food than would be predicted based on traditional ME values.

CONCLUSION

Conventional ME systems have considerable errors, and these are especially relevant when applied to specific foods designed for having special relevance to body weight loss. Such foods tend to be formulated to have relatively high levels of protein, resistant starch, and sugar and fat substitutes. The calorific value of such foods needs to be determined with greater accuracy. The model of Coles *et al.* (2013c), discussed here, gives a means of ranking foods for their "available" energy contents under a defined standardized metabolic condition and provides the food industry with a useful tool to develop and evaluate such greatly needed weight-loss food options. There is considerable scope for the industry to restructure foods and food ingredients, to manipulate structure interactions during digestion, and to design foods that supply altered nutrient profiles and lower amounts of available energy for unit food dry matter. It is important, however, that for such foods, available energy is described with accuracy.

REFERENCES

Addy, R., 2012. Momentum builds to overhaul global calorie system. Cited Food Navigator, http://www.foodnavigator.com/Science-Nutrition/Momentum-builds-to-overhaul-global-calorie-system.

Butts, C.A., Munro, J.A., Moughan, P.J., 2012. *In vitro* determination of protein and amino acid digestibility for humans. Br. J. Nutr. 108, S282−S287.

Coles, L.T., Moughan, P.J., Darragh, A.J., 2005. *In vitro* digestion and fermentation methods, including gas production techniques, as applied to nutritive evaluation of foods in the hindgut of humans and other simple-stomached animals. Anim. Feed Sci. Technol. 123, 421−444.

Coles, L.T., 2010. Prediction of cellular ATP generation from foods in the adult human - application to developing specialist weight-loss foods. PhD Thesis, Massey University, Palmerston north.

Coles, L.T., Moughan, P.J., Awati, A., Darragh, A.J., Zou, M.L., 2010. Predicted apparent digestion of energy-yielding nutrients differs between the upper and lower digestive tracts in rats and humans. J. Nutr. 140, 469−476.

Coles, L.T., Moughan, P.J., Awati, A., Darragh, A.J., 2011. Influence of assay conditions on the *in vitro* hindgut digestibility of dry matter. Food Chem. 125, 1351−1358.

Coles, L.T., Moughan, P.J., Awati, A., Darragh, A.J., 2013a. Optimisation of inoculum concentration and incubation duration for an *in vitro* hindgut dry matter digestibility assay. Food Chem. 136, 624−631.

Coles, L.T., Moughan, P.J., Awati, A., Darragh, A.J., 2013b. Validation of a dual *in vivo/in vitro* assay for predicting the digestibility of dietary energy in humans. J. Sci. Food Agric. 93, 2637−2645.

Coles, L.T., Rutherfurd, S.M., Moughan, P.J., 2013c. A model to predict the ATP equivalents of macronutrients absorbed from food. Food Funct. 4, 432−442.

Deglaire, A., Bos, C., Tomé, D., Moughan, P.J., 2009. Ileal digestibility of dietary protein in the growing pig and adult human. Br. J. Nutr. 102, 1752−1759.

Deglaire, A., Moughan, P.J., 2012. Animal models for determining amino acid digestibility in humans—a review. Br. J. Nutr. 108, S273−S281.

Edwards, C.A., Gibson, G., Champ, M., Jensen, B.-B., Mathers, J.C., Nagengast, I., 1996. *In vitro* method for quantification of the fermentation of starch by human faecal bacteria. J. Sci. Food Agric. 71, 209−217.

Flatt, J.P., Tremblay, A., 1997. Energy expenditure and substrate oxidation. In: Bray, G.A., Bouchard, C., James, W.P.T. (Eds.), Handbook of Obesity. Marcel Dekker, New York, pp. 513–537.

Henare, S.J., Rutherfurd, S.M., Drummond, L.N., Borges, V., Boland, M.J., Moughan, P.J., 2012. Digestible nutrients and available (ATP) energy contents of two varieties of kiwifruit (*Actinidia deliciosa* and *Actinidia chinensis*). Food Chem. 130, 67–72.

Hendriks, W.H., van Baal, J., Bosch, G., 2012. Ileal and faecal protein digestibility measurement in humans and other non-ruminants—a comparative species view. Br. J. Nutr. 108, S247–S257.

Merrill, A.L., Watt, B.K., 1973. Energy value of foods: basis and derivation. Agriculture Handbook 74. US Department of Agriculture, Agricultural Research Service, Washington DC.

Miller, E.R., Ullrey, D.E., 1987. The pig as a model for human nutrition. Annu. Rev. Nutr. 7, 361–382.

Moughan, P.J., Rowan, A.M., 1989. The pig as a model for human nutrition research. Proc. Nutr. Soc. NZ 14, 116–123.

Novotny, J.A., Gebauer, S.K., Baer, D.J., 2012. Discrepancy between the Atwater factor predicted and empirically measured energy values of almonds in human diets. Am. J. Clin. Nutr. 96, 296–301.

Zou, M.L., Moughan, P.J., Awati, A., Livesey, G., 2007. Accuracy of the Atwater factors and related food energy conversion factors with low-fat, high-fiber diets when energy intake is reduced spontaneously. Am. J. Clin. Nutr. 86, 1649–1656.

Combined Phytosterol and Fish Oil Therapy for Lipid Lowering and Cardiovascular Health

Melinda Phang[1] and Manohar Garg[2]

[1]*University of Newcastle, Nutraceuticals Research Group, Newcastle, NSW, Australia,*
[2]*School of Biomedical Sciences & Pharmacy, University of Newcastle, Callaghan, NSW, Australia;
Riddet Institute, Massey Univesity, Palmerston North, New Zealand*

CONTENTS

INTRODUCTION

Hyperlipidemia is widely recognized as a major risk factor in the development of cardiovascular disease (CVD). Hallmark characteristics are increased flux of free fatty acids (FFA), elevated circulating levels of low-density lipoprotein (LDL) cholesterol, triglycerides (TG), and apolipoprotein B (apoB), as well as reduced plasma levels of high density lipoprotein (HDL)

Food Structures, Digestion and Health. http://dx.doi.org/10.1016/B978-0-12-404610-8.00016-5

437

cholesterol (Kolovou et al., 2005). The lipid abnormality involved in hyper-lipidemia is an increase in circulating FFA originating from adipose tissue as well as inadequate esterification and FFA metabolism (Bernard, 2008). Consequently, reduced retention of fatty acids by adipose tissue leads to an increased flux of FFA returning to the liver (Kolovou et al., 2005), which in turn stimulates hepatic triglyceride synthesis, promoting the production of apoB and the assembly and secretion of very low density lipoprotein (VLDL) (Funatusu et al., 2001). When plasma TG concentrations are increased, TG-rich HDL particles are formed and undergo catabolism, hence HDL cholesterol is reduced in the presence of elevated FFA. The elevated VLDL particles are lipolyzed and hence are unable to efficiently bind to LDL receptors, while the exchange of cholesterol esters with TGs forms TG-rich lipoproteins resulting in the generation of small, dense, LDL choles-terol particles (Micallef and Garg, 2009). The importance of LDL cholesterol in the disease process is paramount as the oxidation of LDL results in the migration of LDL into the intima of the artery and the genesis of an atheroscle-rotic plaque (Matsuura et al., 2008). Oxidized LDL is more rapidly taken up by macrophages in intima than native LDL, resulting in cholesterol accumu-lation in the cells, the formation of foam cells, and atherogenesis (Matsuura et al., 2008). Subsequently, elevated LDL cholesterol concentrations are a known risk factor for atherosclerosis and coronary heart disease (CHD) (Kiechl and Willeit, 1999), which has led to the recommendation of decreasing dietary cholesterol intake in order to decrease circulating choles-terol levels. However, recent research has indicated that dietary cholesterol does not significantly contribute to atherosclerosis or the risk of CHD in the general population, as it increases both the LDL fraction of cholesterol and the HDL (anti-atherogenic) fraction (McNamara et al., 1998).

A curvilinear relationship between total cholesterol levels and deaths from coronary heart disease is apparent (Stamler et al., 1986), while increasing levels of HDL are associated with reducing the relative risk of CHD (Gotto and Brinton, 2004). Therefore, elevated LDL cholesterol and low HDL cholesterol levels both serve as independent risk factors for CHD. Further-more, the risk with each parameter is compounded when they are present in combination, for example hypertriglyceridemia coupled with raised LDL cholesterol increases the risk of a coronary event six-fold compared to hypertriglyceridemia alone (Cullen, 2000).

Current therapeutic preventive action focuses on lowering the LDL choles-terol concentration. Pharmacologically, statins or 3-hydroxy-3-methyl-glutaryl-CoA reductase (HMG-CoA reductase) inhibitors are used to decrease cholesterol synthesis within the body by inhibiting the activity of HMG-CoA reductase, the rate-limiting enzyme in cholesterol synthesis.

While drug therapies involving statins, ezetimibe, or fenofibrates are effective in modulating and correcting hyperlipidemia, other means are required for both low-risk individuals and patients experiencing adverse effects of drugs. The high rate of adverse effects with statin use including liver function abnormalities and elevations in creatine kinase (Kastelein *et al.*, 2008) highlight the need for a safe and efficacious alternative.

Naturally occurring compounds and lifestyle modifications are increasingly used in the management of hyperlipidemia. Although drug treatment is effective, diet-based monotherapies or combination, dietary approaches to manage hyperlipidemia will be integral to sustainable health care systems of the future. Emerging research has demonstrated that the fatty acid composition of the diet is a more important lifestyle modulator of LDL cholesterol levels, as high levels of saturated fatty acid and trans fatty acid intakes have been shown to result in elevated plasma LDL cholesterol concentrations (Oh *et al.*, 2005). Dietary lipid-lowering therapies, whether based on alterations in nutrient intake or functional foods, play an increasingly recognized role as modulators of blood lipoprotein metabolism. A range of functional foods (i.e., flaxseed, policosanols, red yeast rice, viscous fiber, garlic, almonds, macadamia, and soy proteins) have been assessed as potential complementary alternatives to manage hyperlipidemia; however, while these functional foods effectively reduce both total and LDL cholesterol, they have no effect on TGs or HDL cholesterol (Fletcher *et al.*, 2005). Thus, it is important to appreciate the need to focus not only on LDL cholesterol reduction, but also on the incorporation of HDL cholesterol raising therapy. Diet may affect HDL in two ways: those caused by changes in the fatty acid composition of the diet, and those that affect plasma TG levels. In this regard, the combination of phytosterols and long-chain omega-3 polyunsaturated fatty acids is of particular interest with their demonstrated ability to modulate cholesterol and triglyceride metabolism in different fashions. This chapter will explore the efficacy of phytosterols and omega-3 fatty acids against cardiovascular risk factors such as lipid aberrations, hypertension, inflammation, and platelet aggregation. An introduction of their structure and biochemistry will be followed by a review of human clinical trials. The complementary and synergistic efficacy of the combined phytosterols and omega-3 fatty acids will be discussed with a focus on the mechanisms involved.

PHYTOSTEROLS
Structure and Derivatives

Plant sterols and stanols, referred to as phytosterols, are naturally occurring compounds which are structurally analogous to cholesterol. They serve to provide the same basic functions in plants as cholesterol in animals for the regulation of membrane fluidity and other physiological functions associated with plant biology (Brufau *et al.*, 2008). However, they are not synthesized by animals and enter the body only via the diet. Phytosterols share structural similarities with cholesterol, having a steroid nucleus and a hydroxyl group at the C3 position. Conversely, they are differentiated by their degree of saturation and side-chain configuration at the C24 position (Clifton *et al.*, 2002). In general, they are thought to stabilize plant membranes with an increase in the sterol:phospholipid ratio leading to membrane rigidification (Itzhaki *et al.*, 1990). However, individual phytosterols differ in their effect on membrane stability (Marsan *et al.*, 1996). Phytostanols are a fully saturated subgroup of phytosterols and are less abundant, while most phytosterols contain 28 or 29 carbons and one or two carbon–carbon double bonds, typically one in the sterol nucleus and sometimes a second in the alkyl side chain (Tikkanen, 2005). More than 250 different phytosterols have been reported in plant species with the most common belonging to the 4-desmethyl sterol family, made up of three compounds and accounting for most of the phytosterol mass. The most common are β-sitosterol (additional ethyl group at C24), campesterol (additional methyl group at C24), and stigmasterol (additional ethyl group at C24 and a double bond at C22), which account for 65%, 30%, and 3%, respectively, of total dietary phytosterol intake (Weihrauch and Gardner 1978; Moreau *et al.*, 2002) (Figure 16.1).

Natural Phytosterol Intake

Phytosterols are found in all plant-based foods; plants, legumes, nuts, and seeds have relatively high concentrations of phytosterols, whereas cereal

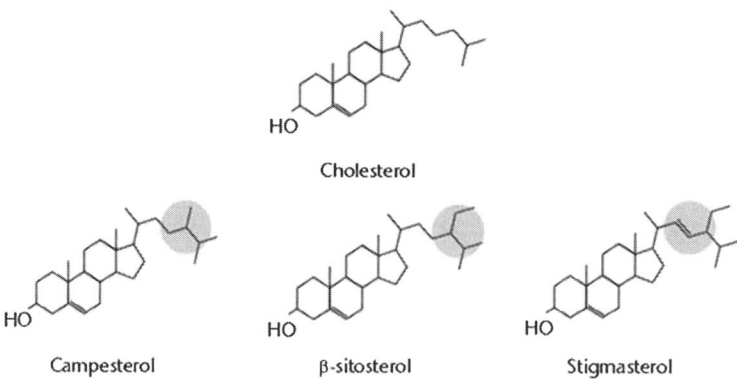

Cholesterol

Campesterol   β-sitosterol   Stigmasterol

■ **FIGURE 16.1** Structure of cholesterol and common phytosterols. Phytosterols are structurally related to cholesterol with a modified side chain configuration at carbon C-24. Campesterol and sitosterol and have a methyl and an ethyl substitute at C-24, respectively, whereas stigmasterol contains a double bond at C-22.

grains, fruits, and vegetables contain only modest amounts (Phillips *et al.*, 2005; Jimenez-Escrig *et al.*, 2006). The most concentrated source can be found in vegetable oils including corn oil (800–1500 mg/100 g) and palm oil (70–100 mg/100 g) (Phillips *et al.*, 2005). However, because of the high intake of cereal products in humans, cereals are quantitatively an important source of phytosterols (Normen *et al.*, 2001; Valsta *et al.*, 2004). Total intake of phytosterols from natural sources is estimated to average 100–300 mg daily in various populations and is similar to cholesterol intake, whereas vegetarians and Asian populations consume at least double the amount of phytosterols (Nair, 1984; Carr and Jesch, 2006). Phytostanols are much less abundant in nature than phytosterols, and consequently we typically consume much lower amounts (25 mg/day) in our diets (Cerqueira *et al.*, 1979). Evidence suggests that phytosterols consumed in normal habitual dietary quantities can influence cholesterol metabolism. It has been demonstrated that phytosterols naturally present in wheat germ and corn oil significantly decreased intestinal cholesterol absorption (Ostlund *et al.*, 2002a, 2003). In addition, the amount of phytosterols consumed in habitual diets has been reported to be inversely correlated with serum cholesterol concentration (Andersson *et al.*, 2004; Klingberg *et al.*, 2008). The absorption of phytosterols is largely dependent on the nature of the C24 side chain where increasing complexity increases hydrophobicity therefore reducing absorption (Salen *et al.*, 1970; Heinemann *et al.*, 1993; Samders *et al.*, 2000). The overall net absorption of phytosterols is <5% of that consumed as most is rapidly excreted by the liver (Andersson *et al.*, 2004). Hence, one of the hallmarks of dietary phytosterols is their ability to lower cholesterol while being poorly absorbed themselves. In supplemental studies, phytosterol absorption ranged from 0.9 to 19% for campesterol (Lutjohann *et al.*, 1995; Ostlund *et al.*, 2002b). Collectively, these findings suggest that phytosterol intake at levels found naturally in the food supply may exert a minimal cholesterol lowering effect.

Absorption and Metabolism

Cholesterol in the intestinal lumen comes from both dietary and biliary sources, whereas phytosterols are not synthesized by animals and enter the body only via the diet (Carr and Jesch, 2006). Phytosterols or cholesterol must be incorporated into micelles for absorption. Biliary cholesterol is secreted into the intestine in the form of mixed micelles composed primarily of bile salts, phosphatidylcholine and cholesterol (Hofmann and Borgstroem, 1964).

Biliary secretions are important for efficient micellarization of dietary sterols. Micellarization of sterols in turn results in more efficient absorption with smaller donor particles (Haikal *et al.*, 2008), by interacting with the intestinal brush border and penetrating the unstirred water layer of the enterocyte membrane. Non-esterified dietary phytosterols compete with dietary cholesterol for micellarization to displace micellarized biliary cholesterol. The displacement of biliary cholesterol from micelles is quantitatively more important because biliary cholesterol secretion contributes a greater amount of cholesterol to the intestinal lumen than the diet (Carr and Jesch, 2006). The phytosterol esters must be hydrolyzed to compete for micellarization. While the exact mechanism by which phytosterol absorption occurs is still uncertain, the striking difference in absorption with cholesterol implicates protein-mediated sterol transporters. A number of proteins have been shown to be involved in intestinal uptake, including Niemann-Pick C1 like 1 (NPC1LI), scavenger receptor class B type 1 (SR-B1), ATP-binding cassette (ABC) transporters ABCA1, ABCG1, and the ABCG5/G8 heterodimer (Hui *et al.*, 2008; Fitzgerald *et al.*, 2010).

Within the enterocyte, cholesterol and a small percentage of phytosterols are esterified by acyl cholesterol acyltransferase (ACAT), packaged by chylomicrons and secreted into the lymphatic system. The unesterified cholesterol and phytosterols are incorporated into VLDL or secreted via the bile. The low affinity of ACAT for sitosterol in the liver results in a greater secretion and higher hepatic clearance rate than that for campesterol (Sudhop *et al.*, 2002). NPC1L1 has been implicated as being essential for cholesterol absorption of higher capacity, with only 15% of cholesterol absorption occurring in NPC1L1 knockout mice (Altmann *et al.*, 2004). Another study demonstrated that NPC1L1 facilitated sitosterol uptake at approximately 40% of the rate in comparison to cholesterol (Yamanashi *et al.*, 2007).

Mechanism of Action

The proposed mechanisms for the cholesterol-lowering effects of phytosterols have centered on metabolic events shared by cholesterol and phytosterols, including intestinal solubility, cellular uptake, interaction with digestive enzymes, and gene regulation. Essentially, phytosterols are able to reduce the absorption of both dietary and biliary cholesterol from the intestinal tract by displacing cholesterol from micelles, hence limiting intestinal solubility of cholesterol and decreasing the hydrolysis of cholesterol esters in the small intestine (Subbiah *et al.*, 1971; Ikeda *et al.*, 1989; Ling and Jones, 1995). They have the potential to reduce cholesterol absorption by 30 to 50% (Jones *et al.*, 2000).

A proposed mechanism by which phytosterols may decrease cholesterol solubility in the intestinal lumen is via co-crystallization, thereby making them insoluble. Competitive solubilization experiments of cholesterol micro-emulsion droplets, which mimic the functionality of micelles, demonstrated that the presence of phytosterols within the micelle significantly lowers the relative solubility of cholesterol (Garti *et al.*, 2006). This is further supported by studies suggesting that β-sitosterol has an increased affinity for biliary micelles, compared with cholesterol, demonstrated at high sterol concentrations (Ling and Jones, 1995).

Another potential mechanism is through transporters that are important for sterol flux; ABCA1, ABCG1, and ABCG5/G8. The hyperaccumulation of phytosterols, commonly called sitosterolemia (named as it was originally detected by hyperaccretion of sitosterol (Bhattacharyya and Connor, 1974)), has been attributed to a mutation in the sterol efflux heterodimer of ABCG5/G8 (Salen *et al.*, 2002). Individuals with malfunctioning ABCG5/G8 have increased levels of phytosterols. This initially led to the conclusion that this cotransporter was the phytosterol-specific efflux protein, but later research demonstrated that it also transported sterols. ABCG5/G8 is also important in sterol secretion into bile (Wang *et al.*, 2007).

ABCA1 is present in multiple tissues and is regarded as specifically donating sterols to ApoAI as part of the process of developing nascent HDL (Wang *et al.*, 2004) while ABCG1 and ABCG4 deliver cholesterol to mature HDL (Wang *et al.*, 2001). It is thought that phytosterols induce the expression of the ABCA1 transporters, which are unable to differentiate between cholesterol and β-sitosterol, thus increasing the efflux of cholesterol (Plat and Mensink, 2002a). *In vitro* studies have also shown that ACAT-mediated esterification is less efficient for phytosterols compared to cholesterol (Field and Mathur, 1983). ACAT converts free sterols to sterol esters, which are the storage form of the sterols and the primary form in which cholesterol is incorporated into chylomicrons. It has been suggested that phytosterols may suppress ACAT activity, consequently reducing intestinal cholesterol uptake (Krauss *et al.*, 2000).

Clearly, phytosterols encompass a wide variety of biological interactions. Above all, they are known for their efficacious cholesterol-lowering properties (Eisenberg, 2003). The mechanisms by which phytosterols reduce cholesterol absorption and consequently reduce LDL cholesterol will be further explored throughout this review.

The dose—response effect of phytosterols and omega-3 fatty acids is also an important issue to address when considering the efficacy of such nutritional therapies. A large majority of studies provide evidence that both phytosterols and omega-3 fatty acids are effective lipid-lowering agents over the short term; however, long-term studies are needed to establish sustained, continued efficacy of such functional foods.

Phytosterol-Enriched Foods, Dose, and Food Matrix

For several years, the existence and dietary effects of phytosterols were largely ignored and poorly understood. Sterol chemists and biochemists focused their efforts on cholesterol as elevated serum cholesterol levels were shown to be a prominent risk factor for CVD. Recent strategies for lowering serum cholesterol utilize dietary restrictions to limit cholesterol intake and/or require the use of drugs such as statins which inhibit cholesterol biosynthesis in humans. The prospect of lowering cholesterol levels by consuming "functional" foods fortified with natural phytonutrients would seem more ideal than pharmacological means or dietary restrictions. For this purpose, there has been emerging and considerable interest for commercial marketing of phytosterol products. Interestingly, the use of phytosterols to lower serum cholesterol levels is not a new idea. In the 1950s, sitosterol was used as a supplement and as a drug (Cytellin, marketed by Eli Lilly) to lower serum cholesterol levels in hypercholesterolemic individuals (Pollak, 1953). However, due to poor solubility and bioavailability of the free phytosterols, the serum cholesterol-lowering effects were inconsistent, and very high doses (up to $25-50$ g per day) were sometimes required for efficacy. This problem of solubility and bioavailability led to many confounding results in early clinical studies, thus when the more predictable and effective statins were introduced, the use of these phytosterol products diminished rapidly.

With the growing interest in functional foods in recent years, the use of phytosterols for reducing serum cholesterol levels has regained considerable momentum. This can be largely attributed to the work of scientists in Finland, who esterified phytostanols with fatty acids (stanyl esters) in order to improve their solubility. This allowed the first practical, commercial-scale production of phytosterol-containing foods such as margarines (Moreau *et al.*, 2002). Due to the success with esterified forms of stanols, many researchers lost interest in free phytosterols and phytostanols and it appeared that stanyl esters were superior to all other sterol forms for cholesterol reduction. However, subsequent research has shown that fatty acid esters of sterols are also easily formulated in food products and can be useful in functional foods.

The daily doses, considered optimal for the purpose of lowering blood cholesterol levels, are $2-3$ g of phytosterols, which translates to $3.4-5.2$ g in esterified form. This recommended daily dose is typically divided in

1–3 portions of food providing 1.7–5.2 g ester, which equals 1–3 g phytosterol equivalents. The general consensus is that phytosterol therapy is most effective at lowering LDL cholesterol with a maximum efficacy at ≈ 3 g/day (Law, 2000; Demonty *et al.*, 2009). Other studies have shown that ≈ 2 g/day consumption of phytosterol esters can achieve reductions of 10–15% in LDL cholesterol in approximately 90% of individuals (Wolfs *et al.*, 2006; Madsen *et al.*, 2007).

Demonty *et al.* (2009) developed a continuous dose– response equation using 141 trial arms (from 84 studies) and concluded that a 3 g/day dose reduces LDL cholesterol by 11%, whereas a dose of 2 g/day lowers LDL cholesterol by 9%. Thus, phytosterol consumption of 2 g/day is nearly as effective and probably represents a more reasonable and achievable dose. Consequently, the US National Cholesterol Education Program recommends a phytosterol dose of 2 g/day to enhance lowering of LDL cholesterol (NCEP, 2002).

Phytosterols have been incorporated into various food matrices, although the results have been variable. They have significantly lowered serum LDL cholesterol when incorporated into margarine spreads, cream cheese, yoghurt, milk, cereal bars, bread, chocolate, capsules, and tablets (Demonty *et al.*, 2009). Despite the variability of phytosterol efficacy in various food matrices, the meta-analysis by Demonty *et al.* (2009) suggested that neither the fat content nor the nature of the food as dairy or non-dairy significantly affected the product's efficacy to lower LDL cholesterol. In contrast, a meta-analysis by Abumweis *et al.* (Abumweis *et al.*, 2008) concluded that reductions in LDL cholesterol were greater when phytosterols were incorporated into fat spreads, mayonnaise, milk, and yoghurt compared with muffins, non-fat beverages, cereal bars, and chocolate.

A convenient means of delivering phytosterol esters in the diet has been through their dispersion in fat spreads. Given that fats are needed to solubilize sterols, margarines are an ideal vehicle to increase the lipid solubility and facilitate the incorporation of phytosterols into the micelles. Nestel *et al.* (2001) reported that consumption of 20 g/day of a sterol ester-fortified margarine (2.4 g/day) for 4 weeks reduced total cholesterol and LDL cholesterol by 12.2% and 13.63%, respectively (Nestel *et al.*, 2001). These reductions were similar to those reported by Weststrate and Meijer (1998) and Gylling and Miettinen (1999).

A systematic review by Moruisi and colleagues (Moruisi *et al.*, 2006) concluded that consumption of phytosterol-enriched fat spreads (2.3 ± 0.5 g/day) for 6.5 ± 1.9 weeks significantly reduced total cholesterol (7–11%) with a mean decrease of 65 mmol/L, and LDL cholesterol

(10—15%) with a mean decrease of 0.64 mmol/L in hyperlipidemic subjects compared to controls. Triglycerides and HDL cholesterol were not affected.

In a long-term study (12 months), consumption of 20 g/day of spread (providing 1.6 g/day phytosterols) resulted in a consistent reduction of 4 and 6% in total and LDL cholesterol in mildly hyperlipidemic subjects. Phytosterol supplementation did not alter hormone levels, hematological parameters, or fat-soluble vitamin concentrations. However, α- and β-carotene levels were adjusted by 15—25% lower relative to control subjects (Hendriks et al., 2003). Changes in carotenoid concentration can be simply prevented by an increase of fruit and vegetable consumption during phytosterol supplementation (Noakes et al., 2002).

A community intervention program in the Netherlands assessed the voluntary use and effectiveness of phytosterol/stanol-enriched fat spreads during the time of their introduction to the Dutch market (de Jong et al., 2007). Blood samples that were collected in 1999 and then in 2003 revealed significant changes in total cholesterol concentration in nonusers ($+2\%$), enriched-spread users (-4%), cholesterol-lowering drug users (-17%), and combination users (spread + statin) (-29%).

Studies that have assessed the effects of other phytosterol-enriched products have also reported reductions in cholesterol concentrations. Noakes et al. (2005) compared the effects of supplementation with phytosterol-enriched milk (300 ml/day; 2.0 g phytosterols/day), and yoghurt (300 g/day; 1.8 g phytosterols/day) in hypocholesterolemic subjects for 3 weeks. Both the milk and yoghurt effectively reduced serum cholesterol by 6—8% and 6%, respectively.

While plant sterols are generally considered not effective in reducing plasma triglyceride concentrations, several clinical investigations have reported consistent yet variable reductions in circulating triglycerides in response to phytosterol consumption (Clifton et al., 2008; Plana et al., 2008). Though previous meta-analyses have failed to establish this relationship (Berger et al., 2004; Abumweis et al., 2008), another study reported a 9% reduction in plasma triglyceride in hypertriglycerolemic patients consuming plant sterol-enriched soy milk (Rideout et al., 2009). Thus, there is speculation that modulation of triglycerides in response to phytosterol consumption may be more readily detected in study populations with high baseline triglycerides.

Genetic factors may also affect the response to phytosterol supplementation. However, with the exception of those that carry the sitosterolemia gene, it is reported that many of the common polymorphisms do not alter the beneficial cholesterol-lowering effects of phytosterol supplementation (Plat and Mensink, 2002b). It has also become increasingly clear that some individuals respond to phytosterol consumption with major shifts in lipid profiles, whereas others are much less responsive. This variability of responsiveness is likely to be associated with single nuclear polymorphisms (SNPs) in Niemann-Pick C1-like 1 (NPC1L1) and ATP binding cassette proteins G5 (ABCG5) and G8 (ABCG8), genes that regulate intestinal sterol absorption and efflux (Rudkowska *et al.*, 2008; Izar *et al.*, 2011).

OMEGA-3 POLYUNSATURATED FATTY ACIDS
Structure, Biochemistry, and Dietary Sources

It has been long established that atherosclerosis is affected by both the type and quantity of fatty acids in dietary lipids (Hu *et al.*, 2001). Such effects are distinguished by the chain length, degree of unsaturation, as well as the position on the triglyceride backbone, which all contribute to the manner of the digestion and absorption of the fatty acid and how it affects lipoprotein metabolism (Karupaiah and Sundram, 2007). Fatty acids form the main constituents of dietary fats and oils including those contained in plants and animals. They are mainly esterified to glycerol, as triacylglycerols, although some are present as esterified components of phospholipids, glycolipids, and other lipids. Thus, they are seldom present as free fatty acids, but commonly as ester-linked groups within various types of lipid molecules, the triacylglycerols being the most common and of greatest importance (Eisenberg, 1983). Although dietary fat has been generally considered as deleterious to human health, fatty acids play vital biochemical and physiological roles and some of the most important fatty acids must be obtained from the diet. Among these are the essential polyunsaturated fatty acids that cannot be synthesized by mammalian cells. In the absence of sufficient dietary intake of these fatty acids, deficiency symptoms occur, and hence they are classified as essential fatty acids.

The polyunsaturated fatty acids can be divided into two categories, namely the n-3 (omega-3) and the n-6 (omega-6) polyunsaturated fatty acids. This classification is based on the position of the first double bond from the methyl terminal, being located at the third carbon in the n-3 and at the sixth carbon in the n-6PUFA. The n-3PUFA, α-linoleic acid (ALA; 18:3n-3), and the n-6PUFA, linoleic acid (LA; 18:2n-6), are the predominant essential fatty acids that are derived from a variety of animal and plant foods. LA is desaturated and elongated to arachidonic acid (AA; 20:4n-6) while ALA is desaturated and elongated to eicosapentaenoic acid (EPA; 20:5n-3) and docosahexaenoic acid (DHA; 22:6n-3). The metabolites derived from the n-6 AA or the n-3 EPA are the 2- and 3-series eicosanoids, respectively (Yaqoob, 2004). The major dietary sources of LA are corn and sunflower oils while ALA is found predominantly in green leafy vegetables, flaxseed, canola, and soya bean oils. The long-chain omeg-3 fatty acids (EPA, DPA, and DHA) are predominantly found in oily fish such as salmon, tuna, and mackerel (Grundy and Denke, 1990). There is now constant growth in the availability of foods that are fortified with omega-3 fatty acids in the consumer market. Such fortified sources include bread, margarine spreads, milk, and eggs that are produced by chickens fed omega-3-rich diets or fish oils (Ruxton, 2004).

Mechanism of Action

In general, eicosanoids derived from AA are pro-inflammatory and pro-aggregatory agonists while those derived from EPA have anti-inflammatory, anti-arrhythmic, anti-aggregatory, and hypotriglyceridemic responses (Salmon and Terano, 1985; Calder, 2004). These fatty acids could play a vital role in the management of hypertension and hyperlipidemia, and in the prevention of several diseases including coronary heart disease and type 2 diabetes (Lombardo and Chicco, 2006). These effects are mediated by alterations in circulating plasma lipids, eicosanoids, cytokines, and physicochemical properties in the phospholipid membrane.

Mammalian cell membranes consist of a lipid bilayer composed primarily of phospholipids and cholesterol. Proteins that drive critical cellular functions, such as receptors, transporters, and enzymes are embedded in the lipid bilayer (Roshanai et al., 1981). Dietary intake of both EPA and DHA present in fish oil can alter the membrane phospholipid composition of the cells, alter eicosanoid synthesis, and regulate transcription factor activity (Calder, 2004). The hypolipidemic effects of fish oils may be mediated by several mechanisms incorporating reduction of triglyceride synthesis and chylomicron secretion from intestinal cells and suppression of hepatic fatty acid synthesis and triglyceride production, thereby limiting VLDL secretion (Nestel et al., 1984). Numerous studies have demonstrated that supplementation with omega-3 fatty acids favorably modifies serum and tissue lipid profiles. The most consistent finding is a drastic reduction in fasting and postprandial serum triglycerides and FFA (Weintraub et al., 1988). These changes have been observed with EPA and DHA alone (Woodman et al., 2003) and with their combination in fish oil. The reduced VLDL production by the liver (Nestel et al., 1984) largely results from (1) decreased availability of FFA released from adipose stores, (2) suppression of lipogenic activity, (3) an increase in the activity of triglyceride-synthesizing enzymes (DGAT, diacylglycerol acyltransferase; and PAP, phosphatidate phosphohydrolase), (4) the induction of genes involved in fatty acid oxidation, and (5) an increase in phospholipid synthesis (Clarke, 2004). This regulation of gene expression proceeds through the inhibition of SREBP-1 (sterol regulatory element-binding transcription factor 1) and the activation of peroxisome proliferator receptors (PPAR) (Seo *et al.*, 2005). SREBP-1 is a major transcription factor controlling fatty acid synthesis, both *de novo* lipogenesis and PUFA synthesis (Jump *et al.*, 2008). Dietary omega-3 fatty acids can suppress SREBP-1, which decreases the expression of enzymes involved in fatty acid synthesis and PUFA synthesis (Jump *et al.*, 2005; Jump, 2008). In this way, dietary polyunsaturated fatty acids function as feedback inhibitors of all fatty acid synthesis. There is evidence that omega-3 fatty acids also reduce apoprotein synthesis, as the production of apoB is diminished, and the daily flux or transport of VLDL apoB is reduced ((Brown *et al.*, 1999). An increased lipolytic activity of lipoprotein lipase (LPL), which catalyzes the degradation of VLDLs and chylomicrons in extrahepatic tissues, completes the hypotriglyceridemic effect observed with omega-3 fatty acid supplementation (Davidson, 2006). As a consequence of triglyceride and VLDL reduction, HDL synthesis is indirectly affected. In the absence of circulating triglycerides, the cholesteryl ester transfer protein is reduced thereby reducing the amount of triglycerides being transferred from VLDL to HDL. This results in a modest increase in triglyceride-poor HDL and possibly apoA-1 concentration (Bruce *et al.*, 1998; Borggreve *et al.*, 2007). The collective actions of dietary omega-3 fatty acids involve more than alteration of plasma lipids and lipoproteins; growing evidence suggests that omega-3 fatty acids may exert several beneficial effects via modulation of eicosanoid synthesis and functions of various cells associated with the cardiovascular system (Kinsella *et al.*, 1990).

Role in Eicosanoid Synthesis

The essentiality of LA and ALA is that they are precursors for some of the more important highly unsaturated fatty acids that are AA, DHA, and EPA. Conceivably, the most potent effects of PUFAs are related to their enzymatic conversion to a series of oxygenated derivatives, i.e., eicosanoids that include prostanoids and leukotrienes. These possess potent biological activity and their homeostatic functions in regulating platelet and endothelial vessel–wall interactions and monocyte and macrophage behavior are relevant to the initiation and progress of atherogenesis (Marcus, 1984; Mahley, 1985).

EPA and DHA compete with AA for prostaglandin and leukotriene synthesis at the cyclooxygenase (COX) and lipoxygenase (LOX) level to produce eicosanoids of 3-series (prostaglandin E_3, thromboxane A_3, and leukotriene B_5) or

2-series (prostaglandin E_2, thromboxane A_2, and Leukotriene B_4), respectively (Prichard et al., 1995). Eicosanoids that are derived from AA have opposing effects to those derived from EPA and a dietary imbalance in favor of n-6PUFA may contribute to detrimental effects on cardiovascular health (Harris et al., 2009). Therefore, obtaining preformed EPA and DHA directly from the diet is important for adequate incorporation into cell membranes and uptake in body tissues (Burdge and Calder, 2005).

Epidemiological studies indicate that humans evolved on a diet with a ratio of omega-6:omega-3 polyunsaturated fatty acids of approximately 1, compared to the modern typical western diet where the ratio has increased to 10−30:1 (Simopoulos, 2006). This change has been attributed to the increased intake of omega-6 fatty acids coupled with a decreased intake of omega-3 fatty acids, particularly EPA and DHA. In the western diet, AA is the most abundant 20 carbon polyunsaturated fatty acid available as a substrate for eicosanoid synthesis due to the greater abundance of LA over ALA. Consequently, the tissue PLs are enriched in AA. Unfortunately, a high omega-6:omega-3 ratio promotes the pathogenesis of many diseases, including CVD, cancer, and inflammatory diseases, whereas increased levels of long-chain omega-3 fatty acids (lower omega-6:omega-3), with an optimal ratio of 2−4:1, exert suppressive effects due to eicosanoid function (Simopoulos, 2002). For CVD prevention, the National Heart Foundation (NHF) of Australia and the American Heart Association (AHA) recommend two to three servings of oily fish a week or 500 mg/day of EPA/DHA for adults (Colquhoun et al., 2008). However, the Australian National Health and Medical Research Council (NHMRC) estimate that consuming at least 160 mg/day for males and 90 mg/day for females is adequate, while individuals with documented CHD or with lipid abnormalities are recommended by the AHA to consume 1000 mg/day of EPA/DHA (Colquhoun et al., 2008).

Omega-3 Fatty Acid Supplementation and Human Health

A plethora of evidence rising from epidemiological and experimental studies suggests that the consumption of omega-3 fatty acids is associated with a reduced risk of CVD, certain types of cancer, inflammatory disease, diabetes mellitus, multiple sclerosis, and clinical depression (Ginsberg and Toal, 2009). Observational, clinical, animal, and in vitro studies have reported the overall cardioprotective role of omega-3 fatty acids owing to their anti-aggregatory, anti-arrhythmic, anti-hypertensive, and anti-inflammatory effects (von Schacky and Weber, 1985; Knapp et al., 1986; Gaudette and Holub, 1990; Kinsella et al., 1990; Axelrod et al., 1994; Prisco et al.,

1995; Mori *et al.*, 1997; Wensing *et al.*, 1999; Akiba *et al.*, 2000; Grundt and Nilsen, 2008; Harris *et al.*, 2008). Secondary prevention trials have reported that supplementation with omega-3 fatty acids reduces premature mortality from coronary heart disease, reduces risk-impaired glucose tolerance and diabetes, as well as exerting beneficial effects on thrombosis and arterial compliance (Mozaffarian and Rimm, 2006; Zhoa *et al.*, 2007). Such favorable effects of omega-3 fatty acids are mediated by alterations in circulating plasma lipids, eicosanoids, cytokines and physicochemical properties in the membrane (Mori *et al.*, 2000).

It is well accepted that high doses of omega-3 fatty acids reduce triglycerides and improve HDL cholesterol concentrations, both EPA and DHA have a long-standing clinical use for lowering plasma triacylglycerols which are reduced approximately in parallel with the unsaturation index of fatty acids (Nestel *et al.*, 1984). The effect of omega-3 fatty acid supplementation on plasma triglycerides is sustained with continued feeding (Saynor, 1984; Saynor *et al.*, 1984), is dose responsive (Sanders and Hochland, 1983), and is reversible (Goodnight *et al.*, 1982). Connor (1986) reported that in hyperlipidemic patients, fish oil at 30 g/day decreased plasma triglyceride concentrations by $32 \pm 7.5\%$. The decrease in triglyceride concentrations was dose dependent with a 41% and 52% decrease in total plasma triglycerides, respectively.

Grimsgaard and colleagues (Grimsgaard *et al.*, 1997) reported on the effects of supplementation with highly purified EPA (3.8 g/day) or DHA (3.6 g/day) for 7 weeks in healthy male subjects. They reported that both EPA and DHA markedly reduced plasma triglycerides by 21% and 26%, respectively. In addition, a rise in HDL was apparent in the DHA group only. These findings provided convincing evidence that EPA and DHA are equally effective at reducing serum triglycerides, and that DHA may also raise HDL cholesterol as well as LDL cholesterol particle size (both anti-atherogenic outcomes). Omega-3 fatty acids have contrasting effects on LDL cholesterol where slight increases of LDL cholesterol are observed; however, the size of the LDL molecule is also increased, which is considered to be less atherogenic (Connor, 1968).

The mechanism by which this occurs remains unknown but many speculations have been proposed. It is thought that rates of synthesis of LDL are increased, rather than the number of receptors themselves; however, the potential associated cardiovascular risk is largely compensated for by the incorporation of omega-3 fatty acids into the surface phospholipids of small, dense lipoprotein particles, which change their structure to exhibit antioxidant properties (Harris, 1989). Given that omega-3 fatty acids exert beneficial effects on other CVD risk factors such as HDL and VLDL cholesterol, triglycerides, CETP, triglyceride-synthesizing enzymes (DGAT and PAP), fatty acid oxidation, lipogenesis, blood pressure, and inflammatory biomarkers, this may offset any potentially negative effects on changes in LDL cholesterol.

PHYTOSTEROL AND OMEGA-3 COMBINATION THERAPY

There is considerable evidence to suggest that omega-3 fatty acid supplementation retards the atherogenic process with improvements in vascular function, thrombogenicity, and lipoprotein profile. Additionally, reductions in inflammatory cytokines, adhesion molecules, and vasoconstrictive eicosanoids have been reported with the consumption of omega-3 fatty acids (De Caterina and Zampolli, 2001). Numerous epidemiological studies have found inverse associations between the intake of omega-3 fatty acids and plasma concentrations of C-reactive protein (CRP), tumor necrosis factor α (TNFα), E-selectin, and intercellular adhesion molecule-1 (ICAM-1) (Lopez-Garcia et al., 2004; Hjerkinn et al., 2005; Zampelas et al., 2005). Studies also investigated the effects of omega-3 fatty acids in the secondary prevention of myocardial infarction (MI). In the Diet and Reinfarction Trial (DART), patients recovered from MI were randomly allocated dietary advice to either reduce the ratio of polyunsaturated to saturated fat, increase fatty fish intake, or increase cereal fiber intake. Patients who were advised to increase their fish to at least two fish meals a week had a 29% decrease in 2-year all-cause mortality compared to those not advised (Burr et al., 1989). Similar findings were also reported in the GISSI-Prevenzione trial where post-MI patients were randomized to receive 1 g/day of EPA plus a DHA supplement or placebo for 3.5 years. Patients receiving treatment of omega-3 fatty acids experienced significantly lowered combined risk of mortality, non-fatal myocardial infarction, and stroke by 10–15% (Marchioli, 1999; Marchioli et al., 2002). These clinical trials provide evidence-based medicine, which underpins the validity of epidemiological findings and physiological plausibility of the cardioprotective effects of omega-3 fatty acids. Comparatively, minimal work has been done to investigate the cardioprotective effects of phytosterols in human nutrition, apart from its hypocholesterolemic effect.

A study conducted by Madsen and coworkers (Madsen et al., 2007) reported that the consumption of 2 g/day of phytosterols had no effect on Apo A-1, Lp(a), or CRP, but significantly decreased apoB, which is a strong predictor of coronary events. Ridker (2003) and Yusuf et al. (2004) also reported similar findings.

Given that phytosterols and fish oil improve the blood lipid profile via different mechanisms, it is plausible to suggest that combination of phytosterols and omega-3 may further reduce cardiovascular risk factors. Plant sterols are known to reduce plasma and LDL cholesterol, while the triglyceride-lowering properties of long-chain n-3PUFA are well established. This is particularly important for individuals who cannot achieve LDL cholesterol target goals with diet and phytosterols alone; certain combination treatments may offer further advantages. Both phytosterols and omega-3 fatty acids are normal dietary components, and as such their use would be expected to present minimal side effects and safety concerns.

Since the 1970s, phytosterols have been esterified to optimize their functionality, solubility, and incorporation into commercial food (Mattson *et al.*, 1977, 1982). Esterification of phytosterols with long-chain fatty acids increases fat solubility 10-fold and allows delivery of several grams daily in fatty foods such as margarine. Phytosterols can also be dispersed in water after emulsification with lecithin and reduce cholesterol absorption when added to non-fat foods (Jandacek *et al.*, 1977).

Most randomized clinical trials have tested the effect on total cholesterol and LDL-C levels of food containing plant sterols while only a limited number of studies have tested the effect of plant sterols in conjunction with omega-3 fatty acids. Results from animal models suggest that phytosterols esterified to omega-3 fatty acids reduce both plasma cholesterol and TG concentrations. In male hamsters, when fed phytosterol esters esterified to fish oil, a significant reduction in non-HDL cholesterol was observed (Ewart *et al.*, 2002). Following the same phytosterol—fish oil ester diet, insulin-resistant rats also experienced significant reductions in serum triglyceride and total cholesterol levels (Russell *et al.*, 2002). Unfortunately, these studies did not allow for a comparison between a phytosterol-only group and a fish oil-only treatment group. Therefore, it cannot be determined whether combined phytosterol—fish oil esters were the most effective compared with phytosterol esters or fatty acid esters alone.

A study by Jones and coworkers (Demonty *et al.*, 2006; Jones *et al.*, 2007) compared the effects of phytosterols that were esterified to fatty acids derived from fish oil versus sunflower oil or olive oil. The study was a crossover design in 21 hypercholesterolemic subjects supplemented for 28 days. Each treatment contained 1.7 g/day phytosterols, and the phytosterol—fish oil treatment contained 5.4 g/day fish oil (EPA + DHA). Their results indicated that phytosterols esterified with fish oil fatty acids did not alter LDL cholesterol levels compared with the other phytosterol esters, but fasting and postprandial triglyceride levels were significantly decreased in the fish oil phytosterol treatment. Therefore, a combination of phytosterols and omega-3 fatty acids, whether consumed as fish oil or esterified directly to phytosterols, may be more cardioprotective owing of their anti-inflammatory and triglyceride-lowering properties. This also suggests that the phytosterol carrier may play a role in the effects of various phytosterol preparations.

Indeed, we have previously shown the benefits of combined therapy in our research unit. A study was designed to directly compare the efficacy of a combination treatment with phytosterols and omega-3 fatty acids to treat hyperlipidemia. The study was a 3-week randomized double-blind, placebo-controlled, 2×2 factorial design involving 60 hyperlipidemic subjects (Micallef and Garg, 2008). Participants were randomized to receive either sunola oil capsules alone or in combination with 25 g/day of a spread (containing 2 g/day phytosterols) or 1.4 g/day of omega-3 fatty acid (high DHA) capsules alone or in combination with 25 g/day of a phytosterol-enriched spread. The plasma lipid profile was significantly improved with a 13.3% reduction in total cholesterol, a 12.5% reduction in LDL cholesterol, a 25.9% reduction in triglycerides, and an 8.6% increase in HDL cholesterol in the combined phytosterol and fish-oil group, and these changes were more prominent compared to those in other groups. Furthermore, significant reductions were observed in plasma-CRP, IL-6, TNFα, and LTB$_4$ with the consumption of phytosterols in combination with omega-3 fatty acid supplementation. Thus, phytosterol supplementation counteracted the LDL-raising effects of high-DHA fish oil. Combined therapy with phytosterols and omega-3 fatty acids exerted not only complementary, but also synergistic effects on circulating lipid levels. In conclusion, combination therapy could lower triglycerides and LDL cholesterol concentrations while raising HDL cholesterol levels thereby allowing optimum reduction in CVD risk in addition to anti-inflammatory effects and other general health benefits.

CONCLUDING REMARKS AND FUTURE DIRECTIONS

Phytosterols lower plasma cholesterol and LDL cholesterol without any considerable effects in HDL cholesterol or triglyceride levels. Conversely, fish oil lowers triglyceride and increases HDL cholesterol, which is accompanied by a slight increase in LDL. In addition, supplementation with marine omega-3 fatty acids has been shown to elicit anti-inflammatory, anti-arrhythmic, and anti-aggregatory actions. Therefore, dietary supplementation of both phytosterols and omega-3 fatty acids can be expected to optimize the blood lipid profile for maximum protection against CVD, i.e., reduce plasma cholesterol, LDL cholesterol, and triglyceride and raise

HDL cholesterol. Moreover, the effects of combined supplementation are not only complementary but also synergistic. It should be noted that the combined treatment provided a greater reduction. Evidently, phytosterols when esterified to the omega-3 fatty acids do not reduce LDL cholesterol. Therefore, to retain the synergistic effects, they should be taken individually but simultaneously in the same meal/supplement, but not in a chemically combined form. Using the Framingham equation to calculate the 10-year cardiovascular risk, a significantly greater reduction in overall cardiovascular risk in individuals consuming the combination treatment (22.6%) is apparent compared to the phytosterol and fish oil groups alone (15.1% and 15.3%, respectively) (Micallef and Garg, 2009).

Consumption of plant sterols (2 g/day) is advocated by current guidelines as part of an optimum dietary therapy to reduce LDL-C levels. The consumption of 1 g/day of EPA/DHA in individuals with lipid abnormalities is also recommended. Given these guidelines, a single functional food incorporating both phytosterols and omega-3 fatty acids should be developed as a comprehensive strategy not only for improving the lipid profile but also for the additional cardiovascular benefits. Most importantly, such a convenient means may also allow improved compliance.

REFERENCES

Abumweis, S.S., Barake, R., Jones, P.J., 2008. Plant sterols/stanols as cholesterol lowering agents: a meta-analysis of randomized controlled trials. Food Nutr. Res. 52.

Akiba, S., Murata, T., Kitatani, K., Sato, T., 2000. Involvement of lipoxygenase pathway in docosapentaenoic acid-induced inhibition of platelet aggregation. Biol. Pharm. Bull. 23, 1293−1297.

Altmann, S.W., Davis Jr., H.R., Zhu, L.J., Yao, X., Hoos, L.M., Tetzloff, G., et al., 2004. Niemann-Pick C1 Like 1 protein is critical for intestinal cholesterol absorption. Science 303, 1201−1204.

Andersson, S.W., Skinner, J., Ellegard, L., Welch, A.A., Bingham, S., Mulligan, A., et al., 2004. Intake of dietary plant sterols is inversely related to serum cholesterol concentration in men and women in the EPIC Norfolk population: a cross-sectional study. Eur. J. Clin. Nutr. 58, 1378−1385.

Axelrod, L., Camuso, J., Williams, E., Kleinman, K., Briones, E., Schoenfeld, D., 1994. Effects of a small quantity of omega-3 fatty acids on cardiovascular risk factors in NIDDM. A randomized, prospective, double-blind, controlled study. Diabetes Care 17, 37−44.

Berger, A., Jones, P.J., Abumweis, S.S., 2004. Plant sterols: factors affecting their efficacy and safety as functional food ingredients. Lipids Health Dis. 3, 5.

Bernard, J., 2008. Free fatty acid receptor family: novel targets for the treatment of diabetes and dyslipidemia. Curr. Opin. Investig. Drugs 9, 1078−1083.

Bhattacharyya, A.K., Connor, W.E., 1974. Beta-sitosterolemia and xanthomatosis. A newly described lipid storage disease in two sisters. J. Clin. Invest. 53, 1033−1043.

Borggreve, S.E., Hillege, H.L., Dallinga-Thie, G.M., de Jong, P.E., Wolffenbuttel, B.H., Grobbee, D.E., et al., 2007. High plasma cholesteryl ester transfer protein levels may favour reduced incidence of cardiovascular events in men with low triglycerides. Eur. Heart J. 28, 1012−1018.

Brown, A.M., Castle, J., Hebbachi, A.M., Gibbons, G.F., 1999. Administration of n-3 fatty acids in the diets of rats or directly to hepatocyte cultures results in different effects on hepatocellular ApoB metabolism and secretion. Arterioscler. Thromb. Vasc. Biol. 19, 106−114.

Bruce, C., Sharp, D.S., Tall, A.R., 1998. Relationship of HDL and coronary heart disease to a common amino acid polymorphism in the cholesteryl ester transfer protein in men with and without hypertriglyceridemia. J. Lipid Res. 39, 1071−1078.

Brufau, G., Canela, M.A., Rafecas, M., 2008. Phytosterols: physiologic and metabolic aspects related to cholesterol-lowering properties. Nutr. Res. 28, 217−225.

Burdge, G.C., Calder, P.C., 2005. Conversion of alpha-linolenic acid to longer-chain polyunsaturated fatty acids in human adults. Reprod. Nutr. Dev. 45, 581−597.

Burr, M.L., Fehily, A.M., Gilbert, J.F., Rogers, S., Holliday, R.M., Sweetnam, P.M., et al., 1989. Effects of changes in fat, fish, and fibre intakes on death and myocardial reinfarction: diet and reinfarction trial (DART). Lancet 2, 757−761.

Calder, P.C., 2004. n-3 Fatty acids and cardiovascular disease: evidence explained and mechanisms explored. Clin. Sci. (Lond.) 107, 1−11.

Carr, T.P., Jesch, E.D., 2006. Food components that reduce cholesterol absorption. Adv. Food Nutr. Res. 51, 165−204.

Cerqueira, M.T., Fry, M.M., Connor, W.E., 1979. The food and nutrient intakes of the Tarahumara Indians of Mexico. Am. J. Clin. Nutr. 32, 905−915.

Clarke, S.D., 2004. The multi-dimensional regulation of gene expression by fatty acids: polyunsaturated fats as nutrient sensors. Curr. Opin. Lipidol. 15, 13−18.

Clifton, P., 2002. Plant sterol and stanols—comparison and contrasts. Sterols versus stanols in cholesterol lowering: is there a difference? Atheroscler. Suppl. 3, 5−9.

Clifton, P.M., Mano, M., Duchateau, G.S., van der Knaap, H.C., Trautwein, E.A., 2008. Dose−response effects of different plant sterol sources in fat spreads on serum lipids and C-reactive protein and on the kinetic behavior of serum plant sterols. Eur. J. Clin. Nutr. 62, 968−977.

Colquhoun, D., Ferreira-Jardim, A., Udell, T., Eden, B., 2008. Fish, fish oils, n-3 polyunsaturated fatty acids and cardiovascular health. Review of Evidence. National Heart Foundation of Australia, 1−54.

Connor, W.E., 1968. Dietary sterols: their relationship to atherosclerosis. J. Am. Diet. Assoc. 52, 202−208.

Connor, W.E., 1986. Hypolipidemic effects of dietary w-3 fatty acids in normal and hyperlipidemic humans: effectiveness and mechanisms. In: Simopoulos, A.P., Kifer, R.R., Martin, R.E. (Eds.), Health Effects of Polyunsaturated Fatty Acids in Seafoods. Academic Press, New York, p. 173.

Cullen, P., 2000. Evidence that triglycerides are an independent coronary heart disease risk factor. Am. J. Cardiol. 86, 943−949.

Davidson, M.H., 2006. Mechanisms for the hypotriglyceridemic effect of marine omega-3 fatty acids. Am. J. Cardiol. 98, 27i−33i.

De Caterina, R., Zampolli, A., 2001. Lipids. n-3 fatty acids: antiatherosclerotic effects. Lipids 36 (Suppl.), S69—S78.

de Jong, N., Zuur, A., Wolfs, M.C., Wendel-Vos, G.C., van Raaij, J.M., Schuit, A.J., 2007. Exposure and effectiveness of phytosterol/-stanol-enriched margarines. Eur. J. Clin. Nutr. 61, 1407—1415.

Demonty, I., Chan, Y.M., Pelled, D., Jones, P.J., 2006. Fish-oil esters of plant sterols improve the lipid profile of dyslipidemic subjects more than do fish-oil or sunflower oil esters of plant sterols. Am. J. Clin. Nutr. 84, 1534—1542.

Demonty, I., Ras, R.T., van der Knaap, H.C., Duchateau, G.S., Meijer, L., Zock, P.L., et al., 2009. Continuous dose—response relationship of the LDL-cholesterol-lowering effect of phytosterol intake. J. Nutr. 139, 271—284.

Eisenberg, D., 2003. Naturally available oils contain phytosterols that affect cholesterol absorption. Curr. Atheroscler. Rep. 5, 55.

Eisenberg, S., 1983. Lipoproteins and lipoprotein metabolism. A dynamic evaluation of the plasma fat transport system. Klin. Wochenschr. 61, 119—132.

Ewart, H.S., Cole, L.K., Kralovec, J., Layton, H., Curtis, J.M., Wright, J.L., Murphy, M.G., 2002. Fish oil containing phytosterol esters alters blood lipid profiles and left ventricle generation of thromboxane a(2) in adult guinea pigs. J. Nutr. 132, 1149—1152.

Field, F.J., Mathur, S.N., 1983. Beta-sitosterol: esterification by intestinal acylcoenzyme A: cholesterol acyltransferase (ACAT) and its effect on cholesterol esterification. J. Lipid Res. 24, 409—417.

Fitzgerald, M.L., Mujawar, Z., Tamehiro, N., 2010. ABC transporters, atherosclerosis and inflammation. Atherosclerosis 211, 361—370.

Fletcher, B., Berra, K., Ades, P., Braun, L.T., Burke, L.E., Durstine, J.L., et al., 2005. Managing abnormal blood lipids: a collaborative approach. Circulation 112, 3184—3209.

Funatsu, T., Suzuki, K., Goto, M., Arai, Y., Kakuta, H., Tanaka, H., et al., 2001. Prolonged inhibition of cholesterol synthesis by atorvastatin inhibits apo B-100 and triglyceride secretion from HepG2 cells. Atherosclerosis 157, 107—115.

Garti, N., Avrahami, M., Aserin, A., 2006. Improved solubilization of Celecoxib in U-type nonionic microemulsions and their structural transitions with progressive aqueous dilution. J. Colloid Interface Sci. 299, 352—365.

Gaudette, D.C., Holub, B.J., 1990. Albumin-bound docosahexaenoic acid and collagen-induced human platelet reactivity. Lipids 25, 166—169.

Ginsberg, G.L., Toal, B.F., 2009. Quantitative approach for incorporating methylmercury risks and omega-3 fatty acid benefits in developing species-specific fish consumption advice. Environ. Health Perspect. 117, 267—275.

Goodnight Jr., S.H., Harris, W.S., Connor, W.E., Illingworth, D.R., 1982. Polyunsaturated fatty acids, hyperlipidemia, and thrombosis. Arteriosclerosis 2, 87—113.

Gotto Jr., A.M., Brinton, E.A., 2004. Assessing low levels of high-density lipoprotein cholesterol as a risk factor in coronary heart disease: a working group report and update. J. Am. Coll. Cardiol. 43, 717—724.

Grimsgaard, S., Bonaa, K.H., Hansen, J.B., Nordoy, A., 1997. Highly purified eicosapentaenoic acid and docosahexaenoic acid in humans have similar triacylglycerol-lowering effects but divergent effects on serum fatty acids. Am. J. Clin. Nutr. 66, 649—659.

Grundt, H., Nilsen, D.W., 2008. n-3 fatty acids and cardiovascular disease. Haematologica 93, 807–812.

Grundy, S.M., Denke, M.A., 1990. Dietary influences on serum lipids and lipoproteins. J. Lipid Res. 31, 1149–1172.

Gylling, H., Miettinen, T.A., 1999. Cholesterol reduction by different plant stanol mixtures and with variable fat intake. Metabolism 48, 575–580.

Haikal, Z., Play, B., Landrier, J.F., Giraud, A., Ghiringhelli, O., Lairon, D., Jourdheuil-Rahmani, D., 2008. NPC1L1 and SR-BI are involved in intestinal cholesterol absorption from small-size lipid donors. Lipids 43, 401–408.

Harris, W.S., 1989. Fish oils and plasma lipid and lipoprotein metabolism in humans: a critical review. J. Lipid Res. 30 (6), 785–807.

Harris, W.S., Miller, M., Tighe, A.P., Davidson, M.H., Schaefer, E.J., 2008. Omega-3 fatty acids and coronary heart disease risk: clinical and mechanistic perspectives. Atherosclerosis 197, 12–24.

Harris, W.S., Mozaffarian, D., Rimm, E., Kris-Etherton, P., Rudel, L.L., Appel, L.J., et al., 2009. Omega-6 fatty acids and risk for cardiovascular disease: a science advisory from the American Heart Association Nutrition Subcommittee of the Council on Nutrition, Physical Activity, and Metabolism; Council on Cardiovascular Nursing; and Council on Epidemiology and Prevention. Circulation 119, 902–907.

Heinemann, T., Axtmann, G., von Bergmann, K., 1993. Comparison of intestinal absorption of cholesterol with different plant sterols in man. Eur. J. Clin. Invest. 23, 827–831.

Hendriks, H.F., Brink, E.J., Meijer, G.W., Princen, H.M., Ntanios, F.Y., 2003. Safety of long-term consumption of plant sterol esters-enriched spread. Eur. J. Clin. Nutr. 57, 681–692.

Hjerkinn, E.M., Seljeflot, I., Ellingsen, I., Berstad, P., Hjermann, I., Sandvik, L., Arnesen, H., 2005. Influence of long-term intervention with dietary counseling, long-chain n-3 fatty acid supplements, or both on circulating markers of endothelial activation in men with long-standing hyperlipidemia. Am. J. Clin. Nutr. 81, 583–589.

Hofmann, A.F., Borgstroem, B., 1964. The intraluminal phase of fat digestion in man: the lipid content of the micellar and oil phases of intestinal content obtained during fat digestion and absorption. J. Clin. Invest. 43, 247–257.

Hu, F.B., Manson, J.E., Willett, W.C., 2001. Types of dietary fat and risk of coronary heart disease: a critical review. J. Am. Coll. Nutr. 20, 5–19.

Hui, D.Y., Labonte, E.D., Howles, P.N., 2008. Development and physiological regulation of intestinal lipid absorption. III. Intestinal transporters and cholesterol absorption. Am. J. Physiol. Gastrointest. Liver Physiol. 294, G839–G843.

Ikeda, I., Tanabe, Y., Sugano, M., 1989. Effects of sitosterol and sitostanol on micellar solubility of cholesterol. J. Nutr. Sci. Vitaminol. (Tokyo) 35, 361–369.

Itzhaki, H., Borochov, A., Mayak, S., 1990. Age-related changes in petal membranes from attached and detached rose flowers. Plant Physiol. 94, 1233–1236.

Izar, M.C., Tegani, D.M., Kasmas, S.H., Fonseca, F.A., 2011. Phytosterols and phytosterolemia: gene-diet interactions. Genes Nutr. 6, 17–26.

Jandacek, R.J., Webb, M.R., Mattson, F.H., 1977. Effect of an aqueous phase on the solubility of cholesterol in an oil phase. J. Lipid Res. 18, 203–210.

Jimenez-Escrig, A., Santos-Hidalgo, A.B., Saura-Calixto, F., 2006. Common sources and estimated intake of plant sterols in the Spanish diet. J. Agric. Food Chem. 54, 3462–3471.

Jones, P.J., Demonty, I., Chan, Y.M., Herzog, Y., Pelled, D., 2007. Fish-oil esters of plant sterols differ from vegetable-oil sterol esters in triglycerides lowering, carotenoid bioavailability and impact on plasminogen activator inhibitor-1 (PAI-1) concentrations in hypercholesterolemic subjects. Lipids Health Dis. 6, 28.

Jones, P.J., Raeini-Sarjaz, M., Ntanios, F.Y., Vanstone, C.A., Feng, J.Y., Parsons, W.E., 2000. Modulation of plasma lipid levels and cholesterol kinetics by phytosterol versus phytostanol esters. J. Lipid Res. 41, 697—705.

Jump, D.B., 2008. N-3 polyunsaturated fatty acid regulation of hepatic gene transcription. Curr. Opin. Lipidol. 19, 242—247.

Jump, D.B., Botolin, D., Wang, Y., Xu, J., Christian, B., Demeure, O., 2005. Fatty acid regulation of hepatic gene transcription. J. Nutr. 135, 2503—2506.

Jump, D.B., Botolin, D., Wang, Y., Xu, J., Demeure, O., Christian, B., 2008. Docosa-hexaenoic acid (DHA) and hepatic gene transcription. Chem. Phys. Lipids 153, 3—13.

Karupaiah, T., Sundram, K., 2007. Effects of stereospecific positioning of fatty acids in triacylglycerol structures in native and randomized fats: a review of their nutritional implications. Nutr. Metab. (Lond.) 4, 16.

Kastelein, J.J., Akdim, F., Stroes, E.S., Zwinderman, A.H., Bots, M.L., Stalenhoef, A.F., et al., 2008. Simvastatin with or without ezetimibe in familial hypercholesterolemia. N. Engl. J. Med. 358, 1431—1443.

Kiechl, S., Willeit, J., 1999. The natural course of atherosclerosis. Part I: incidence and progression. Arterioscler. Thromb. Vasc. Biol. 19, 1484—1490.

Kinsella, J.E., Lokesh, B., Stone, R.A., 1990. Dietary n-3 polyunsaturated fatty acids and amelioration of cardiovascular disease: possible mechanisms. Am. J. Clin. Nutr. 52, 1—28.

Klingberg, S., Ellegard, L., Johansson, I., Hallmans, G., Weinehall, L., Andersson, H., Winkvist, A., 2008. Inverse relation between dietary intake of naturally occurring plant sterols and serum cholesterol in northern Sweden. Am. J. Clin. Nutr. 87, 993—1001.

Knapp, H.R., Reilly, I.A., Alessandrini, P., FitzGerald, G.A., 1986. In vivo indexes of platelet and vascular function during fish-oil administration in patients with atherosclerosis. N. Engl. J. Med. 314, 937—942.

Kolovou, G.D., Anagnostopoulou, K.K., Cokkinos, D.V., 2005. Pathophysiology of dyslipidaemia in the metabolic syndrome. Postgrad. Med. J. 81, 358—366.

Krauss, R.M., Eckel, R.H., Howard, B., Appel, L.J., Daniels, S.R., Deckelbaum, R.J., et al., 2000. AHA Dietary Guidelines: revision 2000: a statement for healthcare professionals from the Nutrition Committee of the American Heart Association. Circulation 102, 2284—2299.

Law, M.R., 2000. Plant sterol and stanol margarines and health. West. J. Med. 173, 43—47.

Ling, W.H., Jones, P.J., 1995. Dietary phytosterols: a review of metabolism, benefits and side effects. Life Sci. 57, 195—206.

Lombardo, Y.B., Chicco, A.G., 2006. Effects of dietary polyunsaturated n-3 fatty acids on dyslipidemia and insulin resistance in rodents and humans. A review. J. Nutr. Biochem. 17, 1—13.

Lopez-Garcia, E., Schulze, M.B., Manson, J.E., Meigs, J.B., Albert, C.M., Rifai, N., et al., 2004. Consumption of (n-3) fatty acids is related to plasma biomarkers of inflammation and endothelial activation in women. J. Nutr. 134, 1806—1811.

Lutjohann, D., Bjorkhem, I., Beil, U.F., von Bergmann, K., 1995. Sterol absorption and sterol balance in phytosterolemia evaluated by deuterium-labeled sterols: effect of sitostanol treatment. J. Lipid Res. 36, 1763—1773.

Madsen, M.B., Jensen, A.M., Schmidt, E.B., 2007. The effect of a combination of plant sterol-enriched foods in mildly hypercholesterolemic subjects. Clin. Nutr. 26, 792—798.

Mahley, R.W., 1985. Atherogenic lipoproteins and coronary artery disease: concepts derived from recent advances in cellular and molecular biology. Circulation 72, 943−948.

Marchioli, R., 1999. [Results of GISSI Prevenzione: diet, drugs, and cardiovascular risk. Researchers of GISSI Prevenzione]. Cardiologia 44 (Suppl. 1(Pt 2)), 745−746.

Marchioli, R., Barzi, F., Bomba, E., Chieffo, C., Di Gregorio, D., Di Mascio, R., et al., 2002. Early protection against sudden death by n-3 polyunsaturated fatty acids after myocardial infarction: time-course analysis of the results of the Gruppo Italiano per lo Studio della Sopravvivenza nell'Infarto Miocardico (GISSI)-Prevenzione. Circulation 105, 1897−1903.

Marcus, A.J., 1984. The eicosanoids in biology and medicine. J. Lipid Res. 25, 1511−1516.

Marsan, M.P., Warnock, W., Muller, I., Nakatani, Y., Ourisson, G., Milon, A., 1996. Synthesis of deuterium-labeled plant sterols and analysis of their side-chain mobility by solid state deuterium NMR. J. Org. Chem. 61, 4252−4257.

Matsuura, E., Hughes, G.R., Khamashta, M.A., 2008. Oxidation of LDL and its clinical implication. Autoimmun. Rev. 7, 558−566.

Mattson, F.H., Grundy, S.M., Crouse, J.R., 1982. Optimizing the effect of plant sterols on cholesterol absorption in man. Am. J. Clin. Nutr. 35, 697−700.

Mattson, F.H., Volpenhein, R.A., Erickson, B.A., 1977. Effect of plant sterol esters on the absorption of dietary cholesterol. J. Nutr. 107, 1139−1146.

McNamara, J.R., Shah, P.K., Nakajima, K., Cupples, L.A., Wilson, P.W., Ordovas, J.M., Schaefer, E.J., 1998. Remnant lipoprotein cholesterol and triglyceride reference ranges from the Framingham Heart Study. Clin. Chem. 44, 1224−1232.

Micallef, M.A., Garg, M.L., 2008. The lipid-lowering effects of phytosterols and (n-3) polyunsaturated fatty acids are synergistic and complementary in hyperlipidemic men and women. J. Nutr. 138, 1086−1090.

Micallef, M.A., Garg, M.L., 2009. Beyond blood lipids: phytosterols, statins and omega-3 polyunsaturated fatty acid therapy for hyperlipidemia. J. Nutr. Biochem. 20, 927−939.

Moreau, R.A., Whitaker, B.D., Hicks, K.B., 2002. Phytosterols, phytostanols, and their conjugates in foods: structural diversity, quantitative analysis, and health-promoting uses. Prog. Lipid Res. 41, 457−500.

Mori, T.A., Beilin, L.J., Burke, V., Morris, J., Ritchie, J., 1997. Interactions between dietary fat, fish, and fish oils and their effects on platelet function in men at risk of cardiovascular disease. Arterioscler. Thromb. Vasc. Biol. 17, 279−286.

Mori, T.A., Burke, V., Puddey, I.B., Watts, G.F., O'Neal, D.N., Best, J.D., Beilin, L.J., 2000. Purified eicosapentaenoic and docosahexaenoic acids have differential effects on serum lipids and lipoproteins, LDL particle size, glucose, and insulin in mildly hyperlipidemic men. Am. J. Clin. Nutr. 71, 1085−1094.

Moruisi, K.G., Oosthuizen, W., Opperman, A.M., 2006. Phytosterols/stanols lower cholesterol concentrations in familial hypercholesterolemic subjects: a systematic review with meta-analysis. J. Am. Coll. Nutr. 25, 41−48.

Mozaffarian, D., Rimm, E.B., 2006. Fish intake, contaminants, and human health: evaluating the risks and the benefits. JAMA 296, 1885−1899.

Nair, P.P., 1984. Diet, nutrition intake, and metabolism in populations at high and low risk for colon cancer. Introduction: correlates of diet, nutrient intake, and metabolism in relation to colon cancer. Am. J. Clin. Nutr. 40 (4 Suppl.), 880−886.

NCEP, 2002. Third Report of the National Cholesterol Education Program (NCEP) Expert Panel on Detection, Evaluation, and Treatment of High Blood Cholesterol in Adults (Adult Treatment Panel III) final report. Circulation 106, 3143−3421.

Nestel, P., Cehun, M., Pomeroy, S., Abbey, M., Weldon, G., 2001. Cholesterol-lowering effects of plant sterol esters and non-esterified stanols in margarine, butter and low-fat foods. Eur. J. Clin. Nutr. 55, 1084–1090.

Nestel, P.J., Connor, W.E., Reardon, M.F., Connor, S., Wong, S., Boston, R., 1984. Suppression by diets rich in fish oil of very low density lipoprotein production in man. J. Clin. Invest. 74, 82–89.

Noakes, M., Clifton, P., Ntanios, F., Shrapnel, W., Record, I., McInerney, J., 2002. An increase in dietary carotenoids when consuming plant sterols or stanols is effective in maintaining plasma carotenoid concentrations. Am. J. Clin. Nutr. 75, 79–86.

Noakes, M., Clifton, P.M., Doornbos, A.M., Trautwein, E.A., 2005. Plant sterol ester-enriched milk and yoghurt effectively reduce serum cholesterol in modestly hypercholesterolemic subjects. Eur. J. Nutr. 44, 214–222.

Normen, A.L., Brants, H.A., Voorrips, L.E., Andersson, H.A., van den Brandt, P.A., Goldbohm, R.A., 2001. Plant sterol intakes and colorectal cancer risk in the Netherlands Cohort Study on Diet and Cancer. Am. J. Clin. Nutr. 74, 141–148.

Oh, K., Hu, F.B., Manson, J.E., Stampfer, M.J., Willett, W.C., 2005. Dietary fat intake and risk of coronary heart disease in women: 20 years of follow-up of the nurses' health study. Am. J. Epidemiol. 161, 672–679.

Ostlund Jr., R.E., McGill, J.B., Zeng, C.M., Covey, D.F., Stearns, J., Stenson, W.F., Spilburg, C.A., 2002a. Gastrointestinal absorption and plasma kinetics of soy Delta(5)-phytosterols and phytostanols in humans. Am. J. Physiol. Endocrinol. Metab. 282, E911–916.

Ostlund Jr., R.E., Racette, S.B., Okeke, A., Stenson, W.F., 2002b. Phytosterols that are naturally present in commercial corn oil significantly reduce cholesterol absorption in humans. Am. J. Clin. Nutr. 75, 1000–1004.

Ostlund Jr., R.E., Racette, S.B., Stenson, W.F., 2003. Inhibition of cholesterol absorption by phytosterol-replete wheat germ compared with phytosterol-depleted wheat germ. Am. J. Clin. Nutr. 77, 1385–1389.

Phillips, K.M., Ruggio, D.M., Ashraf-Khorassani, M., 2005. Phytosterol composition of nuts and seeds commonly consumed in the United States. J. Agric. Food Chem. 53, 9436–9445.

Plana, N., Nicolle, C., Ferre, R., Camps, J., Cos, R., Villoria, J., Masana, L., 2008. Plant sterol-enriched fermented milk enhances the attainment of LDL-cholesterol goal in hypercholesterolemic subjects. Eur. J. Nutr. 47, 32–39.

Plat, J., Mensink, R.P., 2002a. Increased intestinal ABCA1 expression contributes to the decrease in cholesterol absorption after plant stanol consumption. FASEB J. 16, 1248–1253.

Plat, J., Mensink, R.P., 2002b. Relationship of genetic variation in genes encoding apolipoprotein A-IV, scavenger receptor BI, HMG-CoA reductase, CETP and apolipoprotein E with cholesterol metabolism and the response to plant stanol ester consumption. Eur. J. Clin. Invest. 32, 242–250.

Pollak, O.J., 1953. Reduction of blood cholesterol in man. Circulation 7, 702–706.

Prichard, B.N., Smith, C.C., Ling, K.L., Betteridge, D.J., 1995. Fish oils and cardiovascular disease. BMJ 310, 819–820.

Prisco, D., Filippini, M., Francalanci, I., Paniccia, R., Gensini, G.F., Serneri, G.G., 1995. Effect of n-3 fatty acid ethyl ester supplementation on fatty acid composition of the single platelet phospholipids and on platelet functions. Metabolism 44, 562–569.

Rideout, T.C., Chan, Y.M., Harding, S.V., Jones, P.J., 2009. Low and moderate-fat plant sterol fortified soymilk in modulation of plasma lipids and cholesterol kinetics in subjects with normal to high cholesterol concentrations: report on two randomized crossover studies. Lipids Health Dis. 8, 45.

Ridker, P.M., 2003. Clinical application of C-reactive protein for cardiovascular disease detection and prevention. Circulation 107, 363−369.

Roshanai, F., Sanders, T., Howarth, D., Haines, A., 1981. Prostaglandin precursors in platelet phosphoglycerides following acute myocardial infarction. Thrombosis Research 24, 169−173.

Rudkowska, I., AbuMweis, S.S., Nicolle, C., Jones, P.J., 2008. Association between non-responsiveness to plant sterol intervention and polymorphisms in cholesterol metabolism genes: a case−control study. Appl. Physiol. Nutr. Metab. 33, 728−734.

Russell, J.C., Ewart, H.S., Kelly, S.E., Kralovec, J., Wright, J.L., Dolphin, P.J., 2002. Improvement of vascular dysfunction and blood lipids of insulin-resistant rats by a marine oil-based phytosterol compound. Lipids 37, 147−152.

Ruxton, C., 2004. Health benefits of omega-3 fatty acids. Nurs. Stand. 18, 38−42.

Salen, G., Ahrens Jr., E.H., Grundy, S.M., 1970. Metabolism of beta-sitosterol in man. J. Clin. Invest. 49, 952−967.

Salen, G., Patel, S., Batta, A.K., 2002. Sitosterolemia. Cardiovasc. Drug Rev. 20, 255−270.

Salmon, J.A., Terano, T., 1985. Supplementation of the diet with eicosapentaenoic acid: a possible approach to the treatment of thrombosis and inflammation. Proc. Nutr. Soc. 44, 385−389.

Sanders, T.A., Hochland, M.C., 1983. A comparison of the influence on plasma lipids and platelet function of supplements of omega 3 and omega 6 polyunsaturated fatty acids. Br. J. Nutr. 50, 521−529.

Saynor, R., 1984. Effects of omega-3 fatty acids on serum lipids. Lancet 2, 696−697.

Saynor, R., Verel, D., Gillott, T., 1984. The effect of MaxEPA on the serum lipids, platelets, bleeding time and GTN consumption. Br. J. Clin. Pract. Suppl. 31, 70−74.

Seo, T., Blaner, W.S., Deckelbaum, R.J., 2005. Omega-3 fatty acids: molecular approaches to optimal biological outcomes. Curr. Opin. Lipidol. 16, 11−18.

Simopoulos, A.P., 2002. Omega-3 fatty acids and cardiovascular disease: The epidemiological evidence. Environ. Health Prevent. Med. 6, 203−209.

Simopoulos, A.P., 2006. Evolutionary aspects of diet, the omega-6/omega-3 ratio and genetic variation: nutritional implications for chronic diseases. Biomedicine and Pharmacotherapy = Biomedecine and Pharmacotherapie 60, 502−507.

Stamler, J., Wentworth, D., Neaton, J.D., 1986. Is relationship between serum cholesterol and risk of premature death from coronary heart disease continuous and graded? Findings in 356,222 primary screenees of the Multiple Risk Factor Intervention Trial (MRFIT). JAMA 256, 2823−2828.

Subbiah, M.T., Kottke, B.A., Carlo, I.A., 1971. Uptake of campesterol in pigeon intestine. Biochim. Biophys. Acta 249, 643−646.

Sudhop, T., Sahin, Y., Lindenthal, B., Hahn, C., Luers, C., Berthold, H.K., von Bergmann, K., 2002. Comparison of the hepatic clearances of campesterol, sitosterol, and cholesterol in healthy subjects suggests that efflux transporters controlling intestinal sterol absorption also regulate biliary secretion. Gut 51, 860−863.

Tikkanen, M.J., 2005. Plant sterols and stanols. Handb. Exp. Pharmacol. 170, 215−230.

Valsta, L.M., Lemstrom, A., Ovaskainen, M.L., Lampi, A.M., Toivo, J., Korhonen, T., Piironen, V., 2004. Estimation of plant sterol and cholesterol intake in Finland: quality of new values and their effect on intake. Br. J. Nutr. 92, 671−678.

von Schacky, C., Weber, P.C., 1985. Metabolism and effects on platelet function of the purified eicosapentaenoic and docosahexaenoic acids in humans. J. Clin. Invest. 76, 2446−2450.

Wang, H.H., Patel, S.B., Carey, M.C., Wang, D.Q., 2007. Quantifying anomalous intestinal sterol uptake, lymphatic transport, and biliary secretion in Abcg8(−/−) mice. Hepatology 45, 998−1006.

Wang, N., Lan, D., Chen, W., Matsuura, F., Tall, A.R., 2004. ATP-binding cassette transporters G1 and G4 mediate cellular cholesterol efflux to high-density lipoproteins. Proc. Natl. Acad. Sci. USA 101, 9774−9779.

Wang, N., Silver, D.L., Thiele, C., Tall, A.R., 2001. ATP-binding cassette transporter A1 (ABCA1) functions as a cholesterol efflux regulatory protein. J. Biol. Chem. 276, 23742−23747.

Weihrauch, J.L., Gardner, J.M., 1978. Sterol content of foods of plant origin. J. Am. Diet. Assoc. 73, 39−47.

Weintraub, M.S., Zechner, R., Brown, A., Eisenberg, S., Breslow, J.L., 1988. Dietary polyunsaturated fats of the W-6 and W-3 series reduce postprandial lipoprotein levels. Chronic and acute effects of fat saturation on postprandial lipoprotein metabolism. J. Clin. Invest. 82, 1884−1893.

Wensing, A.G., Mensink, R.P., Hornstra, G., 1999. Effects of dietary n-3 polyunsaturated fatty acids from plant and marine origin on platelet aggregation in healthy elderly subjects. Br. J. Nutr. 82, 183−191.

Weststrate, J.A., Meijer, G.W., 1998. Plant sterol-enriched margarines and reduction of plasma total- and LDL-cholesterol concentrations in normocholesterolaemic and mildly hypercholesterolaemic subjects. Eur. J. Clin. Nutr. 52, 334−343.

Wolfs, M., de Jong, N., Ocke, M.C., Verhagen, H., Monique Verschuren, W.M., 2006. Effectiveness of customary use of phytosterol/-stanol enriched margarines on blood cholesterol lowering. Food Chem. Toxicol. 44, 1682−1688.

Woodman, R.J., Mori, T.A., Burke, V., Puddey, I.B., Barden, A., Watts, G.F., Beilin, L.J., 2003. Effects of purified eicosapentaenoic acid and docosahexenoic acid on platelet, fibrinolytic and vascular function in hypersensitive type 2 diabetic patients. Antherosclerosis 166, 85−93.

Yamanashi, Y., Takada, T., Suzuki, H., 2007. Niemann-Pick C1-like 1 overexpression facilitates ezetimibe-sensitive cholesterol and beta-sitosterol uptake in CaCo-2 cells. J. Pharmacol. Exp. Ther. 320, 559−564.

Yaqoob, P., 2004. Fatty acids and the immune system: from basic science to clinical applications. Proc. Nutr. Soc. 63, 89−104.

Yusuf, S., Hawken, S., Ounpuu, S., Dans, T., Avezum, A., Lanas, F., et al., 2004. Effect of potentially modifiable risk factors associated with myocardial infarction in 52 countries (the INTERHEART study): case−control study. Lancet 364, 937−952.

Zampelas, A., Panagiotakos, D.B., Pitsavos, C., Das, U.N., Chrysohoou, C., Skoumas, Y., Stefanadis, C., 2005. Fish consumption among healthy adults is associated with decreased levels of inflammatory markers related to cardiovascular disease: the ATTICA study. J. Am. Coll. Cardiol. 46, 120−124.

Dairy Materials as Delivery Tools for Bioactive Components in Dairy Platforms

Anilda Guri[1,2] and Milena Corredig[1]

[1]*Department of Food Science, University of Guelph, Ontario, Canada,*
[2]*Canadian Research Institute for Food Safety, University of Guelph, Ontario, Canada*

CONTENTS

INTRODUCTION

MILK AS A DELIVERY PLATFORM OF BIOACTIVE COMPONENTS

Milk is an aqueous solution of lactose, minerals, and small-molecular-weight components, in which fat globules and proteins are dispersed. The

Food Structures, Digestion and Health. http://dx.doi.org/10.1016/B978-0-12-404610-8.00017-7

structure of the colloidal particles present in milk, both at the molecular and supramolecular level, is well conserved between species, and serves multiple biological functionalities. Indeed, milk is an excellent example of a food matrix having both nutritional and physiological roles. The milk components can be either assembled into novel structures, or they can be disrupted during digestion, resulting in a variety of fragments with distinct biological functionalities (Ward and German, 2004).

The proteins present in the serum phase of milk (whey proteins) exist as monomers or oligomers, and some of their functions are yet to be elucidated. They are considered of high nutritional quality because of their amino acid composition. Casein proteins constitute about 80% of the total protein in milk, and are critical to the growth and development of the neonate (Holt, 2004). Caseins are present as an assembly of colloidal particles with diameters ranging from 80 to 500 nm, and their structure represents a great example of nature's solution to nutrient delivery. The casein micelles are destabilized during digestion by the gastric acidic environment and partial hydrolysis of the polyelectrolyte layer of κ-casein, present on the surface, creating a structure that plays a role in modulating stomach emptying of the proteins and fat globules. In addition to the protein, triglycerides in milk are also present as colloidal supramolecular structures, covered by a trilayer membrane composed mainly of phospholipids and proteins. It is becoming increasingly evident (Argov *et al.*, 2008; Evers *et al.*, 2008) that the structural organization of the milk fat globule plays a major role in biological functions related to gut health and development.

In recent years there has been an increasing interest in understanding the role of these structures beyond supplying the nutritional components necessary to the growth of the neonate (Macierzanka *et al.*, 2011; Michalski *et al.*, 2013; Barbe *et al.*, 2013). In addition, dairy components have been employed to create functional structures in foods that are not only nutritionally dense, but also deliver beneficial molecules not originally present in milk. In this chapter we will give an overview of our current understanding of milk as a platform for delivery of beneficial components; however, this is not intended by any means to be an exhaustive review, as new information continues to become available on the central place that milk components can play in the development of functional foods. In this chapter we will focus attention on the main milk colloids, proteins, and milk fat globules, which have shown the potential to act as a platform for the delivery of bioactive molecules.

MILK MACROMOLECULES, STRUCTURE, AND DELIVERY FUNCTIONS

Proteins and lipid aggregates not only encode bioactive structures that will become important once digested, and will serve protective roles, for example in the regulation of the immune system or by showing symbiotic functions with beneficial microorganisms; they will also modify their structure, aggregate, and form larger assemblies, which play a major role in the formation of the structure and texture of dairy products.

Increasing scientific and commercial interest has been shown in the multiple functional properties of milk proteins (Clare *et al.*, 2003; Pihlanto and Korhonen, 2003; Zimecki and Kruzel, 2007). Whey proteins and caseins have been widely studied for their ability to bind hydrophobic molecules, interact with other biopolymers, stabilize emulsions, self-assemble, and form gels. The use of whey and casein ingredients is now widespread in the food industry.

Lipids are present in milk as fat globules varying in size from less than 100 nm to about 15 μm in diameter (Argov *et al.*, 2008). Triglycerides, phospholipids, and glycosphingolipids are the main lipid constituents in milk (Keenan and Dylewski, 1994). A membrane composed of phospholipids, glycosphingolipids, and proteins surrounds the fat globules and protects them from instability. The composition and changes of the milk fat globule membrane (MFGM) with processing have been widely reviewed in the literature (Parodi, 1997; Spitsberg, 2005; Evers *et al.*, 2008; Dewettinck *et al.*, 2008a; Jimenez-Flores *et al.*, 2013).

Whey Proteins

The whey fraction of milk contains a wide variety of proteins that differ from each other in their chemical structure, processing properties, and biological functions. Whey proteins represent about 20% of the total proteins in milk. They are globular in nature. The major proteins in whey are β-lactoglobulin, α-lactalbumin, bovine serum albumin, immunoglobulins, proteose peptones, lactoferrin, and lactoperoxidase. Whey protein isolates and concentrates are widespread in the food industry as ingredients, because of their nutritive value and versatile processing functionalities, such as solubility, viscosity, gel-forming capacity, emulsification, and also for their ability to form nanoparticles (Mehravar *et al.*, 2009; Arroyo-Maya *et al.*, 2012). The main protein in whey is β-lactoglobulin, which is a globular protein composed of 162 amino acid residues with two disulfide groups and one free sulfhydryl group. This protein has been shown to bind to hydrophobic molecules (Wang *et al.*, 1997), and its aggregates have been suggested as nanovehicles for delivery of nutraceutical molecules (Zimet and Livney, 2009). α-Lactalbumin comprises about 20% of the total whey proteins, and it serves a role in lactose biosynthesis (Farrell *et al.*, 2004; Arroyo-Maya *et al.*, 2012). Other components of great biological value, such as immunoglobulins, lactoferrin, and growth factors, are also present in the milk serum. These components are found in colostrum in much greater concentrations than in milk, thus reflecting their importance to the health of the neonate (Korhonen, 1977; Pakkanen and Aalto, 1997; Scammell, 2001). In particular, lactoferrin, a glycoprotein (80 kDa) member of the transferrin family in milk, consists of 673 amino acid residues. Lactoferrin and its peptides play an important role in innate defense against antimicrobial and antiviral activities (Kitts and Weiler, 2003), antioxidant activities (Korhonen and Pihlanto, 2007), and immunomodulating and cell growth effects (Swart et al., 1998; Mercier et al., 2004). Because of its positive charge at neutral pH, lactoferrin has been shown to form nanoparticles and coacervates (Shimoni et al., 2013) that could be utilized as delivery systems. In addition, the stability of lactoferrin to gastric digestion may be improved by its interaction with milk phospholipids (Liu et al., 2013).

In sum, the excellent nutritional value of whey proteins, coupled with their functional properties such as abilities to form gels and stabilize emulsions have extensively increased their utilization as delivery systems.

Caseins

Casein micelles contain four main phosphoproteins (α_{s1}-, α_{s2}-, β-, and κ-casein) with molecular weights between 19 and 25 kDa. These proteins contain a large number of proline residues, which makes them rheomorphic structures (Holt, 2004). These proteins are often used as ingredients in food products, for their ability to gel, self-assemble, and adsorb at oil—water and air—water interfaces. In milk, the caseins are present as casein micelles, colloidal particles between 80 and 500 nm in diameter, which are highly hydrated. Six percent of the total weight of these protein assemblies is composed of colloidal calcium phosphate. The casein micelles are conserved structures within mammalian species, and have evolved to carry essential nutrients such as calcium, phosphate, and biologically significant protein to the neonate. The composition of native micelles seems to vary depending on their size, with smaller micelles relatively rich in κ-casein and having a lower content of β-casein (Dalgleish and Corredig, 2012). Casein micelles are stabilized electrostatically and sterically by a polyelectrolyte layer of κ-casein, mainly present on the surface of the particles. The micelles can be readily destabilized by the hydrolysis of κ-casein carried out by gastric proteases (Dalgleish and Corredig, 2012).

Caseins have been shown to be retained longer than whey proteins before their amino acids are detected in the bloodstream (Barbe et al., 2013). This difference in the kinetics of amino acid absorption has been attributed to the structuring of the casein micelles in the stomach, causing a slower postprandial release (Barbe *et al.*, 2013). This is due to their ability to form a gel under acidic conditions and in the presence of specific stomach proteases. During acidification, the casein micelles gradually solubilize calcium phosphate until completion at pH of about 5, hence providing an excellent pH-triggered release system for calcium and phosphate. In addition, at a pH close to their isoelectric point (pH 4.6) casein micelles decrease their overall charge and undergo isoelectric precipitation (Fox and Brodkorb, 2008), forming a self-supporting network. The casein curd formed in the stomach will have characteristics of gel formed by a mixed enzymatic (rennet) and acid coagulation, as the pH of aggregation of the casein micelles significantly increases with partial hydrolysis of the κ-casein layer (Doublier *et al.*, 2000; Li and Dalgleish, 2006). Casein micelles are also prone to phase separation in the presence of small concentrations of soluble polysaccharides, because of thermodynamic incompatibility (Doublier *et al.*, 2000). This may be of great significance during digestion of a mixed meal containing dairy proteins and soluble fiber.

Casein proteins can be obtained in non-micellar form after precipitation with acid and resuspension at neutral pH. Under these conditions, these proteins are mostly present as monomers or nanoparticles (<50 nm in size), and the processing properties of caseinates are quite different from those of the original casein micelles.

Casein micelles as well as casein protein aggregates have been suggested as nanodelivery systems (Haham *et al.*, 2012) and their nanoencapsulation seems to have minimal impact in the processing behavior of the milk (Livney, 2010), although this is yet a source of debate (see, for example, O'Connell *et al.*, 1998; Haratifar and Corredig, 2013). In addition, the ability of milk components to form gels, protein aggregates, and phase-separated domains in the presence of other components is an opportunity to design matrices that will modify their structure during gastrointestinal transit.

Milk Lipids

Milk lipids are mostly present as fat globules surrounded with a membrane protecting them from coalescence. The fat globules are a main source of energy for the neonate. They contain a wide range of fatty acids, and they play an important role as a delivery system for hydrophobic molecules. As in the case of milk proteins, the fat globules and their supramolecular structure are a great example of nature's solution to deliver health benefits through food.

The membrane surrounding the fat globules (MFGM) has a heterogeneous composition, and a unique structure (Dewettinck *et al.*, 2008b). In its native form, the MFGM contains an inner polar lipid monolayer, surrounding the triglyceride core of the fat droplets, a proteinaceous coat in between, and a bilayer membrane of phospholipids and proteins on the outside. The phospholipid composition of the MFGM is quite different from that derived from soy or egg extracts (Burling and Graverholt, 2008; Farhang and Corredig, 2011). The glycerophospholipids (phosphatidylcholine and phosphatidylethanolamine) are the most abundant species in milk phospholipids, together with sphingolipids (sphingomyelin, gangliosides, and cerebrosides) (Rombaut *et al.*, 2006). The MFGM also contains a large variety of proteins. A recent study of the MFGM proteome identified up to 120 proteins, of which 71% are membrane-associated proteins and 24% are cytoplasmic in origin (Cavaletto *et al.*, 2008). Almost 50% of the protein is associated with cell signaling and immune response (Spitsberg, 2005; Fong and Norris, 2009).

Processing milk or cream (agitation, heating, cooling, homogenization, evaporation, spray drying) can modify the size of the fat globules as well as the composition of the MFGM, and this may affect to different extents the bioactivity and the behavior of the fat globules during gastrointestinal transit (Michalski *et al.*, 2013). Milk contains as little as 2 g/L of MFGM material. In spite of the small concentration in milk, this material has received much attention in recent years due to both its health-beneficial properties and technological functionalities (Evers *et al.*, 2008). In particular, the MFGM seems to contribute greatly to gastrointestinal health functions.

Recent research has highlighted the presence in human milk of very small fat globules, and in these globules the MFGM constitutes a large percentage of the total weight, resulting in much higher relative concentrations of polar versus non-polar lipids (Argov *et al.*, 2008). The authors concluded that a large portion of the milk fat globules are secreted into milk not for delivery of fat but most likely for metabolic and nutritional functions. It may be hypothesized that some of the very small vesicles may not necessarily have a lipid core, but they could have different organization of the lipid membrane, and resemble the structure of liposomes, with a hydrophilic core.

MILK COMPONENTS DURING DIGESTION

Proteins and lipid aggregates in milk encode bioactive structures that will become important once digested, and will serve functional roles, for instance in the regulation of the immune system or important body conditions, such as appetite control or blood pressure modulation. Milk protein sequences are rich in physiologically important amino acids, such as tryptophan, leucine, and arginine, often employed in supplemental nutrition because of their known roles in stimulating muscle protein synthesis or being involved in neural functions (Etzel, 2004).

In milk, many of the components have demonstrated multiple functions (German *et al.*, 2002). The need for better understanding of the structures in milk and not only their assembly during processing and storage, but also their disruption and reassembly during digestion continues to grow. It is becoming increasingly clear that the fate of bioactives encoded or embedded in the structures formed by proteins is highly dependent on the matrix composition, and not only on the gastrointestinal conditions (Turgeon and Rioux, 2011).

Increased understanding of the bioactive role played by the milk components and the effect of the structure of the milk matrix on the delivery of additional biological functionality will result, on the one hand, in the isolation and fractionation of key components from milk for food and pharmaceutical applications, but on the other, in a clear identification of which food processes should be employed to extend the nutritional and biological value of dairy products.

Milk Protein Digestion and Bioactive Peptides

The proteins in milk represent functional building blocks for many food matrices. In addition to encoding within their primary structures a number of bioactive peptides (Clare and Swaisgood, 2000; Korhonen and Pihlanto, 2006; Haque *et al.*, 2009; Mills *et al.*, 2011), they assemble to form aggregates, polymers, and nanotubes. The organization of such structures will define the function of the matrix they form and how it will be disrupted during digestion.

It is becoming increasingly clear that the absorption of dietary amino acids by the gut does not only depend on the type of protein ingested (Boirie *et al.*, 1997), but also on the processing history and the structure of the matrix (Turgeon and Rioux, 2011; Rioux and Turgeon, 2012; Le Feunteun *et al.*, 2013). Dairy proteins can be ingested in a great variety of structures, and recent work has demonstrated that the kinetics of amino acids absorption can be predicted taking into account the structure modifications that these components are subjected to within the stomach (Le Feunteun *et al.*, 2013).

Milk proteins contain a high level of nutritionally important amino acids, and their high bioavailability is usually maintained after processing, as there is a limited amount of lysine damage (Rutherfurd and Moughan, 2005). However, the processing history will affect the kinetics of digestion and absorption (Almaas *et al.*, 2006; Dupont *et al.*, 2010). For example, when ingesting untreated milk, a mixed coagulation (pepsin + acid, in adults, and rennin + acid in infants) of the casein micelles may cause the gel to coagulate in the stomach, with whey proteins resisting digestion (β-lactoglobulin may resist digestion in its native form (Almaas *et al.*, 2006; Barbe *et al.*, 2013)) and remaining soluble. On the other hand, very different rates of absorption may occur when milk has been heated, and whey proteins are denatured and form aggregates with the caseins at low pH (Rioux and Turgeon, 2012).

It has been demonstrated that the macrostructure of a milk matrix (fluid or gel) has a strong impact on digestion (Barbe *et al.*, 2013). The gelling of the milk in the stomach slows down the digestion of the protein and the availability of the amino acids in the blood (Barbe *et al.*, 2013). Gastric emptying rates are affected also by the structure of the gel (Le Feunteun *et al.*, 2013), with rennet gels being more retained within the stomach, compared to pasteurized milk or acid gels. The reasons for these differences are not fully understood.

Depending on the composition and matrix of the food, there will be different satiety responses, not only because of differences in gastric emptying rates, but also because of the stimulation of synthesis of gastrointestinal hormones. Protein-induced satiety seems to be of vital importance for managing weight loss through suppression of signaling molecules and hormones (Veldhorst *et al.*, 2008). Milk proteins (casein and whey proteins) have been shown to have an impact on satiety and suppress appetite (Bowen *et al.*, 2008; Westerterp-Plantenga *et al.*, 2009). It has been reported that whey proteins elicit a higher secretion of gut hormones and greater satiety effects than caseins (Hall *et al.*, 2003), and these results are in full agreement with the understanding that there are different digestion kinetics for the two protein types, as mentioned above (Boiric *et al.*, 1997).

Recent *in vivo* digestion studies (Boutrou *et al.*, 2013) have demonstrated that a large variety of casein fragments can be found at physiologically significant concentrations in the jejunum of healthy humans after ingestion of caseins. The majority of the bioactive peptides still unabsorbed seem to derive from β-casein proteolysis, and among the most abundant peptides isolated in the study, there were some with recognized bioactivity, such as β-casomorphins and antihypertensive peptides.

Whey proteins seem to have limited gastric digestibility, and they are mostly hydrolyzed in the upper intestine. The same study (Boutrou *et al.*, 2013) showed that the size of the whey protein peptides isolated in the intestine is generally greater than for those derived from caseins (9−15 amino acid residues compared to 6−9 residues, respectively). Hence it can be concluded that caseins are more susceptible to hydrolysis, but their peptides are still present unabsorbed in the jejunum. The peptidic fractions isolated seem to derive from the regions more resistant to hydrolysis, rich in proline and glycine. These findings were recently supported by an *in vivo* clinical trial using piglets (Bouzerzour *et al.*, 2012; Boutrou *et al.*, 2013).

Digestion of caseins causes the release of phosphorylated peptides, casein-ophosphopeptides. Numerous studies both *in vitro* and *in vivo* have demonstrated that casein phosphopeptides, formed after trypsinolysis, are still found unabsorbed in the intestine where they may play an important physiological role in mineral absorption (Naito and Suzuki, 1974; Ono *et al.*, 1994).

The binding capacity of phosphopeptides to minerals like iron, calcium, and zinc have been investigated (Argyri *et al.*, 2007, 2009; Garcia-Nebot *et al.*, 2009). Cell culture models have been used to study the iron, zinc, and calcium uptake triggered by casein bioactive peptides. The association between minerals and caseins is the basis of the effect of dairy products on mineral bioavailability.

Many reports are available on the bioactive peptides derived from milk proteins (Korhonen and Pihlanto, 2003; FitzGerald *et al.*, 2004). The bioactive peptide sequences will remain inactive as long as they are encrypted in the native protein sequence, until their release during protein hydrolysis (Clare and Swaisgood, 2000; Power *et al.*, 2013). To fully understand the formation of bioactive peptides in the gut from milk matrices is a very difficult task, as structural changes to the proteins, interactions with other ingredients, processing history, and macromolecular structure will affect their release.

Peptides may also be generated during manufacturing of dairy products, with a careful control of the processing conditions as, in this case also, various factors will affect the rate and extent of hydrolysis. In general, hydrolysis-derived peptides have been shown to have an impact on a variety of biological and physiological activities, with effects such as antihypertensive, immunomodulatory, antimicrobial, antioxidative, and antithrombotic (Clare and Swaisgood, 2000; Haque *et al.*, 2009; Power *et al.*, 2013). Some antimicrobial peptides released from milk are considered as multifunctional because of other physiological functions that they can exert (Clare and Swaisgood, 2000; Hartmann and Meisel, 2007; Lopez-Exposito *et al.*, 2007). When obtained through processing and incorporated in the dairy matrix, the peptides may play multiple functionality roles, first techno-functional, such as their ability to impart foaming, emulsifying, and structuring properties, and later, physiological roles, once ingested. In particular, fermentation of milk by probiotics has been shown to produce molecules of interest for control of growth or virulence inhibition of pathogens and for modulation of the gut microbiota (Tomita *et al.*, 2002; Zeinhom *et al.*, 2012). There are then a number of advantages in using fermented dairy matrices as carriers for biological molecules, as the matrix itself is rich in components with multiple functionality (Ehlers *et al.*, 2011; Tellez *et al.*, 2011).

Fat Globule Digestion and Milk Fat Globule Membrane Bioactivity

Fat globules in milk are a sophisticated delivery system, targeted not only to release fatty acids, but also to impart additional gut health. Fat globules are only partly digested in the gastric phase. Gastric lipase in human adults results in the release of 5−30% of the fatty acids. Most of the lipids are digested in the duodenum, once mixed with bile salts and pancreatic secretions (Armand *et al.*, 1999; Phan and Tso, 2001). The kinetics of digestion of milk fat globules are highly dependent on the size of the fat globules, but also on the quality of the interface and its structure. The size and supramolecular structure of the fat globules will affect the kinetics of lipolysis (Armand *et al.*, 1999; Berton *et al.*, 2012).

Processing of milk and cream can alter lipid digestion, as both size and membrane structure will be affected. A human intervention study with intragastrically delivered emulsions showed that a lower initial fat droplet size facilitated lipolysis, although fat assimilation overall was not affected, because of the efficient digestion in the small intestine (Armand *et al.*, 1999). In addition to the structural organization of the triglycerides and their liquid or solid state at body temperature, the supramolecular organization of the membrane carrying the lipids can affect the digestion of the fatty acids and their absorption (Michalski *et al.*, 2013). The size of the lipid droplet also seems to be critical for milk fat globules, and it has been reported that the catalytic efficiency of pancreatic lipase will be higher on small fat globules prepared by high pressure homogenization compared to native globules (Berton *et al.*, 2012). These research findings will certainly lead to novel developments of dairy matrices with fat globules designed with specific hydrolysis behaviors.

In the past few years there has been increasing evidence that the milk fat globule membrane may play a very important role in gastrointestinal health. The MFGM fractions have beneficial properties such as inhibition of growth of cancer cells (Snow *et al.*, 2010), inhibition of pathogen adhesion (Guri *et al.*, 2012), or antimicrobial properties (Spitsberg, 2005; Sanchez-Juanes *et al.*, 2009). For example, fat globules from both bovine and goat milk inhibit the adhesion of *Salmonella enteritidis* to HT-29 human adenocarcinoma cells in a cultured cell experiment (Guri *et al.*, 2012). Fractions of MFGM also showed an inhibitory effect on the virulence (expression of

Shiga toxin gene) of *E. coli* O157:H7 (Tellez *et al.*, 2012). Heat treatment of the cream does not seem to affect these properties of the MFGM (Guri *et al.*, 2012; Tellez *et al.*, 2012).

Many of the studies on the health benefits associated with the MFGM have focused on single components present in the membranes, mostly some of the major proteins and the phospholipids (Karlsson, 1989; Parodi, 1997; Hancock *et al.*, 2002). Glycosphingolipids are one of the most studied bioactive components of the MFGM. These molecules and their catabolites have effects on biological functions such as cell–cell interactions, differentiation, apoptosis, and immune recognition (Peguet-Navarro *et al.*, 2003; Morales *et al.*, 2004). Phospholipids represent the most abundant species of polar lipids in the MFGM and are known to beneficially affect diverse cell functions including absorption of nutrients, molecular transport systems, growth, and development and regulation of the nervous system (Spitsberg, 2005). The MFGM is also rich in glycosylated proteins, shown to prevent physical attachment of bacteria and pathogens to the gastrointestinal mucosa (Schroten *et al.*, 1992; Sanchez-Juanes *et al.*, 2009).

In addition to single components, research has been conducted on MFGM isolates. These isolates are effective in the inhibition of *Helicobacter pylori* infection in mice, and hemagglutination and adhesion of *H. pylori* in cell culture models (Wang *et al.*, 2001). The authors observed the same effect in defatted and non-defatted fractions from buttermilk (the by-product of butter making, containing MFGM), suggesting that the biofunctionality is related to proteinaceous components present in MFGM. In addition to inhibiting attachment of pathogenic microorganisms, there are some reports of the MFGM binding to beneficial microorganisms (Guillaume *et al.*, 2010).

In addition to antimicrobial activity and effects on bacterial attachment, another important bioactivity of MFGM is related to colon carcinogenesis. Recent reports showed the role of native MFGM isolates in inducing apoptosis in HT-29 human colon cancer cells (Zanabria *et al.*, 2013). Similarly, other reports (Snow *et al.*, 2011) demonstrated that feeding rats with an MFGM fraction from buttermilk can reduce significantly the incidence of aberrant crypt foci and provide protection against gastrointestinal leakiness caused by lipopolysaccharide (Snow *et al.*, 2010, 2011). Most studies suggest that the supramolecular structure of the MFGM plays a major role in the delivery of the bioactive molecules contained in the MFGM, such as the unique polar lipids, immune stimulating peptides, and signaling molecules (Dewettinck *et al.*, 2008a; Jimenez-Flores and Brisson, 2008).

MILK COMPONENTS AS DELIVERY SYSTEMS

Supramolecular structures such as those of casein micelles, fat globules, and "lactosomes" (very small fat globules (Argov *et al.*, 2008)) are highly conserved among mammals and have been studied as examples of nature's delivery vehicles. Studying these structures and modulating the aggregation of the milk components will lead to a better understanding of how to develop more efficacious delivery systems.

Milk Proteins Assemblies as Delivery Systems

Milk proteins may form structures that could be utilized to deliver bioactive compounds (Livney, 2010). These proteins are used because of their surface activity, leading them to readily adsorb at interfaces. They can also form aggregates that can be modified by processing as well as changes in environmental conditions such as pH and temperature. Hence, they have been looked at as a means to deliver certain bioactive molecules normally having low bioaccessibility, for example, polyphenols such as curcumin or tea catechins.

Whey protein microparticles can be obtained and employed to protect and transport bioactive molecules through the digestive tract and ensure their sustained release (Mercier *et al.*, 2004; Saint-Sauveur *et al.*, 2008; Hebrard *et al.*, 2013). For example, whey protein gel particles containing alginates have been shown as a successful model for encapsulation of vitamin D, retinol, and probiotic cultures (Hebrard *et al.*, 2013). Cold gelation methods have often been utilized to protect heat-sensitive components or living microorganisms (Martin and de Jong, 2012; Hebrard *et al.*, 2013). Cold-set gelation of whey proteins has been successfully used for iron fortification of foods (Martin and de Jong, 2012). Acid-induced or salt-induced gelation, desolvation, as well as enzymatically-induced cross-linking have been developed to fabricate milk proteins' nano- and microparticles for various applications (Rosenberg and Lee, 2004; Gunasekaran *et al.*, 2007). The usage of whey protein-based gels can be advantageous for triggered release of bioactives, for instance by causing modifications of the gel structure once reaching particular environments during gastrointestinal transit. For example, whey protein-based microspheres incorporating micronized calcium alginate may be activated only at the acidic pH of the stomach (Rosenberg and Lee, 2004).

A number of studies have been reported on the interactions of β-lactoglobulin with physiologically relevant molecules. Often the interactions have been quantified using fluorescence techniques (Liang and Subirade, 2010; Augustin *et al.*, 2011). The benefits of the presence of a complex have been shown, for example, in the case of folic acid and β-lactoglobulin, whereby folic acid binds to a hydrophobic pocket of the protein in a groove between the α-helix and the β-barrel and the complex improves the photostability of the bioactive molecule against UV radiation (Liang and Subirade, 2010). Retinol has also been shown to bind in the calyx cavity of β-lactoglobulin causing quenching of the intrinsic fluorescence of tryptophan amino acids (Cho *et al.*, 1994). On the other hand, binding studies with resveratrol, a water soluble molecule, showed that this polyphenol associates with the surface of β-lactoglobulin (Liang *et al.*, 2008).

Heat-treated β-lactoglobulin has been employed to form nanoparticles (<50 nm) containing tea polyphenols. These nanoparticles can protect water-sensitive compounds, and, in the case of the tea polyphenols, the encapsulation significantly suppresses bitterness and astringency of the mixture (Shpigelman *et al.*, 2010). Such small particles could be used in clear beverage products. Whey proteins also form coacervates with polysaccharides, and these more complex particles could also be used as encapsulating agents (Aberkane *et al.*, 2012). For example, β-lactoglobulin, in combination with pectin, can form aggregates that carry hydrophobic molecules such as omega-3 fatty acids (Zimet and Livney, 2009) or polyphenols (Liang *et al.*, 2008; Shpigelman *et al.*, 2010). Lactoferrin also has been shown to form different types of nanoparticles depending on temperature, pH, or the presence of polysaccharides (Peinado *et al.*, 2010). These nanoparticles have been employed to stabilize oil in water emulsions (Shimoni *et al.*, 2013). Whey protein nanoparticles also can be prepared using desolvation methods, using ethanol, and can encapsulate hydrophilic molecules or appreciable concentrations of minerals. These systems are unique as they are stable under acidic conditions (Gülseren *et al.*, 2012a, b).

Whey proteins can form fibrils under specific environmental and processing conditions (Loveday *et al.*, 2009). Those structures can be formed by heating at low pH or under static high pressure processing. It has been shown that heat-induced β-lactoglobulin fibrils are fully digested by pepsin (Bateman *et al.*, 2010). Nanotubes are also formed with α-lactalbumin, for example, during the partial hydrolysis by a protease extracted from *Bacillus licheniformis*. These nanotubes consist of heterogeneous self-assembled structures with molar masses of approximately 11 kDa (de Kruif, 2007; Graveland-Bikker *et al.*, 2009).

In addition to whey proteins, casein micelles are recognized as natural delivery vehicles. The caseins are assembled in nature as highly hydrated particles, and their structure can be modulated by changes in environmental conditions and processing (Dalgleish and Corredig, 2012). In recent years, the potential to capitalize on the ability of casein proteins to bind bioactive molecules has been extensively studied. For example, caseinate-resveratrol complexes can be used to overcome the limitations related to the low solubility of resveratrol in water (Chen, 2010). Another important bioactive polyphenol, curcumin, has been studied in relation to its interactions with isolated caseins (Sneharani *et al.*, 2009). In addition to the association of curcumin with casein monomers (sodium caseinate or isolated caseins) (Rahimi Yazdi and Corredig, 2012), curcumin has also been shown to penetrate the core of the casein micelles (Sahu *et al.*, 2008; Rahimi Yazdi and Corredig, 2012), and its binding behavior is very similar to that with casein micelles after their dissociation with calcium chelating agents (Sahu *et al.*, 2008).

The entrapment of hydrophobic molecules within casein micelles has also been demonstrated using vitamin D_2 (Semo *et al.*, 2007). The binding caused no changes in the size and morphology of the casein micelles, as measured by dynamic light scattering or electron microscopy. More importantly, the incorporation of these bioactive molecules in the casein micelle seems to be beneficial for the bioefficacy of the molecules. In the case of curcumin (Sahu *et al.*, 2008) higher solubility can be achieved, and in the case of vitamin D_2, complex formation has a protective effect against photochemical degradation (Semo *et al.*, 2007). However, it has also been suggested that, due to the high affinity with phenolic compounds, caseins may bind with them in the lumen during digestion, perhaps decreasing their bioaccessibility (Alexandropoulou *et al.*, 2006). This has been disputed by current studies demonstrating similar uptake of complexed and free polyphenols using cell culture models (Sahu *et al.*, 2008; Elzoghby *et al.*, 2011). The binding of caseins to these compounds, however, may affect the processing functionality of the casein micelles, and, in turn, processing may affect the association of polyphenols with the proteins. Heating and static high pressure have been shown to change the affinity of the caseins for these bioactive molecules (Rahimi Yazdi and Corredig, 2012; Rahimi Yazdi *et al.*, 2013). In the case of heat-treated milk, the whey proteins associated with the surface of casein micelles significantly increased the capacity of milk proteins to bind curcumin (Rahimi Yazdi and Corredig, 2012).

Cell culture models have been used to evaluate bioavailability, metabolism, and biological activity. The encapsulation of curcumin in casein micelles does not impair the cytotoxic effects on HeLa cells compared to an equal dose of free curcumin (Sahu *et al.*, 2008). The ability of curcumin to remain dispersed in solution increases substantially when encapsulated with casein proteins: for example, an increase in solubility for curcumin of 2500-fold occurs when associated with β-casein nanoparticles (Esmaili *et al.*, 2011). Furthermore, the bioavailability assessed using a human leukemia cell line showed an enhanced cytotoxicity of curcumin encapsulated in β-casein micelles (Esmaili *et al.*, 2011).

Casein micelles were also employed to encapsulate green tea catechins (Shukla *et al.*, 2009; Haratifar and Corredig, 2013). Casein micelles offer protection of tea catechins, and the bioefficacy was well preserved when assessed using HT-29 adenocarcinoma cells (Haratifar, 2012; Guri *et al.*, 2013). The authors suggested that tea polyphenols encapsulated within casein micelles do not alter the breakdown of casein proteins during *in vitro* digestion. The nanoencapsulated epigallocatechin gallate (EGCG) retains its bioefficacy confirming casein micelles as a proper delivery vehicle for polyphenols (Guri *et al.*, 2013).

Milk Phospholipid Liposomes

Only recently, because of an increased availability of commercial fractions of milk phospholipids, has it been possible to study the physicochemical and delivery properties of milk phospholipid liposomes (Thompson and Singh, 2006; Thompson *et al.*, 2009; Farhang *et al.*, 2012). Liposomes are often employed as delivery systems in the pharmaceutical and cosmetic industry, and are well regarded by consumers because of the recognized bioactivity (Thompson and Singh, 2006; Thompson *et al.*, 2009) and biocompatibility of phospholipids (Spitsberg, 2005).

Liposomes prepared with milk phospholipids show different physico-chemical properties compared to those prepared with soy phospholipids (Thompson and Singh, 2006). In particular, the membrane of milk liposomes is thicker and less permeable than that of soy-derived liposomes (Thompson and Singh, 2006). They also show higher entrapment efficiencies for β-carotene and potassium chromate (hydrophobic and hydrophilic tracer molecules, respectively) than those measured for soy liposomes (Thompson and Singh, 2006; Thompson *et al.*, 2009). Hence, it could be hypothesized that these vesicles, most probably also present in nature, are well-suited delivery systems for both hydrophobic and hydrophilic compounds in milk.

Small phospholipid vesicles have demonstrated controlled release properties. Soy liposomes have been shown to be effective in protecting enzymes during processing and can be used, for example, to encapsulate β-galactosidase and aid in the digestion of lactose in dairy products (Rodriguez-Nogales and Lopez, 2006; Hermida *et al.*, 2009). However, less work has been carried out with milk phospholipid liposomes. Their degradation during gastrointestinal transit has been reported only recently (Liu *et al.*, 2012, 2013), and it is clear that a better understanding of the physicochemical properties of these naturally occurring vesicles in various environments may initiate the development of more sophisticated designs of dairy products.

Liposomes can entrap enzymes and proteins that would be otherwise digested in the gastric phase. The entrapped enzymes may be then released in the upper intestine, because of the decreased stability of the liposomes in the presence of bile salts, and specific lipases. For example, it has been shown that β-galactosidase in soy liposomes can be released in the upper intestine and could be employed to aid in lactose digestion (Rodriguez-Nogales and Lopez, 2006). However, it is important to point out that the composition of the bilayer and the physical properties of the liposomes may affect their pH stability and their permeability during digestion (Rowland and Woodley, 1980). For example, the difference between a fluid and a solid membrane may affect not only colloidal stability and encapsulation, but also pH permeation. Hence, liposomes may be colloidally stable during *in vitro* digestion, and retain a high ratio of the entrapped bioactive, but they may not be able to protect acid-sensitive bioactives during gastrointestinal transit (Rowland and Woodley, 1980; Hermida *et al.*, 2009).

Recent work on *in vitro* digestion of vesicles prepared with milk phospholipids demonstrated that these liposomes were little affected by simulated gastric fluids containing pepsin, and protected lactoferrin from gastric digestion (Liu *et al.*, 2013). The liposomes made with milk phospholipids were more stable to duodenal digestion than soybean-based liposomes, showing less change in their average diameter, surface charge, and free fatty acid release (Liu *et al.*, 2012). These results may suggest, once more, that milk phospholipid vesicles in milk may aid in the delivery of bioactives to the intestinal cells.

CONCLUSIONS AND OUTLOOK

The structure of dairy products is quite complex, containing multiple structural features, such as polymeric and particulate gel structures, phase-separated domains, emulsion droplets, lipid vesicles, and air bubbles. At a molecular level, the same component may be present in multiple structural forms, with differences in susceptibility to hydrolysis during digestion. Supramolecular structures are of great importance in dairy matrices, and this work has attempted to describe some of Nature's ways of providing multiple biological functions within a structure. Milk is a great example of how the matrix embeds a number of delivery structures that will be released during digestion, to serve multiple functionalities.

In the past, research efforts have focused on how to create different structures in dairy products, modulating the properties and assembling behavior of the various molecules present in the matrix. We are now at a turning point, as increasing evidence is brought forward that the organization, the physical state of the components, and the structure of the matrix will affect digestion, absorption, and metabolism of nutrients. This will certainly modify the scope of food material science research. The disruption, disassembly, and reassembly of the structures during gastrointestinal processing are fundamental aspects related to the delivery of health benefits using food matrices (Marciani *et al.*, 2007; Golding *et al.*, 2011; Michalski *et al.*, 2013). A better understanding of these processes will result in the development of novel formulations.

The delivery of bioactive molecules through food is particularly challenging, as the bioactives may be prone to degradation, may react with other ingredients, and may compromise the sensory properties of the food. In addition, the delivery of the bioactives will be affected by the structure of the food matrix. Hence, we will need a combination of novel processing approaches, tailored to specific biological functionality, and new understanding of material science and how material science may affect human sensorial and physiological functions. So, we may expect in the future that biology, nutrition, and material sciences will all be taken into consideration in the design of dairy products, and food material science will be specifically employed to predict the disruption and recombination of the building blocks in the food matrix, to ensure delivery of biological functionality during digestion.

REFERENCES

Aberkane, L., Jasniewski, J., Gaiani, C., Hussain, R., Scher, J., Sanchez, C., 2012. Structuration mechanism of β-lactoglobulin—acacia gum assemblies in presence of quercetin. Food Hydrocoll. 29, 9—20.

Alexandropoulou, I., Komaitis, M., Kapsokefalou, M., 2006. Effects of iron, ascorbate, meat and casein on the antioxidant capacity of green tea under conditions of in vitro digestion. Food Chem. 94, 359—365.

Almaas, H., Cases, A.L., Devold, T.G., Holm, H., Langsrud, T., Aabakken, L., et al., 2006. In vitro digestion of bovine and caprine milk by human gastric and duodenal enzymes. Int. Dairy J. 16, 961—968.

Argov, N., Lemay, D.G., German, J.B., 2008. Milk fat globule structure and function: nanoscience comes to milk production. Trends Food Sci. Technol. 19, 617—623.

Argyri, K., Miller, D.D., Glahn, R.P., Zhu, L., Kapsokefalou, M., 2007. Peptides isolated from in vitro digests of milk enhance iron uptake by Caco-2 cells. J. Agric. Food Chem. 55, 10221—10225.

Argyri, K., Tako, E., Miller, D.D., Glahn, R.P., Komaitis, M., Kapsokefalou, M., 2009. Milk peptides increase iron dialyzability in water but do not affect DMT-1 expression in Caco-2 cells. J. Agric. Food Chem. 57, 1538—1543.

Armand, M., Pasquier, B., Andre, M., Borel, P., Senft, M., Peyrot, J., et al., 1999. Digestion and absorption of two fat emulsions with different droplet sizes in the human digestive tract. Am. J. Clin. Nutr. 70, 1096—1106.

Arroyo-Maya, I.J., Rodiles-López, J.O., Cornejo-Mazón, M., Gutiérrez-López, G.F., Hernández-Arana, A., Toledo-Núñez, C., et al., 2012. Effect of different treatments on the ability of α-lactalbumin to form nanoparticles. J. Dairy Sci. 95, 6204—6214.

Augustin, M.A., Abeywardena, M.Y., Patten, G., Head, R., Lockett, T., Luca, A.D., Sanguansri, L., 2011. Effects of microencapsulation on the gastrointestinal transit and tissue distribution of a bioactive mixture of fish oil, tributyrin and resveratrol. J. Funct. Foods 3, 25—37.

Barbe, F., Menard, O., Le Gouar, Y., Buffiere, C., Famelart, M., Laroche, B., et al., 2013. The heat treatment and the gelation are strong determinants of the kinetics of milk proteins digestion and of the peripheral availability of amino acids. Food Chem. 136, 1203—1212.

Bateman, L., Ye, A., Singh, H., 2010. In vitro digestion of β-lactoglobulin fibrils formed by heat treatment at low pH. J. Agric. Food Chem. 58, 9800—9808.

Berton, A., Rouvellac, S., Robert, B., Rousseau, F., Lopez, C., Crenon, I., 2012. Effect of the size and interface composition of milk fat globules on their in vitro digestion by the human pancreatic lipase: native versus homogenized milk fat globules. Food Hydrocoll. 29, 123—134.

Boirie, Y., Dangin, M., Gachon, P., Vasson, M.P., Maubois, J.L., Beaufrere, B., 1997. Slow and fast dietary proteins differently modulate postprandial protein accretion. Proc. Natl. Acad. Sci. USA 94, 14930—14935.

Boutrou, R., Gaudichon, C., Dupont, D., Jardin, J., Airinei, G., Marsset-Baglieri, A., et al., 2013. Sequential release of milk protein-derived bioactive peptides in the jejunum in healthy humans. Am. J. Clin. Nutr. 97, 1314—1323.

Bouzerzour, K., Morgan, F., Cuinet, I., Bonhomme, C., Jardin, J., Le Huerou-Luron, I., Dupont, D., 2012. In vivo digestion of infant formula in piglets: protein digestion kinetics and release of bioactive peptides. Br. J. Nutr. 108, 2105—2114.

Bowen, J., Noakes, M., Clifton, P., 2008. Role of protein and carbohydrate sources on acute appetite responses in lean and overweight men. Nutr. Diet. 65, S71—S78.

Burling, H., Graverholt, G., 2008. Milk—a new source for bioactive phospholipids for use in food formulations. Lipid Technol. 20, 229—231.

Cavaletto, M., Giuffrida, M.G., Conti, A., 2008. Milk fat globule membrane components—a proteomic approach. Bioactive Components of Milk 606, 129—141.

Chen, C., 2010. Casein complexes. US Patent 2010/0099607 A1.

Cho, Y.J., Batt, C.A., Sawyer, L., 1994. Probing the retinol-binding site of bovine beta-lactoglobulin. J. Biol. Chem. 269, 11102—11107.

Clare, D., Catignani, G., Swaisgood, H., 2003. Biodefense properties of milk: the role of antimicrobial proteins and peptides. Curr. Pharm. Design 9, 1239—1255.

Clare, D.A., Swaisgood, H.E., 2000. Bioactive milk peptides: a prospectus. J. Dairy Sci. 83, 1187—1195.

Dalgleish, D.G., Corredig, M., 2012. The structure of the casein micelle of milk and its changes during processing. Annu. Rev. Food Sci. Technol. 3, 449—467.

de Kruif, C.G., 2007. AGFD 34-Milk protein nanotubes: formation, structure and stability of α-lactalbumin nanotubes for application in food and non-food systems. Abstr. Pap. Am. Chem. Soc. 233, 140.

Dewettinck, K., Rombaut, R., Thienpont, N., Le, T.T., Messens, K., Van Camp, J., 2008a. Nutritional and technological aspects of milk fat globule membrane material. Int. Dairy J. 18, 436–457.

Dewettinck, K., Rombaut, R., Thienpont, N., Le, T.T., Messens, K., Van Camp, J., 2008b. Nutritional and technological aspects of milk fat globule membrane material. Int. Dairy J. 18, 436–457.

Doublier, J.L., Garnier, C., Renard, D., Sanchez, C., 2000. Protein-polysaccharide interactions. Curr. Opin. Colloid Interface Sci. 5, 202–214.

Dupont, D., Mandalari, G., Molle, D., Jardin, J., Rolet-Repecaud, O., Duboz, G., et al., 2010. Food processing increases casein resistance to simulated infant digestion. Mol. Nutr. Food Res. 54, 1677–1689.

Ehlers, P.I., Kivimaki, A.S., Turpeinen, A.M., Korpela, R., Vapaatalo, H., 2011. High blood pressure-lowering and vasoprotective effects of milk products in experimental hypertension. Br. J. Nutr. 106, 1353–1363.

Elzoghby, A.O., Abo El-Fotoh, W.S., Elgindy, N.A., 2011. Casein-based formulations as promising controlled release drug delivery systems. J. Controlled Release 153, 206–216.

Esmaili, M., Ghaffari, S.M., Moosavi-Movahedi, Z., Atri, M.S., Sharifizadeh, A., Farhadi, M., et al., 2011. Beta casein-micelle as a nano vehicle for solubility enhancement of curcumin; food industry application. LWT—Food Sci. Technol. 44, 2166–2172.

Etzel, M.R., 2004. Manufacture and use of dairy protein fractions. J. Nutr. 134, 996S–1002S.

Evers, J.M., Haverkamp, R.G., Holroyd, S.E., Jameson, G.B., Mackenzie, D.D.S., McCarthy, O.J., 2008. Heterogeneity of milk fat globule membrane structure and composition as observed using fluorescence microscopy techniques. Int. Dairy J. 18, 1081–1089.

Farhang, B., Corredig, M., 2011. Milk phospholipids: a nanocarrier system for delivery of bioactive compounds. In: Ahmad, M. (Ed.), Lipids in Nanotechnology, pp. 53–68. AOCS.

Farhang, B., Kakuda, Y., Corredig, M., 2012. Encapsulation of ascorbic acid in liposomes prepared with milk fat globule membrane-derived phospholipids. Dairy Sci. Technol. 92, 353–366.

Farrell Jr. H., Jimenez-Flores, R., Bleck, G., Brown, E., Butler, J., Creamer, L., et al., 2004. Nomenclature of the proteins of cows' milk—sixth revision. J. Dairy Sci. 87, 1641–1674.

FitzGerald, R.J., Murray, B.A., Walsh, D.J., 2004. Hypotensive peptides from milk proteins. J. Nutr. 134, 980S–988S.

Fong, B.Y., Norris, C.S., 2009. Quantification of milk fat globule membrane proteins using selected reaction monitoring mass spectrometry. J. Agric. Food Chem. 57, 6021–6028.

Fox, P.F., Brodkorb, A., 2008. The casein micelle: historical aspects, current concepts and significance. Int. Dairy J. 18, 677–684.

Garcia-Nebot, M.J., Alegria, A., Barbera, R., Clemente, G., Romero, F., 2009. Does the addition of caseinophosphopeptides or milk improve zinc in vitro bioavailability in fruit beverages? Food Res. Int. 42, 1475–1482.

German, J.B., Dillard, C.J., Ward, R.E., 2002. Bioactive components in milk. Curr. Opin. Clin. Nutr. Metab. Care 5, 653–658.

Golding, M., Wooster, T.J., Day, L., Xu, M., Lundin, L., Keogh, J., Clifton, P., 2011. Impact of gastric structuring on the lipolysis of emulsified lipids. Soft Matter 7, 3513–3523.

Graveland-Bikker, J.F., Koning, R.I., Koerten, H.K., Geels, R.B.J., Heeren, R.M.A., de Kruif, C.G., 2009. Structural characterization of alpha-lactalbumin nanotubes. Soft Matter 5, 2020–2026.

Gülseren, I., Fang, Y., Corredig, M., 2012a. Whey protein nanoparticles prepared with desolvation with ethanol: characterization, thermal stability and interfacial behavior. Food Hydrocoll. 29, 258–264.

Gülseren, I., Fang, Y., Corredig, M., 2012b. Zinc incorporation capacity of whey protein nanoparticles prepared with desolvation with ethanol. Food Chem. 135, 770–774.

Guillaume, B., Hannah, F.P., John, P.S., Rafael, J., 2010. Characterization of Lactobacillus reuteri interaction with milk fat globule membrane components in dairy products. J. Agric. Food Chem. 58, 5612–5619.

Gunasekaran, S., Ko, S., Xiao, L., 2007. Use of whey proteins for encapsulation and controlled delivery applications. J. Food Eng. 83, 31–40.

Guri, A., Griffiths, M., Khursigara, C.M., Corredig, M., 2012. The effect of milk fat globules on adherence and internalization of Salmonella Enteritidis to HT-29 cells. J. Dairy Sci. 95, 6937–6945.

Guri, A., Haratifar, S., Corredig, M., 2013. Bioavailability of tea catechins encapsulated in milk caseins during different digestion models. Proc. 2nd Int. Conf. on Food Digestion 2, 44.

Haham, M., Ish-Shalom, S., Nodelman, M., Duek, I., Segal, E., Kustanovich, M., Livney, Y.D., 2012. Stability and bioavailability of vitamin D nanoencapsulated in casein micelles. Food Funct. 3, 737–744.

Hall, W.L., Millward, D.J., Rogers, P.J., Morgan, L.M., 2003. Physiological mechanisms mediating aspartame-induced satiety. Physiol. Behav. 78, 557–562.

Hancock, J.T., Salisbury, V., Ovejero-Boglione, M.C., Cherry, R., Hoare, C., Eisenthal, R., Harrison, R., 2002. Antimicrobial properties of milk: dependence on presence of xanthine oxidase and nitrite. Antimicrob. Agents Chemother. 46, 3308–3310.

Haque, E., Chand, R., Kapila, S., 2009. Biofunctional properties of bioactive peptides of milk origin. Food Rev. Int. 25, 28–43.

Haratifar, S., 2012. Nanoencapsulation of tea catechins in casein micelles: effects on processing and biological functionalities. PhD thesis. University of Guelph.

Haratifar, S., Corredig, M., 2014. Interactions between tea catechins and casein micelles affect renneting. Food Chem. 143, 27–32.

Hartmann, R., Meisel, H., 2007. Food-derived peptides with biological activity: from research to food applications. Antimicrob. Agents Chemother. 18, 163–169.

Hebrard, G., Hoffart, V., Cardot, J., Subirade, M., Beyssac, E., 2013. Development and characterization of coated-microparticles based on whey protein/alginate using the Encapsulator device. Drug Devel. Indus. Pharm. 39, 128–137.

Hermida, L.G., Sabes-Xamani, M., Barnadas-Rodriguez, R., 2009. Combined strategies for liposome characterization during in vitro digestion. J. Liposome Res. 19, 207–219.

Holt, C., 2004. An equilibrium thermodynamic model of the sequestration of calcium phosphate by casein micelles and its application to the calculation of the partition of salts in milk. Eur. Biophys. J. Biophys. Lett. 33, 421–434.

Jimenez-Flores, R., Brisson, G., 2008. The milk fat globule membrane as an ingredient: why, how, when? Dairy Sci. Technol. 88, 5–18.

Karlsson, K.A., 1989. Animal glycosphingolipids as membrane attachment sites for bacteria. Annu. Rev. Biochem. 58, 309–350.

Keenan, T., Dylewski, D., 1994. Intracellular origin of milk fat globules and the nature and the structure of the milk lipid globule membrane. In: Fox, P.F. (Ed.), Advanced Dairy Chemistry-2—Lipids. Chapman and Hall, London, pp. 89–130.

Kitts, D.D., Weiler, K., 2003. Bioactive proteins and peptides from food sources. Applications of bioprocesses used in isolation and recovery. Curr. Pharm. Des. 9, 1309–1323.

Korhonen, H., 1977. Antimicrobial factors in bovine colostrum [dairy cattle, milk and cream]. J. Sci. Agric. Soc. Finland 49, 434–447.

Korhonen, H., Pihlanto, A., 2003. Bioactive peptides: novel applications for milk proteins. Appl. Biotechnol. Food Sci. Policy 1, 133–144.

Korhonen, H., Pihlanto, A., 2006. Bioactive peptides: production and functionality. Int. Dairy J. 16, 945–960.

Korhonen, H., Pihlanto, A., 2007. Technological options for the production of health-promoting proteins and peptides derived from milk and colostrum. Curr. Pharm. Des. 13, 829–843.

Le Feunteun, S., Barbé, F., Rémond, D., Ménard, O., Le Gouar, Y., Dupont, D., Laroche, B., 2014. Impact of the dairy matrix structure on milk protein digestion kinetics: mechanistic modelling based on mini-pig *in vivo* data. Food Bioprocess Technol. 7, 1099–1113.

Li, J., Dalgleish, D.G., 2006. Controlled proteolysis and the properties of milk gels. J. Agric. Food Chem. 54, 4687–4695.

Liang, L., Subirade, M., 2010. β-Lactoglobulin/folic acid complexes: formation, characterization, and biological implication. J. Phys. Chem. B. 114, 6707–6712.

Liang, L., Tajmir-Riahi, H.A., Subirade, M., 2008. Interaction of β-lactoglobulin with resveratrol and its biological implications. Biomacromolecules 9, 50–56.

Liu, W., Ye, A., Liu, C., Liu, W., Singh, H., 2012. Structure and integrity of liposomes prepared from milk- or soybean-derived phospholipids during in vitro digestion. Food Res. Int. 48, 499–506.

Liu, W., Ye, A., Liu, W., Liu, C., Singh, H., 2013. Stability during in vitro digestion of lactoferrin-loaded liposomes prepared from milk fat globule membrane-derived phospholipids. J. Dairy Sci. 96, 2061–2070.

Livney, Y.D., 2010. Milk proteins as vehicles for bioactives. Curr. Opin. Colloid Interface Sci. 15, 73–83.

Lopez-Exposito, I., Quiros, A., Amigo, L., Recio, I., 2007. Casein hydrolysates as a source of antimicrobial, antioxidant and antihypertensive peptides. Lait 87, 241–249.

Loveday, S.M., Rao, M.A., Creamer, L.K., Singh, H., 2009. Factors affecting rheological characteristics of fibril gels: the case of β-lactoglobulin and α-lactalbumin. J. Food Sci. 74, R47–R55.

Macierzanka, A., Rigby, N.M., Corfield, A.P., Wellner, N., Böttger, F., Mills, E.N.C., Mackie, A.R., 2011. Adsorption of bile salts to particles allows penetration of intestinal mucus. Soft Matter 7, 8077–8084.

Marciani, L., Wickham, M., Singh, G., Bush, D., Pick, B., Cox, E., et al., 2007. Enhancement of intragastric acid stability of a fat emulsion meal delays gastric emptying and increases cholecystokinin release and gallbladder contraction. Am. J. Physiol.—Gastrointest. Liver Physiol. 292, G1607–G1613.

Martin, A.H., de Jong, G.A.H., 2012. Impact of protein pre-treatment conditions on the iron encapsulation efficiency of whey protein cold-set gel particles. Eur. Food Res. Technol. 234, 995–1003.

Mehravar, R., Jahanshahi, M., Saghatoleslami, N., 2009. Production of biological nanoparticles from α-lactalbumin for drug delivery and food science application. African J. Biotechnol. 8, 6822–6827.

Mercier, A., Gauthier, S.F., Fliss, I., 2004. Immunomodulating effects of whey proteins and their enzymatic digests. Int. Dairy J. 14, 175–183.

Michalski, M.C., Genot, C., Gayet, C., Lopez, C., Fine, F., Joffre, F., et al., 2013. Multiscale structures of lipids in foods as parameters affecting fatty acid bioavailability and lipid metabolism. Prog. Lipid Res. 52, 354–373.

Mills, S., Ross, R.P., Hill, C., Fitzgerald, G.F., Stanton, C., 2011. Milk intelligence: mining milk for bioactive substances associated with human health. Int. Dairy J. 21, 377–401.

Morales, A., Colell, A., Mari, M., Garcia-Ruiz, C., Fernandez-Checa, J.C., 2004. Glycosphingolipids and mitochondria: Role in apoptosis and disease. Glycoconjugate J. 20, 579–588.

Naito, H., Suzuki, H., 1974. Further evidence for the formation in vivo of phosphopeptide in the intestinal lumen from dietary beta-casein. Agric. Biol. Chem. 38, 1543–1545.

O'Connell, J.E., Fox, P.D., Tan-Kintia, R., Fox, P.F., 1998. Effects of tea, coffee and cocoa extracts on the colloidal stability of milk and concentrated milk. Int. Dairy J. 8, 689–693.

Ono, T., Ohotawa, T., Takagi, Y., 1994. Complexes of casein phosphopeptide and calcium-phosphate prepared from casein micelles by tryptic digestion. Biosci. Biotechnol. Biochem. 58, 1376–1380.

Pakkanen, R., Aalto, J., 1997. Growth factors and antimicrobial factors of bovine colostrum. Int. Dairy J. 7, 285–297.

Parodi, P.W., 1997. Cows' milk fat components as potential anticarcinogenic agents. J. Nutr. 127, 1055–1060.

Peguet-Navarro, J., Sportouch, M., Popa, I., Berthier, O., Schmitt, D., Portoukalian, J., 2003. Gangliosides from human melanoma tumors impair dendritic cell differentiation from monocytes and induce their apoptosis. J. Immunol. 170, 3488–3494.

Peinado, I., Lesmes, U., Andres, A., McClements, D.J., 2010. Fabrication and morphological characterization of biopolymer particles formed by electrostatic complexation of heat treated lactoferrin and anionic polysaccharides. Langmuir 26, 9827–9834.

Phan, C.T., Tso, P., 2001. Intestinal lipid absorption and transport. Frontiers Biosci. 6, D299–D319.

Pihlanto, A., Korhonen, H., 2003. Bioactive peptides and proteins. Adv. Food Nutr. Res. 47, 175–276.

Power, O., Jakeman, P., FitzGerald, R.J., 2013. Antioxidative peptides: enzymatic production, in vitro and in vivo antioxidant activity and potential applications of milk-derived antioxidative peptides. Amino Acids 44, 797–820.

Rahimi Yazdi, S., Corredig, M., 2012. Heating of milk alters the binding of curcumin to casein micelles. A fluorescence spectroscopy study. Food Chem. 132, 1143–1149.

Rahimi Yazdi, S., Bonomi, F., Iametti, S., Miriani, M., Brutti, A., Corredig, M., 2014. Binding of curcumin to casein micelles increases after static high pressure treatment of skim milk. J. Dairy Res. 80, 152−158.

Rioux, L.E., Turgeon, S.L., 2012. The ratio of casein to whey protein impacts yogurt digestion in vitro. Food Digestion 3, 25−35.

Rodriguez-Nogales, J.M., Lopez, A.D., 2006. A novel approach to develop beta-galactosidase entrapped in liposomes in order to prevent an immediate hydrolysis of lactose in milk. Int. Dairy J. 16, 354−360.

Rombaut, R., Dejonckheere, V., Dewettinck, K., 2006. Microfiltration of butter serum upon casein micelle destabilization. J. Dairy Sci. 89, 1915−1925.

Rosenberg, M., Lee, S.J., 2004. Calcium-alginate coated, whey protein-based microspheres: preparation, some properties and opportunities. J. Microencapsulation 21, 263−281.

Rowland, R.N., Woodley, J.F., 1980. The stability of liposomes *in vitro* to pH, bile-salts and pancreatic lipase. Biochim. Biophys. Acta 620, 400−409.

Rutherfurd, S.M., Moughan, P.J., 2005. Digestible reactive lysine in selected milk-based products. J. Dairy Sci. 88, 40−48.

Sahu, A., Kasoju, N., Bora, U., 2008. Fluorescence study of the curcumin-casein micelle complexation and its application as a drug nanocarrier to cancer cells. Biomacromolecules 9, 2905−2912.

Saint-Sauveur, D., Gauthier, S.F., Boutin, Y., Montoni, A., 2008. Immunomodulating properties of a whey protein isolate, its enzymatic digest and peptide fractions. Int. Dairy J. 18, 260−270.

Sanchez-Juanes, F., Alonso, J.M., Zancada, L., Hueso, P., 2009. Distribution and fatty acid content of phospholipids from bovine milk and bovine milk fat globule membranes. Int. Dairy J. 19, 273−278.

Scammell, A., 2001. Production and uses of colostrum. Aust. J. Dairy Technol 56, 74−82.

Schroten, H., Hanisch, F.G., Plogmann, R., Hacker, J., Uhlenbruck, G., Nobis-Bosch, R., Wahn, V., 1992. Inhibition of adhesion of S-fimbriated Escherichia coli to buccal epithelial cells by human milk fat globule membrane components: a novel aspect of the protective function of mucins in the non-immunoglobulin fraction. Infect. Immun 60, 2893−2899.

Semo, E., Kesselman, E., Danino, D., Livney, Y.D., 2007. Casein micelle as a natural nano-capsular vehicle for nutraceuticals. Food Hydrocoll. 21, 936−942.

Shimoni, G., Shani Levi, C., Levi Tal, S., Lesmes, U., 2013. Emulsions stabilization by lactoferrin nano-particles under in vitro digestion conditions. Food Hydrocoll. 33, 264−272.

Shpigelman, A., Israeli, G., Livney, Y.D., 2010. Thermally-induced protein-polyphenol co-assemblies: beta lactoglobulin-based nanocomplexes as protective nanovehicles for EGCG. Food Hydrocoll. 24, 735−743.

Shukla, A., Narayanan, T., Zanchi, D., 2009. Structure of casein micelles and their complexation with tannins. Soft Matter 5, 2884−2888.

Sneharani, A.H., Singh, S.A., Rao, A.G.A., 2009. Interaction of α_{s1}-casein with curcumin and its biological implications. J. Agric. Food Chem. 57, 10386−10391.

Snow, D.R., Ward, R.E., Olsen, A., Jimenez-Flores, R., Hintze, K.J., 2011. Membrane-rich milk fat diet provides protection against gastrointestinal leakiness in mice treated with lipopolysaccharide. J. Dairy Sci. 94, 2201−2212.

Snow, D.R., Jimenez-Flores, R., Ward, R.E., Cambell, J., Young, M.J., Nemere, I., Hintze, K.J., 2010. Dietary milk fat globule membrane reduces the incidence of aberrant crypt foci in Fischer-344 rats. J. Agric. Food Chem. 58, 2157—2163.

Spitsberg, V.L., 2005. Invited review: bovine milk fat globule membrane as a potential nutraceutical. J. Dairy Sci. 88, 2289—2294.

Swart, P.J., Kuipers, E.M., Smit, C., Van Der Strate, B., Harmsen, M.C., Meijer, D., 1998. Antiviral activity of lactoferrin. Adv. Exp. Med. Biol. 443, 205—213.

Tellez, A., Corredig, M., Guri, A., Zanabria, R., Griffiths, M.W., Delcenserie, V., 2012. Bovine milk fat globule membrane affects virulence expression in Escherichia coli O157:H7. J. Dairy Sci. 95, 6313—6319.

Tellez, A., Corredig, M., Turner, P.V., Morales, R., Griffiths, M., 2011. A peptidic fraction from milk fermented with Lactobacillus helveticus protects mice against Salmonella infection. Int. Dairy J. 21, 607—614.

Thompson, A.K., Singh, H., 2006. Preparation of liposomes from milk fat globule membrane phospholipids using a microfluidizer. J. Dairy Sci. 89, 410—419.

Thompson, A.K., Couchoud, A., Singh, H., 2009. Comparison of hydrophobic and hydrophilic encapsulation using liposomes prepared from milk fat globule-derived phospholipids and soya phospholipids. Dairy Sci. Technol. 89, 99—113.

Tomita, M., Wakabayashi, H., Yamauchi, K., Teraguchi, S., Hayasawa, H., 2002. Bovine lactoferrin and lactoferricin derived from milk: production and applications. Biochem. Cell Biol. 80, 109—112.

Turgeon, S.L., Rioux, L.E., 2011. Food matrix impact on macronutrients nutritional properties. Food Hydrocoll. 25, 1915—1924.

Veldhorst, M., Smeets, A., Soenen, S., Hochstenbach-Waelen, A., Hursel, R., Diepvens, K., et al., 2008. Protein-induced satiety: effects and mechanisms of different proteins. Physiol. Behav. 94, 300—307.

Wang, Q.W., Allen, J.C., Swaisgood, H.E., 1997. Binding of vitamin D and cholesterol to beta-lactoglobulin. J. Dairy Sci. 80, 1054—1059.

Wang, X., Hirmo, S., Willen, R., Wadstrom, T., 2001. Inhibition of Helicobacter pylori infection by bovine milk glycoconjugates in a BALB/cA mouse model. J. Med. Microbiol. 50, 430—435.

Ward, R.E., German, J.B., 2004. Understanding milk's bioactive components: a goal for the Genomics toolbox. J. Nutr. 134, 962S—967S.

Westerterp-Plantenga, M.S., Nieuwenhuizen, A., Tome, D., Soenen, S., Westerterp, K.R., 2009. Dietary protein, weight loss, and weight maintenance. Annu. Rev. Nutr. 29, 21—41.

Zanabria, R., Tellez, A.M., Griffiths, M., Corredig, M., 2013. Milk fat globule membrane isolate induces apoptosis in HT-29 human colon cancer cells. Food Funct. 4, 222—230.

Zeinhom, M., Tellez, A.M., Delcenserie, V., El-Kholy, A.M., El-Shinawy, S.H., Griffiths, M.W., 2012. Yogurt containing bioactive molecules produced by Lactobacillus acidophilus La-5 exerts a protective effect against enterohemorrhagic Escherichia coli in mice. J. Food Prot. 75, 1796—1805.

Zimecki, M., Kruzel, M.L., 2007. Milk-derived proteins and peptides of potential therapeutic and nutritive value. J. Exp. Ther. Oncol. 6, 89—106.

Zimet, P., Livney, Y.D., 2009. Beta-lactoglobulin and its nanocomplexes with pectin as vehicles for omega-3 polyunsaturated fatty acids. Food Hydrocoll. 23, 1120—1126.

The Importance of Microbiota and Host Interactions Throughout Life

N.C. Roy,[1,2,3] S.A. Bassett,[1] W. Young,[1] C. Thum,[1,2] W.C. McNabb[2,3,4]

[1]*Food Nutrition & Health Team, Food & Bio-based Products Group, AgResearch Grasslands, Palmerston North, New Zealand,* [2]*Riddet Institute, Massey University, Palmerston North, New Zealand,* [3]*Gravida, National Centre for Growth and Development, The University of Auckland, Auckland, New Zealand,* [4]*AgResearch Grasslands, Palmerston North, New Zealand*

CONTENTS

INTRODUCTION

There has been a shift in the cause of mortality in the last two centuries. The predominant causes of death have moved from communicable diseases (Beaglehole and Bonita, 2009) to non-communicable diseases such as cancer and metabolic diseases (Beaglehole and Yach, 2003; Popkin *et al.*, 2012). This shift has been driven largely by improved sanitation and cleaner water supplies, superior hygiene practices, and the availability of antimicrobial agents (Curtis *et al.*, 2011) combined with an increase in caloric intake and reduction in physical activity (Uauy and Diaz, 2005), at least in some societies. Other lifestyle factors and genetics also play a major role in the development of non-communicable diseases (Phillips, 2013).

Over the course of the past few decades, growing scientific evidence has supported the hypothesis that interventions during critical periods of

Food Structures, Digestion and Health. http://dx.doi.org/10.1016/B978-0-12-404610-8.00018-9

development may "program" later health outcomes by altering organ/tissue structure and function in a way that predisposes the organism to diseases in later life (Barker, 1966, 1988). One of the most influential, but modifiable, factors that affects fetal growth and organ development is the quality of the maternal diet, and its intake during pregnancy and lactation (Canani *et al.*, 2011; Sebert *et al.*, 2011). Studies of fetal over- and undernourishment have demonstrated that developmental plasticity allows a range of different phenotypes to emerge from a single genotype, largely in response to the environmental cues received during pregnancy (Gluckman and Hanson, 2004). A mismatch between the environment "predicted" during the peri-natal period and the postnatal environment encountered is the likely cause for increased susceptibility of non-communicable diseases in later life (Gluckman and Hanson, 2004; Gluckman *et al.*, 2005). Most organs/tissues, including the adipose tissue, pancreas, kidney, skeletal muscle, brain, and gastrointestinal tract (GIT), have been reported to be imprinted by early disturbances (Warner and Ozanne, 2010).

The GIT plays a major role in health. It is where food and drink are digested and nutrients absorbed (and recycled) to support the body's growth, maintenance, and repair mechanisms (including the GIT's). The human GIT is home to an estimated 100 trillion microorganisms collectively referred to as the microbiota. They are thought to outnumber human cells 10-fold, and represent by far the largest microbial community associated with the human body (Mackie *et al.*, 1999). This complex ecosystem represents a huge reservoir of metabolic capability and plays a crucial role in a number of developmental and nutritional processes in the GIT. The GIT microbiota in humans is highly complex and comprises many hundreds of bacterial species (or phylotypes) (Mackie *et al.*, 1999). The density of the microbiota varies among the different parts of the GIT. The stomach and proximal small intestine contain relatively low numbers of microbes (10^3-10^5 bacteria/g or mL content), because of low pH, bile acids (Inagaki *et al.*, 2006), and rapid digesta flow in this region. The predominant bacteria in the jejunum and ileum are the acid-tolerant lactobacilli and streptococci (Mackie *et al.*, 1999). The ileum, however, maintains a more diverse microbiota and higher bacterial numbers (10^8/g or mL content) than the upper small intestine and is considered a transition zone preceding the colon (Mackie *et al.*, 1999). The colon is characterized by slow turnover, large numbers of bacteria ($10^{10}-10^{11}$/g or mL content), low redox potential, and relatively high short-chain fatty acid (SCFA) concentrations, and it is also the first site of microbial colonization (Mackie *et al.*, 1999).

Birth has been traditionally viewed as transitioning from a sterile intrauterine environment, based on rapid colonization by the maternal microbiota and ecological succession, progressing through to eventual stabilization of the complex adult microbiota (Favier *et al.*, 2002; Palmer *et al.*, 2007). The pattern of colonization of the GIT can have both short-term and long-term health effects (Bager *et al.*, 2008; Kalliomaki *et al.*, 2008; McLoughlin and Mills, 2011; Roberts *et al.*, 2011; Cho *et al.*, 2012). Bacteria in the GIT carry out a variety of functions; they provide essential nutrients such as vitamins and SCFAs, stimulate the development of the immune system, especially adaptive responses, and provide general protection against pathogen colonization (Hooper *et al.*, 2012; Nicholson *et al.*, 2012). Changes in the composition and reduced diversity of the GIT microbiota have been linked to increased risk of the onset of over 25 diseases or syndromes (de Vos and de Vos, 2012). While the microbiota in healthy adults is resilient to compositional changes (Lozupone *et al.*, 2012), the GIT core microbiota of elderly individuals (65 years old or more) is distinct from that of younger adults (Claesson *et al.*, 2011). This raises important questions about the dynamism and stability of the GIT microbiota and its interactions with the host and consequential effects on health.

This chapter discusses the evolution of the microbial communities in the GIT, and focuses mainly on interactions between the microbiota and host mucosal metabolism and how these differ throughout life from birth to old age (summarized in Figure 18.1).

MICROBIOTA AND HOST INTERACTIONS IN EARLY POSTNATAL LIFE

Newborn infants are exposed to a range of microbial challenges once they leave the intrauterine environment. The major cellular components of the GIT mucosal immune system are present at birth. It is recognized, however, that the development of the protective functions of the GIT requires stimulation by the microbiota following colonization with the maternal inoculum during passage through the birth canal (Guarner and Malagelada, 2003; Forchielli and Walker, 2005). Initially, the microbiota of the neonatal GIT is predominantly composed of facultative anaerobes such as *Escherichia coli* and *Streptococci* sp. (Hooper, 2004). A linear increase in microbiota diversity has been reported from birth to weaning (Koenig *et al.*, 2011). In contrast, Pantoja-Feliciano *et al.* (2013) reported a biphasic microbial colonization of the GIT, which is probably linked to the transition of nutrition from breast milk to solid foods. They showed that newborn mice (as a commonly used model to study changes in the GIT microbiota) acquired a colonic microbiota that resembled maternal vaginal communities, but at 3 and 9 days of age, the mice had a substantial loss of microbial diversity

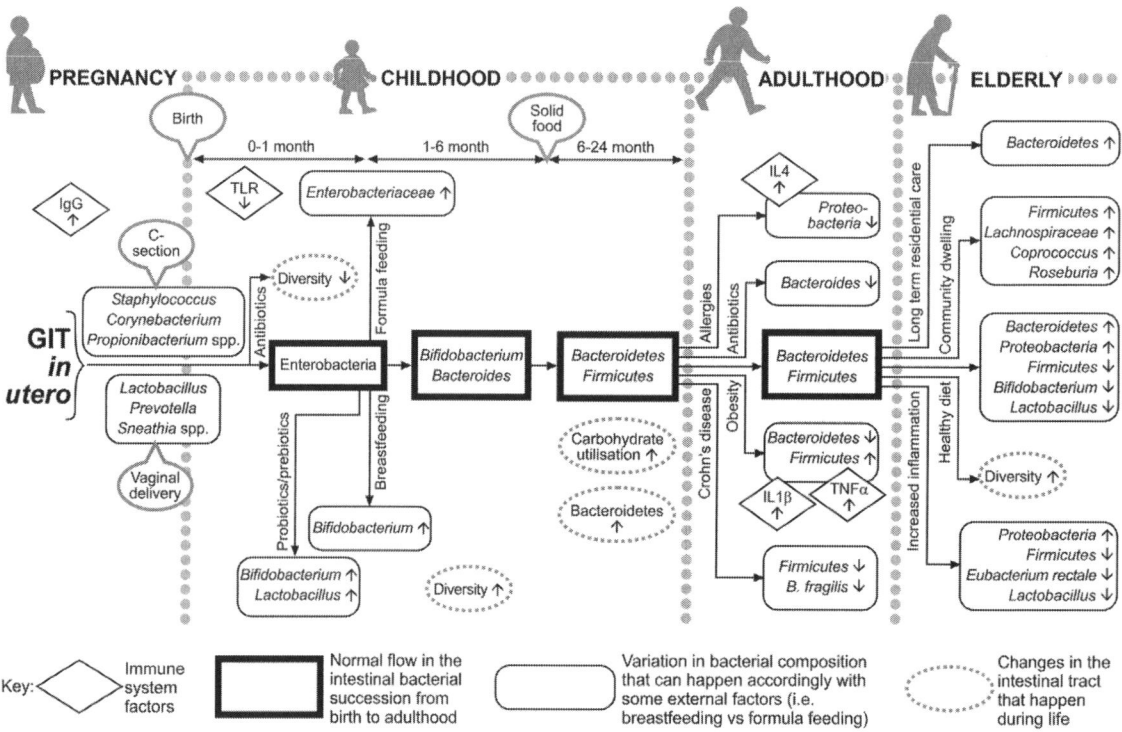

■ **FIGURE 18.1** Development of the GIT microbiota from birth through to elderly life. This diagram shows the dominant bacterial taxa through different stages of life and how they change under different health and diet conditions. Initial community structure is influenced by delivery mode and type of feeding. Diversity increases as the infant grows and begins eating solid food. Different disease states, diets, and environmental conditions all play a role in shaping the microbiota throughout life. *Figure adapted from Pantoja-Feliciano* et al. *(2013).*

and increased prevalence of lactic acid bacteria was observed (Pantoja-Feliciano *et al.*, 2013). The dominance of lactic acid bacteria reflects their ability to ferment milk lactose and casein into lactic acid leading to the inhibition of the growth of obligate anaerobic bacteria. Leading up to and after weaning, the microbiota becomes increasingly complex as new bacterial species become established and the proportions of the different bacterial populations change radically. These include obligate anaerobes such as *Bacteroides*, *Bifidobacterium*, and *Clostridium* species (Fallani *et al.*, 2011). In humans, commencement of dietary supplementation with solids is associated with a decline in *Bifidobacterium* numbers. By 12 months of age, *Bacteroides* species and *Clostridia* become more prevalent in the infant GIT microbiota, which begins to resemble that of the adult (Stark and Lee, 1982; Palmer *et al.*, 2007).

The increased diversity of the microbial community around weaning also coincides with dramatic changes in the morphology and function of the GIT. Changes in the absorptive capability of the small intestine enable the gradual transition from a high-fat milk diet with lactose as the primary source of carbohydrate to solid diets containing plant-based carbohydrates (Sabat and Veloso, 2003; Muncan *et al.*, 2011). During the weaning period, withdrawal of maternal immunoglobulins present in the milk accompanies important changes in the mucosal immune system, such as an increase in intraepithelial lymphocytes (IELs) and lamina propria lymphocytes (LPLs) in the villi of the small intestine (Hooper, 2004). At weaning, the type of dietary complex carbohydrates that pass undigested into the large intestine changes; carbohydrates (oligosaccharides) found in breast milk (Newburg, 2000) are replaced by complex polysaccharides of plant origin as a major source of carbon and nitrogen for bacterial fermentation (De Filippo *et al.*, 2010). This results in expansion of non-pathogenic commensal bacterial populations in the large intestine (Fallani *et al.*, 2011). Differences in fecal microbiota composition and reduced microbiota diversity of infants has been linked to increased incidence of atopic eczema and obesity in later life (Kalliomaki *et al.*, 2008; Wang *et al.*, 2008; Reinhardt *et al.*, 2009). These findings support the concept that manipulation of the GIT microbiota in early life may provide a window of opportunity for long-lasting effects on community composition and activity, and consequently, host health.

In parallel with microbiota alterations in the GIT, the mucosal immune system undergoes numerous changes from early postnatal life to adulthood. T cell numbers increase dramatically in the small and large intestine of mice, with the largest expansion occurring between 2 and 4 weeks of age, while expression of the T cell activation marker CD69 on $\alpha\beta$ and $\gamma\delta$ IELs in the large intestine increases between the ages of 3 and 8 weeks (Kuo *et al.*, 2001). Changes in lymphocyte cytokine expression over time are also apparent, with increasing production of interferon gamma (IFN-γ) and tumor necrosis factor alpha (TNF-α) with advancing age observed in humans between 0 and 18 years of age (Wiegering *et al.*, 2009). Furthermore, thymic activity is generally thought to be greatest during early life (Shanker, 2004), leading to maturation and selection of various T cell populations, including regulatory T cells, which are thought to play a crucial role in maintaining GIT homeostasis (Thompson and Powrie, 2004).

Products of bacterial fermentation (SCFAs) in the large intestine are also known to affect host metabolism. For example, butyrate has been shown to stimulate mucosal cell proliferation (Kripke *et al.*, 1989; Scheppach, 1994) and to be a major energy substrate of the mucosa (Roediger, 1980; Scheppach, 1994). Furthermore, SCFAs (in particular butyrate) have been shown to induce beneficial immunomodulatory effects, such as reducing *in vitro* pro-inflammatory cytokine production (Saemann *et al.*, 2000; Niers *et al.*, 2005), inhibiting intracellular signaling mechanisms following pro-inflammatory cytokine stimulation (Inan *et al.*, 2000; Zapolska-Downar *et al.*, 2004), and stimulating mucosal repair (Segain *et al.*, 2000).

In addition to changes induced by SCFA production resulting from dietary complex carbohydrate fermentation, increased direct bacterial contact with the mucosa due to proliferation of *Lactobacillus* and *Bifidobacterium* species can also affect GIT immune function. The interaction of cellular pattern recognition receptors with extracellular and intracellular bacterial structures, such as cell wall peptidoglycans, can trigger various signaling cascades that alter the profile of cytokine secretion in monocytes and intestinal epithelial cells (IECs) (Netea *et al.*, 2004a; Haller, 2006). Expression of the major pattern recognition receptors, Toll-like receptors (TLRs), and nucleotide-binding oligomerization domain (NOD) protein receptors, occurs primarily in IECs and antigen-presenting cells such as macrophages and dendritic cells (Strober *et al.*, 2006). IECs are, in most cases, the first line of contact between the commensal bacterial population in the GIT and the host defense system. Not only do they provide an essential barrier function (Fasano and Shea-Donohue, 2005), they are also thought to influence immune functions by directly or indirectly modifying dendritic cells, IEL, and LPL activity (Rimoldi *et al.*, 2004; Ferrero, 2005; Kelly *et al.*, 2005).

Stimulation of pattern recognition receptors contributes to either mucosal inflammation or tolerance via induction of intracellular signaling pathways that modulate the activity of nuclear factor kappa-light-chain-enhancer of activated B cells (NF-κB), a multifunctional transcription factor involved in regulation of the immune response (Netea *et al.*, 2004b; Tong *et al.*, 2004; Watanabe *et al.*, 2004, 2005). Commensal bacteria play a crucial role in immune system function. Numerous studies have shown that bacteria, including some *Lactobacillus* and *Bifidobacterium* species, can modify T cell expression of cytokines and reduce host susceptibility to pathogens (von der Weid *et al.*, 2001; Niers *et al.*, 2005; Kim *et al.*, 2006). Commensal bacteria are able to affect these changes through a variety of mechanisms. For example, two species of *Lactobacillus* have been shown to modulate dendritic cell suppression of T cell proliferation via binding with dentritic cell-SIGN, a cell surface structure involved in both microbial internalization and dendritic cell migration (Smits *et al.*, 2005). The GIT obligate anaerobe *Bacteroides thetaiotaomicron* is capable of attenuating GIT inflammation by promoting the nuclear translocation of NF-κB, thereby suppressing expression of pro-inflammatory cytokines such as TNF-α and IL-1α (Kelly *et al.*, 2004).

The question of whether these changes are inherent processes programmed by the host DNA, or if they are driven by exposure to the resident microbiota, has been examined by studies using germ-free animals. Findings from these studies have highlighted the profound effects of the microbiota on intestinal morphology and physiology. Bacterial colonization of germ-free rodents and rabbits resulted in thickening of the mucosa, increased colon crypt depth, stimulation of crypt cell proliferation, and shorter and broader ileal villi (Boot *et al.*, 1985; Sharma *et al.*, 1995). Other effects of bacterial exposure include increased epithelial cell turnover, a thicker colonic epithelial mucus layer, and increased numbers of mucus-containing goblet cells (Goodlad *et al.*, 1989; Meslin *et al.*, 1993; Kleessen *et al.*, 2003). Furthermore, presence of the microbiota has also been shown to modify the structure of mucin secreted by the host. The mucus layer produced by goblet cells is an integral component of the mucosal barrier. Epithelial glycans expressed by the host prior to weaning are primarily terminated by sialic acid (Hooper, 2004). However, during weaning there is a shift towards expression of fucose-terminated glycans, facilitated by a corresponding change in expression of fucosyltransferases that modify the terminal residues (Bry *et al.*, 1996; Hooper, 2004). This shift in glycan structure is not observed in germ-free mice, but can be initiated by subsequent colonization with *B. thetaiotaomicron* (Bry *et al.*, 1996).

The plasticity of the microbiota during early stages of life results in compositional and associated biochemical changes that may have long-lasting impacts on the adult host. However, the mechanisms by which the microbiota influence host physiology in adulthood are still ill-defined.

RESILIENCE OF MICROBIOTA AND HOST INTERACTIONS IN ADULTS

Understanding the dynamics and stability of the human GIT microbiota in adulthood is essential to assess its potential as a marker of health and disease risk. Long-term stability of normal GIT microbiota in healthy adults is ill-defined. Many factors contribute to the composition of the microbiota in adult humans (host genotype, disease, diet, etc., reviewed elsewhere (Lagier *et al.*, 2012a, b)). In the absence of perturbations, composition of the fecal microbiota is host specific and fairly stable, and the within-subject variability over time is much smaller than between subjects (Durban *et al.*, 2012). In a recent study, mice fed high-fat diets that were given Lactobacillus salivarius, modified to produce a bacteriocin, initially showed an altered fecal microbiota composition and lowered weight gain compared to mice given wild-type Lactobacillus salivarius (Clarke *et al.*, 2013).

However, by the end of the experimental period, differences between body weights of treated and control mice were no longer apparent, and the fecal microbiota composition was no longer distinguishable between the groups (Clarke *et al.*, 2013). Likewise, the human throat and GIT microbiota have also been shown to recover over time even after antibiotic treatment (Jakobsson *et al.*, 2010). The stability of the GIT microbiota over time suggests the existence of a core group of microbes that remain with an individual over time (Zoetendal *et al.*, 2008), or that the ecosystem is resilient (De La Cochetiere *et al.*, 2005; Jernberg *et al.*, 2007; Dethlefsen *et al.*, 2008). It has been estimated that humans share a core microbiota (40% of the phylotypes) (Jalanka-Tuovinen *et al.*, 2011), which is substantially larger than the ranges (0−2%−30%) previously reported (Tap *et al.*, 2009; Turnbaugh *et al.*, 2009a; Qin *et al.*, 2010; Willing *et al.*, 2010). A recent study using a combination of high-throughput methods and sequencing of the genomes of anaerobic bacteria showed that 60% of strains remained stable over 5 years; members of the Bacteroidetes and Actinobacteria were more stable components than other members of the microbiota (Faith *et al.*, 2013).

Increasing evidence points to the fact that despite microbiota stability over time, the metabolic functions of the microbiota might be subject to fluctuations caused by substrate availability, external factors, stress, and host physiology (Durban *et al.*, 2012). The use of antibiotics was also shown to destabilize the microbiota, perhaps leading to the rapid loss of formerly ubiquitous organisms such as *Helicobacter pylori* over a span of two or three generations (Blaser and Falkow, 2009). Thus, despite the characteristic resilience found in complex ecosystems, the persistent presence of perturbations may contribute to the loss in recovery, giving way to the establishment of new organisms (Petchey *et al.*, 2008; Dethlefsen and Relman, 2011). If important species are lost, the effects may cascade resulting in secondary extinctions that may have important implications for human health (Cho and Blaser, 2012). High biodiversity may diminish this risk. The substantial non-linear interactions in complex, co-evolved systems ensure that ecological networks are sufficiently robust to withstand random removals (Sole and Montoya, 2001). However, with repeated perturbations, the effects of gene loss can be amplified by downstream effects on co-colonizing microbes and on the host. Because many of the GIT microorganisms produce one or more biochemical compounds that influence the growth, survival, and reproduction of other organisms, the effects of extinctions may magnify (Borrvall and Ebenman, 2006; Cho and Blaser, 2012). In the short term, functional redundancy may mask extinction effects, but in the long run, extinctions lead to loss of contingency responses that can cause ecological crashes

(Sole and Montoya, 2001; Cho and Blaser, 2012). Considering the importance of bacterial association exploiting parallel and sequential metabolic pathways, the contents of the collective genomes of the entire microbiota, often referred to as the metagenome, may have a significant impact on human health.

Several metabolic diseases, such as obesity (Ley *et al.*, 2006), diabetes (Cani *et al.*, 2008), and inflammatory bowel diseases (Yu and Huang, 2013), are associated with changes in the GIT microbiota. Diet is the main factor known to modulate the composition of the GIT microbiota in humans and mice; both short-term dietary interventions and long-term dietary habits may have a considerable effect on the human GIT microbiota. Rural African children, for example, who consumed high amounts of plant polysaccharides, had low levels of Firmicutes and increased levels of Bacteroidetes in their fecal microbiota compared with Italian children, who had high levels of Enterobacteriaceae (De Filippo *et al.*, 2010). Changes in the microbiota were also observed following a 10-day dietary intervention with high-fat/low-fiber or low-fat/high-fiber diets (Wu *et al.*, 2011). However, this change in the microbiota composition was not sufficient to alter the enterotype of an individual (Wu *et al.*, 2011), suggesting that a long-term change may be required to provoke a major shift in GIT microbiota composition. Changes in daily carbohydrate intake, however, may affect specific groups of colonic bacteria over a short period of time. Consumption of the prebiotic inulin for 16 days has been shown to increase levels of *Faecalibacterium prausnitzii* and *Bifidobacterium* sp. in humans (Ramirez-Farias *et al.*, 2009). The GIT microbiota may also be affected by dietary fat. Mice fed on high-fat diets have reduced numbers of Bacteroidetes, and increased numbers of Firmicutes and Proteobacteria within 24 hours (Turnbaugh *et al.*, 2008, 2009b).

Probiotics, commonly used as dietary supplements, are living microorganisms that affect the GIT and associated immune system, and have numerous beneficial effects on GIT function and immune responses (Behnsen *et al.*, 2013). Administration of *F. prausnitzii* to mice with colitis, for example, led to the restoration of GIT microbiota to a normal state, and decreased colitis (Sokol *et al.*, 2008).

It is important to recognize, however, that the most compelling evidence concerning modulation of the GIT microbiota by diet comes from animal rather than human studies (Park *et al.*, 2013). Extrapolating data from animal studies to humans, especially in predicting modulation of the GIT microbiota by diet, has its limitations. The physiological effects of food are influenced in part by an individual's GIT microbiota and the genes of both the human host and the microbiota. The interrelationships between dietary type and eating frequency, the composition of the GIT microbiota, and harvest of nutrients and energy are further impacted by variation in human environmental exposure and genotype. In order to isolate some of these factors, Turnbaugh *et al.* (Turnbaugh *et al.*, 2009b) created humanized gnotobiotic mouse as a well-defined, representative animal model of the human GIT ecosystem. Germ-free adult C57BL/6J mice were transplanted with fresh or frozen fecal microbial communities from adult humans. The authors followed the patterns of bacterial colonization using culture-independent metagenomic analyses to confirm that these humanized mice were stably and heritably colonized with representative bacteria. They were also able to demonstrate the effects of moving from a low-fat, high-plant polysaccharide diet to a high-fat, high-sugar "Western-style" diet. Within a single day, this change in diet shifted the composition of the microbiota and various metabolic pathways in the microbiome, and altered microbiota gene expression. Although the nature of the donor affected the initial structure of the microbial community, the composition of this community could be rapidly shifted by diet. Such models permit a detailed study of the complex interactions between the GIT microbiota and immune system, and the way in which they contribute to several metabolic diseases.

In summary, the GIT microbiota, as with other complex ecosystems that populate specific human anatomical niches, possess a community characterized by a high organizational level. Despite its inherent resilience, there are factors that may perturb this organizational integrity, potentially leading to disease and accelerated aging.

MICROBIOTA AND HOST INTERACTIONS DURING AGING

The aging process is influenced by a number of factors including genetics, infection, diet, physical activity, and the environment (Pray *et al.*, 2010; Cusack *et al.*, 2011). These age-related changes can impact on the function and efficacy of the GIT and thus on the maintenance of health (Cusack *et al.*, 2011). A wide range of physiological changes occur along the GIT with ageing; however, the structure and absorptive functions of the small intestine are generally well preserved in healthy, older people (Russell and Murray, 1992; Fernandez-Banares *et al.*, 2009; Tiihonen *et al.*, 2010). There is a reduction in the absorption of some nutrients associated with aging, e.g., vitamin B_{12}, lactose, calcium, vitamin D, iron, zinc, and magnesium; whereas the absorption of other nutrients is unchanged (Russell and Murray, 1992; Fernandez-Banares *et al.*, 2009; Tiihonen *et al.*, 2010). In contrast, the colon undergoes significant structural changes such as decreased neuronal density and rectal elasticity (Wade and Cowen, 2004). This results in an increase in digesta transit time, which, combined with a general decrease in physical activity, changing diet (including low dietary fiber intake), and certain medications, contributes to increased bloating and constipation. Increased fecal retention can lead to a shift from beneficial saccharolytic fermentation towards increased levels of putrefaction that can result in a build-up of harmful by-products such as ammonia and phenols (Woodmansey, 2007) (as described in Cusack *et al.*, 2011). The longer colonic transit time observed in the elderly results in a decrease in the proportions of slow growing species (e.g., methanogens), and an increase in sulfate reducing bacteria (El Oufir *et al.*, 1996), which are associated with increased inflammation (Rowan *et al.*, 2010; Devkota *et al.*, 2012). These changes, coupled with an increase in total SCFAs, lead to a reduction in the fecal pH (El Oufir *et al.*, 1996; Jeffery and O'Toole, 2013), and in conjunction with dietary intake, will determine the eventual composition of the microbiota (Jeffery and O'Toole, 2013).

Increasing age is also associated with a progressive decline in the integrity and function of both the innate and adaptive immune systems. This is termed immunosenescence (Shanley *et al.*, 2009) (as described in Biagi *et al.*, 2012), and can lead to greater susceptibility to infection, malignancy, autoimmune disease, and delayed wound healing (Biagi *et al.*, 2010; Cusack *et al.*, 2011). This progressive increase in chronic, low-grade pro-inflammatory status has been termed "inflamm-aging" (Franceschi, 2007; Franceschi *et al.*, 2007) and results in a localized, persistent inflammation of the GIT mucosa which can contribute to systemic inflammation (Biagi *et al.*, 2012). The inflammation associated with immunosenescence favors the growth of potentially pathogenic species (or "pathobionts") which can outnumber the healthy GIT microbiota to support inflammation and consolidate a pro-inflammatory status in the host (Biagi *et al.*, 2013). Medications such as antibiotics, commonly administered to the elderly for a variety of age-related infections, also perturb the composition of the microbiota. Well known for causing antibiotic-associated diarrhea, *Clostridium difficile* can cause a severe, even life-threatening disease in the elderly. Although an increased risk of infection has been reported with increasing age, it is unclear whether this is a result of changes associated with aging or an increased frequency of antimicrobial use in older adults (Simor *et al.*, 2002).

Age-related changes in the GIT combined with changes in diet and immune system reactivity affect the composition of the microbiota, leading to increased numbers of facultative anaerobes (including enterobacteria), and decreased numbers of beneficial organisms such as lactobacilli and bifidobacteria (Donini *et al.*, 2009; Biagi *et al.*, 2012). An increased proportion of opportunistic bacteria is one of the most prominent (and most frequently reported) characteristics of an aged-associated GIT microbiota. Any increase in their population numbers, combined with the physiological changes associated with aging, means that these pathobionts can overgrow and cause infections. However, results obtained using molecular methods, such as pyrosequencing, have questioned the long-held belief that a decrease in health-promoting species such as bifidobacteria is associated with aging. Recent studies (Rajilić-Stojanović *et al.*, 2009; Biagi *et al.*, 2010; Lahtinen *et al.*, 2012) contradicted these results except in extremely old people (centenarians) (Biagi *et al.*, 2010). These differences may be explained by geographical differences and/or the temporal instability of these microbial populations (Claesson *et al.*, 2011). The microbiota may also lose its resilience with aging (Biagi *et al.*, 2011).

The consequence of age-related GIT microbial changes is often a shift in dominant bacterial species (Mariat *et al.*, 2009; Cusack *et al.*, 2011). However, comparing studies from different nationalities and age ranges has led to conflicting results (Magrone *et al.*, 2013). While most studies have reported changes to the dominant microbiota including a decrease in Clostridium cluster XIVa (Hayashi *et al.*, 2003; Biagi *et al.*, 2010), an elderly German cohort (60 years old or more) showed the opposite (Mueller *et al.*, 2006). Likewise, several European studies reported a decrease in *F. prausnitzii* species (Italians 60 years old or more and Italian centenarians (Mueller *et al.*, 2006; Biagi *et al.*, 2010)) while other European studies reported the opposite (Mueller *et al.*, 2006; Claesson *et al.*, 2011). Similarly, studies have reported that the Firmicutes and Bacteroidetes ratio was lower in elderly people compared to young adults (Mariat *et al.*, 2009; Claesson *et al.*, 2011; Biagi *et al.*, 2012), whereas another study (Biagi *et al.*, 2010) found no significant differences in these ratios between Italian centenarians, elderly, and young adults. Even the definition of "elderly" used between different studies makes it difficult to ascertain when the GIT is first influenced by the aging process, particularly when there is a difference in the age and health status of those considered "young elderly" compared to those who are "extremely old" (Biagi *et al.*, 2012). The number of striking variances observed in the composition of the GIT microbiota between countries also suggests that dietary and lifestyle diversity of the elderly may play a significant role in determining age-related modifications (Hayashi *et al.*, 2003; Mueller *et al.*, 2006; Mariat *et al.*, 2009; Biagi *et al.*, 2010, 2012; Claesson *et al.*, 2011, 2012).

One of the most important modulators of the GIT microbiota is diet, an important consideration given that malnutrition is highly prevalent among the elderly (Kinross and Nicholson, 2012). The groundbreaking work of Claesson *et al.* (Claesson *et al.*, 2012) provided evidence to suggest that diet influences the composition of the microbiota in older populations, which subsequently affects health. This large cohort study showed that the microbiota of Irish elderly living in long-term care was significantly less diverse than those living in the community, and that this change in microbiota correlated with increased frailty, co-morbidity, nutritional status, markers of inflammation, and microbial-produced metabolites. Elderly living in long-term care had higher levels of Bacteriodetes, while community dwellers and younger adults had higher levels of Firmicutes and unknown bacterial species. Those living in the community also had higher levels of SCFAs (Claesson *et al.*, 2012) which have been shown to influence the immune system (Maslowski *et al.*, 2009); consequently, not only was the composition of the microbiota different, but the microbiota had different types of activity. Overall, their results revealed that the "healthiest" elderly lived in the community, consumed a different diet, and had a distinct microbiota from those in long-term care. This work indicates a clear link between diet, microbiota composition, health, and the rate of aging in older people. This suggests that dietary intervention could provide beneficial changes in the microbiota to maintain better health and promote healthy aging.

Biagi *et al.* (Biagi *et al.*, 2010) were the first to study the microbiota of an extremely old population, which consisted of Italian centenarians. Their study showed that the compositional stability of the adult GIT microbiota may last longer than expected, because only the centenarians showed a compromised microbiota. The metabolic phenotype of extreme longevity was investigated (Collino *et al.*, 2013) in an attempt to identify biological markers of longevity and thus gain an understanding of the specific mechanisms and processes of aging. Using a mostly female cohort of centenarians from northern Italy, this study showed a number of changes in the lipidome that the authors suggested could "reflect the centenarians' unique capability to respond to accumulating oxidative and chronic inflammatory processes characteristic of their extreme aging phenotype." They also identified cellular detoxification mechanisms that they believed could result in activation of an anti-oxidative response. Metabolic profiling of urine identified three markers of longevity; phenylacetylglutamine (PAG) and p-cresol sulfate (PCS), both of which are produced by bacteria, and 2-hydroxybenzoate (2HB), which they correlated with phylogenetic bacterial groups. They concluded that the longevity process affects the structure and composition of the human GIT microbiota, particularly characterized by an increase in Proteobacteria. However, these longevity markers should be validated in other geographical and cultural backgrounds. This work supports that of Biagi *et al.* (Biagi *et al.*, 2010) who showed that while centenarians have less Clostridium cluster XIVa (these bacteria are negatively correlated with PAG and PCS and, as butyrate-producing species, have been shown to reduce intestinal inflammation), they have an increase in facultative anaerobes such as Proteobacteria.

The aging process is one that is occurring throughout life. Healthy aging starts with healthy behaviors in earlier stages of life. Diet, physical exercise, environment, and other lifestyle choices have all been shown to affect health. Diet, in particular, has been shown to affect the composition of the GIT microbiota (Brinkworth *et al.*, 2009), thus specific foods or diets could be used to engineer the GIT microbiota to delay or prevent the inflamm-aging process and promote healthy aging. Improving the wellness of the aged through modulation of the microbiota may be an important consideration when designing foods for both the aged and those wishing to promote their own healthy longevity. Those working in this field argue that dietary interventions for the elderly should be distinct from those developed for the general population because of the physiological and microbial differences observed between these groups (Cusack *et al.*, 2011). A number of probiotic foods have been, or are currently being, tested for their ability to modulate the intestinal microbiota of the elderly. Probiotic cheese containing *Lactobacillus rhamnosus* HN001 and *Lactobacillus acidophilus* NCFM™ fed to healthy elderly volunteers (living in a Finnish nursing home) was associated with lower levels of *Clostridium difficile*, particularly in those who had *C. difficile* at the beginning of the study, but did not alter levels of the major bacterial groups (Lahtinen *et al.*, 2012). Likewise, a biscuit containing the probiotics *Bifidobacterium longum* Bar33 and *Lactobacillus helveticus* Bar13, fed to elderly Italian volunteers, was effective in rectifying some of the age-related dysbioses of the GIT microbiota, particularly in reducing the number of opportunistic pathogens (e.g., Clostridium species) associated with aging (Rampelli *et al.*, 2013). Studies such as these support the use of functional foods to address the microbial dysbioses of the aging GIT. However, given the transient nature of probiotic strains, these foods will require regular, ongoing consumption in order to achieve the maximum benefits.

CONCLUDING REMARKS

There can no longer be any doubt that the microbiota plays a major role in human health and disease. There is an ever-increasing body of evidence for an array of essential functions performed by the GIT microbiota, such as assisting the normal development of the immune system and many of the other key metabolic processes required for host health. Conversely, disruption of the microbiota composition can induce changes in the symbiotic interactions between the microbiota and its host resulting in a large number of non-communicable diseases such as diabetes, obesity, and inflammatory bowel diseases. As a consequence, the role of the intestinal microbiota is valued not only in the pathogenesis of chronic diseases, but also in determining host function and homeostasis beyond the GIT.

The developmental trajectory of the human GIT microbiota primarily begins at birth and is influenced throughout life by a series of complex and dynamic interactions, including diet, lifestyle, disease, and antibiotic use. Together with the host genome, these environmental factors continually influence the diversity and function of the GIT microbiota. While initially low, the diversity of the microbiota increases with age as it moves towards the stable core community structure observed in healthy adults. However, microbial diversity is reduced in old age, accompanied by low-level systemic inflammation ("inflamm-aging") and increased frailty.

One of the most important modulators of the GIT microbiota is diet; thus, dietary modulation to manipulate specific microbial species may offer new therapeutic approaches to promote health throughout the lifespan of the host. Preventing or delaying age-related diseases and maintaining good health through nutrition, particularly functional foods and dietary supplements, are therefore becoming increasingly important strategies. However, the physiological and microbial disparities associated with the different stages of human development will require specific nutritional strategies. In particular, it is more difficult for both the very young (e.g., formula-fed babies) and the elderly to obtain the nutrients they need from their diet, so functional foods can play a role in improving nutrient intake. Overall, the most success in this area has been achieved using probiotics (often in association with prebiotics such as inulin) as these can be added to a wide variety of foods across the age spectrum.

Significant growth and competition in the global foods for health market, along with rising consumer awareness in the role of nutrition for health and well-being, has resulted in a need for increased regulation around health and nutritional claims on food. To maintain consumer confidence in foods with value-added benefits, many international food safety authorities have introduced rules governing the labeling of foods with claims to health benefits. Health claims must now be backed by solid scientific evidence to reduce the risk of misleading or deceptive claims being made about the health properties of food. However, there are currently significant limitations encountered when attempting to gain regulatory approval for beneficial health claims, particularly for probiotic strains, by the European Food Safety Administration (EFSA). The applications for health and nutrition claim approval have to specify the quantities that should be consumed to achieve the claim, the target population, any relevant health risks if consumed in excess, and warnings about unsuitability for individuals with certain profiles (Regulation 1924/2006). To date, EFSA has primarily accepted Article 13.1 (health claims that refer to vitamins or minerals) and Article 14 (disease risk reduction claims that mainly refer to specific nutrients), for example, iodine for children's growth, as well as those that refer to plant sterols and stanols. For other claim areas, it may, for example, be possible to label foods as "containing probiotics"; a strategy adopted by the manufacturers of certain infant formulas and which appears to be a somewhat effective selling point in the minds of many consumers. It is also important to note that just as there is variation in the composition of the GIT microbiota between healthy individuals, some probiotic strains may be more effective than others and individual responses may vary.

REFERENCES

Bager, P., Wohlfahrt, J., Westergaard, T., 2008. Caesarean delivery and risk of atopy and allergic disease: meta-analyses. Clin. Exp. Allergy 38, 634–642.

Barker, D.J., 1966. Low intelligence. Its relation to length of gestation and rate of foetal growth. Br. J. Prev. Soc. Med. 20, 58–66.

Barker, D.J., 1988. Childhood causes of adult diseases. Arch. Dis. Child. 63, 867–869.

Beaglehole, R., Bonita, R., 2009. A scorecard for assessing progress in global public health. J. Epidemiol. Comm. Health 63, 507–508.

Beaglehole, R., Yach, D., 2003. Globalisation and the prevention and control of non-communicable disease: the neglected chronic diseases of adults. Lancet 362, 903–908.

Behnsen, J., Deriu, E., Sassone-Corsi, M., Raffatellu, M., 2013. Probiotics: properties, examples, and specific applications Cold Spring Harb. Perspect. Med. 3 a010074.

Biagi, E., Candela, M., Fairweather-Tait, S., Franceschi, C., Brigidi, P., 2012. Aging of the human metaorganism: the microbial counterpart. Age (Dordrecht, Netherlands) 34, 247–267.

Biagi, E., Candela, M., Franceschi, C., Brigidi, P., 2011. The aging gut microbiota: new perspectives. Ageing Res. Rev. 10, 428–429.

Biagi, E., Candela, M., Turroni, S., Garagnani, P., Franceschi, C., Brigidi, P., 2013. Ageing and gut microbes: perspectives for health maintenance and longevity. Pharmacol. Res. 69, 11–20.

Biagi, E., Nylund, L., Candela, M., Ostan, R., Bucci, L., Pini, E., et al., 2010. Through ageing, and beyond: gut microbiota and inflammatory status in seniors and centenarians. PLoS One 5.

Blaser, M.J., Falkow, S., 2009. What are the consequences of the disappearing human microbiota? Nat. Rev. Microbiol. 7, 887–894.

Boot, R., Koopman, J.P., Kruijt, B.C., Lammers, R.M., Kennis, H.M., Lankhorst, A., et al., 1985. The "normalization" of germ-free rabbits with host-specific caecal microflora. Lab. Anim. 19, 344–352.

Borrvall, C., Ebenman, B., 2006. Early onset of secondary extinctions in ecological communities following the loss of top predators. Ecol. Lett. 9, 435–442.

Brinkworth, G.D., Noakes, M., Clifton, P.M., Bird, A.R., 2009. Comparative effects of very low-carbohydrate, high-fat and high-carbohydrate, low-fat weight-loss diets on bowel habit and faecal short-chain fatty acids and bacterial populations. Br. J. Nutr. 101, 1493−1502.

Bry, L., Falk, P.G., Midtvedt, T., Gordon, J.I., 1996. A model of host-microbial interactions in an open mammalian ecosystem. Science 273, 1380−1383.

Canani, R.B., Costanzo, M.D., Leone, L., Bedogni, G., Brambilla, P., Cianfarani, S., et al., 2011. Epigenetic mechanisms elicited by nutrition in early life. Nutr. Res. Rev. 24, 198−205.

Cani, P.D., Bibiloni, R., Knauf, C., Waget, A., Neyrinck, A.M., Delzenne, N.M., Burcelin, R., 2008. Changes in gut microbiota control metabolic endotoxemia-induced inflammation in high-fat diet-induced obesity and diabetes in mice. Diabetes 57, 1470−1481.

Cho, I., Blaser, M.J., 2012. The human microbiome: at the interface of health and disease. Nat. Rev. Genet. 13, 260−270.

Cho, I., Yamanishi, S., Cox, L., Methe, B.A., Zavadil, J., Li, K., et al., 2012. Antibiotics in early life alter the murine colonic microbiome and adiposity. Nature 488, 621−626.

Claesson, M.J., Cusack, S., O'Sullivan, O., Greene-Diniz, R., De Weerd, H., Flannery, E., et al., 2011. Composition, variability, and temporal stability of the intestinal microbiota of the elderly. Proc. Natl. Acad. Sci. USA 108, 4586−4591.

Claesson, M.J., Jeffery, I.B., Conde, S., Power, S.E., O'Connor, E.M., Cusack, S., et al., 2012. Gut microbiota composition correlates with diet and health in the elderly. Nature 488, 178−184.

Clarke, S.F., Murphy, E.F., O'Sullivan, O., Ross, R.P., O'Toole, P.W., Shanahan, F., Cotter, P.D., 2013. Targeting the microbiota to address diet-induced obesity: a time dependent challenge. PLoS One 8, e65790.

Collino, S., Montoliu, I., Martin, F.P., Scherer, M., Mari, D., Salvioli, S., et al., 2013. Metabolic signatures of extreme longevity in northern Italian centenarians reveal a complex remodeling of lipids, amino acids, and gut microbiota metabolism. PLoS One 8.

Curtis, V., Schmidt, W., Luby, S., Florez, R., Toure, O., Biran, A., 2011. Hygiene: new hopes, new horizons. Lancet Infect. Dis. 11, 312−321.

Cusack, S., Claesson, M.J., O'Toole, P.W., 2011. How beneficial is the use of probiotic supplements for the aging gut? Aging Health 7, 179−186.

De Filippo, C., Cavalieri, D., Di Paola, M., Ramazzotti, M., Poullet, J.B., Massart, S., et al., 2010a. Impact of diet in shaping gut microbiota revealed by a comparative study in children from Europe and rural Africa. Proc. Natl. Acad. Sci. USA 107, 14691−14696.

De La Cochetiere, M.F., Durand, T., Lepage, P., Bourreille, A., Galmiche, J.P., Dore, J., 2005. Resilience of the dominant human fecal microbiota upon short-course antibiotic challenge. J. Clin. Microbiol. 43, 5588−5592.

de Vos, W.M., de Vos, E.A., 2012. Role of the intestinal microbiome in health and disease: from correlation to causation. Nutr. Rev. 70 (Suppl. 1), S45−S56.

Dethlefsen, L., Huse, S., Sogin, M.L., Relman, D.A., 2008. The pervasive effects of an antibiotic on the human gut microbiota, as revealed by deep 16S rRNA sequencing. PLoS Biol. 6, e280.

Dethlefsen, L., Relman, D.A., 2011. Incomplete recovery and individualized responses of the human distal gut microbiota to repeated antibiotic perturbation. Proc. Natl. Acad. Sci. USA 108 (Suppl. 1), 4554−4561.

Devkota, S., Wang, Y., Musch, M.W., Leone, V., Fehlner-Peach, H., Nadimpalli, A., et al., 2012. Dietary-fat-induced taurocholic acid promotes pathobiont expansion and colitis in Il10−/− mice. Nature 487, 104−108.

Donini, L.M., Savina, C., Cannella, C., 2009. Nutrition in the elderly—the role of fiber. Arch. Gerontol. Geriatrics Suppl. 1, 61−69.

Durban, A., Abellan, J.J., Jimenez-Hernandez, N., Latorre, A., Moya, A., 2012. Daily follow-up of bacterial communities in the human gut reveals stable composition and host-specific patterns of interaction. FEMS Microbiol. Ecol. 81, 427−437.

El Oufir, L., Flourie, B., Bruley des Varannes, S., Barry, J.L., Cloarec, D., Bornet, F., Galmiche, J.P., 1996. Relations between transit time, fermentation products, and hydrogen consuming flora in healthy humans. Gut 38, 870−877.

Faith, J.J., Guruge, J.L., Charbonneau, M., Subramanian, S., Seedorf, H., Goodman, A.L., et al., 2013. The long-term stability of the human gut microbiota. Science 341, 1237439.

Fallani, M., Amarri, S., Uusijarvi, A., Adam, R., Khanna, S., Aguilera, M., et al., 2011. Determinants of the human infant intestinal microbiota after the introduction of first complementary foods in infant samples from five European centres. Microbiology 157, 1385−1392.

Fasano, A., Shea-Donohue, T., 2005. Mechanisms of disease: the role of intestinal barrier function in the pathogenesis of gastrointestinal autoimmune diseases. Nat. Clin. Pract. Gastroenterol. Hepatol. 2, 416−422.

Favier, C.F., Vaughan, E.E., De Vos, W.M., Akkermans, A.D., 2002. Molecular monitoring of succession of bacterial communities in human neonates. Appl. Environ. Microbiol. 68, 219−226.

Fernandez-Banares, F., Monzon, H., Forne, M., 2009. A short review of malabsorption and anemia. World J. Gastroenterol. 15, 4644−4652.

Ferrero, R.L., 2005. Innate immune recognition of the extracellular mucosal pathogen, Helicobacter pylori. Mol. Immunol. 42, 879−885.

Forchielli, M.L., Walker, W.A., 2005. The role of gut-associated lymphoid tissues and mucosal defence. Br. J. Nutr. 93 (Suppl. 1), S41−S48.

Franceschi, C., 2007. Inflammaging as a major characteristic of old people: can it be prevented or cured? Nutr. Rev. 65, S173−S176.

Franceschi, C., Capri, M., Monti, D., Giunta, S., Olivieri, F., Sevini, F., et al., 2007. Inflammaging and anti-inflammaging: a systemic perspective on aging and longevity emerged from studies in humans. Mech. Ageing Devel. 128, 92−105.

Gluckman, P.D., Cutfield, W., Hofman, P., Hanson, M.A., 2005. The fetal, neonatal, and infant environments—the long-term consequences for disease risk. Early Hum. Dev. 81, 51−59.

Gluckman, P.D., Hanson, M.A., 2004. Living with the past: evolution, development, and patterns of disease. Science 305, 1733−1736.

Goodlad, R.A., Ratcliffe, B., Fordham, J.P., Wright, N.A., 1989. Does dietary fibre stimulate intestinal epithelial cell proliferation in germ free rats? Gut 30, 820−825.

Guarner, F., Malagelada, J.R., 2003. Gut flora in health and disease. Lancet 361, 512−519.

Haller, D., 2006. Intestinal epithelial cell signalling and host-derived negative regulators under chronic inflammation: to be or not to be activated determines the balance towards commensal bacteria. Neurogastroenterol. Motil. 18, 184−199.

Hayashi, H., Sakamoto, M., Kitahara, M., Benno, Y., 2003. Molecular analysis of fecal microbiota in elderly individuals using 16S rDNA library and T-RFLP. Microbiol. Immunol. 47, 557–570.

Hooper, L.V., 2004. Bacterial contributions to mammalian gut development. Trends Microbiol. 12, 129–134.

Hooper, L.V., Littman, D.R., Macpherson, A.J., 2012. Interactions between the microbiota and the immune system. Science 336, 1268–1273.

Inagaki, T., Moschetta, A., Lee, Y.K., Peng, L., Zhao, G., Downes, M., et al., 2006. Regulation of antibacterial defense in the small intestine by the nuclear bile acid receptor. Proc. Natl. Acad. Sci. USA 103, 3920–3925.

Inan, M.S., Rasoulpour, R.J., Yin, L., Hubbard, A.K., Rosenberg, D.W., Giardina, C., 2000. The luminal short-chain fatty acid butyrate modulates NF-kappaB activity in a human colonic epithelial cell line. Gastroenterology 118, 724–734.

Jakobsson, H.E., Jernberg, C., Andersson, A.F., Sjolund-Karlsson, M., Jansson, J.K., Engstrand, L., 2010. Short-term antibiotic treatment has differing long-term impacts on the human throat and gut microbiome. PLoS One 5, e9836.

Jalanka-Tuovinen, J., Salonen, A., Nikkila, J., Immonen, O., Kekkonen, R., Lahti, L., et al., 2011. Intestinal microbiota in healthy adults: temporal analysis reveals individual and common core and relation to intestinal symptoms. PLoS One 6, e23035.

Jeffery, I.B., O'Toole, P.W., 2013. Diet-microbiota interactions and their implications for healthy living. Nutrients 5, 234–252.

Jernberg, C., Lofmark, S., Edlund, C., Jansson, J.K., 2007. Long-term ecological impacts of antibiotic administration on the human intestinal microbiota. ISME J. 1, 56–66.

Kalliomaki, M., Collado, M.C., Salminen, S., Isolauri, E., 2008. Early differences in fecal microbiota composition in children may predict overweight. Am. J. Clin. Nutr. 87, 534–538.

Kelly, D., Campbell, J.I., King, T.P., Grant, G., Jansson, E.A., Coutts, A.G., et al., 2004. Commensal anaerobic gut bacteria attenuate inflammation by regulating nuclear-cytoplasmic shuttling of PPAR-gamma and RelA. Nat. Immunol. 5, 104–112.

Kelly, D., Conway, S., Aminov, R., 2005. Commensal gut bacteria: mechanisms of immune modulation. Trends Immunol. 26, 326–333.

Kim, Y.G., Ohta, T., Takahashi, T., Kushiro, A., Nomoto, K., Yokokura, T., et al., 2006. Probiotic Lactobacillus casei activates innate immunity via NF-kappaB and p38 MAP kinase signaling pathways. Microbes Infect. 8, 994–1005.

Kinross, J., Nicholson, J.K., 2012. Dietary and social modulation of gut microbiota in the elderly. Nat. Rev. Gastroenterol. Hepatol. 9, 563–564.

Kleessen, B., Hartmann, L., Blaut, M., 2003. Fructans in the diet cause alterations of intestinal mucosal architecture, released mucins and mucosa-associated bifidobacteria in gnotobiotic rats. Br. J. Nutr. 89, 597–606.

Koenig, J.E., Spor, A., Scalfone, N., Fricker, A.D., Stombaugh, J., Knight, R., et al., 2011. Succession of microbial consortia in the developing infant gut microbiome. Proc. Natl. Acad. Sci. 108, 4578–4585.

Kripke, S.A., Fox, A.D., Berman, J.M., Settle, R.G., Rombeau, J.L., 1989. Stimulation of intestinal mucosal growth with intracolonic infusion of short-chain fatty acids. JPEN J. Parenter. Enteral. Nutr. 13, 109–116.

Kuo, S., El Guindy, A., Panwala, C.M., Hagan, P.M., Camerini, V., 2001. Differential appearance of T cell subsets in the large and small intestine of neonatal mice. Pediatr. Res. 49, 543–551.

Lagier, J.C., Armougom, F., Million, M., Hugon, P., Pagnier, I., Robert, C., et al., 2012a. Microbial culturomics: paradigm shift in the human gut microbiome study. Clin. Microbiol. Infect. 18, 1185–1193.

Lagier, J.C., Million, M., Hugon, P., Armougom, F., Raoult, D., 2012b. Human gut microbiota: repertoire and variations. Front. Cell. Infect. Microbiol. 2, 136.

Lahtinen, S., Forssten, S., Aakko, J., Granlund, L., Rautonen, N., Salminen, S., et al., 2012. Probiotic cheese containing Lactobacillus rhamnosus HN001 and Lactobacillus acidophilus NCFM® modifies subpopulations of fecal lactobacilli and Clostridium difficile in the elderly. Age 34, 133–143.

Ley, R.E., Turnbaugh, P.J., Klein, S., Gordon, J.I., 2006. Microbial ecology: human gut microbes associated with obesity. Nature 444, 1022–1023.

Lozupone, C.A., Stombaugh, J.I., Gordon, J.I., Jansson, J.K., Knight, R., 2012. Diversity, stability and resilience of the human gut microbiota. Nature 489, 220–230.

Mackie, R.I., Sghir, A., Gaskins, H.R., 1999. Developmental microbial ecology of the neonatal gastrointestinal tract. Am. J. Clin. Nutr. 69, 1035S–1045S.

Magrone, T., de Heredia, F.P., Jirillo, E., Morabito, G., Marcos, A., Serafini, M., 2013. Functional foods and nutraceuticals as therapeutic tools for the treatment of diet-related diseases. Can. J. Physiol. Pharmacol. 91, 387–396.

Mariat, D., Firmesse, O., Levenez, F., Guimarães, V.D., Sokol, H., Doré, J., et al., 2009. The firmicutes/bacteroidetes ratio of the human microbiota changes with age. BMC Microbiol. 9.

Maslowski, K.M., Vieira, A.T., Ng, A., Kranich, J., Sierro, F., Yu, D., et al., 2009. Regulation of inflammatory responses by gut microbiota and chemoattractant receptor GPR43. Nature 461, 1282–1286.

McLoughlin, R.M., Mills, K.H., 2011. Influence of gastrointestinal commensal bacteria on the immune responses that mediate allergy and asthma. J. Allergy Clin. Immunol. 127, 1097–1107; quiz 1108–1099.

Meslin, J.C., Andrieux, C., Sakata, T., Beaumatin, P., Bensaada, M., Popot, F., et al., 1993. Effects of galacto-oligosaccharide and bacterial status on mucin distribution in mucosa and on large intestine fermentation in rats. Br. J. Nutr. 69, 903–912.

Mueller, S., Saunie, R.K., Hanisch, C., Norin, E., Alm, L., Midtvedt, T., et al., 2006. Differences in fecal microbiota in different European study populations in relation to age, gender, and country: a cross-sectional study. Appl. Environ. Microbiol. 72, 1027–1033.

Muncan, V., Heijmans, J., Krasinski, S.D., Buller, N.V., Wildenberg, M.E., Meisner, S., et al., 2011. Blimp1 regulates the transition of neonatal to adult intestinal epithelium. Nat. Commun. 2, 452.

Netea, M.G., Kullberg, B.J., de Jong, D.J., Franke, B., Sprong, T., Naber, T.H., et al., 2004a. NOD2 mediates anti-inflammatory signals induced by TLR2 ligands: implications for Crohn's disease. Eur. J. Immunol. 34, 2052–2059.

Netea, M.G., Van der Meer, J.W., Kullberg, B.J., 2004b. Toll-like receptors as an escape mechanism from the host defense. Trends Microbiol. 12, 484–488.

Newburg, D.S., 2000. Oligosaccharides in human milk and bacterial colonization. J. Pediatr. Gastroenterol. Nutr. 30 (Suppl. 2), S8–S17.

Nicholson, J.K., Holmes, E., Kinross, J., Burcelin, R., Gibson, G., Jia, W., Pettersson, S., 2012. Host—gut microbiota metabolic interactions. Science 336, 1262—1267.

Niers, L.E., Timmerman, H.M., Rijkers, G.T., van Bleek, G.M., van Uden, N.O., Knol, E.F., et al., 2005. Identification of strong interleukin-10 inducing lactic acid bacteria which down-regulate T helper type 2 cytokines. Clin. Exp. Allergy 35, 1481—1489.

Palmer, C., Bik, E.M., Digiulio, D.B., Relman, D.A., Brown, P.O., 2007. Development of the human infant intestinal microbiota. PLoS Biol. 5, e177.

Pantoja-Feliciano, I.G., Clemente, J.C., Costello, E.K., Perez, M.E., Blaser, M.J., Knight, R., Dominguez-Bello, M.G., 2013. Biphasic assembly of the murine intestinal microbiota during early development. ISME J. 7, 1112—1115.

Park, D.Y., Ahn, Y.T., Park, S.H., Huh, C.S., Yoo, S.R., Yu, R., et al., 2013. Supplementation of Lactobacillus curvatus HY7601 and Lactobacillus plantarum KY1032 in diet-induced obese mice is associated with gut microbial changes and reduction in obesity. PLoS One 8, e59470.

Petchey, O.L., Eklof, A., Borrvall, C., Ebenman, B., 2008. Trophically unique species are vulnerable to cascading extinction. Am. Nat. 171, 568—579.

Phillips, C.M., 2013. Nutrigenetics and metabolic disease: current status and implications for personalised nutrition. Nutrients 5, 32—57.

Popkin, B.M., Adair, L.S., Ng, S.W., 2012. Global nutrition transition and the pandemic of obesity in developing countries. Nutr. Rev. 70, 3—21.

Pray, L., Boon, C., Miller, E.A., Pillsbury, L., 2010. Providing Healthy and Safe Foods As We Age: Workshop Summary. National Academies Press, Washington, DC.

Qin, J., Li, R., Raes, J., Arumugam, M., Burgdorf, K.S., Manichanh, C., et al., 2010. A human gut microbial gene catalogue established by metagenomic sequencing. Nature 464, 59—65.

Rajilić-Stojanović, M., Heilig, H.G.H.J., Molenaar, D., Kajander, K., Surakka, A., Smidt, H., De Vos, W.M., 2009. Development and application of the human intestinal tract chip, a phylogenetic microarray: analysis of universally conserved phylotypes in the abundant microbiota of young and elderly adults. Environ. Microbiol. 11, 1736—1751.

Ramirez-Farias, C., Slezak, K., Fuller, Z., Duncan, A., Holtrop, G., Louis, P., 2009. Effect of inulin on the human gut microbiota: stimulation of Bifidobacterium adolescentis and Faecalibacterium prausnitzii. Br. J. Nutr. 101, 541—550.

Rampelli, S., Candela, M., Severgnini, M., Biagi, E., Turroni, S., Roselli, M., et al., 2013. A probiotics-containing biscuit modulates the intestinal microbiota in the elderly. J. Nutr. Health Aging 17, 166—172.

Reinhardt, C., Reigstad, C.S., Backhed, F., 2009. Intestinal microbiota during infancy and its implications for obesity. J. Pediatr. Gastroenterol. Nutr. 48, 249—256.

Rimoldi, M., Chieppa, M., Vulcano, M., Allavena, P., Rescigno, M., 2004. Intestinal epithelial cells control dendritic cell function. Ann. N.Y. Acad. Sci. 1029, 66—74.

Roberts, S.E., Wotton, C.J., Williams, J.G., Griffith, M., Goldacre, M.J., 2011. Perinatal and early life risk factors for inflammatory bowel disease. World J. Gastroenterol. 17, 743—749.

Roediger, W.E., 1980. Role of anaerobic bacteria in the metabolic welfare of the colonic mucosa in man. Gut 21, 793—798.

Rowan, F., Docherty, N.G., Murphy, M., Murphy, B., Calvin Coffey, J., O'Connell, P.R., 2010. Desulfovibrio bacterial species are increased in ulcerative colitis. Dis. Colon Rectum 53, 1530—1536.

Russell, L.C., Murray, J., 1992. Patient education: recommendations regarding sunscreens, drugs, and diet. Ann. Plast. Surg. 28, 14−16.

Sabat, P., Veloso, C., 2003. Ontogenic development of intestinal disaccharidases in the precocial rodent Octodon degus (Octodontidae). Comp. Biochem. Physiol. Part A: Mol. Integr. Physiol. 134, 393−397.

Saemann, M.D., Bohmig, G.A., Osterreicher, C.H., Burtscher, H., Parolini, O., Diakos, C., et al., 2000. Anti-inflammatory effects of sodium butyrate on human monocytes: potent inhibition of IL-12 and up-regulation of IL-10 production. FASEB J. 14, 2380−2382.

Scheppach, W., 1994. Effects of short chain fatty acids on gut morphology and function. Gut 35, S35−S38.

Sebert, S., Sharkey, D., Budge, H., Symonds, M.E., 2011. The early programming of metabolic health: is epigenetic setting the missing link? Am. J. Clin. Nutr. 94, 1953S−1958S.

Segain, J.P., Raingeard de la Bletiere, D., Bourreille, A., Leray, V., Gervois, N., Rosales, C., et al., 2000. Butyrate inhibits inflammatory responses through NFkappaB inhibition: implications for Crohn's disease. Gut 47, 397−403.

Shanker, A., 2004. Is thymus redundant after adulthood? Immunol. Lett. 91, 79−86.

Shanley, D.P., Aw, D., Manley, N.R., Palmer, D.B., 2009. An evolutionary perspective on the mechanisms of immunosenescence. Trends Immunol. 30, 374−381.

Sharma, R., Schumacher, U., Ronaasen, V., Coates, M., 1995. Rat intestinal mucosal responses to a microbial flora and different diets. Gut 36, 209−214.

Simor, A.E., Bradley, S.F., Strausbaugh, L.J., Crossley, K., Nicolle, L.E., 2002. Shea Position Paper: *Clostridium difficile* in long-term care facilities for the elderly. Infect. Control Hosp. Epidemiol. 23, 696−703.

Smits, H.H., Engering, A., van der Kleij, D., de Jong, E.C., Schipper, K., van Capel, T.M., et al., 2005. Selective probiotic bacteria induce IL-10-producing regulatory T cells in vitro by modulating dendritic cell function through dendritic cell-specific intercellular adhesion molecule 3-grabbing nonintegrin. J. Allergy Clin. Immunol. 115, 1260−1267.

Sokol, H., Pigneur, B., Watterlot, L., Lakhdari, O., Bermudez-Humaran, L.G., Gratadoux, J.J., et al., 2008. Faecalibacterium prausnitzii is an anti-inflammatory commensal bacterium identified by gut microbiota analysis of Crohn disease patients. Proc. Natl. Acad. Sci. USA 105, 16731−16736.

Sole, R.V., Montoya, J.M., 2001. Complexity and fragility in ecological networks. Proc. Biol. Sci. 268, 2039−2045.

Stark, P.L., Lee, A., 1982. The microbial ecology of the large bowel of breast-fed and formula-fed infants during the first year of life. J. Med. Microbiol. 15, 189−203.

Strober, W., Murray, P.J., Kitani, A., Watanabe, T., 2006. Signalling pathways and molecular interactions of NOD1 and NOD2. Nat. Rev. Immunol. 6, 9−20.

Tap, J., Mondot, S., Levenez, F., Pelletier, E., Caron, C., Furet, J.P., et al., 2009. Towards the human intestinal microbiota phylogenetic core. Environ. Microbiol. 11, 2574−2584.

Thompson, C., Powrie, F., 2004. Regulatory T cells. Curr. Opin. Pharmacol. 4, 408−414.

Tiihonen, K., Ouwehand, A.C., Rautonen, N., 2010. Human intestinal microbiota and healthy ageing. Ageing Res. Rev. 9, 107−116.

Tong, X., Yin, L., Washington, R., Rosenberg, D.W., Giardina, C., 2004. The p50-p50 NF-kappaB complex as a stimulus-specific repressor of gene activation. Mol. Cell. Biochem. 265, 171−183.

Turnbaugh, P.J., Backhed, F., Fulton, L., Gordon, J.I., 2008. Diet-induced obesity is linked to marked but reversible alterations in the mouse distal gut microbiome. Cell Host Microbe 3, 213–223.

Turnbaugh, P.J., Hamady, M., Yatsunenko, T., Cantarel, B.L., Duncan, A., Ley, R.E., et al., 2009a. A core gut microbiome in obese and lean twins. Nature 457, 480–484.

Turnbaugh, P.J., Ridaura, V.K., Faith, J.J., Rey, F.E., Knight, R., Gordon, J.I., 2009b. The effect of diet on the human gut microbiome: a metagenomic analysis in humanized gnotobiotic mice. Sci. Transl. Med. 1, 6ra14.

Uauy, R., Diaz, E., 2005. Consequences of food energy excess and positive energy balance. Public Health Nutr. 8, 1077–1099.

von der Weid, T., Bulliard, C., Schiffrin, E.J., 2001. Induction by a lactic acid bacterium of a population of CD4(+) T cells with low proliferative capacity that produce transforming growth factor beta and interleukin-10. Clin. Diagn. Lab. Immunol. 8, 695–701.

Wade, P.R., Cowen, T., 2004. Neurodegeneration: a key factor in the ageing gut. Neurogastroenterol. Motil. 16 (Suppl. 1), 19–23.

Wang, M., Karlsson, C., Olsson, C., Adlerberth, I., Wold, A.E., Strachan, D.P., et al., 2008. Reduced diversity in the early fecal microbiota of infants with atopic eczema. J. Allergy Clin. Immunol. 121, 129–134.

Warner, M.J., Ozanne, S.E., 2010. Mechanisms involved in the developmental programming of adulthood disease. Biochem. J. 427, 333–347.

Watanabe, T., Kitani, A., Murray, P.J., Strober, W., 2004. NOD2 is a negative regulator of Toll-like receptor 2-mediated T helper type 1 responses. Nat. Immunol. 5, 800–808.

Watanabe, T., Kitani, A., Strober, W., 2005. NOD2 regulation of Toll-like receptor responses and the pathogenesis of Crohn's disease. Gut 54, 1515–1518.

Wiegering, V., Eyrich, M., Wunder, C., Gunther, H., Schlegel, P.G., Winkler, B., 2009. Age-related changes in intracellular cytokine expression in healthy children. Eur. Cytokine Netw. 20, 75–80.

Willing, B.P., Dicksved, J., Halfvarson, J., Andersson, A.F., Lucio, M., Zheng, Z., et al., 2010. A pyrosequencing study in twins shows that gastrointestinal microbial profiles vary with inflammatory bowel disease phenotypes. Gastroenterology 139, 1844–1854. e1841.

Woodmansey, E.J., 2007. Intestinal bacteria and ageing. J. Appl. Microbiol. 102, 1178–1186.

Wu, G.D., Chen, J., Hoffmann, C., Bittinger, K., Chen, Y.Y., Keilbaugh, S.A., et al., 2011. Linking long-term dietary patterns with gut microbial enterotypes. Science 334, 105–108.

Yu, C.G., Huang, Q., 2013. Recent progress on the role of gut microbiota in the pathogenesis of inflammatory bowel disease. J. Dig. Dis. 14, 513–517.

Zapolska-Downar, D., Siennicka, A., Kaczmarczyk, M., Kolodziej, B., Naruszewicz, M., 2004. Butyrate inhibits cytokine-induced VCAM-1 and ICAM-1 expression in cultured endothelial cells: the role of NF-kappaB and PPARalpha. J. Nutr. Biochem. 15, 220–228.

Zoetendal, E.G., Rajilic-Stojanovic, M., de Vos, W.M., 2008. High-throughput diversity and functionality analysis of the gastrointestinal tract microbiota. Gut 57, 1605–1615.

Index

Note: Page numbers with "*f*" denote figures; "*t*" tables.